Simulation of Flight

Simulation of Flight

Douglas F. Hunsaker

Associate Professor
Mechanical and Aerospace Engineering
Utah State University

BARTLETT PEAK
PRESS

Printed by Bartlett Peak Press, Logan Utah.

For permissions, inquiries, additional information, or if you find errors or inaccuracies in this book, please contact: doug.hunsaker@usu.edu

Companion website: https://doughunsaker.com/simulation_of_flight

ISBN: 979-8-951325-01-3

Printed in the United States of America

10 9 8 7 6 5 4 3 2 1

Dedicated to my high-school calculus teacher

Curtis Vandenbosch,

who believed in us enough to challenge us.

Quotes to Live By

The following quotes have guided me in my career.

The noblest question in the world is, 'What good may I do in it?'
- Benjamin Franklin

The difference between successful people and unsuccessful people is
the successful people do all the things the unsuccessful people don't want to do.
- John DeJoria

The best dividends on the labor invested have invariably come from
seeking more knowledge rather than more power.
- Orville and Wilbur Wright, 1906

Simple can be harder than complex;
you have to work hard to get your thinking clean to make it simple.
- Steve Jobs

The success or otherwise of a design therefore depends to a large extent
on the designer's knowledge of the physics of the flow,
and no improvements in numerical and experimental design tools
are ever likely to dispose of the need for physical insight.
- D. Küchemann, 1978

We are what we repeatedly do.
Excellence, then is not an act, but a habit.
- Aristotle

Repetition is the first principle of all learning.
- Robert Bruner

There is so much satisfaction that comes from knowing that
you can change and become whatever you want.
- Whitney McCoy

You must do the thing that you think you cannot do.
- Eleanor Roosevelt

What does it take to get to the moon?
The will to do it.
- Wernher Von Braun

A picture is worth a thousand words;
an equation is worth a thousand pictures.
- Warren F. Phillips

All models are wrong, but some are useful.
- George E. P. Box

No formulation has ever been completely general.
- Warren F. Phillips

If we let ourselves, we shall always be waiting for some distraction or other to end before we can really get down to our work. The only people who achieve knowledge are those who seek it while the conditions are still unfavorable. Favorable conditions never come.
- C. S. Lewis

The heights by great men reached and kept
were not attained by sudden flight,
but they while their companions slept,
were toiling upward in the night.
- Henry W. Longfellow

Once you have tasted flight,
you will forever walk the earth with your eyes turned skyward,
for there you have been,
and there you will always long to return.
- Leonardo Da Vinci

It appears that the computer has done for our minds
what the automobile did for our bodies.
- Warren F. Phillips

If you would not be forgotten,
as soon as you are dead and rotten,
either write things worth reading,
or do things worth writing.
- Benjamin Franklin

I don't believe good work is ever done in a hurry.
- C. S. Lewis

Have the courage to follow your heart and intuition.
They somehow know what you truly want to become.
- Steve Jobs

Conquering the universe one has to solve two problems:
gravity and red tape.
We could have mastered gravity.
- Wernher Von Braun

The truth will set you free,
but first it will [tick] you off.
- Joe Klaas

All good research is performed in the context of a paper.
All good research programs are performed in the context of a book.
- Douglas F. Hunsaker

If you simply try to tell the truth
(without caring two pence how often it has been told before)
you will, nine times out of ten,
become original without ever having noticed it.
- C. S. Lewis

Just write every day of your life. Read intensely. Then see what happens.
Most of my friends who are put on that diet have very pleasant careers.
- Ray Bradbury

I slept and dreamt that life was joy.
I awoke and saw that life was service.
I acted and behold, service was joy.
- Rabindranath Tagore

Every man is my superior in some way.
- Warren F. Phillips

Our life is a mission, not a career.
- Stephen Covey

Without great solitude, no serious work is possible.
- Pablo Picasso

Where is the Life we have lost in living?
Where is the wisdom we have lost in knowledge?
Where is the knowledge we have lost in information?
- Thomas Stearns Eliot

Wisdom is the principal thing; therefore get wisdom:
and with all thy getting get understanding.
- Proverbs 4:7

Contents

Preface

This book is intended for undergraduate and graduate-level courses in flight dynamics and simulation. It also serves as a reference for professionals working in related fields. To get the most out of this book, readers should have prior knowledge of aerodynamics, flight mechanics, numerical methods, and fundamental engineering principles. A solid background in calculus, linear algebra, differential equations, statics, and dynamics is assumed. This book builds on that foundation to provide a cohesive understanding of how to accurately simulate the dynamics of flight.

The focus of the text is on modeling the six-degree-of-freedom motion of a vehicle moving through a fluid. Requisite numerical methods are introduced and applied in worked examples and exercises. Through the process, students are guided in building a full six-degree-of-freedom flight simulator, incorporating aerodynamics, propulsion, atmospheric disturbances, actuator dynamics, control effects, geographic transformations, and graphics. Guidance is also provided on code structure and techniques to increase computational efficiency.

One distinguishing feature of this book is its treatment of attitude representation using both Euler angles and quaternions. Euler angles are relatively easy to visualize, whereas quaternions offer greater computational efficiency and avoid singularities that hinder simulation robustness. This dual treatment shows how both representations can be used together to create a user-friendly and computationally efficient simulation framework.

Unique topics in this text include a closed-form solution for estimating the mass and inertia of finite wings and rotors, a section about modeling electric propulsion, an extensive chapter about common trim conditions, detailed methods for implementing atmospheric disturbances, examples of real-world communication methods between various software packages, and an introduction to graphics and visualization. Many of these topics are not addressed in detail in other simulation texts.

The aerodynamic modeling in this book is also more comprehensive than in many traditional simulation texts. Methods for estimating aerodynamic forces and moments for both symmetric and asymmetric aircraft are developed directly from aerodynamic theory. The text also presents a unique nonlinear sub-stall aerodynamic model for fixed-wing aircraft, together with a post-stall model based on Newtonian theory. Together, these models enable accurate aerodynamic predictions over a wide range of angles of attack and compare favorably with experimental data. A section on dataset structures and use is also included for high-fidelity aerodynamic modeling.

Worked examples throughout the book demonstrate how to apply theoretical concepts. Over 80 example problems are included, most of which are worked to a level that can be used for verifying a student's code and implementation. Based on years of teaching experience, the author emphasizes clear, detailed examples to aid debugging and ensure accuracy. Each chapter concludes with two sets of problems: one focused on theory, the other on coding. Theory problems help reinforce analytical understanding, while coding problems guide the student through the development of a working simulation computer code.

Throughout the text, the term *vehicle* is used rather than *aircraft* to reflect the broad applicability of the material. The principles covered in the text apply not only to fixed-wing aircraft in Earth's atmosphere, but also to rotorcraft, underwater vehicles, natural fliers, or vehicles in other planetary environments.

A companion website (`https://doughunsaker.com/simulation_of_flight`) is available, which contains additional resources and code validation tools to support students as they study and implement fundamental concepts requisite for the simulation of flight.

Douglas F. Hunsaker

Acknowledgments

First I would like to thank Utah State University and the academic institutional setup of this nation for making works like this possible. The structure and purpose of universities in society has shifted and adapted throughout history. Although the value of universities is often questioned, I appreciate that such institutions exist as places of higher learning and aspiration of knowledge, somewhat removed from the economic demands of society. I believe that without such organizations, mankind could not continue to advance at the rate it has through the past century. I consider it an obligation of academics to organize the understanding of mankind in ways that are digestible and teachable to the next generation. As so eloquently expressed in the poem *The Bridge Builder* by Will Allen Dromgoole, it is our opportunity and duty to build bridges for those who come after us. This book was written as a labor of love out of sincere hope to support the next generation of engineers in the exciting field of flight simulation, and would not have been possible without the current academic structure.

I also want to thank the students with whom I have had the chance to interface and teach, who have in unique ways contributed to the outcome of this book. This includes many students, but in particular, Ben Moulton, Sabrina Snow, and Christian Bolander who assisted with typesetting support through their individual publications, Spencer Adams who formatted the NASA 1979 aerodynamic model to that provided in this text, Austin Kohler who allowed me to use his HUD graphics, Ashton Gilbert who helped generate the cover image, Zach Montgomery who checked some mathematical developments and solutions with his code, and Tony Daniels who spent significant time checking solutions and editing citations. Additionally, I'd like to thank the Fall 2025 cohort of students in the MAE 6510 Flight Simulation course for being the first to use the book in a classroom setting.

Many professionals in the aerospace engineering community have also supported me on this journey. I would like to extend a special thanks to Todd Moon, Nick Alley, Kevin Bowcutt, Wayne Goodrich, and Tyler Fean who reviewed drafts of the book and provided helpful feedback. I would also specifically like to thank my graduate advisors Warren Phillips and Deryl Snyder, for their mentorship and immense impact on my understanding of flight dynamics and simulation.

Most of all, I want to express my deepest gratitude to my wife, Lindsey, for her sincere support and patience through the development of this book. I had no idea how challenging the road would be, and could not have done it without her constant love and encouragement.

Douglas F. Hunsaker

Notes on Code Development

One purpose of this textbook is to guide the reader through the process of developing a computer code for vehicle flight simulation. Because of this, many of the examples in the chapters and problems at the end of each chapter are geared towards developing computer code. Additionally, example functions and code structure are included in examples and in the appendices. Though some might find the actual code syntax included in the book to be useful, it is likely that the usefulness of the code syntax will rapidly decline as coding languages mature and change with new technologies. The inclusion of code serves to clearly document the underlying algorithms and suggested code structure. These algorithms can be implemented in any coding language, and can be adapted as best coding practices and languages evolve.

Choice of Computer Language

There are several main-stream computer languages currently in use that can be used to develop flight simulation software. There are benefits and drawbacks to each language.

Fortran is a compiled language and perhaps the oldest language still widely used in the aeronautics industry. The benefit of Fortran is that it is very fast and efficient. Additionally, it has inherent array-handling capabilities that make data manipulation straightforward. However, most graphics and communication packages run in C or C++, and therefore, to include graphics in a simulation, function bindings are typically required.

C/C++ is a compiled language that is much more widely used than Fortran across industries. Because of this, there are many open-source libraries for array handling, linear algebra, graphics, and communication available for free on the internet. One drawback of C/C++ is that it does not have intrinsic array handling built in without the use of additional libraries. The user has complete control in how information is passed between functions or how memory is allocated. Although this is a powerful property of the language, it can result in unintentional memory leaks and can cause problems when accessing or writing memory to arrays if the user is not extremely careful.

Python is an open-source scripted language, and perhaps the most common language in use today. Because Python is scripted, it is typically orders of magnitude slower than compiled languages such as C/C++ and Fortran. Python is often used for rapid code development and data manipulation, since it has incredible amounts of functionality inherent to the language. Once a code is developed, it can be profiled and the most computationally-intensive functions can be converted to a compiled language to increase the computational speed.

Matlab® is a commercial scripted language that is widely used for simulation and controls in the aerospace community. Like Python, it is much slower than compiled languages. However, this tool has incredible functionality built-in pertaining to system dynamics analysis, controls, and simulation.

The information contained in this book is general enough to be implemented by the reader in any computer language of choice.

Code Structure

The current best-practice structure for code is object-oriented programming. Object orientation is extremely useful for developing sustainable code that can interface with other software packages and libraries, and every language listed above can be used to develop object-oriented code. This textbook will use object-oriented code format for some aspects, but efforts are made to keep the code as accessible and readable as possible.

For some problems and examples, a suggested code structure is included. Best practice for handling input and output from functions and classes will vary with coding language and use case. For example, it is sometimes best to pass information in and out of a function through the function call (within parentheses), including the solution. The return statement of the function can then be used to return error flags such as an integer or boolean, which the code calling the function can check to test whether the function performed as expected. Such coding practices are largely left to the developer's preferences and best practices.

Input and Output Structure

Two types of input and output files are commonly employed in the examples. These are Comma-Separated Value (CSV) files, and JavaScript Object Notation (JSON) files.

CSV files are widely used across many industries and can be read and/or written by many software packages. In the examples associated with this book, CSV files are typically used for data sets that must be read into the code or written by the code.

The JSON file format is a form of markup language commonly employed in modern web development, and is extremely useful for reading and writing structured data. JSON is used for most input files, including configuration parameters or simulation settings, and has significant advantages over traditional ASCII-based input-file structures common to older software. In most coding languages, when JSON files are read in, they are converted directly to dictionaries and can be accessed or edited as such. Open-source JSON libraries are available online for both Fortran and C++. For example, in the Fortran language, the JSON module written by Jacob Williams and available at http://github.com/jacobwilliams/json-fortran can be used. In the C++ language, the header-only library "JSON for Modern C++" maintained by Niels Lohmann and available at https://github.com/nlohmann/json can be used. Some additional useful JSON functions that call the Lohmann JSON library are included on the companion website. In

Python, JSON functionality can be used by simply including "import json" in the header. Matlab has similar functionality to that of Python.

Naming Convention

It is good coding practice to select and adhere to a naming convention for variables, functions, classes, structs, etc. Two main naming conventions have emerged over the last 60 years of code development. These are typically called *camel case* and *snake case*.

Camel case is a naming convention in which capitalization is used to denote each word in a multi-word name. Some conventions capitalize the first letter of each word and other capitalize the first letter of each word except the first. For example, a variable representing wing sweep might be named `WingSweep` or `wingSweep`. This is a common naming convention used in older languages but is still widely used today.

The snake naming convention has gained popularity more recently especially among developers using the Python language. This naming convention uses lower-case letters for all letters in a name, and an underscore between words. In this convention, the variable for wing sweep might be named `wing_sweep`. Most example code presented in this book will follow the snake naming convention.

Numerical Accuracy

Many examples are included throughout the text. Although some are rudimentary and straightforward, they can be useful for checking computer code that implements the relationships presented. Identifying errors in a code in the early stages by testing the code on problems with known solutions can significantly reduce debugging time.

Modern computers typically use double precision for most computations. This results in an accuracy of about 15 to 16 significant digits. Of course, current flight simulators are not accurate to that level of precision. Inherent inaccuracies are included in the modeling due to the limited accuracy of the mass and inertia models, predicted aerodynamic behavior, or gravitational models of the Earth. However, many computations included in this text will be shown to 12 or 13 digits of accuracy, or as high a precision as is practical due to space. Although this level of precision exceeds the accuracy of many physical models, the student is encouraged to match all computations out to the precision shown. This practice will help the student find errors in their implementation and can greatly assist in the learning process.

Units

The units of measure given or available in a problem or situation are not always the units of interest. Hence, it is important to be able to convert measurements in one unit system to another. The most common measurement system is the International System of Units, also known as SI units. However, in the aeronautics community, US Customary units, also known as English units, are often used.

The development in this textbook is general such that code can be developed using any consistent unit system. Since English units are widely used in the aeronautics community, most of the worked examples are in English units. However, some examples are also worked using SI units.

Typically, computer codes are written such that they employ a standard unit system for the computations. Other units of measure can be accepted from the user when initializing the code, but these are immediately converted to the main unit system at run time. The user can also sometimes specify what units the results will be given in, which again simply requires a conversion from the base unit system when displaying the results. For example, a code could be written to employ meters as the base unit of length in all computations of the code. However, the user could specify input parameters in units of centimeters, kilometers, miles, feet, etc. The code would recognize each of these and immediately convert the given length into a length in meters for all subsequent computations in the code. The reverse conversion can be performed when results are to be displayed.

Many computations in the textbook require the use of angles. It is good coding practice to store and use any angle quantities in radians within the code. User inputs can be accepted in degrees and immediately converted to radians within the code. Code outputs can also be converted to degrees for ease of interpreting results.

Constants and Conversion Factors

The exact definitions and values of constants and units of measurement have changed over time as science, measurement techniques, and standards have improved. For example, the definition of a kilogram was redefined in 2018 by the International Bureau of Weights and Measures [1]. Appendix A includes values for the constants and conversion factors used in this textbook. Variations in the accepted values for these constants and conversion factors are typically very small, and well within the accuracy of concern for most flight simulation scenarios. Therefore, the exact values of these constants is not of great importance, as long as the values are sufficiently accurate for the simulation.

In order to facilitate learning and code debugging, the student is encouraged to use the values for the constants and conversion factors exactly as given in Appendix A throughout the development of his or her individual code. This will significantly help with debugging. Once the student is confident in the code implementation, other constants or conversion factors can be used to modify the simulation if needed.

It is worth considering one constant as an example. The standard gravitational constant at sea level is internationally accepted to be 9.80665 m/s^2. Of course, gravity at sea level varies as a function of location on Earth. However, this constant is widely accepted as the standard, and is the standard used in this textbook. Since the currently accepted standard conversion from meters to feet is 1 ft = 0.3048 m, the gravitational constant at sea level used in this book out to double precision is 32.1740485564304 ft/s^2. Many engineering textbooks shorten this to simply 32.174 or 32.2. However, for the examples worked in English units in this textbook, double precision for this constant is retained. By strictly adhering to the constants and conversion factors listed in Appendix A, code developed by students can be accurately compared to the example solutions given in the text.

atan2 Function

The `atan2` function is used in several algorithms to preserve the quadrant when computing the solution to the mathematical expression $\tan^{-1}(y/x)$. In most programming languages, the function call accepts two input parameters in the form `atan2(y,x)` where y represents the numerator value and x represents the denominator value. Note that the value for y is the first parameter, and the value for x is the second parameter accepted in these function calls. This structure is used for most mathematical libraries in C++, Fortran, Matlab®, and Python. However, it is sometimes helpful to check solutions in Excel®. It is important to note that in Excel®, the `atan2` function accepts the two parameters in the opposite order, i.e. `atan2(x,y)`. In the present text, any reference to `atan2` will use the more common notation with the numerator value y as the first input, and the denominator value x as the second input to the function.

Douglas F. Hunsaker

Chapter 1

Coordinate Systems and Attitude Descriptors

1.1 Introduction

Flight dynamics is the study of the motion of aircraft, and flight simulation is the prediction of that motion as a function of time. At first glance, it may seem that the value of flight simulation lies primarily in visualization of vehicle dynamics, performance prediction, and pilot training. While these are important applications, they represent only a small part of the role that accurate and modular flight simulation tools play in the development of modern aerospace vehicles. Because hardware and software development, flight testing, and data collection are both expensive and time consuming, flight simulation has become an indispensable engineering tool throughout the vehicle development process. It enables engineers to evaluate concepts rapidly, explore design tradeoffs, reduce technical risk, and investigate system behavior long before a prototype is built or flown. Common applications include design trade studies, Monte-Carlo analysis, control-system development, autopilot tuning, system identification, flight-safety risk assessment, and root-cause investigation of flight incidents. As aerospace systems become increasingly complex and development schedules continue to accelerate, the importance of accurate, modular, and well-validated flight simulation continues to grow. This book develops the mathematical models, numerical methods, and software architecture required to build such simulations from first principles.

While this book will focus on vehicles traveling through the Earth's atmosphere, the principles presented in this text can be applied to a larger class of problems than aircraft. For example, the motion of submarines in water or other objects in an atmosphere of any planet could be studied and simulated with these principles. Hence, in this text we will refer to the aircraft as a *vehicle* and give special attention to vehicles traveling through the Earth's atmosphere.

The motion of any vehicle depends on the mass properties of the vehicle as well as the forces and moments exerted on the vehicle. The forces and moments exerted on an aircraft at any point in time come from the pressure and shear distribution on the surface, thrust from a propulsion unit, and gravity. Other pseudo forces and moments can result from gyroscopic effects of objects attached to the vehicle such as propellers and turbine blades, or Coriolis forces from the rotation of the aircraft.

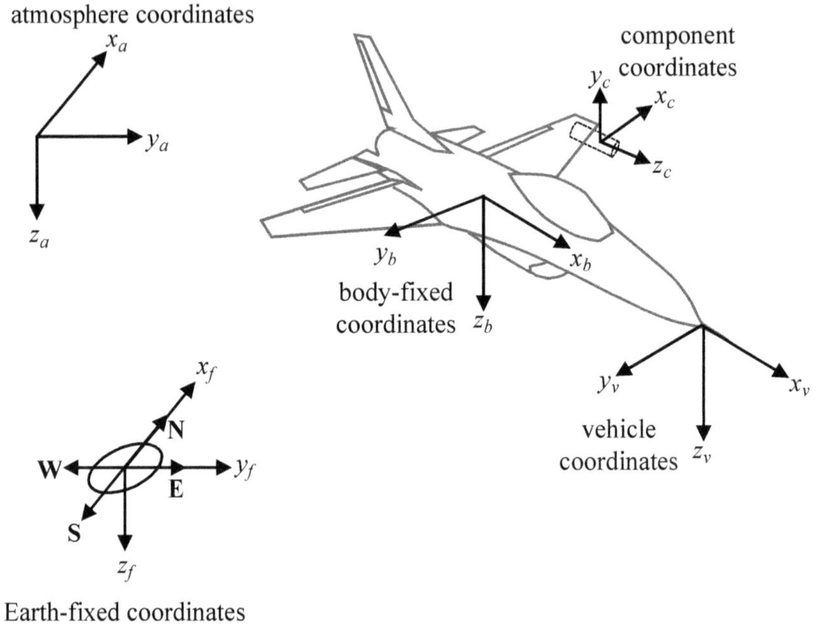

Figure 1.1.1. Coordinate systems used in this textbook.

In order to study the dynamics of any vehicle, a coordinate system must be chosen. The choice of this coordinate system can have a significant effect on the complexity of the analysis. For example, the inertia tensor of the vehicle is most naturally defined in a coordinate system fixed to the vehicle, whereas the gravity vector is most naturally defined in a coordinate system fixed to the Earth. Additionally, the dynamics of a vehicle are most easily described about a coordinate system with an origin at the vehicle center-of-gravity (CG), which can change with time, whereas components onboard the vehicle are most easily defined relative to a fixed coordinate system on the vehicle. Hence, we will use multiple coordinate systems throughout our analyses. In this text we will use the Cartesian coordinate systems shown in Fig. 1.1.1. These include the Earth-fixed, atmosphere, vehicle, body-fixed, and component coordinate systems.

The **Earth-fixed coordinate system** is a flat-Earth coordinate system fixed to the surface of the Earth with its origin at sea level. The x-axis points to the north, y-axis to the east, and z towards the center of the Earth. This coordinate system will be used as the inertial coordinate system for physics based on flat-Earth relations. The subscript f will be used to denote components in this coordinate system.

The **atmosphere coordinate system** has the same orientation as the Earth-fixed coordinate system, and moves relative to the Earth-fixed coordinate system at a constant velocity equal to the average wind speed. Because it moves at a constant velocity relative to the Earth, it can be considered an inertial coordinate system. The subscript a will be used to denote components in this coordinate system.

The **vehicle coordinate system** is fixed to the vehicle. The origin location relative to the vehicle is arbitrary. The coordinate system is oriented such that the x-axis points out the nose, the y-axis points out the right, and the z-axis points out the bottom of the vehicle. The subscript v will be used to denote components in this coordinate system.

The **body-fixed coordinate system** is fixed to the vehicle with the origin at the vehicle center-of-gravity. This coordinate system has the same orientation as the vehicle coordinate system. The subscript b will be used to denote components in this coordinate system.

The **component coordinate system** is a local coordinate system of any individual component of the vehicle. For example, this can be a local coordinate system of a propeller, wing, nacelle, or payload. The subscript c will be used to denote components in this coordinate system.

Of course, other useful coordinate systems could be included and used in the analysis as long as the appropriate physics is also added to the analysis. For example, a coordinate system fixed to the solar system could be added to model the motion of the Earth relative to the Sun, and a coordinate system fixed to a sub-component could be used to define the relative location of a sub-component to a component.

Information in one coordinate system can be expressed in terms of a second coordinate system if the orientation of the second coordinate system is known relative to the first coordinate systems. The orientation of one coordinate system relative to another is often referred to as the *attitude*.

In the following analysis, we will use the subscripts 1 and 2 to represent vector components in two different coordinate systems. The first (1) will be referred to as the *base* coordinate system. The second (2) will be referred to as the *dependent* coordinate system. The orientation, or attitude, of any dependent coordinate system relative to a base coordinate system can be expressed several ways. Two of the most common methods include Euler angles and quaternions. Both of these methods are presented in this section, which is a summary of work presented by Phillips [2, 3].

1.2 Euler Angles

The orientation of one coordinate system relative to another can be expressed using three consecutive rotations about coordinate-system axes. The set of angles associated with the rotations are known as the *Euler angles*. There is one requirement for these rotations, which is that two consecutive rotations may not occur about the same axis. With this restriction, there are twelve possible combinations. Six are symmetric and include

$$
\begin{aligned}
&\varphi_x \to \varphi_y \to \varphi_x, &&\varphi_x \to \varphi_z \to \varphi_x,\\
&\varphi_y \to \varphi_x \to \varphi_y, &&\varphi_y \to \varphi_z \to \varphi_y,\\
&\varphi_z \to \varphi_x \to \varphi_z, &&\varphi_z \to \varphi_y \to \varphi_z
\end{aligned}
$$

Table 1.2.1. Euler angle definitions typically used in the field of aeronautics.

Name	Rotation Axis	Symbol	Limits	Rotation Order
Bank	φ_x	ϕ	$-180^\circ < \phi \leq 180^\circ$	3
Elevation	φ_y	θ	$-90^\circ \leq \theta \leq 90^\circ$	2
Azimuth	φ_z	ψ	$0^\circ \leq \psi < 360^\circ$	1

Six are asymmetric and include

$$\begin{aligned} &\varphi_x \to \varphi_y \to \varphi_z, && \varphi_x \to \varphi_z \to \varphi_y, \\ &\varphi_y \to \varphi_x \to \varphi_z, && \varphi_y \to \varphi_z \to \varphi_x, \\ &\varphi_z \to \varphi_x \to \varphi_y, && \varphi_z \to \varphi_y \to \varphi_x \end{aligned}$$

Any of these twelve sets of Euler angles can be used, and the choice of which set to use varies by field. For example, in the field of orbital mechanics, it is common to use the rotations defined by $\varphi_z \to \varphi_x \to \varphi_z$. However, in the field of aeronautics, it is common to use the rotations defined by $\varphi_z \to \varphi_y \to \varphi_x$. In this text we will exclusively use the rotations customary in the field of aeronautics. The angles for each rotation are typically called the *azimuth angle* φ_z, *elevation angle* φ_y, and *bank angle* φ_x. The symbols typically used in the field of aeronautics for each rotation are given in Table 1.2.1. In order to make the definition of any one orientation unique, the limits included in the table are typically applied to these angles.

The order in which the Euler angle rotations are applied is very important and also included in the table. To demonstrate this process, we begin with an arbitrary vector expressed in the base coordinate system $(v_{x_1}, v_{y_1}, v_{z_1})$. The traditional Euler angles used in the field of aeronautics are applied by first rotating this vector about the z axis of the base coordinate system by the azimuth angle $\psi = \varphi_z$. This gives a vector in a new intermediate coordinate system we will refer to with a subscript A and results in

$$\begin{bmatrix} v_{x_A} \\ v_{y_A} \\ v_{z_A} \end{bmatrix} = \begin{bmatrix} C_\psi & S_\psi & 0 \\ -S_\psi & C_\psi & 0 \\ 0 & 0 & 1 \end{bmatrix} \begin{bmatrix} v_{x_1} \\ v_{y_1} \\ v_{z_1} \end{bmatrix} \tag{1.2.1}$$

where the capital S and C stand for sine and cosine respectively of the angle in subscript. The resulting vector is then rotated about the y axis of the A coordinate system by the elevation angle $\theta = \varphi_y$ to yield a vector in a new intermediate coordinate system we will refer to as B, which results in

$$\begin{bmatrix} v_{x_B} \\ v_{y_B} \\ v_{z_B} \end{bmatrix} = \begin{bmatrix} C_\theta & 0 & -S_\theta \\ 0 & 1 & 0 \\ S_\theta & 0 & C_\theta \end{bmatrix} \begin{bmatrix} v_{x_A} \\ v_{y_A} \\ v_{z_A} \end{bmatrix} \tag{1.2.2}$$

Finally, the resulting vector is rotated about the x axis of the B coordinate system by the bank angle $\phi = \varphi_x$ to yield a vector in the final dependent coordinate system, which results in

$$\begin{bmatrix} v_{x2} \\ v_{y2} \\ v_{z2} \end{bmatrix} = \begin{bmatrix} 1 & 0 & 0 \\ 0 & C_\phi & S_\phi \\ 0 & -S_\phi & C_\phi \end{bmatrix} \begin{bmatrix} v_{xB} \\ v_{yB} \\ v_{zB} \end{bmatrix} \tag{1.2.3}$$

Combining Eqs. (1.2.1)–(1.2.3) gives

$$\begin{bmatrix} v_{x2} \\ v_{y2} \\ v_{z2} \end{bmatrix} = \begin{bmatrix} 1 & 0 & 0 \\ 0 & C_\phi & S_\phi \\ 0 & -S_\phi & C_\phi \end{bmatrix} \begin{bmatrix} C_\theta & 0 & -S_\theta \\ 0 & 1 & 0 \\ S_\theta & 0 & C_\theta \end{bmatrix} \begin{bmatrix} C_\psi & S_\psi & 0 \\ -S_\psi & C_\psi & 0 \\ 0 & 0 & 1 \end{bmatrix} \begin{bmatrix} v_{x1} \\ v_{y1} \\ v_{z1} \end{bmatrix} \tag{1.2.4}$$

The three rotation matrices can be multiplied to yield a single transformation equation

$$\begin{bmatrix} v_{x2} \\ v_{y2} \\ v_{z2} \end{bmatrix} = \begin{bmatrix} C_\theta C_\psi & C_\theta S_\psi & -S_\theta \\ S_\phi S_\theta C_\psi - C_\phi S_\psi & S_\phi S_\theta S_\psi + C_\phi C_\psi & S_\phi C_\theta \\ C_\phi S_\theta C_\psi + S_\phi S_\psi & C_\phi S_\theta S_\psi - S_\phi C_\psi & C_\phi C_\theta \end{bmatrix} \begin{bmatrix} v_{x1} \\ v_{y1} \\ v_{z1} \end{bmatrix} \tag{1.2.5}$$

A similar process can be followed to compute the inverse transformation, which yields

$$\begin{bmatrix} v_{x1} \\ v_{y1} \\ v_{z1} \end{bmatrix} = \begin{bmatrix} C_\theta C_\psi & S_\phi S_\theta C_\psi - C_\phi S_\psi & C_\phi S_\theta C_\psi + S_\phi S_\psi \\ C_\theta S_\psi & S_\phi S_\theta S_\psi + C_\phi C_\psi & C_\phi S_\theta S_\psi - S_\phi C_\psi \\ -S_\theta & S_\phi C_\theta & C_\phi C_\theta \end{bmatrix} \begin{bmatrix} v_{x2} \\ v_{y2} \\ v_{z2} \end{bmatrix} \tag{1.2.6}$$

Note that the transformation matrix in Eq. (1.2.6) is the inverse and transpose of the transformation matrix in Eq. (1.2.5). These equations can be used to transform the components of a vector in one coordinate system to another coordinate system, as long as the Euler angles for the transformation are known.

EXAMPLE 1.2.1 A vehicle with a weight of 100 lbf is oriented relative to the Earth-fixed coordinate system with Euler angles $\phi = 60$ deg, $\theta = 10$ deg, and $\psi = 45$ deg. Compute the components of the weight vector in the body-fixed coordinate system.

Solution: The weight vector of any vehicle is known in the Earth-fixed coordinate system as $(W_{x_f}, W_{y_f}, W_{z_f}) = (0, 0, W)$. From Eq. (1.2.5), the weight vector from the Earth-fixed coordinate system can be expressed in the body-fixed coordinate system as

$$\begin{bmatrix} W_{x_b} \\ W_{y_b} \\ W_{z_b} \end{bmatrix} = \begin{bmatrix} C_\theta C_\psi & C_\theta S_\psi & -S_\theta \\ S_\phi S_\theta C_\psi - C_\phi S_\psi & S_\phi S_\theta S_\psi + C_\phi C_\psi & S_\phi C_\theta \\ C_\phi S_\theta C_\psi + S_\phi S_\psi & C_\phi S_\theta S_\psi - S_\phi C_\psi & C_\phi C_\theta \end{bmatrix} \begin{bmatrix} 0 \\ 0 \\ W \end{bmatrix}$$

Simplifying the right-hand side gives

$$\begin{bmatrix} W_{x_b} \\ W_{y_b} \\ W_{z_b} \end{bmatrix} = W \begin{bmatrix} -S_\theta \\ S_\phi C_\theta \\ C_\phi C_\theta \end{bmatrix}$$

Notice that this relationship is independent of the azimuth angle. Hence, the weight vector in the body-fixed coordinate system is independent of the azimuth direction the vehicle is pointing. Evaluating this relationship using the values given in the problem statement gives

$$\begin{bmatrix} W_{x_b} \\ W_{y_b} \\ W_{z_b} \end{bmatrix} = \begin{bmatrix} -17.36481776669 \\ 85.28685319524 \\ 49.24038765061 \end{bmatrix} \text{lbf}$$

The change in Euler angles as a function of time is related to the rotation rate of the dependent coordinate system. The rotation rate of a coordinate system can be expressed as a rotation vector with components $(\omega_x, \omega_y, \omega_z)$. Using a similar transformation method, the time derivative of the Euler angles can be found as a function of the current orientation and rotation rates of the dependent coordinate system [4]

$$\begin{bmatrix} \dot{\phi} \\ \dot{\theta} \\ \dot{\psi} \end{bmatrix} = \begin{bmatrix} 1 & S_\phi S_\theta / C_\theta & C_\phi S_\theta / C_\theta \\ 0 & C_\phi & -S_\phi \\ 0 & S_\phi / C_\theta & C_\phi / C_\theta \end{bmatrix} \begin{bmatrix} \omega_{x_2} \\ \omega_{y_2} \\ \omega_{z_2} \end{bmatrix} \tag{1.2.7}$$

The inverse of this transformation is [4]

$$\begin{bmatrix} \omega_{x_2} \\ \omega_{y_2} \\ \omega_{z_2} \end{bmatrix} = \begin{bmatrix} 1 & 0 & -S_\theta \\ 0 & C_\phi & S_\phi C_\theta \\ 0 & -S_\phi & C_\phi C_\theta \end{bmatrix} \begin{bmatrix} \dot{\phi} \\ \dot{\theta} \\ \dot{\psi} \end{bmatrix} \tag{1.2.8}$$

Note that the inverse of Eq. (1.2.7) is not its transpose.

One benefit of using Euler angles to define the orientation of one coordinate system relative to another is the fairly intuitive nature of the formulation. Given a set of Euler angles, one can relatively easily visualize the orientation of one coordinate system relative to another. Although Euler angles are widely used in the field of aeronautics, the result shown in Eq. (1.2.7) highlights a significant drawback of employing Euler angles for describing the orientation of vehicles. Note that when $\cos(\theta) = 0$ (i.e. $\theta = \pm 90$ deg), Eq. (1.2.7) is singular. This is referred to as *gimbal lock*. For this formulation, gimbal lock occurs when the vehicle is oriented with the vehicle x axis pointed straight up or straight down relative to the Earth-fixed coordinate system. This is problematic for many aircraft, and even more problematic for launch vehicles. Gimbal lock is a characteristic inherent to all Euler-angle formulations. The exact Euler-angle values at which gimbal lock occurs varies with the formulation. However, every Euler-angle formulation suffers from gimbal lock at some orientation. For this reason, the Euler-angle formulation is not typically used in flight simulation.

Euler angles are not the only method for describing the orientation of one coordinate system relative to another. Two alternate common methods include the *direction-cosine* and *quaternion* formulations. Although the direction-cosine formulation is widely used and gimbal-lock free, it is not as computationally efficient as the quaternion formulation. Hence, in this text we will not include the direction-cosine formulation, and will instead focus on the use of quaternions to represent orientation. To understand the development and definition of the quaternion formulation, the reader must first understand a geometric construct known as the Euler axis.

1.3 Euler Axis

The orientation of one coordinate system relative to another can be defined by a single rotation angle about a single axis as suggested by Euler [5]. The axis about which the rotation occurs is often termed the *Euler axis*, but is also sometimes referred to as the *eigen axis*, or *characteristic axis*.

Whereas the Euler angle formulation is defined by only three components (the three Euler angles ϕ, θ, and ψ), the Euler axis formulation is defined by four. These include the three components of a vector that points along the Euler axis $\mathbf{E} = (E_x, E_y, E_z)$ and a rotation angle Θ about that axis.

Very simple examples of this are easy to visualize. For example, the orientation specified by the Euler angles $\phi = 45$ deg, $\theta = 0$, and $\psi = 0$ is a simple rotation about the x axis. Hence the Euler axis is $E_x = 1$, $E_y = 0$, $E_z = 0$, and the associated angle for the Euler-axis formulation is $\Theta = 45$ deg. Similar examples are straightforward for the other two Euler angles in the case that only one Euler angle is nonzero.

The orientation of one coordinate system relative to another involves only three degrees of freedom. Therefore, there is some redundancy in the Euler-axis formulation. Since the length of the Euler axis is arbitrary, an additional equation is often specified by requiring the length of the Euler axis to be unity

$$E_x^2 + E_y^2 + E_z^2 \equiv 1 \tag{1.3.1}$$

One noteworthy characteristic of this transformation is that the Euler axis has the same components in both coordinate systems. Therefore,

$$\begin{bmatrix} E_{x_1} \\ E_{y_1} \\ E_{z_1} \end{bmatrix} = \begin{bmatrix} E_{x_2} \\ E_{y_2} \\ E_{z_2} \end{bmatrix} \equiv \begin{bmatrix} E_x \\ E_y \\ E_z \end{bmatrix} \tag{1.3.2}$$

The Euler axis formulation results in a transformation often called *Euler's formula.* This can be written as [3]

$$\begin{bmatrix} v_{x_2} \\ v_{y_2} \\ v_{z_2} \end{bmatrix} = \begin{bmatrix} E_{xx} + C_\Theta & E_{xy} + E_z S_\Theta & E_{xz} - E_y S_\Theta \\ E_{xy} - E_z S_\Theta & E_{yy} + C_\Theta & E_{yz} + E_x S_\Theta \\ E_{xz} + E_y S_\Theta & E_{yz} - E_x S_\Theta & E_{zz} + C_\Theta \end{bmatrix} \begin{bmatrix} v_{x_1} \\ v_{y_1} \\ v_{z_1} \end{bmatrix} \tag{1.3.3}$$

where $E_{ij} = E_i E_j (1 - C_\Theta)$. The inverse transformation can be found by simply changing the sign of the rotation angle Θ. This gives

$$\begin{bmatrix} v_{x_1} \\ v_{y_1} \\ v_{z_1} \end{bmatrix} = \begin{bmatrix} E_{xx} + C_\Theta & E_{xy} - E_z S_\Theta & E_{xz} + E_y S_\Theta \\ E_{xy} + E_z S_\Theta & E_{yy} + C_\Theta & E_{yz} - E_x S_\Theta \\ E_{xz} - E_y S_\Theta & E_{yz} + E_x S_\Theta & E_{zz} + C_\Theta \end{bmatrix} \begin{bmatrix} v_{x_2} \\ v_{y_2} \\ v_{z_2} \end{bmatrix} \tag{1.3.4}$$

EXAMPLE 1.3.1 Use Euler's formula to develop an Euler-axis expression for the components of the weight vector in the vehicle coordinate system for an arbitrary orientation of the vehicle relative to the Earth-fixed coordinate system.

Solution: The weight vector of the vehicle in the Earth-fixed coordinate system is $(W_{x_f}, W_{y_f}, W_{z_f}) = (0, 0, W)$. Using this in Eq. (1.3.3), the weight vector from the Earth-fixed coordinate system can be expressed in the vehicle coordinate system as

$$\begin{bmatrix} W_{x_v} \\ W_{y_v} \\ W_{z_v} \end{bmatrix} = \begin{bmatrix} E_{xx} + C_\Theta & E_{xy} + E_z S_\Theta & E_{xz} - E_y S_\Theta \\ E_{xy} - E_z S_\Theta & E_{yy} + C_\Theta & E_{yz} + E_x S_\Theta \\ E_{xz} + E_y S_\Theta & E_{yz} - E_x S_\Theta & E_{zz} + C_\Theta \end{bmatrix} \begin{bmatrix} 0 \\ 0 \\ W \end{bmatrix}$$

Simplifying the right-hand side gives an expression for the weight vector in the vehicle coordinate system

$$\begin{bmatrix} W_{x_v} \\ W_{y_v} \\ W_{z_v} \end{bmatrix} = W \begin{bmatrix} E_{xz} - E_y S_\Theta \\ E_{yz} + E_x S_\Theta \\ E_{zz} + C_\Theta \end{bmatrix}$$

The change in the Euler axis and rotation angle with respect to time depends on the rotation rate of the dependent coordinate system and can be expressed as [3]

$$\begin{bmatrix} \dot{\Theta} \\ \dot{E}_x \\ \dot{E}_y \\ \dot{E}_z \end{bmatrix} = \frac{1}{2} \begin{bmatrix} 2E_x & 2E_y & 2E_z \\ E'_{xx} + \frac{C_{\Theta/2}}{S_{\Theta/2}} & E'_{xy} - E_z & E'_{xz} + E_y \\ E'_{xy} + E_z & E'_{yy} + \frac{C_{\Theta/2}}{S_{\Theta/2}} & E'_{yz} - E_x \\ E'_{xz} - E_y & E'_{yz} + E_x & E'_{zz} + \frac{C_{\Theta/2}}{S_{\Theta/2}} \end{bmatrix} \begin{bmatrix} \omega_{x_2} \\ \omega_{y_2} \\ \omega_{z_2} \end{bmatrix} \tag{1.3.5}$$

where $E'_{ij} = -E_i E_j C_{\Theta/2}/S_{\Theta/2}$. Notice that Eq. (1.3.5) has a singularity when $\Theta = 0$. Although this characteristic is not typically referred to as gimbal lock in the traditional sense, it represents a similar limitation in this formulation. Because of this, the Euler axis formulation is not commonly used in flight simulation. However, it forms the basis of an alternate formulation that does not suffer from singularities. Before discussing this alternate formulation, it is helpful to understand the concept of a quaternion, and the related algebra.

1.4 Quaternion Algebra

The concept of a quaternion and the associated algebraic system was developed by William R. Hamilton [6–8]. A quaternion consists of four components, which include a scalar component, and three vector components. Here we will use the brackets {} to denote a quaternion. A quaternion $\{\mathbf{Q}\}$ can be written in the form

$$\{\mathbf{Q}\} \equiv Q_0 + Q_x \mathbf{i}_x + Q_y \mathbf{i}_y + Q_z \mathbf{i}_z \tag{1.4.1}$$

where Q_0, Q_x, Q_y, and Q_z are scalars, and $\mathbf{i}_x$, $\mathbf{i}_y$, and $\mathbf{i}_z$ are unit vectors in the Cartesian coordinate system.

Because a quaternion includes both scalar and vector components, the associated algebra includes both scalar and vector-like characteristics. Quaternion algebra also includes characteristics similar to complex algebra. In fact, scalars, vectors, and complex numbers can all be thought of as special cases of the more general mathematical construct of a quaternion. For example, a scalar is simply a quaternion for which $Q_x = Q_y = Q_z = 0$. A vector is a quaternion for which $Q_0 = 0$. A complex number is a quaternion for which two of the vector components are zero. For example, $Q_y = Q_z = 0$.

Quaternion addition and subtraction follow the commutative law. This yields

$$\{\mathbf{A}\} + \{\mathbf{B}\} = \{\mathbf{B}\} + \{\mathbf{A}\} = (A_0 + B_0) + (A_x + B_x)\mathbf{i}_x + (A_y + B_y)\mathbf{i}_y + (A_z + B_z)\mathbf{i}_z \tag{1.4.2}$$

In order to perform multiplication, a *quaternion product* is used. This product is different than a cross product or dot product. The quaternion product of unit vectors in the Cartesian coordinate system is defined as

$$\begin{array}{lll} \mathbf{i}_x \otimes \mathbf{i}_x \equiv -1, & \mathbf{i}_x \otimes \mathbf{i}_y \equiv \mathbf{i}_z, & \mathbf{i}_x \otimes \mathbf{i}_z \equiv -\mathbf{i}_y, \\ \mathbf{i}_y \otimes \mathbf{i}_x \equiv -\mathbf{i}_z, & \mathbf{i}_y \otimes \mathbf{i}_y \equiv -1, & \mathbf{i}_y \otimes \mathbf{i}_z \equiv \mathbf{i}_x, \\ \mathbf{i}_z \otimes \mathbf{i}_x \equiv \mathbf{i}_y, & \mathbf{i}_z \otimes \mathbf{i}_y \equiv -\mathbf{i}_x, & \mathbf{i}_z \otimes \mathbf{i}_z \equiv -1 \end{array} \tag{1.4.3}$$

Note that the quaternion product of unit vectors follows the right hand rule, except that the quaternion product of a unit vector with itself is -1. Hence, the quaternion product has characteristics similar to both the vector cross product and complex numbers. The quaternion product follows the distributive law, which yields

$$\begin{aligned} \{\mathbf{A}\} \otimes \{\mathbf{B}\} &= (A_0 + A_x\mathbf{i}_x + A_y\mathbf{i}_y + A_z\mathbf{i}_z) \otimes (B_0 + B_x\mathbf{i}_x + B_y\mathbf{i}_y + B_z\mathbf{i}_z) \\ &= (A_0B_0 - A_xB_x - A_yB_y - A_zB_z) \\ &\quad + (A_0B_x + A_xB_0 + A_yB_z - A_zB_y)\mathbf{i}_x \\ &\quad + (A_0B_y - A_xB_z + A_yB_0 + A_zB_x)\mathbf{i}_y \\ &\quad + (A_0B_z + A_xB_y - A_yB_x + A_zB_0)\mathbf{i}_z \end{aligned} \tag{1.4.4}$$

The quaternion product is not commutative. Therefore, in general, $\{\mathbf{A}\} \otimes \{\mathbf{B}\} \neq \{\mathbf{B}\} \otimes \{\mathbf{A}\}$. Note that the quaternion product of two scalars (i.e. the vector components of both quaternions are zero) results in the multiplication of the two scalars. Also note that the quaternion product of two vectors (i.e. the scalar component of both quaternions is zero) does not result in the vector dot product or the vector cross product. Rather it results in

$$\mathbf{A} \otimes \mathbf{B} = -\mathbf{A} \cdot \mathbf{B} + \mathbf{A} \times \mathbf{B} \tag{1.4.5}$$

The magnitude of a quaternion is defined similar to that of a vector or complex number

$$|\{\mathbf{Q}\}| \equiv \sqrt{Q_0^2 + Q_x^2 + Q_y^2 + Q_z^2} \tag{1.4.6}$$

and the conjugate is defined as

$$\{\mathbf{Q}\}^* \equiv Q_0 - Q_x\mathbf{i}_x - Q_y\mathbf{i}_y - Q_z\mathbf{i}_z \tag{1.4.7}$$

where the asterisk denotes a conjugate. The quaternion product of a quaternion with its conjugate results in

$$\{\mathbf{Q}\} \otimes \{\mathbf{Q}\}^* = Q_0^2 + Q_x^2 + Q_y^2 + Q_z^2 = |\{\mathbf{Q}\}|^2 \tag{1.4.8}$$

which is similar to that of a complex number.

With a basic understanding of the concept of a quaternion and quaternion algebra, we now turn to the development of an alternate form of coordinate transformations called the Euler-Rodrigues quaternion.

1.5 Euler-Rodrigues Quaternion

The Euler-Rodrigues transformation can be developed from the Euler axis definition through the change of variables

$$\begin{bmatrix} e_0 \\ e_x \\ e_y \\ e_z \end{bmatrix} \equiv \begin{bmatrix} \cos(\Theta/2) \\ E_x \sin(\Theta/2) \\ E_y \sin(\Theta/2) \\ E_z \sin(\Theta/2) \end{bmatrix} \tag{1.5.1}$$

These four components are commonly known as the *Euler-Rodrigues symmetric parameters* and represent four components of a quaternion that can be used for transformations. This formulation is widely used in many fields of study. However, there is a great deal of variability in the order in which the parameters are listed and how they are labeled. Here we will use the subscript 0 to denote the scalar component, and the subscripts x, y, and z to denote the three vector components.

Using the definitions given in Eq. (1.5.1), the transformation from Eq. (1.3.3) known as Euler's formula can be expressed in terms of the Euler-Rodrigues quaternion components as

$$\begin{bmatrix} V_{x_2} \\ V_{y_2} \\ V_{z_2} \end{bmatrix} = \begin{bmatrix} e_x^2 + e_0^2 - e_y^2 - e_z^2 & 2(e_x e_y + e_z e_0) & 2(e_x e_z - e_y e_0) \\ 2(e_x e_y - e_z e_0) & e_y^2 + e_0^2 - e_x^2 - e_z^2 & 2(e_y e_z + e_x e_0) \\ 2(e_x e_z + e_y e_0) & 2(e_y e_z - e_x e_0) & e_z^2 + e_0^2 - e_x^2 - e_y^2 \end{bmatrix} \begin{bmatrix} V_{x_1} \\ V_{y_1} \\ V_{z_1} \end{bmatrix} \tag{1.5.2}$$

The inverse transformation is

$$\begin{bmatrix} V_{x_1} \\ V_{y_1} \\ V_{z_1} \end{bmatrix} = \begin{bmatrix} e_x^2 + e_0^2 - e_y^2 - e_z^2 & 2(e_x e_y - e_z e_0) & 2(e_x e_z + e_y e_0) \\ 2(e_x e_y + e_z e_0) & e_y^2 + e_0^2 - e_x^2 - e_z^2 & 2(e_y e_z - e_x e_0) \\ 2(e_x e_z - e_y e_0) & 2(e_y e_z + e_x e_0) & e_z^2 + e_0^2 - e_x^2 - e_y^2 \end{bmatrix} \begin{bmatrix} V_{x_2} \\ V_{y_2} \\ V_{z_2} \end{bmatrix} \tag{1.5.3}$$

The transformation given in Eq. (1.5.2) can also be written in terms of quaternion products in the form

$$\begin{bmatrix} V_{x_2} \\ V_{y_2} \\ V_{z_2} \end{bmatrix} = \begin{bmatrix} e_0 \\ -e_x \\ -e_y \\ -e_z \end{bmatrix} \otimes \left(\begin{bmatrix} 0 \\ V_{x_1} \\ V_{y_1} \\ V_{z_1} \end{bmatrix} \otimes \begin{bmatrix} e_0 \\ e_x \\ e_y \\ e_z \end{bmatrix} \right) \tag{1.5.4}$$

Likewise, the transformation given in Eq. (1.5.3) can be written in terms of quaternion products by applying the conjugate of the quaternion, which yields

$$\begin{bmatrix} v_{x_1} \\ v_{y_1} \\ v_{z_1} \end{bmatrix} = \begin{bmatrix} e_0 \\ e_x \\ e_y \\ e_z \end{bmatrix} \otimes \left(\begin{bmatrix} 0 \\ v_{x_2} \\ v_{y_2} \\ v_{z_2} \end{bmatrix} \otimes \begin{bmatrix} e_0 \\ -e_x \\ -e_y \\ -e_z \end{bmatrix} \right) \tag{1.5.5}$$

Because there is a zero in the quaternion multiplication shown in Eq. (1.5.4), this equation can be expanded and simplified to a form that is more computationally efficient than that given in Eq. (1.5.2) or (1.5.4). Using a temporary quaternion $\{\mathbf{T}\}$, the transformation can be computed more efficiently using a two-step process suggested by Phillips [2]

$$\{\mathbf{T}\} = \begin{bmatrix} T_0 \\ T_x \\ T_y \\ T_z \end{bmatrix} = \begin{bmatrix} 0 \\ v_{x_1} \\ v_{y_1} \\ v_{z_1} \end{bmatrix} \otimes \begin{bmatrix} e_0 \\ e_x \\ e_y \\ e_z \end{bmatrix} = \begin{bmatrix} -v_{x_1}e_x - v_{y_1}e_y - v_{z_1}e_z \\ v_{x_1}e_0 + v_{y_1}e_z - v_{z_1}e_y \\ -v_{x_1}e_z + v_{y_1}e_0 + v_{z_1}e_x \\ v_{x_1}e_y - v_{y_1}e_x + v_{z_1}e_0 \end{bmatrix} \tag{1.5.6}$$

$$\begin{bmatrix} v_{x_2} \\ v_{y_2} \\ v_{z_2} \end{bmatrix} = \begin{bmatrix} e_0 \\ -e_x \\ -e_y \\ -e_z \end{bmatrix} \otimes \begin{bmatrix} T_0 \\ T_x \\ T_y \\ T_z \end{bmatrix} = \begin{bmatrix} e_0T_x - e_xT_0 - e_yT_z + e_zT_y \\ e_0T_y + e_xT_z - e_yT_0 - e_zT_x \\ e_0T_z - e_xT_y + e_yT_x - e_zT_0 \end{bmatrix} \tag{1.5.7}$$

A similar process can be used to efficiently compute transformations given by Eq. (1.5.5)

EXAMPLE 1.5.1 Develop the quaternion expression for the components of the weight vector in the body-fixed coordinate system for an arbitrary orientation of the vehicle relative to the Earth-fixed coordinate system.

Solution: The weight vector of the vehicle in the Earth-fixed coordinate system is $(W_{x_f}, W_{y_f}, W_{z_f}) = (0, 0, W)$. Using this in Eq. (1.5.2), the weight vector from the Earth-fixed coordinate system can be expressed in the body-fixed coordinate system as

$$\begin{bmatrix} W_{x_b} \\ W_{y_b} \\ W_{z_b} \end{bmatrix} = \begin{bmatrix} e_x^2 + e_0^2 - e_y^2 - e_z^2 & 2(e_xe_y + e_ze_0) & 2(e_xe_z - e_ye_0) \\ 2(e_xe_y - e_ze_0) & e_y^2 + e_0^2 - e_x^2 - e_z^2 & 2(e_ye_z + e_xe_0) \\ 2(e_xe_z + e_ye_0) & 2(e_ye_z - e_xe_0) & e_z^2 + e_0^2 - e_x^2 - e_y^2 \end{bmatrix} \begin{bmatrix} 0 \\ 0 \\ W \end{bmatrix}$$

Simplifying the right-hand side gives an expression for the weight vector in the vehicle coordinate system

$$\begin{bmatrix} W_{x_b} \\ W_{y_b} \\ W_{z_b} \end{bmatrix} = W \begin{bmatrix} 2(e_xe_z - e_ye_0) \\ 2(e_ye_z + e_xe_0) \\ e_z^2 + e_0^2 - e_x^2 - e_y^2 \end{bmatrix}$$

Differentiating Eq. (1.5.1) with respect to time gives a relationship between the time rate of change of the Euler axis components and the quaternion components

$$\begin{bmatrix} \dot{e}_0 \\ \dot{e}_x \\ \dot{e}_y \\ \dot{e}_z \end{bmatrix} = \frac{\dot{\Theta}}{2} \begin{bmatrix} -\sin(\Theta/2) \\ E_x \cos(\Theta/2) \\ E_y \cos(\Theta/2) \\ E_z \cos(\Theta/2) \end{bmatrix} + \begin{bmatrix} 0 \\ \dot{E}_x \sin(\Theta/2) \\ \dot{E}_y \sin(\Theta/2) \\ \dot{E}_z \sin(\Theta/2) \end{bmatrix} \tag{1.5.8}$$

Using this result along with Eq. (1.5.1) in Eq. (1.3.5) gives the time rate of change of the quaternion as a function of the quaternion and rotation rate of the dependent coordinate system

$$\begin{bmatrix} \dot{e}_0 \\ \dot{e}_x \\ \dot{e}_y \\ \dot{e}_z \end{bmatrix} = \frac{1}{2} \begin{bmatrix} -e_x & -e_y & -e_z \\ e_0 & -e_z & e_y \\ e_z & e_0 & -e_x \\ -e_y & e_x & e_0 \end{bmatrix} \begin{bmatrix} \omega_{x_2} \\ \omega_{y_2} \\ \omega_{z_2} \end{bmatrix} \tag{1.5.9}$$

This time rate-of-change of the quaternion can also be written in terms of a quaternion product in the form

$$\begin{bmatrix} \dot{e}_0 \\ \dot{e}_x \\ \dot{e}_y \\ \dot{e}_z \end{bmatrix} = \frac{1}{2} \begin{bmatrix} e_0 \\ e_x \\ e_y \\ e_z \end{bmatrix} \otimes \begin{bmatrix} 0 \\ \omega_{x_2} \\ \omega_{y_2} \\ \omega_{z_2} \end{bmatrix} \tag{1.5.10}$$

However, in terms of computational costs, Eq. (1.5.9) is typically more efficient.

Quaternion Normalization

Because the four components of the quaternion are used to represent only three degrees of freedom, some redundancy is included in the formulation. The final equation to close the formulation includes quaternion normalization. Because the Euler axis **E** is a unit vector, it can be shown that

$$e_0^2 + e_x^2 + e_y^2 + e_z^2 = 1 \tag{1.5.11}$$

Therefore, during the course of any simulation, the length of the quaternion must remain unity.

During the integration procedures common to flight simulation, the differential equation representing the change in the quaternion with rotation rate given in Eqs. (1.5.9) or (1.5.10) will be integrated forward in time. Due to discretization and round-off error, the length of the quaternion may change with time. This results in what is called *orthogonality error*. This error can be significantly reduced by periodically normalizing the quaternion. The normalized quaternion $\{\mathbf{e}\}_r$ can be computed from

$$\{\mathbf{e}\}_r = \frac{\{\mathbf{e}\}}{|\{\mathbf{e}\}|} = \frac{\{\mathbf{e}\}}{\sqrt{e_0^2 + e_x^2 + e_y^2 + e_z^2}} \tag{1.5.12}$$

If the orthogonality error is sufficiently small, the normalized quaternion $\{\mathbf{e}\}_r$ can be very well approximated as

$$\{\mathbf{e}\}_r \approx \{\mathbf{e}\}[1.5 - 0.5(e_0^2 + e_x^2 + e_y^2 + e_z^2)] \tag{1.5.13}$$

Equation (1.5.13) is typically more computationally efficient than Eq. (1.5.12). Application of either equation periodically throughout the simulation is sufficient to keep the orthogonality error at satisfactory levels. For a thorough discussion on error resulting from numerical integration of the quaternion formulation, see Phillips [2]. A condensed discussion is included in Section 5.7 of this book.

1.6 Relations Between Attitude Descriptors

Because Euler angles are generally easier to understand and visualize, user inputs and outputs to software depicting vehicle orientation are often expressed in terms of Euler angles. However, because quaternions are significantly more efficient for computation, the underlying simulation is typically performed using quaternions. In some cases, it is also useful to obtain the Euler axis representation of an orientation. Hence, it is important to be able to convert between the Euler angle, Euler axis, and quaternion representations of an orientation.

The transformation matrix given in Eq. (1.2.5) for the Euler-angle transformation and that given in Eq. (1.5.2) for the quaternion representation must be equal. This gives

$$\begin{bmatrix} e_x^2 + e_0^2 - e_y^2 - e_z^2 & 2(e_x e_y + e_z e_0) & 2(e_x e_z - e_y e_0) \\ 2(e_x e_y - e_z e_0) & e_y^2 + e_0^2 - e_x^2 - e_z^2 & 2(e_y e_z + e_x e_0) \\ 2(e_x e_z + e_y e_0) & 2(e_y e_z - e_x e_0) & e_z^2 + e_0^2 - e_x^2 - e_y^2 \end{bmatrix}$$
$$= \begin{bmatrix} C_\theta C_\psi & C_\theta S_\psi & -S_\theta \\ S_\phi S_\theta C_\psi - C_\phi S_\psi & S_\phi S_\theta S_\psi + C_\phi C_\psi & S_\phi C_\theta \\ C_\phi S_\theta C_\psi + S_\phi S_\psi & C_\phi S_\theta S_\psi - S_\phi C_\psi & C_\phi C_\theta \end{bmatrix} \tag{1.6.1}$$

Equating each term in the transformation matrices and adding the constraint that the length of the quaternion must be unity, it can be shown that [2]

$$\begin{bmatrix} e_0 \\ e_x \\ e_y \\ e_z \end{bmatrix} = \pm \begin{bmatrix} C_{\phi/2} C_{\theta/2} C_{\psi/2} + S_{\phi/2} S_{\theta/2} S_{\psi/2} \\ S_{\phi/2} C_{\theta/2} C_{\psi/2} - C_{\phi/2} S_{\theta/2} S_{\psi/2} \\ C_{\phi/2} S_{\theta/2} C_{\psi/2} + S_{\phi/2} C_{\theta/2} S_{\psi/2} \\ C_{\phi/2} C_{\theta/2} S_{\psi/2} - S_{\phi/2} S_{\theta/2} C_{\psi/2} \end{bmatrix} \tag{1.6.2}$$

Although both the positive and negative solutions are valid, typically the positive solution is used. The inverse relation can also be obtained from Eq. (1.6.1). However, due to gimbal lock, the case when the vehicle is oriented with $\theta = \pm 90$ deg must be addressed specifically. The following algorithm suggested by Phillips [2] can be used for all scenarios

$$\text{if } (e_0 e_y - e_x e_z = 0.5)$$
$$\begin{bmatrix} \phi \\ \theta \\ \psi \end{bmatrix} = \begin{bmatrix} 2\sin^{-1}[e_x/\cos(\pi/4)] + \psi \\ \pi/2 \\ \text{arbitrary} \end{bmatrix}$$
$$\text{if } (e_0 e_y - e_x e_z = -0.5)$$
$$\begin{bmatrix} \phi \\ \theta \\ \psi \end{bmatrix} = \begin{bmatrix} 2\sin^{-1}[e_x/\cos(\pi/4)] - \psi \\ -\pi/2 \\ \text{arbitrary} \end{bmatrix} \tag{1.6.3}$$
$$\text{else}$$
$$\begin{bmatrix} \phi \\ \theta \\ \psi \end{bmatrix} = \begin{bmatrix} \text{atan2}[2(e_0 e_x + e_y e_z), (e_0^2 + e_z^2 - e_x^2 - e_y^2)] \\ \sin^{-1}[2(e_0 e_y - e_x e_z)] \\ \text{atan2}[2(e_0 e_z + e_x e_y), (e_0^2 + e_x^2 - e_y^2 - e_z^2)] \end{bmatrix}$$
$$\text{if } (\psi < 0) \ \psi = \psi + 2\pi$$

The first two if statements in this algorithm test for the case of gimbal lock, or when the x axis of the vehicle is perfectly vertical. Due to machine precision limitations, it is poor coding practice to evaluate whether two real numbers are equal. It is better practice to compare whether two real numbers are sufficiently close in value to be considered equal. For example, the first if statement could be replaced with a comparison statement of the form

$$\text{if } [\text{abs}(e_0 e_y - e_x e_z - 0.5) < \varepsilon_t] \tag{1.6.4}$$

where ε_t is a number on the order of machine precision. Note that in the case of gimbal lock, the azimuth angle is arbitrary. In that case, any value for the azimuth angle may be used within the typical limits $0 \le \psi < 360$ deg. Often zero is used for this special case. However, note that the choice of azimuth angle affects the solution for the bank angle, ϕ. The final if statement in Eq. (1.6.3) ensures the azimuth angle is limited to the limits $0 \le \psi < 360$ deg. This is necessary since the atan2 function returns values from $-\pi$ to π.

By the definition given in Eq. (1.5.1) the Euler-Rodrigues quaternion is related to the Euler-axis formulation according to

$$\begin{bmatrix} e_0 \\ e_x \\ e_y \\ e_z \end{bmatrix} = \begin{bmatrix} \cos(\Theta/2) \\ E_x \sin(\Theta/2) \\ E_y \sin(\Theta/2) \\ E_z \sin(\Theta/2) \end{bmatrix} \tag{1.6.5}$$

The inverse of this yields

$$\begin{bmatrix} \Theta \\ E_x \\ E_y \\ E_z \end{bmatrix} = \begin{bmatrix} 2\cos^{-1}(e_0) \\ e_x/\sin(\Theta/2) \\ e_y/\sin(\Theta/2) \\ e_z/\sin(\Theta/2) \end{bmatrix} \tag{1.6.6}$$

Note that for the case when $\Theta = 0$ (i.e. $e_0 = \pm 1$), Eq. (1.6.6) is indeterminate. However, for this case there is no rotation, making the Euler axis inconsequential.

EXAMPLE 1.6.1 A vehicle is oriented with an attitude of $\phi = 60$ deg, $\theta = 20$ deg, and $\psi = 45$ deg relative to the Earth-fixed coordinate system. Compute the quaternion and Euler axis definitions of this orientation.

Solution: Using the Euler angle values in Eq. (1.6.2) gives the quaternion

$$e_0 = 0.8211739209206, \quad e_x = 0.3973725000478,$$
$$e_y = 0.3273712586793, \quad e_z = 0.2461636585616$$

Using the resulting quaternion values in Eq. (1.6.6) gives the Euler axis terms

$$\Theta = 69.595038053138 \text{ deg}, \quad E_x = 0.6963164740505,$$
$$E_y = 0.5736531856673, \quad E_z = 0.4313529767367$$

1.7 Problems

Theory Problems

1.T1 Prove Eq. (1.2.7) is true by using three rotation matrices to develop the transformation.

1.T2 Find the Euler-axis representation of the orientation given by the Euler angles $\phi = 0$, $\theta = 45$ deg, $\psi = 0$.

1.T3 Find the Euler-axis representation of the orientation given by the Euler angles $\phi = 0$, $\theta = 0$, $\psi = 45$ deg.

1.T4 Develop Eq. (1.5.2) from the definitions given in Eq. (1.5.1) and Euler's formula given in Eq. (1.3.3).

1.T5 Show that Eqs. (1.5.2) and (1.5.4) are equivalent.

1.T6 Use Eqs. (1.3.5), (1.5.1), and (1.5.8) to develop the solution given in Eq. (1.5.9).

1.T7 Show that Eqs. (1.5.9) and (1.5.10) are equivalent.

1.T8 Using the definitions given in Eqs. (1.3.1) and (1.5.1), show that Eq. (1.5.11) is true.

1.T9 Starting with Eq. (1.5.12) and assuming the length of the quaternion is nearly unity, develop the approximation given in Eq. (1.5.13).

1.T10 Starting from Eq. (1.6.1), show that Eq. (1.6.2) is true.

1.T11 Starting from Eq. (1.6.1), show that the general solution given in Eq. (1.6.3) is true.

Coding Problems

1.C1 Write a function named `euler_base_to_dependent` to compute a vector transformation using Euler angles. The function should accept an array of length three containing the three components of a vector in the **base** coordinate system as well as an array of length three containing the three Euler angles $[\phi, \theta, \psi]$ that represent the orientation of the dependent coordinate system. This function should return an array of length three containing the vector components in the **dependent** coordinate system.

1.C2 Write a function named `euler_dependent_to_base` to compute an inverse transformation using Euler angles. The function should accept an array of length three containing the three components of a vector in the **dependent** coordinate system as well as an array of length three containing the three Euler angles $[\phi, \theta, \psi]$ that represent the orientation of the dependent coordinate system. This function should return an array of length three containing the vector components in the **base** coordinate system.

1.C3 Write a function named `quat_mult` that accepts two arrays of length four representing two quaternions and returns the quaternion product.

1.C4 Write a function named `quat_base_to_dependent` to compute a vector transformation using the Euler-Rodrigues quaternion. The function should accept an array of length three containing the three components of a vector in the **base** coordinate system as well as an array of length four containing the four components of the quaternion that represent the orientation of the dependent coordinate system. This function should return an array of length three containing the vector components in the **dependent** coordinate system.

1.C5 Write a function named `quat_dependent_to_base` to compute an inverse transformation using the Euler-Rodrigues quaternion. The function should accept an array of length three containing the three components of a vector in the **dependent** coordinate system as well as an array of length four containing the four components of the quaternion that represent the orientation of the dependent coordinate system. This function should return an array of length three containing the vector components in the **base** coordinate system.

1.C6 Write a function named `quat_norm` that accepts an array of length four containing a quaternion, and returns the normalized quaternion. For this computation, use the exact solution given in Eq. (1.5.12). You may wish to overwrite the quaternion that was passed in to the function with the normalized solution.

1.C7 Write a function named `euler_to_quat` that accepts an array of length three containing the three Euler angles and returns an array of length four containing values for the equivalent quaternion components.

1.C8 Write a function named `quat_to_euler` that accepts an array of length four containing the four components of a quaternion and returns an array of length three containing values for the equivalent Euler angles.

1.C9 The following table shows example solutions for relationships between Euler angles and quaternions.

Case	A	B	C	D
ϕ [deg]	0	45	179	0
θ [deg]	0	45	-60	±90
ψ [deg]	0	45	270	0
e_0	1	0.84462319862	-0.35888381816	0.70710678119
e_x	0	0.19134171618	-0.60926382222	0.0
e_y	0	0.46193976626	0.61543441464	±0.70710678119
e_z	0	0.19134171618	-0.34819603857	0.0

Check your functions written for Problems 1.C7 and 1.C8 by generating the solutions shown in this table. Make sure to check the conversions in both directions.

1.C10 Add all code developed for the problems listed in this chapter to a single "helper" file, which can easily be included in future code development. Modify your code to run as quickly as possible. You can test the speed of your code using the wrapper code available on the book website. The output of this wrapper should produce the results available on the website.

Chapter 2

Mass and Inertia Modeling

2.1 Introduction

The dynamic motion of a vehicle depends on the total mass of the vehicle as well as the distribution of mass within the vehicle. Therefore, in order to accurately simulate vehicle motion, the mass, center of gravity location, and inertial properties of the vehicle must be given. Estimates for each of these vehicle properties can be obtained using various methods.

If the geometry and construction materials of the vehicle are well defined, the mass properties of a complete vehicle can be estimated prior to construction using computer-aided-design (CAD) software. Once the vehicle has been built, the mass properties can be experimentally determined using bifilar and trifilar pendulum measurement methods. These computational and measurement methods are extremely useful in the later stages of vehicle design. During the early phases of design, when the vehicle geometry is still evolving, estimates for the final vehicle mass properties can be obtained using empirical methods or by summing estimated contributions from individual components. This chapter focuses on the latter method, and presents closed-form solutions for common geometries including cuboids, spheres, cylinders, wings, and rotors. All solutions presented in this chapter are exact within the constraints listed. The mass and inertial properties of a complete vehicle can be estimated by summing the contributions of each individual component, as will be shown in Section 2.6.

Note to the reader: If an estimate for the mass and inertial properties of the vehicle you wish to simulate are known, you may choose to skip this chapter. The present chapter is intended to assist those who need a method for estimating the mass and inertial properties of a vehicle.

2.2 General Properties of an Object

The volume of an object will be denoted as $\forall$ and can be obtained by integrating the differential volume over the three-dimensional volume of the object

$$\forall = \iiint_{\forall} d\forall \tag{2.2.1}$$

The total mass of an object $\mathcal{M}$ can be obtained by integrating the density over the three-dimensional volume of the object

$$\mathcal{M} \equiv \iiint_{V} \rho(x, y, z)\, dV \tag{2.2.2}$$

where the density ρ can vary as a function of location within the geometry. In the present development, we shall assume that the density within each component is constant. This may sound limiting, but as will be shown, this formulation can be very useful for estimating very complex structures and mass distributions within a vehicle. For the special case of constant density, Eq. (2.2.2) simplifies to

$$\mathcal{M} = \rho V \tag{2.2.3}$$

The weight W of an object is related to the mass through

$$W = \mathcal{M} g \tag{2.2.4}$$

where g is the acceleration of gravity. At sea level, the standard value for the acceleration of gravity is currently accepted to be 9.80665 m/s^2. Because we are assuming constant density, the mass properties of an object can be completely defined by the object's volume, and any one of the following: mass, density, or sea-level weight.

The center of gravity (CG) of an object depends on the object shape and mass distribution within the object. Because we are limiting our development to components of constant density, the coordinates of the CG location $(\bar{x}, \bar{y}, \bar{z})$ in the component coordinate system can be computed from

$$\bar{x} = \frac{\rho}{\mathcal{M}} \iiint_{V} x\, dV \tag{2.2.5}$$

$$\bar{y} = \frac{\rho}{\mathcal{M}} \iiint_{V} y\, dV \tag{2.2.6}$$

$$\bar{z} = \frac{\rho}{\mathcal{M}} \iiint_{V} z\, dV \tag{2.2.7}$$

The mass moment of inertia of a component can be expressed in tensor form as

$$[\mathbf{I}] = \begin{bmatrix} I_{xx} & -I_{xy} & -I_{xz} \\ -I_{yx} & I_{yy} & -I_{yz} \\ -I_{zx} & -I_{zy} & I_{zz} \end{bmatrix} \tag{2.2.8}$$

where the individual terms for a component of constant density can be computed from the integrals

$$I_{xx} = \rho \iiint_{\forall} (y^2 + z^2)\, d\forall \tag{2.2.9}$$

$$I_{yy} = \rho \iiint_{\forall} (x^2 + z^2)\, d\forall \tag{2.2.10}$$

$$I_{zz} = \rho \iiint_{\forall} (x^2 + y^2)\, d\forall \tag{2.2.11}$$

$$I_{xy} = I_{yx} = \rho \iiint_{\forall} (xy)\, d\forall \tag{2.2.12}$$

$$I_{xz} = I_{zx} = \rho \iiint_{\forall} (xz)\, d\forall \tag{2.2.13}$$

$$I_{yz} = I_{zy} = \rho \iiint_{\forall} (yz)\, d\forall \tag{2.2.14}$$

Here we will refer to the mass moment of inertia as simply the *moment of inertia.*

The moment of inertia tensor $[\mathbf{I}]$ of the object about an arbitrary point $\mathbf{P} = (x, y, z)$ is related to the moment of inertia tensor $[\mathbf{I}]_{CG}$ of the object about the center of gravity $\mathbf{P}_{CG} = (x_{CG}, y_{CG}, z_{CG})$ through the *parallel axis theorem* according to

$$[\mathbf{I}] = [\mathbf{I}]_{CG} + \mathcal{M}\left[(\mathbf{s}\cdot\mathbf{s})[\mathbf{E}] - \mathbf{s}\mathbf{s}^T\right] \tag{2.2.15}$$

where $[\mathbf{E}]$ is the 3×3 identity matrix and $\mathbf{s}$ is the vector from $\mathbf{P}_{CG}$ to $\mathbf{P}$, which can be expressed as

$$\mathbf{s} = \begin{bmatrix} s_x \\ s_y \\ s_z \end{bmatrix} = \mathbf{P} - \mathbf{P}_{CG} = \begin{bmatrix} x - x_{CG} \\ y - y_{CG} \\ z - z_{CG} \end{bmatrix} \tag{2.2.16}$$

The outer product $\mathbf{s}\mathbf{s}^T$ of the column vector $\mathbf{s}$ is defined as

$$\mathbf{s}\mathbf{s}^T = \begin{bmatrix} s_x^2 & s_x s_y & s_x s_z \\ s_y s_x & s_y^2 & s_y s_z \\ s_z s_x & s_z s_y & s_z^2 \end{bmatrix} \tag{2.2.17}$$

Hence, Eq. (2.2.15) can be written as

$$[\mathbf{I}] = [\mathbf{I}]_{CG} + \mathcal{M}\left(\begin{bmatrix} s_x^2 + s_y^2 + s_z^2 & 0 & 0 \\ 0 & s_x^2 + s_y^2 + s_z^2 & 0 \\ 0 & 0 & s_x^2 + s_y^2 + s_z^2 \end{bmatrix} - \begin{bmatrix} s_x^2 & s_x s_y & s_x s_z \\ s_y s_x & s_y^2 & s_y s_z \\ s_z s_x & s_z s_y & s_z^2 \end{bmatrix}\right) \tag{2.2.18}$$

This simplifies to

$$[\mathbf{I}] = [\mathbf{I}]_{CG} + \mathcal{M}\begin{bmatrix} s_y^2 + s_z^2 & -s_x s_y & -s_x s_z \\ -s_y s_x & s_x^2 + s_z^2 & -s_y s_z \\ -s_z s_x & -s_z s_y & s_x^2 + s_y^2 \end{bmatrix} \tag{2.2.19}$$

In component form, this yields

$$\begin{aligned} I_{xx} &= I_{xx_{CG}} + \mathcal{M}(s_y^2 + s_z^2) \\ I_{yy} &= I_{yy_{CG}} + \mathcal{M}(s_x^2 + s_z^2) \\ I_{zz} &= I_{zz_{CG}} + \mathcal{M}(s_x^2 + s_y^2) \\ I_{xy} = I_{yx} &= I_{xy_{CG}} - \mathcal{M}s_x s_y \\ I_{xz} = I_{zx} &= I_{xz_{CG}} - \mathcal{M}s_x s_z \\ I_{yz} = I_{zy} &= I_{yz_{CG}} - \mathcal{M}s_y s_z \end{aligned} \tag{2.2.20}$$

Note that the results of the parallel axis theorem are independent of the sign of s. Hence, the solution is identical whether s is computed from $\mathbf{s} = \mathbf{P} - \mathbf{P}_{CG}$ or $\mathbf{s} = \mathbf{P}_{CG} - \mathbf{P}$. In some cases, the inertia of an object may be known about an arbitrary point **P** instead of the center of gravity. In this case, the inertia about the center of gravity can be computed by rearranging Eq. (2.2.15) to yield

$$[\mathbf{I}]_{CG} = [\mathbf{I}] - \mathcal{M}\left[(\mathbf{s}\cdot\mathbf{s})[\mathbf{E}] - \mathbf{s}\mathbf{s}^T\right] \tag{2.2.21}$$

The mass and moment-of-inertia properties of a complete vehicle can be estimated by summing the influences of the individual components that make up the vehicle. This chapter includes solutions for common simple geometries such as cuboids, cylinders, and spheres, that can be used to model the mass properties of many vehicle components. Additionally, estimates for more complex geometries specific to aircraft such as wings and rotors are also presented.

2.3 Simple Geometries

Rectangular Cuboid

A cuboid is a geometric entity with six faces, also known as a hexahedron. We will limit the current analysis to a rectangular cuboid, which is a cuboid for which all faces are rectangles, and all neighboring faces meet at right angles, as shown in Fig. 2.3.1. The center of gravity of a rectangular cuboid of constant density is located at the center of the object. We define a Cartesian coordinate system such that the origin lies at the object CG and each axis is parallel or perpendicular to the cuboid faces. The cuboid size is defined by a length of the cuboid along each axis, as shown in Fig. 2.3.1.

The volume of the rectangular cuboid is obtained from the integral

$$\forall \equiv \iiint_{\forall} d\forall = \int_{-l_z/2}^{l_z/2}\int_{-l_y/2}^{l_y/2}\int_{-l_x/2}^{l_x/2} dxdydz = l_x l_y l_z \tag{2.3.1}$$

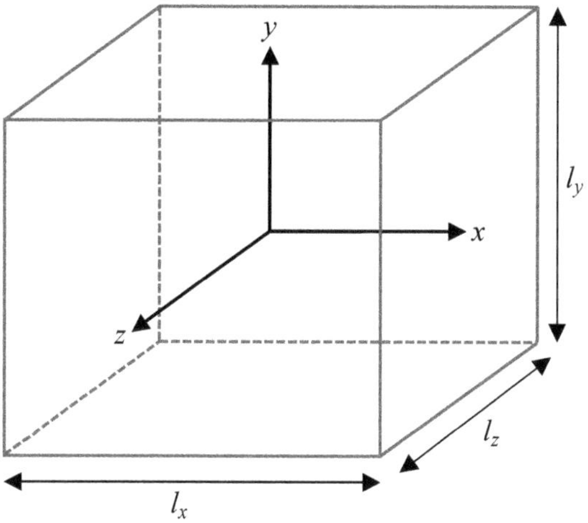

Figure 2.3.1. Rectangular cuboid geometric definition and coordinate system.

Using these same limits of integration in Eqs. (2.2.9)–(2.2.14), performing the integration, and applying the result to Eq. (2.2.8) gives the inertia tensor for a rectangular cuboid of constant density computed about the cuboid origin

$$[\mathbf{I}] = \frac{\mathcal{M}}{12}\begin{bmatrix} (l_y^2 + l_z^2) & 0 & 0 \\ 0 & (l_x^2 + l_z^2) & 0 \\ 0 & 0 & (l_x^2 + l_y^2) \end{bmatrix} \tag{2.3.2}$$

For the case of a hollow rectangular cuboid with inner dimensions l_{x1}, l_{y1}, l_{z1} and outer dimensions l_{x2}, l_{y2}, l_{z2}, it can be shown that the volume is

$$\mathcal{V} = \mathcal{V}_2 - \mathcal{V}_1 \tag{2.3.3}$$

where $\mathcal{V}_1 = l_{x1} l_{y1} l_{z1}$ and $\mathcal{V}_2 = l_{x2} l_{y2} l_{z2}$. The inertia tensor about the origin for the hollow rectangular cuboid of constant density is

$$[\mathbf{I}] = \frac{\mathcal{M}}{12(\mathcal{V}_2 - \mathcal{V}_1)}\begin{bmatrix} A & 0 & 0 \\ 0 & B & 0 \\ 0 & 0 & C \end{bmatrix} \tag{2.3.4}$$

where

$$\begin{aligned} A &= \mathcal{V}_2(l_{y2}^2 + l_{z2}^2) - \mathcal{V}_1(l_{y1}^2 + l_{z1}^2) \\ B &= \mathcal{V}_2(l_{x2}^2 + l_{z2}^2) - \mathcal{V}_1(l_{x1}^2 + l_{z1}^2) \\ C &= \mathcal{V}_2(l_{x2}^2 + l_{y2}^2) - \mathcal{V}_1(l_{x1}^2 + l_{y1}^2) \end{aligned} \tag{2.3.5}$$

EXAMPLE 2.3.1 A vehicle has a fuselage that can be estimated as hollow rectangular cuboid of constant density. It has outer dimensions

$$l_{x_2} = 5.3 \text{ ft}, \ l_{y_2} = 0.8 \text{ ft}, \ l_{z_2} = 1.0 \text{ ft}$$

and inner dimensions of

$$l_{x_1} = 5.0 \text{ ft}, \ l_{y_1} = 0.5 \text{ ft}, \ l_{z_1} = 0.7 \text{ ft}$$

The cuboid is made of expanded polypropylene (EPP) foam, with a density of 0.025 slug/ft^3. Compute the volume of the foam as well as the mass of the fuselage, the weight of the fuselage at sea level, and the inertia tensor of the fuselage about the cuboid origin.

Solution: The volume from the outer cuboid dimensions is

$$\mathcal{V}_2 = l_{x_2} l_{y_2} l_{z_2} = (5.3)(0.8)(1.0) = 4.24 \text{ ft}^3$$

and the volume from the inner dimensions is

$$\mathcal{V}_1 = l_{x_1} l_{y_1} l_{z_1} = (5.0)(0.5)(0.7) = 1.75 \text{ ft}^3$$

Hence, the total volume of the material is

$$\mathcal{V} = \mathcal{V}_2 - \mathcal{V}_1 = 4.24 - 1.75 = 2.49 \text{ ft}^3$$

Using this result in Eq. (2.2.3) along with the density of the foam gives the total mass in slugs

$$\mathcal{M} = \rho \mathcal{V} = (0.025)(2.49) = 0.06225 \text{ slug}$$

The weight of the cuboid at sea level is computed by multiplying the mass by the gravitational constant on the surface of the Earth. This gives

$$W = \mathcal{M} g = 0.06225 \text{ slug} \left(\frac{9.80665 \text{ m/s}^2}{0.3048 \text{ m/ft}} \right) = 2.002834522638 \text{ lbf}$$

Using these results in Eq. (2.3.4) gives the inertia tensor of the hollow cuboid

$$[\mathbf{I}] = \begin{bmatrix} 0.01178875 & 0 & 0 \\ 0 & 0.164029375 & 0 \\ 0 & 0 & 0.161724375 \end{bmatrix} \text{ slug}\cdot\text{ft}^2$$

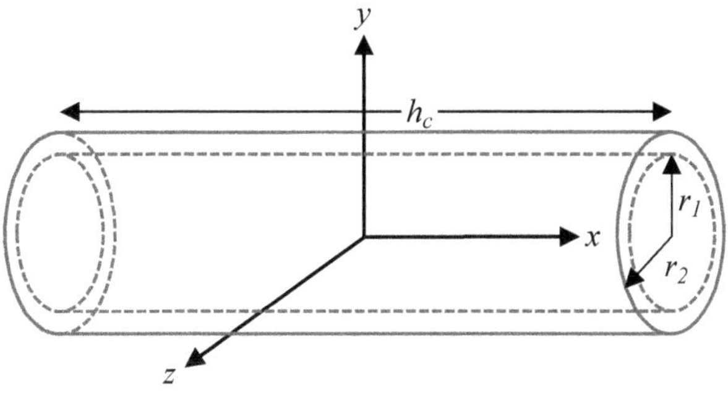

Figure 2.3.2. Cylinder geometric definition and coordinate system.

Cylinder

Consider a hollow cylinder with an inner radius r_1 and outer radius r_2 and having a length h_c as shown in Fig. 2.3.2. Here we will define a coordinate system such that the origin is placed at the cylinder center of gravity, and the x axis is aligned with the axis of the cylinder.

The volume integrals required for Eqs. (2.2.2) and (2.2.9)–(2.2.14) are most easily carried out in cylindrical coordinates (x, r, φ) where $r = y^2 + z^2$. The volume can be obtained from the integral

$$\text{¥} \equiv \iiint_{\text{¥}} d\text{¥} = \int_0^{2\pi} \int_{r_1}^{r_2} \int_{-h_c/2}^{h_c/2} dx r dr d\varphi = \pi h_c (r_2^2 - r_1^2) \tag{2.3.6}$$

The inertia tensor for the hollow cylinder of constant density relative to the component origin is

$$[\mathbf{I}] = \frac{\mathcal{M}}{12} \begin{bmatrix} 6(r_2^2 + r_1^2) & 0 & 0 \\ 0 & 3(r_2^2 + r_1^2) + h_c^2 & 0 \\ 0 & 0 & 3(r_2^2 + r_1^2) + h_c^2 \end{bmatrix} \tag{2.3.7}$$

The case of a solid cylinder of constant density can be computed by simply setting $r_1 = 0$.

EXAMPLE 2.3.2 A vehicle has a tail boom made of a carbon tube that is 3 ft long. It has an outer radius of $r_2 = 0.05$ ft, and an inner radius of $r_1 = 0.025$ ft. The boom weighs 1 lbf at sea level. Find the mass of the boom, the density of the material, and the inertia tensor of the boom.

Solution: The mass of the boom can be found by rearranging Eq. (2.2.4)

$$\mathcal{M} = \frac{W}{g} = 1.0\,\text{lbf}\left(\frac{0.3048\ \text{m/ft}}{9.80665\ \text{m/s}^2}\right) = 0.03108095017157\ \text{slug}$$

In order to compute the density, we need to find the volume of the material. The volume of the material can be found from Eq. (2.3.6)

$$\forall = \pi h_c (r_2^2 - r_1^2) = \pi(3.0)(0.05^2 - 0.025^2) = 0.01767145867644 \text{ ft}^3$$

The density can then be computed by rearranging Eq. (2.2.3)

$$\rho = \frac{\mathcal{M}}{\forall} = \frac{0.03108095017157}{0.01767145867644} = 1.758821993173 \text{ slug/ft}^3$$

Using these results in Eq. (2.3.7) gives the inertia tensor of the boom

$$[\mathbf{I}] = \begin{bmatrix} 4.856398464307\text{E}{-05} & 0 & 0 \\ 0 & 0.023334994621 & 0 \\ 0 & 0 & 0.023334994621 \end{bmatrix} \text{ slug}\cdot\text{ft}^2$$

Sphere

Consider a hollow sphere with an inner radius r_1 and outer radius r_2 as shown in Fig. 2.3.3. The origin is placed at the sphere center of gravity, which, for a sphere of constant density, is at the center of the sphere. Recognizing that a sphere is radially symmetric and the surface area of a sphere is $4\pi r^2$, the volume of the sphere can be obtained from the integral

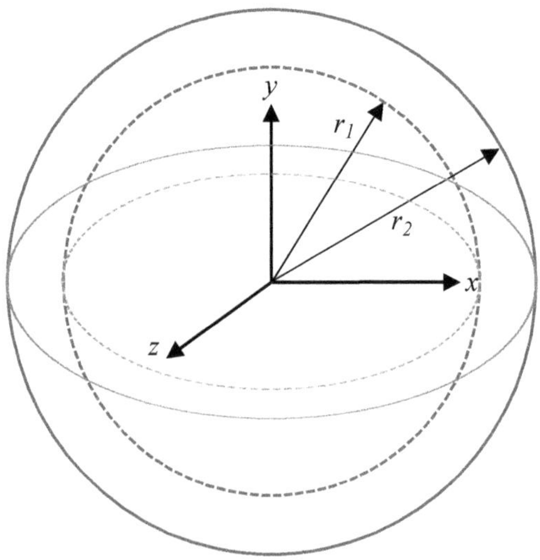

Figure 2.3.3. Sphere geometric definition and coordinate system.

$$\forall \equiv \iiint_{\forall} d\forall = \int_{r_1}^{r_2} 4\pi r^2 dr = \frac{4}{3}\pi(r_2^3 - r_1^3) \tag{2.3.8}$$

The moment of inertia of the sphere of constant density about the object origin is

$$[\mathbf{I}] = \mathcal{M}\frac{2(r_2^5 - r_1^5)}{5(r_2^3 - r_1^3)}\begin{bmatrix} 1 & 0 & 0 \\ 0 & 1 & 0 \\ 0 & 0 & 1 \end{bmatrix} \tag{2.3.9}$$

The case of a solid sphere of constant density can be computed by simply setting $r_1 = 0$.

EXAMPLE 2.3.3 A vehicle has a gimbaled camera that weighs 14 lbf at sea level and is roughly spherical shape with an outer radius of 0.4 ft. Find the average density of materials within the sphere and compute the inertia tensor assuming the density is constant throughout the gimbal.

Solution: The mass of the gimbaled camera can be found by rearranging Eq. (2.2.4)

$$\mathcal{M} = \frac{W}{g} = 14.0\ \text{lbf}\left(\frac{0.3048\ \text{m/ft}}{9.80665\ \text{m/s}^2}\right) = 0.4351333024019\ \text{slug}$$

In order to compute the average density, we need to find the volume of the sphere, which can be found from Eq. (2.3.8)

$$\forall = \frac{4}{3}\pi r_2^3 = \frac{4}{3}\pi 0.4^3 = 0.2680825731063\ \text{ft}^3$$

The density can then be computed by rearranging Eq. (2.2.3)

$$\rho = \frac{\mathcal{M}}{\forall} = \frac{0.4351333024019}{0.2680825731063} = 1.623131624559\ \text{slug/ft}^3$$

Using these results in Eq. (2.3.9), the inertia tensor of the camera is approximately

$$[\mathbf{I}] = \begin{bmatrix} 0.02784853135372 & 0 & 0 \\ 0 & 0.02784853135372 & 0 \\ 0 & 0 & 0.02784853135372 \end{bmatrix} \text{slug}\cdot\text{ft}^2$$

2.4 Wing Segment

Due to the complexity of wing geometries, the mass properties of wings are typically computed using CAD tools. However, simplified approximations have been published by various authors [9, 10] for simple wing segments. The method presented here was developed by Moulton and Hunsaker [10] and can be used to accurately estimate the volume, mass, and inertial properties of a single wing segment of constant density having geometric properties that vary linearly in the spanwise direction. More complex wing geometries can be computed by combining a number of simple wing segments.

The geometries discussed in Section 2.3 were simple geometries, of which the location of the center of gravity was known for each component due to the symmetry of the geometry and the constraint of constant density. Hence, the origin for each component was easily defined at the center of gravity. For the case of a finite wing, we shall define an origin that is not coincident with the wing center of gravity, since the center of gravity of the wing is not known a priori.

In the present analysis, the origin of the wing-segment coordinate frame is placed at the wing root quarter-chord location as shown in Fig. 2.4.1. The x axis extends from the origin toward the root leading edge, the y axis extends in the spanwise direction of a right wing, and the z axis completes the right-hand coordinate system.

A single wing segment can be defined by a span b, root chord c_r, tip chord c_t, root airfoil percent maximum thickness τ_r, tip airfoil percent maximum thickness τ_t, airfoil thickness distribution μ_a, quarter-chord sweep angle $\Lambda_{c/4}$, mounting angle α_0, and dihedral angle Γ. The mounting and dihedral angles are rigid-body rotations of the wing geometry, and will be treated differently from the sweep angle.

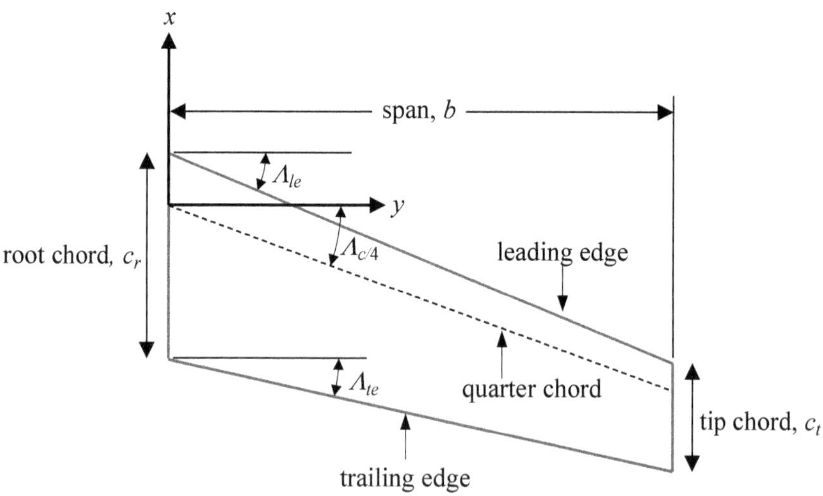

Figure 2.4.1. Planform view of wing geometric definition and coordinate system.

Wing sweep is defined as a chordwise shearing of the wing in the wing x-y plane. Positive sweep skews the wing tip aft, and negative sweep skews the wing tip forward. Three different sweep angles are sometimes referenced. These include the leading-edge, trailing-edge, and quarter-chord sweep angles. Given any one of these sweep angles along with the wing span, root chord, and tip chord, the other sweep angles can be easily computed. For example, the leading-edge and trailing-edge sweep angles are related through

$$\tan \Lambda_{te} = \tan \Lambda_{le} + \frac{c_t - c_r}{b} \tag{2.4.1}$$

The leading-edge and quarter-chord sweep angles are related through

$$\tan \Lambda_{c/4} = \tan \Lambda_{le} + \frac{c_t - c_r}{4b} \tag{2.4.2}$$

The quarter-chord and trailing-edge sweep angles are related through

$$\tan \Lambda_{te} = \tan \Lambda_{c/4} + \frac{3}{4}\frac{c_t - c_r}{b} \tag{2.4.3}$$

In the following analysis, we will use the quarter-chord sweep angle, $\Lambda_{c/4}$, i.e. the angle between the wing y axis and the line that lies along the quarter-chord of the wing.

Wing dihedral and mounting angle are both rigid-body rotations of the wing segment relative to the vehicle coordinate system. Wing dihedral is defined as a rotation of the wing about the vehicle longitudinal axis. Positive dihedral results in a wing tip above the wing root, and negative dihedral or anhedral results in a wing tip below the wing root. The wing mounting angle is defined as a rotation of the wing about the wing y axis. A positive mounting angle increases the angle of attack of the wing. In the present development, the dihedral and mounting angles will be treated as rigid-body rotations of the wing, including the wing coordinate system. With this definition, dihedral and mounting angle do not affect the mass properties of the wing relative to the wing coordinate system. In Section 2.6, the effects of object orientation, including dihedral and mounting angle, relative to the vehicle coordinate system is discussed. Hence, effects of dihedral and mounting angle will not be included in this section, but will be included in the vehicle coordinate system.

In this analysis, the airfoil thickness distribution, sweep angle, and dihedral angle are required to be invariant with spanwise location, whereas the chord and airfoil percent maximum thickness are allowed to vary linearly with spanwise location. This results in the chord distribution

$$c(y) = (c_t - c_r)\frac{y}{b} + c_r \tag{2.4.4}$$

and the airfoil percent maximum thickness distribution

$$\tau_{\max}(y) = (\tau_t - \tau_r)\frac{y}{b} + \tau_r \tag{2.4.5}$$

The dimensional maximum airfoil thickness at any spanwise location can be found from the product of the local airfoil percent maximum thickness $\tau_{\max}$ and the local chord c

$$t_{\max}(y) = \tau_{\max}(y)c(y) \tag{2.4.6}$$

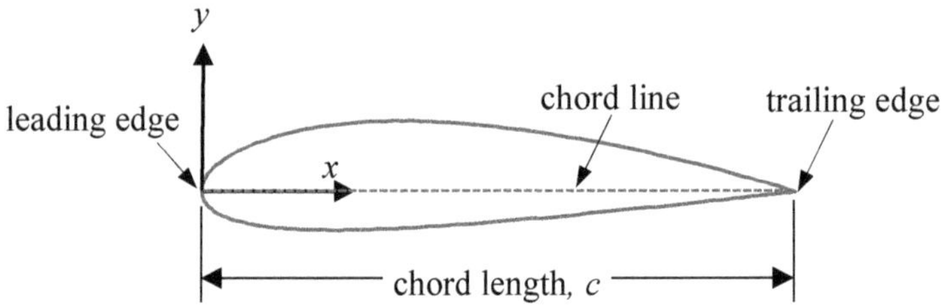

Figure 2.4.2. Airfoil coordinate system.

Note that, since both $\tau_{\max}(y)$ and $c(y)$ are allowed to vary linearly, $t_{\max}(y)$ may be nonlinear. In this development, the influence of airfoil camber and wing twist are neglected, since these generally have very little effect on the mass properties of the wing relative to the other parameters considered.

Effects of Airfoil Geometry

The airfoil used for the wing segment can have a significant effect on the mass properties of the wing. For the purposes of this discussion, it is helpful to define an airfoil coordinate system. Here we use the traditional airfoil coordinate frame, with the origin at the leading edge, the x axis pointing aft, and the y axis pointing up as shown in Fig. 2.4.2. In this development, the ordinates are normalized by the chord length and defined as $\hat{x} \equiv x/c$ and $\hat{y} \equiv y/c$.

In order to evaluate the volume integrals presented in Section 2.2 to compute the volume, CG, and inertia properties of the complete wing, one-dimensional integrals of the airfoil thickness are needed, as shown by Moulton and Hunsaker [10]. To simplify this integration process it is helpful to assume that the mass properties of the airfoil are dominated by the chordwise thickness distribution, and that effects of camber are negligible.

The thickness distribution of an airfoil as a function of percent chord $\tau(\hat{x})$ can be defined as the product of the maximum thickness $\tau_{\max}$ and a dimensionless thickness distribution function $\mu_a(\hat{x})$

$$\tau(\hat{x}) = \tau_{\max}\mu_a(\hat{x}) \tag{2.4.7}$$

In general, the mass of the complete wing segment depends on the following four integrals involving the thickness distribution $\mu_a(\hat{x})$ [10]

$$\upsilon_0 \equiv \frac{\int_0^1 \mu_a(\hat{x})d\hat{x}}{\int_0^1 d\hat{x}} = \int_0^1 \mu_a(\hat{x})d\hat{x} \tag{2.4.8}$$

$$\upsilon_1 \equiv \frac{\int_0^1 \left(\frac{1}{4} - \hat{x}\right)\mu_a(\hat{x})d\hat{x}}{\int_0^1 \left(\frac{1}{4} - \hat{x}\right)d\hat{x}} = -4\int_0^1 \left(\frac{1}{4} - \hat{x}\right)\mu_a(\hat{x})d\hat{x} \tag{2.4.9}$$

$$\upsilon_2 \equiv \frac{\int_0^1 \left(\frac{1}{4}-\hat{x}\right)^2 \mu_a(\hat{x})d\hat{x}}{\int_0^1 \left(\frac{1}{4}-\hat{x}\right)^2 d\hat{x}} = \frac{48}{7}\int_0^1 \left(\frac{1}{4}-\hat{x}\right)^2 \mu_a(\hat{x})d\hat{x} \tag{2.4.10}$$

$$\upsilon_3 \equiv \frac{\int_0^1 \mu_a(\hat{x})^3 d\hat{x}}{\int_0^1 d\hat{x}} = \int_0^1 \mu_a(\hat{x})^3 d\hat{x} \tag{2.4.11}$$

These integrals can be solved for any arbitrary thickness distribution, and the solution to these integrals affects the volume, CG, and mass of the entire wing segment. Here we shall consider how these integrals can be computed with the thickness distribution of a diamond airfoil, as well as a very general thickness distribution that can be used for a wide range of airfoils.

Diamond Airfoil Thickness Distribution

Consider a diamond airfoil with maximum thickness τ_{max} at the nondimensional chordwise location $\hat{x} = \hat{x}_{\text{max}}$. The dimensionless thickness distribution can be expressed as

$$\mu_a(\hat{x}) = \begin{cases} \dfrac{\hat{x}}{\hat{x}_{\text{max}}}, & 0 \le \hat{x} \le \hat{x}_{\text{max}} \\ \dfrac{1-\hat{x}}{1-\hat{x}_{\text{max}}}, & \hat{x}_{\text{max}} \le \hat{x} \le 1 \end{cases} \tag{2.4.12}$$

Using this in Eq. (2.4.8) gives

$$\begin{aligned} \upsilon_0 &= \int_0^{\hat{x}_{\text{max}}} \frac{\hat{x}}{\hat{x}_{\text{max}}} d\hat{x} + \int_{\hat{x}_{\text{max}}}^1 \frac{1-\hat{x}}{1-\hat{x}_{\text{max}}} d\hat{x} \\ &= \frac{1}{2}\frac{\hat{x}^2}{\hat{x}_{\text{max}}}\bigg|_0^{\hat{x}_{\text{max}}} + \frac{\hat{x}-\frac{1}{2}\hat{x}^2}{1-\hat{x}_{\text{max}}}\bigg|_{\hat{x}_{\text{max}}}^1 \\ &= \frac{1}{2} \end{aligned} \tag{2.4.13}$$

Repeating this process for the integrals given in Eqs. (2.4.9)–(2.4.11) gives

$$\upsilon_1 = \frac{4\hat{x}_{\text{max}}+1}{6}, \quad \upsilon_2 = \frac{8\hat{x}_{\text{max}}^2+3}{14}, \quad \upsilon_3 = \frac{1}{4} \tag{2.4.14}$$

A More General Thickness Distribution

An airfoil thickness distribution that can be used for a wide variety of airfoils is given by

$$\mu_a(\hat{x}) = a_0\sqrt{\hat{x}} + a_1\hat{x} + a_2\hat{x}^2 + a_3\hat{x}^3 + a_4\hat{x}^4 \tag{2.4.15}$$

Table 2.4.1. Thickness distribution coefficients of best fit for example airfoils.

Airfoil	a_0	a_1	a_2	a_3	a_4
NACA 4-Digit series	2.969	-1.260	-3.516	2.843	-1.015
NACA 4-Digit series with closed trailing edge [11]	2.980	-1.320	-3.286	2.441	-0.815
Clark Y with $\tau_{\max} = 0.117$	2.947	-1.102	-3.975	3.533	-1.399
Selig S1010 with $\tau_{\max} = 0.06$	4.375	-5.199	2.473	-2.009	0.341

where a_0, a_1, a_2, a_3, and a_4 are arbitrary constants. This is the thickness distribution used by the NACA 4-digit airfoil series, and can be applied to many other airfoils, including those not specifically designed with this thickness distribution, by varying only the parameters a_0, a_1, a_2, a_3, and a_4. Even a diamond airfoil can be reasonably approximated using these coefficients. Table 2.4.1 shows values for each of these coefficients that can be used to match or closely approximate the thickness distribution for a few example airfoils, including some that do not use a NACA 4-digit thickness distribution.

Solutions for the integrals given in Eqs. (2.4.8)–(2.4.11) for the thickness distribution given in Eq. (2.4.15) with arbitrary coefficient values a_0, a_1, a_2, a_3, and a_4 can be found analytically as shown by Moulton and Hunsaker [10]. This results in

$$\upsilon_0 = \frac{1}{60}\left(40a_0 + 30a_1 + 20a_2 + 15a_3 + 12a_4\right) \tag{2.4.16}$$

$$\upsilon_1 = \frac{1}{60}\left(56a_0 + 50a_1 + 40a_2 + 33a_3 + 28a_4\right) \tag{2.4.17}$$

$$\upsilon_2 = \frac{1}{980}\left(856a_0 + 770a_1 + 644a_2 + 553a_3 + 484a_4\right) \tag{2.4.18}$$

$$\begin{aligned}
\upsilon_3 = {} & \frac{2}{5}a_0^3 + a_0^2a_1 + \frac{3}{4}a_0^2a_2 + \frac{3}{5}a_0^2a_3 + \frac{1}{2}a_0^2a_4 + \frac{6}{7}a_0a_1^2 + \frac{4}{3}a_0a_1a_2 \\
& + \frac{12}{11}a_0a_1a_3 + \frac{12}{13}a_0a_1a_4 + \frac{6}{11}a_0a_2^2 + \frac{12}{13}a_0a_2a_3 + \frac{4}{5}a_0a_2a_4 \\
& + \frac{2}{5}a_0a_3^2 + \frac{12}{17}a_0a_3a_4 + \frac{6}{19}a_0a_4^2 + \frac{1}{4}a_1^3 + \frac{3}{5}a_1^2a_2 \\
& + \frac{1}{2}a_1^2a_3 + \frac{3}{7}a_1^2a_4 + \frac{1}{2}a_1a_2^2 + \frac{6}{7}a_1a_2a_3 + \frac{3}{4}a_1a_2a_4 \\
& + \frac{3}{8}a_1a_3^2 + \frac{2}{3}a_1a_3a_4 + \frac{3}{10}a_1a_4^2 + \frac{1}{7}a_2^3 + \frac{3}{8}a_2^2a_3 \\
& + \frac{1}{3}a_2^2a_4 + \frac{1}{3}a_2a_3^2 + \frac{3}{5}a_2a_3a_4 + \frac{3}{11}a_2a_4^2 + \frac{1}{10}a_3^3 \\
& + \frac{3}{11}a_3^2a_4 + \frac{1}{4}a_3a_4^2 + \frac{1}{13}a_4^3
\end{aligned} \tag{2.4.19}$$

Using the values given in Table 2.4.1 in Eqs. (2.4.16)–(2.4.19) yields the results in Table 2.4.2, rounded to 5 significant figures.

Table 2.4.2. Example airfoil integral results for the thickness distribution given in Eq. (2.4.15).

Airfoil	υ_0	υ_1	υ_2	υ_3
NACA 4-Digit series	0.685083	0.467050	0.395795	0.472889
NACA 4-Digit series with closed trailing edge [11]	0.678583	0.452883	0.381336	0.468420
Clark Y with $\tau_{max} = 0.117$	0.692117	0.472483	0.398801	0.488379
Selig S1010 with $\tau_{max} = 0.06$	0.707450	0.453683	0.396377	0.516323

The solution given in Eqs. (2.4.16)–(2.4.19) applies to any airfoil for which the airfoil thickness distribution can be fit to that given in Eq. (2.4.15). As stated previously, the thickness distribution given in Eq. (2.4.15) is quite versatile and can be used for most airfoils, including those outside the NACA 4-digit series. The following example demonstrates how this general solution can be applied to an arbitrary airfoil.

EXAMPLE 2.4.1 Evaluate coefficients that can be used to fit the thickness distribution given in Eq. (2.4.15) to a Göttingen 173 airfoil, and compute the resulting solutions to the integrals given in Eqs. (2.4.8)–(2.4.11) for this thickness distribution.

Solution: Coordinates for the Göttingen 173 airfoil can be obtained from the UIUC Airfoil Database maintained by Michael Selig [12]. The coordinates are given in Table 2.4.3 and the airfoil is shown in Fig. 2.4.3.

Note that the $\hat{x}$ values on the lower surface do not exactly align vertically to those on the upper surface. However, because they are very nearly vertically aligned, we will assume that the corresponding points in the lower and upper surfaces can be used to estimate the camber and thickness. The camber can be roughly estimated as the average between the upper and lower points, while the thickness can be roughly estimated as the difference in the $\hat{y}$ values of the upper and lower points. This results in the values given in Table 2.4.4.

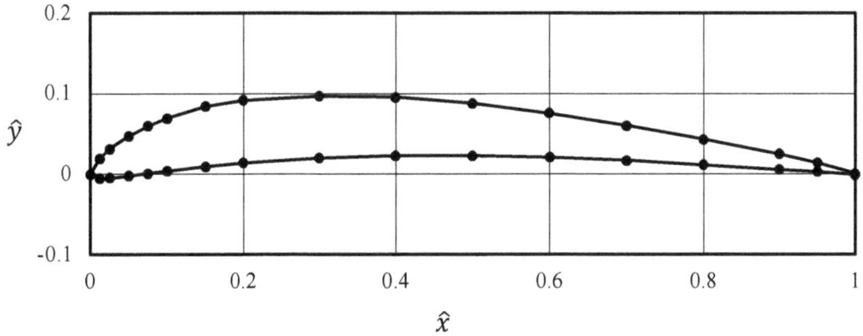

Figure 2.4.3. Göttingen 173 airfoil.

Table 2.4.3. Göttingen 173 airfoil coordinates [12].

Upper Surface		Lower Surface	
$\hat{x}$	$\hat{y}$	$\hat{x}$	$\hat{y}$
0.00000	0.00000	0.00000	0.00000
0.01241	0.01956	0.01253	-0.00544
0.02486	0.03111	0.02502	-0.00438
0.04978	0.04723	0.05001	-0.00197
0.07473	0.05934	0.07500	0.00084
0.09968	0.06896	0.09998	0.00366
0.14962	0.08369	0.14996	0.00919
0.19958	0.09152	0.19994	0.01392
0.29956	0.09648	0.29991	0.01988
0.39956	0.09534	0.39989	0.02284
0.49960	0.08780	0.49989	0.02300
0.59965	0.07556	0.59991	0.02056
0.69972	0.05992	0.69992	0.01672
0.79980	0.04298	0.79995	0.01158
0.89989	0.02484	0.89997	0.00564
0.94994	0.01397	0.94999	0.00287
1.00000	0.00090	1.00000	-0.00090

Table 2.4.4. Göttingen 173 airfoil camber and thickness estimates.

$\hat{x}$	Camber	$\tau(\hat{x})$	Eq. (2.4.15)
0.000000	0.000000	0.000000	0.000000
0.012470	0.007060	0.025000	0.025769
0.024940	0.013365	0.035490	0.035880
0.049895	0.022630	0.049200	0.049182
0.074865	0.030090	0.058500	0.058250
0.099830	0.036310	0.065300	0.064876
0.149790	0.046440	0.074500	0.073413
0.199760	0.052720	0.077600	0.077683
0.299735	0.058180	0.076600	0.078277
0.399725	0.059090	0.072500	0.072687
0.499745	0.055400	0.064800	0.064057
0.599780	0.048060	0.055000	0.054116
0.699820	0.038320	0.043200	0.043557
0.799875	0.027280	0.031400	0.032164
0.899930	0.015240	0.019200	0.018884
0.949965	0.008420	0.011100	0.010983
1.000000	0.000000	0.001800	0.001847

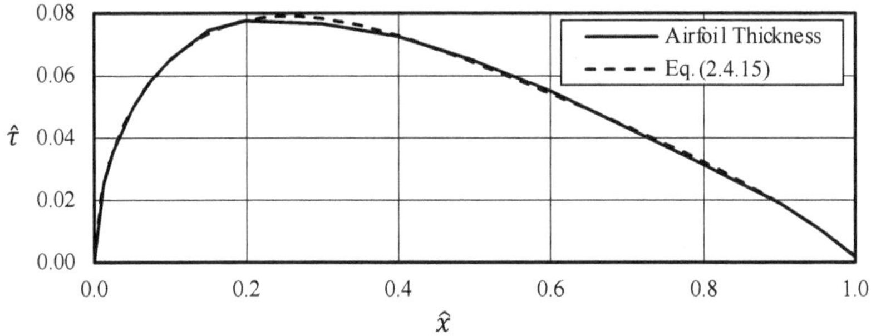

Figure 2.4.4. Göttingen 173 airfoil thickness.

From the estimated thickness distribution given in Table 2.4.4, it appears that this airfoil has a maximum thickness of about 8%. Using $\tau_{\max} = 0.08$ and a nonlinear solver or optimization routine, values for the thickness coefficients can be obtained that minimize the root-mean-square of the error between Eq. (2.4.15) and the estimated thickness data given in the $\tau(\hat{x})$ column of the table. This yields

$$a_0 = 2.95712, \quad a_1 = -0.55554, \quad a_2 = -7.70116,$$
$$a_3 = 9.27685, \quad a_4 = -3.95419$$

Solutions for the thickness distribution given by Eq. (2.4.15) using these coefficients are included in the last column of Table 2.4.4 and plotted in Fig. 2.4.4. As can be seen in the figure, the thickness distribution from Eq. (2.4.15) using the coefficients computed here very closely approximates the thickness of the Göttingen 173 airfoil.

The coefficients computed above can be used in Eqs. (2.4.16)–(2.4.19) to yield

$$\upsilon_0 = 0.65496, \quad \upsilon_1 = 0.41990,$$
$$\upsilon_2 = 0.36760, \quad \upsilon_3 = 0.42197$$

The method presented above can be used to account for the effects of any arbitrary airfoil thickness distribution $\mu_a(\hat{x})$ on the wing-segment mass properties. Results for common thickness distributions have been shown, but the process can be repeated quite simply for other thickness distributions. In the following analysis for wing segments and rotors, the thickness distribution function $\mu_a(\hat{x})$ will be prescribed to be invariant with spanwise location. However, the maximum thickness $\tau_{\max}$ is allowed to vary linearly as a function of spanwise location.

General Integrals

The mass and inertial properties of the wing require solving the volume integrals given in Eqs. (2.2.1), (2.2.5)–(2.2.7), and (2.2.9)–(2.2.14). For the special case of a wing segment with constant density, these integrals can all be written in Cartesian coordinates in the general form

$$S_{i,j,k} = \int_0^b \int_{-\frac{3}{4}c(y)}^{\frac{1}{4}c(y)} \int_{-\frac{1}{2}\tau_{\max}(y)\mu_a(x)c(y)}^{\frac{1}{2}\tau_{\max}(y)\mu_a(x)c(y)} x^i y^j z^k \, dz \, dx \, dy \tag{2.4.20}$$

where i, j, and k are each positive integers. Each of the integrals given in Eqs. (2.2.1), (2.2.5)–(2.2.7), and (2.2.9)–(2.2.14) are a special case of Eq. (2.4.20) with varying values for i, j, and k. To analytically solve Eq. (2.4.20) for a wing segment, Moulton and Hunsaker [10] use a change of variables. Detailed derivations of the solutions included in this section are given by Moulton and Hunsaker [10].

Wing Volume and Center of Gravity

The general volume integral given in Eq. (2.2.1) can be written for the wing segment as a special case of Eq. (2.4.20) where $i = 0$, $j = 0$, $k = 0$. This can be analytically integrated to yield the **volume of a wing segment** [10]

$$\forall = \frac{b}{12}\kappa_a \upsilon_0 \tag{2.4.21}$$

where

$$\kappa_a \equiv \tau_r \left(3c_r^2 + 2c_r c_t + c_t^2\right) + \tau_t \left(c_r^2 + 2c_r c_t + 3c_t^2\right) \tag{2.4.22}$$

The parameter υ_0 depends on the airfoil thickness distribution as discussed previously.

A similar process can be followed to compute the location of the wing-segment CG relative to the wing origin. Because sweep shifts the wing mass in the x direction, the sweep angle must be accounted for in the computation of $\bar{x}$. However, sweep does not affect the y or z ordinates of the center of gravity. To account for sweep, Eq. (2.2.5) can be written as

$$\begin{aligned} \bar{x} &= \frac{\rho}{\mathcal{M}} \iiint_{\forall} \left(x - y \tan \Lambda_{c/4}\right) d\forall \\ &= \frac{\rho}{\mathcal{M}} \iiint_{\forall} x \, d\forall - \frac{\rho}{\mathcal{M}} \tan \Lambda_{c/4} \iiint_{\forall} y \, d\forall \end{aligned} \tag{2.4.23}$$

The integrals given in Eqs. (2.4.23), (2.2.6), and (2.2.7) are each a special case of Eq. (2.4.20) with varying values for i, j, and k. Each integral can be solved analytically to yield the **center of gravity location of the wing segment relative to the wing origin** [10]

$$\bar{x} = -\frac{3\kappa_b \upsilon_1 + 4b\kappa_c \upsilon_0 \tan \Lambda_{c/4}}{20\kappa_a \upsilon_0} \tag{2.4.24}$$

$$\bar{y} = \delta b \frac{\kappa_c}{5\kappa_a} \tag{2.4.25}$$

$$\bar{z} = 0 \tag{2.4.26}$$

where κ_a is given in Eq. (2.4.22),

$$\kappa_b \equiv \tau_r \left(4c_r^3 + 3c_r^2 c_t + 2c_r c_t^2 + c_t^3\right) + \tau_t \left(c_r^3 + 2c_r^2 c_t + 3c_r c_t^2 + 4c_t^3\right) \tag{2.4.27}$$

$$\kappa_c \equiv \tau_r \left(3c_r^2 + 4c_r c_t + 3c_t^2\right) + 2\tau_t \left(c_r^2 + 3c_r c_t + 6c_t^2\right) \tag{2.4.28}$$

and δ denotes whether the wing is a left or right wing according to

$$\delta \equiv \begin{cases} 1, & \text{right wing} \\ -1, & \text{left wing} \end{cases} \tag{2.4.29}$$

The parameters υ_0 and υ_1 depend on the airfoil thickness distribution as discussed previously. Note that because we have neglected the effects of camber and twist, $\bar{z} = 0$.

Wing Inertia Tensor

The volume integrals given in Eqs. (2.2.9)–(2.2.14) can be used to compute the components of the inertia tensor for the wing segment about the origin of the wing-segment coordinate system. Accounting for the effects of sweep, the components of the inertia tensor given in Eq. (2.2.8) about the wing origin can be found from [10]

$$I_{xx} = \rho \iiint_{V} \left(y^2 + z^2\right) dV \tag{2.4.30}$$

$$I_{yy} = \rho \iiint_{V} \left(x^2 + z^2 + y^2 \tan^2 \Lambda_{c/4} - 2xy \tan \Lambda_{c/4}\right) dV \tag{2.4.31}$$

$$I_{zz} = \rho \iiint_{V} \left(x^2 + y^2 + y^2 \tan^2 \Lambda_{c/4} - 2xy \tan \Lambda_{c/4}\right) dV \tag{2.4.32}$$

$$I_{xy} = I_{yx} = \rho \iiint_{V} \left(xy - y^2 \tan \Lambda_{c/4}\right) dV \tag{2.4.33}$$

$$I_{xz} = I_{zx} = \rho \iiint_{V} \left(xz - yz \tan \Lambda_{c/4}\right) dV \tag{2.4.34}$$

$$I_{yz} = I_{zy} = \rho \iiint_{V} yz \, dV \tag{2.4.35}$$

Equations (2.4.30)–(2.4.35) are each a special case of Eq. (2.4.20) with various values for i, j, and k. Each can be analytically integrated to yield [10]

$$I_{xx} = \rho \frac{b}{3360}\left(56b^2\kappa_f \upsilon_0 + \kappa_g \upsilon_3\right) \tag{2.4.36}$$

$$I_{yy} = \rho \frac{b}{10080}\left[84b\left(2b\kappa_f \upsilon_0 \tan^2 \Lambda_{c/4} + \kappa_d \upsilon_1 \tan \Lambda_{c/4}\right) + 49\kappa_e \upsilon_2 + 3\kappa_g \upsilon_3\right] \tag{2.4.37}$$

$$I_{zz} = \rho \frac{b}{1440}\left\{12b\left[2b\left(\tan^2 \Lambda_{c/4} + 1\right)\kappa_f \upsilon_0 + \kappa_d \upsilon_1 \tan \Lambda_{c/4}\right] + 7\kappa_e \upsilon_2\right\} \tag{2.4.38}$$

$$I_{xy} = I_{yx} = -\rho\delta \frac{b^2}{240}\left[4b\kappa_f \upsilon_0 \tan \Lambda_{c/4} + \kappa_d \upsilon_1\right] \tag{2.4.39}$$

$$I_{xz} = I_{zx} = I_{yz} = I_{zy} = 0 \tag{2.4.40}$$

where

$$\kappa_d \equiv \tau_r\left(c_r + c_t\right)\left(2c_r^2 + c_r c_t + 2c_t^2\right) + \tau_t\left(c_r^3 + 3c_r^2 c_t + 6c_r c_t^2 + 10c_t^3\right) \tag{2.4.41}$$

$$\begin{aligned}\kappa_e \equiv\ & \tau_r\left(5c_r^4 + 4c_r^3 c_t + 3c_r^2 c_t^2 + 2c_r c_t^3 + c_t^4\right) \\ & + \tau_t\left(c_r^4 + 2c_r^3 c_t + 3c_r^2 c_t^2 + 4c_r c_t^3 + 5c_t^4\right)\end{aligned} \tag{2.4.42}$$

$$\kappa_f \equiv \tau_r\left(c_r^2 + 2c_r c_t + 2c_t^2\right) + \tau_t\left(c_r^2 + 4c_r c_t + 10c_t^2\right) \tag{2.4.43}$$

$$\begin{aligned}\kappa_g \equiv\ & \tau_r^3\left(35c_r^4 + 20c_r^3 c_t + 10c_r^2 c_t^2 + 4c_r c_t^3 + c_t^4\right) \\ & + \tau_r^2\tau_t\left(15c_r^4 + 20c_r^3 c_t + 18c_r^2 c_t^2 + 12c_r c_t^3 + 5c_t^4\right) \\ & + \tau_r\tau_t^2\left(5c_r^4 + 12c_r^3 c_t + 18c_r^2 c_t^2 + 20c_r c_t^3 + 15c_t^4\right) \\ & + \tau_t^3\left(c_r^4 + 4c_r^3 c_t + 10c_r^2 c_t^2 + 20c_r c_t^3 + 35c_t^4\right)\end{aligned} \tag{2.4.44}$$

The parameters $\upsilon_0, \upsilon_1, \upsilon_2$, and υ_3 depend on the airfoil thickness distribution as discussed previously.

Because the mass is directly related to the volume and density according to Eq. (2.2.3), applying Eq. (2.2.3) along with the relation for volume given in Eq. (2.4.21) gives the tensor components in terms of the wing-segment mass $\mathcal{M}$

$$I_{xx} = \mathcal{M}\left[\frac{56b^2\kappa_f \upsilon_0 + \kappa_g \upsilon_3}{280\kappa_a \upsilon_0}\right] \tag{2.4.45}$$

$$I_{yy} = \mathcal{M}\left[\frac{84b\left(2b\kappa_f \upsilon_0 \tan^2 \Lambda_{c/4} + \kappa_d \upsilon_1 \tan \Lambda_{c/4}\right) + 49\kappa_e \upsilon_2 + 3\kappa_g \upsilon_3}{840\kappa_a \upsilon_0}\right] \tag{2.4.46}$$

$$I_{zz} = \mathcal{M}\left[\frac{12b\left\{2b\left(\tan^2 \Lambda_{c/4} + 1\right)\kappa_f \upsilon_0 + \kappa_d \upsilon_1 \tan \Lambda_{c/4}\right\} + 7\kappa_e \upsilon_2}{120\kappa_a \upsilon_0}\right] \tag{2.4.47}$$

$$I_{xy} = I_{yx} = -\delta b\mathcal{M}\left[\frac{4b\kappa_f \upsilon_0 \tan\Lambda_{c/4} + \kappa_d \upsilon_1}{20\kappa_a \upsilon_0}\right] \tag{2.4.48}$$

$$I_{xz} = I_{zx} = I_{yz} = I_{zy} = 0 \tag{2.4.49}$$

The inertia components given in Eqs. (2.4.36)–(2.4.40) or (2.4.45)–(2.4.49) can be used in Eq. (2.2.8) to give the inertia tensor about the origin of the wing-segment coordinate frame. However, in some cases it can be helpful to know the inertial components of the wing segment about the CG of the wing segment. From the parallel axis theorem given in Eq. (2.2.15), the inertia tensor **[I]** about an arbitrary point is related to the inertia tensor about the center of gravity $[\mathbf{I}]_{CG}$ as given in Eq. (2.2.21) where **s** is the vector from the wing-segment CG to the wing-segment origin

$$\mathbf{s} = \begin{bmatrix} \bar{x} \\ \bar{y} \\ \bar{z} \end{bmatrix} \tag{2.4.50}$$

For any given wing segment, **s** can be computed from Eqs. (2.4.24)–(2.4.26). The reader is reminded that results from the parallel axis theorem are independent of the sign of **s**. Using Eq. (2.4.50) in Eq. (2.2.21) and simplifying gives

$$[\mathbf{I}]_{CG} = [\mathbf{I}] - \mathcal{M}\begin{bmatrix} \bar{y}^2 + \bar{z}^2 & -\bar{x}\bar{y} & -\bar{x}\bar{z} \\ -\bar{x}\bar{y} & \bar{x}^2 + \bar{z}^2 & -\bar{y}\bar{z} \\ -\bar{x}\bar{z} & -\bar{y}\bar{z} & \bar{x}^2 + \bar{y}^2 \end{bmatrix} \tag{2.4.51}$$

Since we are neglecting camber and twist, $\bar{z} = 0$ as shown in Eq. (2.4.26). Hence, Eq. (2.4.51) can be simplified to yield the **inertia tensor of the wing segment about the wing segment CG**

$$[\mathbf{I}]_{CG} = [\mathbf{I}] - \mathcal{M}\begin{bmatrix} \bar{y}^2 & -\bar{x}\bar{y} & 0 \\ -\bar{x}\bar{y} & \bar{x}^2 & 0 \\ 0 & 0 & \bar{x}^2 + \bar{y}^2 \end{bmatrix} \tag{2.4.52}$$

EXAMPLE 2.4.2 Consider a main wing with a semispan of 7.7 ft, a root chord of 1 ft, a tip chord of 0.7 ft, and a quarter-chord sweep angle of 5 deg. The wing is manufactured using a hot-wire technique, and made of expanded polypropylene (EPP) foam which has a density of 0.025 slug/ft^3. The wing uses a traditional NACA 2412 airfoil at the root, and a NACA 0008 airfoil at the tip. Estimate the weight, CG position, and the inertia tensor of the right wing about the right wing CG. Repeat this process to compute the CG position and inertia tensor of the left wing about the left wing CG.

Solution: The parameters depending on the airfoil properties can be computed from Eqs. (2.4.16)–(2.4.19) using the coefficients listed in the first line of Table 2.4.1. This gives

$$\upsilon_0 = 0.685083333, \quad \upsilon_1 = 0.46705,$$
$$\upsilon_2 = 0.395794898, \quad \upsilon_3 = 0.472889489$$

Using the wing parameters in Eqs. (2.4.22) yields

$$\kappa_a = 0.8964$$

and using the results in Eq. (2.4.21) gives the volume

$$\forall = 0.3940530825 \ \text{ft}^3$$

Using the volume along with the density of the foam in Eq. (2.2.3) gives a mass of

$$\mathcal{M} = 0.0098513270625 \ \text{slug}$$

From Eq. (2.2.4) this results in a weight of

$$W = \mathcal{M}g = 0.0098513270625 \ \text{slug}\left(\frac{9.80665 \ \text{m/s}^2}{0.3048 \ \text{m/ft}}\right) = 0.3169570752542 \ \text{lbf}$$

Using the wing parameters in Eqs. (2.4.27) and (2.4.28) gives

$$\kappa_b = 1.31012, \quad \kappa_c = 1.8388$$

Using these results along with the wing sweep, a wing-segment span of 7.7 ft, and $\delta = 1$ for the right wing in Eqs. (2.4.24)–(2.4.26) gives the center of gravity position

$$\bar{x} = -0.425837515 \ \text{ft}, \qquad \bar{y} = 3.159027220 \ \text{ft}, \qquad \bar{z} = 0 \ \text{ft}$$

The additional κ values required to compute the inertia tensor can be found from Eqs. (2.4.41)–(2.4.44), which result in

$$\kappa_d = 1.50832, \quad \kappa_e = 1.738932, \quad \kappa_f = 1.1016, \quad \kappa_g = 0.182986861$$

Using these results in Eqs. (2.4.36)–(2.4.40) or Eqs. (2.4.45)–(2.4.49) and applying this to Eq. (2.2.8) gives the inertia tensor about the origin of the wing

$$[\mathbf{I}] = \begin{bmatrix} 0.143563243 & 0.016910502 & 0 \\ 0.016910502 & 0.002509128 & 0 \\ 0 & 0 & 0.146062456 \end{bmatrix} \ \text{slug·ft}^2$$

Finally, using these results in Eq. (2.4.51) or Eq. (2.4.52) gives the inertia tensor about the wing CG

$$[\mathbf{I}]_{CG} = \begin{bmatrix} 0.045252388 & 0.003658179 & 0 \\ 0.003658179 & 0.000722712 & 0 \\ 0 & 0 & 0.045965185 \end{bmatrix} \ \text{slug·ft}^2$$

Repeating the process for the left wing using $\delta = -1$ gives the center of gravity position

$$\bar{x} = -0.425837515 \ \text{ft}, \ \bar{y} = -3.159027220 \ \text{ft}, \ \bar{z} = 0 \ \text{ft}$$

the inertia tensor about the origin of the wing

$$[\mathbf{I}] = \begin{bmatrix} 0.143563243 & -0.016910502 & 0 \\ -0.016910502 & 0.002509128 & 0 \\ 0 & 0 & 0.146062456 \end{bmatrix} \text{slug}\cdot\text{ft}^2$$

and the inertia tensor about the wing CG

$$[\mathbf{I}]_{CG} = \begin{bmatrix} 0.045252388 & -0.003658179 & 0 \\ -0.003658179 & 0.000722712 & 0 \\ 0 & 0 & 0.045965185 \end{bmatrix} \text{slug}\cdot\text{ft}^2$$

The mass and inertia solutions for the wing segment of constant density given in this section are exact within the approximations used, which include assuming a linear change in chord and airfoil percent thickness as a function of spanwise location, and neglecting airfoil camber and wing twist. For typical wings with non-zero airfoil camber and wing twist, the analytic solutions given in this section typically introduce errors in $\bar{x}$, $\bar{y}$, I_{xx}, I_{yy}, and I_{zz} on the order of 1–2% [10, 13].

2.5 Rotor

Several types of rotors are commonly used on aircraft. These include propellers, jet-engine turbines, helicopter rotor blades, and others. Figure 2.5.1 shows a rotor and the associated local Cartesian coordinates, with the x axis defined along the axis of the rotor. To estimate the mass and inertial properties of a rotor, we will assume that the rotor can be defined by the number of blades N_b, rotor radius r_r, hub radius r_h, hub height h_h, blade root chord c_r, blade tip chord c_t, root airfoil percent thickness τ_r, and tip airfoil percent thickness τ_t. Just as wing twist and airfoil camber were neglected in the analysis of a wing segment, the influence of blade pitch and airfoil camber will be neglected in the present analysis. Therefore, the origin of the rotor coordinate system is also the rotor CG location.

Similar to what was done in the analysis of a wing segment, the chord and maximum airfoil percent thickness will be allowed to vary linearly as a function of radius r according to

$$c(r) = (c_t - c_r)\frac{(r - r_h)}{(r_r - r_h)} + c_r \tag{2.5.1}$$

$$\tau_{\max}(r) = (\tau_t - \tau_r)\frac{(r - r_h)}{(r_r - r_h)} + \tau_r \tag{2.5.2}$$

Hence, the dimensional maximum airfoil thickness at any radius is

$$t(r) = \tau_{\max}(r)c(r) \tag{2.5.3}$$

With these definitions, we will now consider relations for the rotor volume, inertia tensor, and angular momentum.

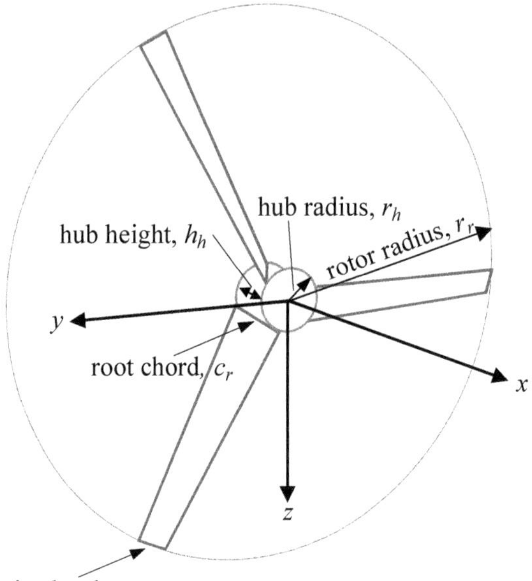

Figure 2.5.1. Rotor geometric definition and coordinate system.

Rotor Volume

The volume of the rotor can be estimated by summing the volume of each blade with the volume of the hub. The volume of the hub can be estimated as the volume of a cylinder with a radius equal to the hub radius, and a height equal to the hub height. Hence, the volume of the hub can be computed from

$$\forall_h = \pi h_h r_h^2 \tag{2.5.4}$$

The volume of an individual rotor blade can be estimated the same way the volume of a wing segment was estimated in the previous section. Replacing the span in Eq. (2.4.21) with the difference in rotor radius and hub radius and multiplying by the number of blades N_b gives to total volume of the blades

$$\forall_b = N_b \frac{r_r - r_h}{12} \kappa_a \upsilon_0 \tag{2.5.5}$$

where κ_a is computed from Eq. (2.4.22) and υ_0 depends on the airfoil thickness as described previously. The **total volume of the rotor** is the sum of the volume of the hub and the volume of the blades

$$\forall = \forall_h + \forall_b \tag{2.5.6}$$

Given a total weight or mass of the rotor, the density of the rotor can be found by rearranging Eqs. (2.2.3) and (2.2.4) and applying the total volume estimated from Eq.

(2.5.6). Once the density is known, the mass and weight of the hub and blades can be found individually from the known individual volumes given in Eqs. (2.5.4) and (2.5.5). This information will be useful in the computation of the inertia tensor.

Rotor Inertia Tensor

The inertia tensor of a rotor can be approximated by considering the inertia tensor of the hub and blades separately. The inertia tensor of the hub can be estimated from Eq. (2.3.7) by assuming a zero radius for the inner radius $r_1 = 0$, setting the outer radius equal to the radius of the hub $r_2 = r_h$, and using a cylinder height equal to the hub height $h = h_h$. This gives

$$[\mathbf{I}]_h = \frac{\mathcal{M}_h}{12}\begin{bmatrix} 6r_h^2 & 0 & 0 \\ 0 & 3r_h^2 + h_h^2 & 0 \\ 0 & 0 & 3r_h^2 + h_h^2 \end{bmatrix} \tag{2.5.7}$$

where $\mathcal{M}_h$ is the mass of the hub.

The inertia tensor of the rotor blades at any point in time depends on the orientation of the blades about the axis of the rotor. Since rotors are commonly spinning during flight, the inertia tensor is a function of time. However, an approximate static inertia tensor for the rotor can be developed by distributing the mass of the rotor blades within a circular disk of the same diameter as the rotor. At each radial distance, we will assume the differential volume of the disk is equal to the differential volume from all the rotor blades at the same radial location. This can be done by setting the area of a thin cylinder A_c equal to the summed cross-sectional area of the rotor blades at any given radial distance. The area of a thin cylinder of radius r and height h_c is

$$A_c(r) = 2\pi r h_c \tag{2.5.8}$$

The total area of the rotor blades passing through a given radius is

$$A_c(r) = N_b \tau_{\max}(r) \upsilon_0 c(r)^2 \tag{2.5.9}$$

Setting Eq. (2.5.8) equal to Eq. (2.5.9) and solving for the disk height gives the height or thickness of the disk as a function of radial position

$$h(r) = \frac{N_b \tau_{\max}(r) \upsilon_0 c(r)^2}{2\pi r} \tag{2.5.10}$$

The components of the inertia tensor are most easily found using the polar coordinates

$$x = x$$

$$y = r\cos\varphi$$

$$z = r\sin\varphi$$

where the angle φ is a right-hand angle about the x axis. Using this change of variables in Eqs. (2.2.9)–(2.2.14) gives the components of the inertia tensor

$$I_{xx} = \rho \iiint_{\forall} r^2 \, d\forall \tag{2.5.11}$$

$$I_{yy} = \rho \iiint_{\forall} x^2 + r^2 \sin^2 \varphi \, d\forall \tag{2.5.12}$$

$$I_{zz} = \rho \iiint_{\forall} x^2 + r^2 \cos^2 \varphi \, d\forall \tag{2.5.13}$$

$$I_{xy} = I_{yx} = \rho \iiint_{\forall} xr \cos \varphi \, d\forall \tag{2.5.14}$$

$$I_{xz} = I_{zx} = \rho \iiint_{\forall} xr \sin \varphi \, d\forall \tag{2.5.15}$$

$$I_{yz} = I_{zy} = \rho \iiint_{\forall} r^2 \sin \varphi \cos \varphi \, d\forall \tag{2.5.16}$$

Using the local disk thickness given in Eq. (2.5.10), these equations can be integrated to yield [10]

$$[\mathbf{I}]_b = \begin{bmatrix} I_{xx} & 0 & 0 \\ 0 & I_{yy} & 0 \\ 0 & 0 & I_{zz} \end{bmatrix} \tag{2.5.17}$$

where

$$I_{xx} = \mathcal{M}_b \left[\frac{r_h^3 \gamma_l + r_h^2 r_r \gamma_m + r_h r_r^2 \gamma_n + r_r^3 \gamma_o}{5 \left(r_r - r_h \right) \kappa_a} \right] \tag{2.5.18}$$

$$I_{yy} = I_{zz} = \frac{\mathcal{M}_b N_b \upsilon_0}{\forall_b} \left[\frac{N_b^2 \upsilon_0^2 A}{13440 \pi^2 r_h r_r (r_h - r_r)^9} + \frac{B}{120} \right]$$

$$I_{xy} = I_{yx} = I_{xz} = I_{zx} = I_{yz} = I_{zy} = 0 \tag{2.5.19}$$

and

$$A = \gamma_a r_h^{10} + \gamma_b r_h^9 r_r + \gamma_c r_h^8 r_r^2 + \gamma_d r_h^7 r_r^3 + \gamma_e r_h^6 r_r^4 + \gamma_f r_h^5 r_r^5 + \gamma_g r_h^4 r_r^6 + \gamma_h r_h^3 r_r^7 + \gamma_i r_h^2 r_r^8 + \gamma_j r_h r_r^9 + \gamma_k r_r^{10} \tag{2.5.20}$$

$$B = r_h^3 \gamma_l + r_h^2 r_r \gamma_m + r_h r_r^2 \gamma_n + r_r^3 \gamma_o \tag{2.5.21}$$

and $\mathcal{M}_b$ is the total mass of the blades. The expressions for the γ parameters in the equations above are [10]

$$\gamma_a = -280 c_t^6 \tau_t^3 \tag{2.5.22}$$

$$
\begin{aligned}
\gamma_b = -\Big[& c_r^6(35\tau_r^3 + 15\tau_r^2\tau_t + 5\tau_r\tau_t^2 + \tau_t^3) \\
&+ 6c_r^5 c_t(5\tau_r^3 + 5\tau_r^2\tau_t + 3\tau_r\tau_t^2 + \tau_t^3) \\
&+ 5c_r^4 c_t^2(5\tau_r^3 + 9\tau_r^2\tau_t + 9\tau_r\tau_t^2 + 5\tau_t^3) \\
&+ 20c_r^3 c_t^3(\tau_r^3 + 3\tau_r^2\tau_t + 5\tau_r\tau_t^2 + 5\tau_t^3) \\
&+ 15c_r^2 c_t^4(\tau_r^3 + 5\tau_r^2\tau_t + 15\tau_r\tau_t^2 + 35\tau_t^3) \\
&+ 2c_r c_t^5(5\tau_r^3 + 45\tau_r^2\tau_t + 315\tau_r\tau_t^2 - 3123\tau_t^3) \\
&+ c_t^6(5\tau_r^3 + 105\tau_r^2\tau_t - 3123\tau_r\tau_t^2 + 4329\tau_t^3)\Big] \\
&- \ln(r_h/r_r)\Big[1680c_r c_t^5\tau_t^3 + 840c_t^6(\tau_r\tau_t^2 - 3\tau_t^3)\Big]
\end{aligned}
\tag{2.5.23}
$$

$$
\begin{aligned}
\gamma_c = 4\Big[& c_r^6(90\tau_r^3 + 40\tau_r^2\tau_t + 14\tau_r\tau_t^2 + 3\tau_t^3) \\
&+ 2c_r^5 c_t(40\tau_r^3 + 42\tau_r^2\tau_t + 27\tau_r\tau_t^2 + 10\tau_t^3) \\
&+ 5c_r^4 c_t^2(14\tau_r^3 + 27\tau_r^2\tau_t + 30\tau_r\tau_t^2 + 20\tau_t^3) \\
&+ 20c_r^3 c_t^3(3\tau_r^3 + 10\tau_r^2\tau_t + 20\tau_r\tau_t^2 + 30\tau_t^3) \\
&+ 5c_r^2 c_t^4(10\tau_r^3 + 60\tau_r^2\tau_t + 270\tau_r\tau_t^2 - 1089\tau_t^3) \\
&+ 2c_r c_t^5(20\tau_r^3 + 270\tau_r^2\tau_t - 3267\tau_r\tau_t^2 + 996\tau_t^3) \\
&+ 3c_t^6(10\tau_r^3 - 363\tau_r^2\tau_t + 332\tau_r\tau_t^2 + 840\tau_t^3)\Big] \\
&+ 1680\ln(r_h/r_r)\Big[5c_r^2 c_t^4\tau_t^3 + 2c_r c_t^5(3\tau_r\tau_t^2 - 4\tau_t^3) + c_t^6(\tau_r^2\tau_t - 4\tau_r\tau_t^2)\Big]
\end{aligned}
\tag{2.5.24}
$$

$$
\begin{aligned}
\gamma_d = -14\Big[& c_r^6(120\tau_r^3 + 56\tau_r^2\tau_t + 21\tau_r\tau_t^2 + 5\tau_t^3) \\
&+ 2c_r^5 c_t(56\tau_r^3 + 63\tau_r^2\tau_t + 45\tau_r\tau_t^2 + 20\tau_t^3) \\
&+ 15c_r^4 c_t^2(7\tau_r^3 + 15\tau_r^2\tau_t + 20\tau_r\tau_t^2 + 20\tau_t^3) \\
&+ 20c_r^3 c_t^3(5\tau_r^3 + 20\tau_r^2\tau_t + 60\tau_r\tau_t^2 - 117\tau_t^3) \\
&+ 5c_r^2 c_t^4(20\tau_r^3 + 180\tau_r^2\tau_t - 1053\tau_r\tau_t^2 - 21\tau_t^3) \\
&+ 6c_r c_t^5(20\tau_r^3 - 351\tau_r^2\tau_t - 21\tau_r\tau_t^2 + 280\tau_t^3) \\
&+ 3c_t^6(-39\tau_r^3 - 7\tau_r^2\tau_t + 280\tau_r\tau_t^2 + 280\tau_t^3)\Big] \\
&- 840\ln(r_h/r_r)\Big[20c_r^3 c_t^3\tau_t^3 + 5c_r^2 c_t^4(9\tau_r\tau_t^2 - 7\tau_t^3) \\
&+ 6c_r c_t^5(3\tau_r^2\tau_t - 7\tau_r\tau_t^2) + c_t^6(\tau_r^3 - 7\tau_r^2\tau_t)\Big]
\end{aligned}
\tag{2.5.25}
$$

$$
\begin{aligned}
\gamma_e =& 28\Big[c_r^6(168\tau_r^3 + 84\tau_r^2\tau_t + 35\tau_r\tau_t^2 + 10\tau_t^3) \\
&\quad + 6c_r^5c_t(28\tau_r^3 + 35\tau_r^2\tau_t + 30\tau_r\tau_t^2 + 20\tau_t^3) \\
&\quad + 5c_r^4c_t^2(35\tau_r^3 + 90\tau_r^2\tau_t + 180\tau_r\tau_t^2 - 174\tau_t^3) \\
&\quad + 20c_r^3c_t^3(10\tau_r^3 + 60\tau_r^2\tau_t - 174\tau_r\tau_t^2 - 33\tau_t^3) \\
&\quad + 15c_r^2c_t^4(20\tau_r^3 - 174\tau_r^2\tau_t - 99\tau_r\tau_t^2 + 70\tau_t^3) \\
&\quad + 2c_rc_t^5(-174\tau_r^3 - 297\tau_r^2\tau_t + 630\tau_r\tau_t^2 + 280\tau_t^3) \\
&\quad + c_t^6(-33\tau_r^3 + 210\tau_r^2\tau_t + 280\tau_r\tau_t^2 + 420\tau_t^3)\Big] \\
&+ 1680\ln(r_h/r_r)\Big[10c_r^4c_t^2\tau_t^3 + 20c_r^3c_t^3(2\tau_r\tau_t^2 - \tau_t^3) \\
&\quad + 15c_r^2c_t^4(2\tau_r^2\tau_t - 3\tau_r\tau_t^2) + 2c_rc_t^5(2\tau_r^3 - 9\tau_r^2\tau_t) - c_t^6\tau_r^3\Big]
\end{aligned}
\tag{2.5.26}
$$

$$
\begin{aligned}
\gamma_f =& -70\Big[c_r^6(126\tau_r^3 + 70\tau_r^2\tau_t + 35\tau_r\tau_t^2 + 15\tau_t^3) \\
&\quad + 10c_r^5c_t(14\tau_r^3 + 21\tau_r^2\tau_t + 27\tau_r\tau_t^2 - 12\tau_t^3) \\
&\quad + 25c_r^4c_t^2(7\tau_r^3 + 27\tau_r^2\tau_t - 36\tau_r\tau_t^2 - 12\tau_t^3) \\
&\quad + 300c_r^3c_t^3(\tau_r^3 - 4\tau_r^2\tau_t - 4\tau_r\tau_t^2 + \tau_t^3) \\
&\quad - 25c_r^2c_t^4(12\tau_r^3 + 36\tau_r^2\tau_t - 27\tau_r\tau_t^2 - 7\tau_t^3) \\
&\quad - 10c_rc_t^5(12\tau_r^3 - 27\tau_r^2\tau_t - 21\tau_r\tau_t^2 - 14\tau_t^3) \\
&\quad + c_t^6(15\tau_r^3 + 35\tau_r^2\tau_t + 70\tau_r\tau_t^2 + 126\tau_t^3)\Big] \\
&+ 4200\ln(r_h/r_r)\Big[-2c_r^5c_t\tau_t^3 - 5c_r^4c_t^2(3\tau_r\tau_t^2 - \tau_t^3) \\
&\quad - 20c_r^3c_t^3(\tau_r^2\tau_t - \tau_r\tau_t^2) - 5c_r^2c_t^4(\tau_r^3 - 3\tau_r^2\tau_t) + 2c_rc_t^5\tau_r^3\Big]
\end{aligned}
\tag{2.5.27}
$$

$$
\begin{aligned}
\gamma_g =& 28\Big[c_r^6(420\tau_r^3 + 280\tau_r^2\tau_t + 210\tau_r\tau_t^2 - 33\tau_t^3) \\
&\quad + 2c_r^5c_t(280\tau_r^3 + 630\tau_r^2\tau_t - 297\tau_r\tau_t^2 - 174\tau_t^3) \\
&\quad + 15c_r^4c_t^2(70\tau_r^3 - 99\tau_r^2\tau_t - 174\tau_r\tau_t^2 + 20\tau_t^3) \\
&\quad + 20c_r^3c_t^3(-33\tau_r^3 - 174\tau_r^2\tau_t + 60\tau_r\tau_t^2 + 10\tau_t^3) \\
&\quad + 5c_r^2c_t^4(-174\tau_r^3 + 180\tau_r^2\tau_t + 90\tau_r\tau_t^2 + 35\tau_t^3) \\
&\quad + 6c_rc_t^5(20\tau_r^3 + 30\tau_r^2\tau_t + 35\tau_r\tau_t^2 + 28\tau_t^3) \\
&\quad + c_t^6(10\tau_r^3 + 35\tau_r^2\tau_t + 84\tau_r\tau_t^2 + 168\tau_t^3)\Big] \\
&+ 1680\ln(r_h/r_r)\Big[c_r^6\tau_t^3 + 2c_r^5c_t(9\tau_r\tau_t^2 - 2\tau_t^3) \\
&\quad + 15c_r^4c_t^2(3\tau_r^2\tau_t - 2\tau_r\tau_t^2) + 20c_r^3c_t^3(\tau_r^3 - 2\tau_r^2\tau_t) - 10c_r^2c_t^4\tau_r^3\Big]
\end{aligned}
\tag{2.5.28}
$$

$$
\begin{aligned}
\gamma_h =14\Big[&3c_r^6(-280\tau_r^3 - 280\tau_r^2\tau_t + 7\tau_r\tau_t^2 + 39\tau_t^3)\\
&+ 6c_r^5c_t(-280\tau_r^3 + 21\tau_r^2\tau_t + 351\tau_r\tau_t^2 - 20\tau_t^3)\\
&+ 5c_r^4c_t^2(21\tau_r^3 + 1053\tau_r^2\tau_t - 180\tau_r\tau_t^2 - 20\tau_t^3)\\
&+ 20c_r^3c_t^3(117\tau_r^3 - 60\tau_r^2\tau_t - 20\tau_r\tau_t^2 - 5\tau_t^3)\\
&+ 15c_r^2c_t^4(-20\tau_r^3 - 20\tau_r^2\tau_t - 15\tau_r\tau_t^2 - 7\tau_t^3)\\
&+ 2c_rc_t^5(-20\tau_r^3 - 45\tau_r^2\tau_t - 63\tau_r\tau_t^2 - 56\tau_t^3)\\
&+ c_t^6(-5\tau_r^3 - 21\tau_r^2\tau_t - 56\tau_r\tau_t^2 - 120\tau_t^3)\Big]\\
- 840\ln(r_h/r_r)\Big[&c_r^6(7\tau_r\tau_t^2 - \tau_t^3) + 6c_r^5c_t(7\tau_r^2\tau_t - 3\tau_r\tau_t^2)\\
&+ 5c_r^4c_t^2(7\tau_r^3 - 9\tau_r^2\tau_t) - 20c_r^3c_t^3\tau_r^3\Big]
\end{aligned}
\tag{2.5.29}
$$

$$
\begin{aligned}
\gamma_i =4\Big[&3c_r^6(840\tau_r^3 + 332\tau_r^2\tau_t - 363\tau_r\tau_t^2 + 10\tau_t^3)\\
&+ 2c_r^5c_t(996\tau_r^3 - 3267\tau_r^2\tau_t + 270\tau_r\tau_t^2 + 20\tau_t^3)\\
&+ 5c_r^4c_t^2(-1089\tau_r^3 + 270\tau_r^2\tau_t + 60\tau_r\tau_t^2 + 10\tau_t^3)\\
&+ 20c_r^3c_t^3(30\tau_r^3 + 20\tau_r^2\tau_t + 10\tau_r\tau_t^2 + 3\tau_t^3)\\
&+ 5c_r^2c_t^4(20\tau_r^3 + 30\tau_r^2\tau_t + 27\tau_r\tau_t^2 + 14\tau_t^3)\\
&+ 2c_rc_t^5(10\tau_r^3 + 27\tau_r^2\tau_t + 42\tau_r\tau_t^2 + 40\tau_t^3)\\
&+ c_t^6(3\tau_r^3 + 14\tau_r^2\tau_t + 40\tau_r\tau_t^2 + 90\tau_t^3)\Big]\\
+ 1680\ln(r_h/r_r)\Big[&4c_r^6\tau_r^2\tau_t - c_r^6\tau_r\tau_t^2 + 8c_r^5c_t\tau_r^3 - 6c_r^5c_t\tau_r^2\tau_t - 5c_r^4c_t^2\tau_r^3\Big]
\end{aligned}
\tag{2.5.30}
$$

$$
\begin{aligned}
\gamma_j = -\Big[&c_r^6(4329\tau_r^3 - 3123\tau_r^2\tau_t + 105\tau_r\tau_t^2 + 5\tau_t^3)\\
&+ 2c_r^5c_t(-3123\tau_r^3 + 315\tau_r^2\tau_t + 45\tau_r\tau_t^2 + 5\tau_t^3)\\
&+ 15c_r^4c_t^2(35\tau_r^3 + 15\tau_r^2\tau_t + 5\tau_r\tau_t^2 + \tau_t^3)\\
&+ 20c_r^3c_t^3(5\tau_r^3 + 5\tau_r^2\tau_t + 3\tau_r\tau_t^2 + \tau_t^3)\\
&+ 5c_r^2c_t^4(5\tau_r^3 + 9\tau_r^2\tau_t + 9\tau_r\tau_t^2 + 5\tau_t^3)\\
&+ 6c_rc_t^5(\tau_r^3 + 3\tau_r^2\tau_t + 5\tau_r\tau_t^2 + 5\tau_t^3)\\
&+ c_t^6(\tau_r^3 + 5\tau_r^2\tau_t + 15\tau_r\tau_t^2 + 35\tau_t^3)\Big]\\
+ 840\ln(r_h/r_r)\Big[&- 3c_r^6\tau_r^3 + c_r^6\tau_r^2\tau_t + 2c_r^5c_t\tau_r^3\Big]
\end{aligned}
\tag{2.5.31}
$$

$$\gamma_k = -280c_r^6\tau_r^3 \tag{2.5.32}$$

$$\gamma_l = -\tau_r\left(10c_r^2 + 4c_r c_t + c_t^2\right) - \tau_t\left(2c_r^2 + 2c_r c_t + c_t^2\right) \tag{2.5.33}$$

$$\gamma_m = \tau_r\left(6c_r^2 - c_t^2\right) - \tau_t\left(2c_r c_t + 3c_t^2\right) \tag{2.5.34}$$

$$\gamma_n = \tau_r\left(3c_r^2 + 2c_r c_t\right) + \tau_t\left(c_r^2 - 6c_t^2\right) \tag{2.5.35}$$

$$\gamma_o = \tau_r\left(c_r^2 + 2c_r c_t + 2c_t^2\right) + \tau_t\left(c_r^2 + 4c_r c_t + 10c_t^2\right) \tag{2.5.36}$$

An alternate form of the γ factors that may be easier to formulate into computer code was published by Moulton and Hunsaker [13]. Code for these γ factors can be found on the book website.

The total inertia of the rotor about the origin of the rotor can be computed by summing the inertia of the hub given in Eq. (2.5.7) with the inertia of the blades given in Eq. (2.5.17) to yield

$$[\mathbf{I}] = [\mathbf{I}]_h + [\mathbf{I}]_b \tag{2.5.37}$$

EXAMPLE 2.5.1 Consider a 4-blade propeller with a diameter of 1.5 ft, a root chord of 0.2 ft, and a tip chord of 0.05 ft. The propeller uses a NACA 4-digit airfoil with 12% thickness. The blades are attached to a hub with a diameter of 0.2 ft and a thickness of 0.1 ft. The total propeller including the hub weighs 5.0 lbf at sea level and is made out of a single material of constant density. Compute the density of the material and the inertia tensor of the propeller.

Solution: From the given geometric properties, the volume of the hub can be found from Eq. (2.5.4)

$$\forall_h = \pi h_h r_h^2 = \pi(0.1)(0.1)^2 = 0.003141593 \text{ ft}^3$$

To compute the volume of the blades, the airfoil coefficient υ_0 and the coefficient κ_a must be obtained. Using Eq. (2.4.16) with the airfoil thickness parameters given in the first row of Table 2.4.1 gives

$$\upsilon_0 = 0.685083333$$

Using the blade geometric properties in Eq. (2.4.22) gives

$$\kappa_a = 0.0252$$

These can be used in Eq. (2.5.5) to obtain the volume of the blades

$$\begin{aligned}\forall_b =& N_b\frac{r_r - r_h}{12}\kappa_a\upsilon_0\\ =&(4)\frac{0.75 - 0.1}{12}(0.0252)(0.685083333) = 0.003740555 \text{ ft}^3\end{aligned}$$

The total volume is then

$$\forall = \forall_h + \forall_b = 0.003141593 + 0.003740555 = 0.006882148 \ \text{ft}^3$$

The total mass of the propeller can be found from

$$\mathcal{M} = \frac{W}{g} = 5.0 \ \text{lbf}\left(\frac{0.3048 \ \text{m/ft}}{9.80665 \ \text{m/s}^2}\right) = 0.1554047508578 \ \text{slug}$$

The density can then be found from

$$\rho = \mathcal{M}/\forall = 0.1554047508578/0.006882148 = 22.5808510192 \ \text{slug/ft}^3$$

The mass of the hub and blades can then be found from

$$\mathcal{M}_h = \rho\forall_h = (22.5808510192)(0.003141593) = 0.07093983567372 \ \text{slug}$$

$$\mathcal{M}_b = \rho\forall_b = (22.5808510192)(0.003740555) = 0.08446491518412 \ \text{slug}$$

The inertia tensor for the hub can be found from Eq. (2.5.7) and yields

$$[\mathbf{I}]_h = \begin{bmatrix} 0.000354699178 & 0 & 0 \\ 0 & 0.000236466119 & 0 \\ 0 & 0 & 0.000236466119 \end{bmatrix} \ \text{slug·ft}^2$$

In order to compute the inertia tensor of the blades, the γ coefficients of the blade must be computed. These can be found by using the geometric parameters in Eqs. (2.5.22)–(2.5.36). This gives

$$\begin{array}{lll} \gamma_a = -7.560\text{E}{-09}, & \gamma_b = -9.49560000\text{E}{-06}, & \gamma_c = \ \ 9.84140399\text{E}{-05} \\ \gamma_d = -4.42064039\text{E}{-04}, & \gamma_e = \ \ 1.08504516\text{E}{-03}, & \gamma_f = -1.46416757\text{E}{-03} \\ \gamma_g = \ \ 8.37803295\text{E}{-04}, & \gamma_h = \ \ 2.82524435\text{E}{-04}, & \gamma_i = -6.38461225\text{E}{-04}, \\ \gamma_j = \ \ 2.81374827\text{E}{-04}, & \gamma_k = -3.0965760\text{E}{-05}, & \gamma_l = -0.0654, \\ \gamma_m = \ \ 0.0252, & \gamma_n = \ \ 0.0198, & \gamma_o = \ \ 0.0204 \end{array}$$

Using these in Eqs. (2.5.17)–(2.5.21) gives

$$[\mathbf{I}]_b = \begin{bmatrix} 0.0101518784 & 0 & 0 \\ 0 & 0.00507643713 & 0 \\ 0 & 0 & 0.00507643713 \end{bmatrix} \ \text{slug·ft}^2$$

Finally, summing the inertia tensor of the hub and blades gives the total inertia tensor of the propeller about the propeller origin

$$[\mathbf{I}] = [\mathbf{I}]_h + [\mathbf{I}]_b = \begin{bmatrix} 0.0105065776 & 0 & 0 \\ 0 & 0.00531290325 & 0 \\ 0 & 0 & 0.00531290325 \end{bmatrix} \ \text{slug·ft}^2$$

The mass and inertia solutions for the rotor of constant density given in this section are exact within the approximations used, which include assuming a linear change in chord and airfoil percent thickness as a function of radius of the blade, approximating the hub as a cylinder, and neglecting airfoil camber and blade pitch. For typical rotors with airfoil camber and blade pitch, the analytic solutions given in this section will likely introduce errors in I_{xx}, I_{yy}, and I_{zz} on the order of 1–2% [13].

Rotor Angular Momentum

As a rotor spins, it creates angular momentum that can affect the motion of the vehicle independent of whether the rotor is generating thrust. The angular momentum of the rotor about its own axis of rotation can be computed from the inertia tensor of the rotor and the rotor angular rate. The inertia tensor is given in Eq. (2.5.37). Due to our choice of a rotor coordinate system, the angular rate is only about the x axis and can be written in vector form as

$$\omega = \begin{bmatrix} \omega_x \\ 0 \\ 0 \end{bmatrix} \tag{2.5.38}$$

The angular momentum vector is the product of the inertia tensor and the angular-rate vector, which gives

$$\mathbf{h} \equiv [\mathbf{I}]\omega = [\mathbf{I}] \begin{bmatrix} \omega_x \\ 0 \\ 0 \end{bmatrix} = \begin{bmatrix} I_{xx}\omega_x \\ 0 \\ 0 \end{bmatrix} \tag{2.5.39}$$

Note that this vector is in the component coordinate system. This will be used in Section 2.7 to evaluate the angular momentum in the vehicle coordinate system due to spinning rotors.

EXAMPLE 2.5.2 The propeller from Example 2.5.1 is a right-handed propeller rotating at a rate of 1000 rpm. Compute the angular momentum of the rotor.

Solution: We first compute the rotation rate in terms of radians per second

$$\omega_x = 1000\left(\frac{\text{rotation}}{\text{minute}}\right)\left(\frac{2\pi\text{ rad}}{1\text{ rotation}}\right)\left(\frac{1\text{ minute}}{60\text{ seconds}}\right) = 104.7197551\text{sec}^{-1}$$

Because the rotor is a right-handed rotor, it is spinning according to the right-hand rule about the x-axis of the rotor coordinate system. Applying Eq. (2.5.39) gives the angular momentum vector

$$\mathbf{h} = [\mathbf{I}] \begin{bmatrix} \omega_x \\ 0 \\ 0 \end{bmatrix} = \begin{bmatrix} I_{xx}\omega_x \\ 0 \\ 0 \end{bmatrix}$$

$$= \begin{bmatrix} (0.01050657755574)(104.7197551) \\ 0 \\ 0 \end{bmatrix} = \begin{bmatrix} 1.10024622878281 \\ 0 \\ 0 \end{bmatrix} \frac{\text{slug}\cdot\text{ft}^2}{\text{s}}$$

2.6 Properties of the Complete Vehicle

The mass and inertial properties of a complete vehicle can be computed by summing the influence of individual components that make up the vehicle. The previous sections presented methods that can be used to estimate the mass and inertial properties of individual components of common geometries. However, the mass and inertial properties of individual components can be determined in many different ways, including using CAD or experimental methods. The analysis shown in this section can be used regardless of the source of the mass and inertial properties of the individual components.

The mass and inertial properties of interest for a complete vehicle typically include the total mass, CG location, and inertia tensor about the vehicle CG. The influence of each component on the global vehicle properties depends on the component position and orientation relative to the vehicle origin and coordinate system.

Component Orientation

Each component has a local coordinate system called the component coordinate system. The position and orientation of a component are typically expressed in terms of its coordinate system relative to the vehicle coordinate system.

In general, the components of a vector in the component coordinate system can be expressed in the vehicle coordinate system using the transformation

$$\begin{bmatrix} v_{x_v} \\ v_{y_v} \\ v_{z_v} \end{bmatrix} = [\mathbf{R}] \begin{bmatrix} v_{x_c} \\ v_{y_c} \\ v_{z_c} \end{bmatrix} \tag{2.6.1}$$

where $[\mathbf{R}]$ is a rotation matrix. The components of the rotation matrix can be computed from any of the formulations given in Chapter 1. For example, the rotation matrix $[\mathbf{R}]$ can be expressed in Euler angles from Eq. (1.2.6) or in quaternions from Eq. (1.5.3) as

$$\begin{aligned}[\mathbf{R}] &= \begin{bmatrix} C_\theta C_\psi & S_\phi S_\theta C_\psi - C_\phi S_\psi & C_\phi S_\theta C_\psi + S_\phi S_\psi \\ C_\theta S_\psi & S_\phi S_\theta S_\psi + C_\phi C_\psi & C_\phi S_\theta S_\psi - S_\phi C_\psi \\ -S_\theta & S_\phi C_\theta & C_\phi C_\theta \end{bmatrix} \\ &= \begin{bmatrix} e_x^2 + e_0^2 - e_y^2 - e_z^2 & 2(e_x e_y - e_z e_0) & 2(e_x e_z + e_y e_0) \\ 2(e_x e_y + e_z e_0) & e_y^2 + e_0^2 - e_x^2 - e_z^2 & 2(e_y e_z - e_x e_0) \\ 2(e_x e_z - e_y e_0) & 2(e_y e_z + e_x e_0) & e_z^2 + e_0^2 - e_x^2 - e_y^2 \end{bmatrix}\end{aligned} \tag{2.6.2}$$

Because Euler angles are simple to visualize, it is good practice to allow the user to specify the orientation of each component relative to the vehicle coordinate system in terms of Euler angles. However, since computing rotations using quaternions is significantly more efficient than using Euler angles, it is good practice to convert the Euler angles to quaternions for computation. This can be done using Eq. (1.6.2).

In general, a component can have nonzero values for all Euler angles and all quaternion components. However, the wing-segment component requires some additional discussion. For most aircraft, the rotation matrix $[\mathbf{R}]$ for the wing is a function of only

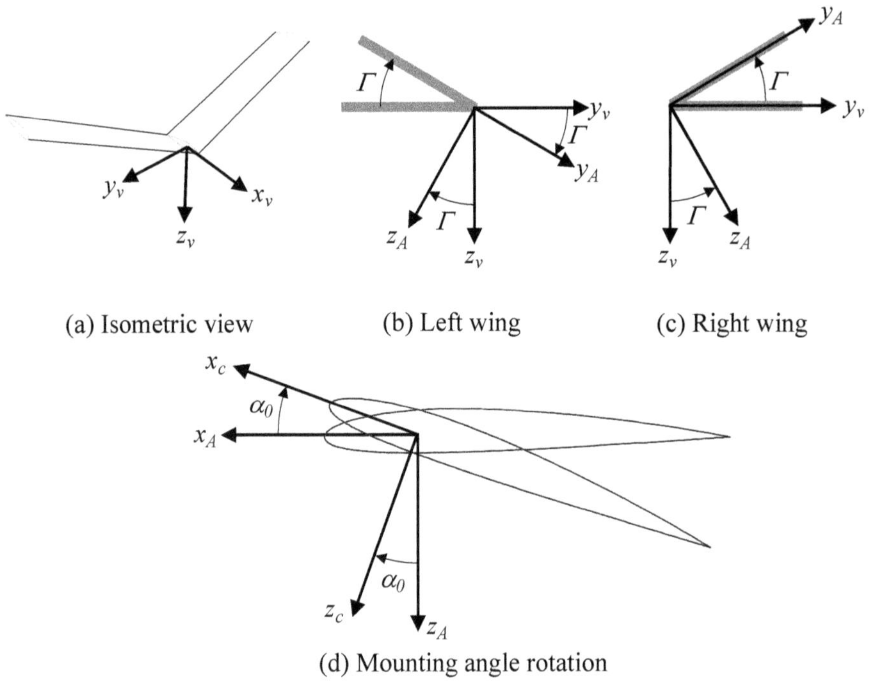

Figure 2.6.1. Wing-segment rigid-body rotations for dihedral and mounting angle.

the dihedral and mounting angles. Dihedral is a rigid-body rotation of the wing segment about the **vehicle** x axis, and the mounting angle is a rigid-body rotation of the wing segment about the **wing** y axis, as depicted in Fig. 2.6.1. The order of rotations is very important, and they can be combined into a single rotation matrix $[\mathbf{R}]$.

We begin with an arbitrary vector $(v_{x_v}, v_{y_v}, v_{z_v})$ expressed in the vehicle coordinate system and apply a rotation about the vehicle x axis by the dihedral angle Γ. Because a positive value for Γ represents a positive dihedral angle and because rotations about any axis are defined according to the right-hand rule, the rotation of the wing about the vehicle x axis for a right wing is $-\Gamma$, and the rotation for a left wing is Γ. Applying Eq. (2.4.29) yields a general rotation angle $-\delta\Gamma$ for dihedral about the vehicle x-axis. Applying this rotation about the x-axis of the vehicle coordinate system gives a vector in an intermediate coordinate system we will refer to with a subscript A

$$\begin{bmatrix} v_{x_A} \\ v_{y_A} \\ v_{z_A} \end{bmatrix} = \begin{bmatrix} 1 & 0 & 0 \\ 0 & C_{\delta\Gamma} & -S_{\delta\Gamma} \\ 0 & S_{\delta\Gamma} & C_{\delta\Gamma} \end{bmatrix} \begin{bmatrix} v_{x_v} \\ v_{y_v} \\ v_{z_v} \end{bmatrix} \tag{2.6.3}$$

The A coordinate system is also shown in Fig. 2.6.1. The resulting vector is then rotated about the y axis of the A coordinate system by the mounting angle α_0. Note that the y

axis of the A coordinate system is equivalent to the y axis of the wing coordinate system. This yields a vector in the wing-component coordinate system

$$\begin{bmatrix} v_{x_c} \\ v_{y_c} \\ v_{z_c} \end{bmatrix} = \begin{bmatrix} C_{\alpha_0} & 0 & -S_{\alpha_0} \\ 0 & 1 & 0 \\ S_{\alpha_0} & 0 & C_{\alpha_0} \end{bmatrix} \begin{bmatrix} v_{x_A} \\ v_{y_A} \\ v_{z_A} \end{bmatrix} \tag{2.6.4}$$

The rotation matrices given in Eqs. (2.6.3) and (2.6.4) can be combined to yield a single rotation matrix

$$\begin{aligned} \begin{bmatrix} v_{x_c} \\ v_{y_c} \\ v_{z_c} \end{bmatrix} &= \begin{bmatrix} C_{\alpha_0} & 0 & -S_{\alpha_0} \\ 0 & 1 & 0 \\ S_{\alpha_0} & 0 & C_{\alpha_0} \end{bmatrix} \begin{bmatrix} 1 & 0 & 0 \\ 0 & C_{\delta\Gamma} & -S_{\delta\Gamma} \\ 0 & S_{\delta\Gamma} & C_{\delta\Gamma} \end{bmatrix} \begin{bmatrix} v_{x_v} \\ v_{y_v} \\ v_{z_v} \end{bmatrix} \\ &= \begin{bmatrix} C_{\alpha_0} & -S_{\alpha_0}S_{\delta\Gamma} & -S_{\alpha_0}C_{\delta\Gamma} \\ 0 & C_{\delta\Gamma} & -S_{\delta\Gamma} \\ S_{\alpha_0} & C_{\alpha_0}S_{\delta\Gamma} & C_{\alpha_0}C_{\delta\Gamma} \end{bmatrix} \begin{bmatrix} v_{x_v} \\ v_{y_v} \\ v_{z_v} \end{bmatrix} \end{aligned} \tag{2.6.5}$$

Since the dihedral and mounting-angle rotations are orthogonal, the inverse transformation can be obtained from the transpose of Eq. (2.6.5), which yields

$$\begin{bmatrix} v_{x_v} \\ v_{y_v} \\ v_{z_v} \end{bmatrix} = \begin{bmatrix} C_{\alpha_0} & 0 & S_{\alpha_0} \\ -S_{\alpha_0}S_{\delta\Gamma} & C_{\delta\Gamma} & C_{\alpha_0}S_{\delta\Gamma} \\ -S_{\alpha_0}C_{\delta\Gamma} & -S_{\delta\Gamma} & C_{\alpha_0}C_{\delta\Gamma} \end{bmatrix} \begin{bmatrix} v_{x_c} \\ v_{y_c} \\ v_{z_c} \end{bmatrix} \tag{2.6.6}$$

Hence, the rotation matrix $[\mathbf{R}]$ for a wing segment can be expressed as

$$[\mathbf{R}] = \begin{bmatrix} C_{\alpha_0} & 0 & S_{\alpha_0} \\ -S_{\alpha_0}S_{\delta\Gamma} & C_{\delta\Gamma} & C_{\alpha_0}S_{\delta\Gamma} \\ -S_{\alpha_0}C_{\delta\Gamma} & -S_{\delta\Gamma} & C_{\alpha_0}C_{\delta\Gamma} \end{bmatrix} \tag{2.6.7}$$

The expression in Eq. (2.6.7) can be used for a wing component rather than the expression for an arbitrary orientation of a component as given in Eq. (2.6.2).

Because quaternion computations are significantly faster than Euler-angle computations, it is advantageous to have a quaternion representation of Eq. (2.6.7) for the development of a computer code. Additionally, because we are applying the Euler-angle formulation commonly used in aeronautics developed in Section 1.2 and understand the relations between this particular Euler-angle formulation and the quaternion formulation as given in Section 1.6, it can be helpful to relate the rotation given in Eq. (2.6.7) to that given in Eq. (2.6.2). At first thought, the reader may assume that dihedral and mounting angle are directly analogous to the Euler bank and elevation angles respectively, commonly used in aeronautics. However, this is not exactly true.

By comparing Eqs. (2.6.2) and (2.6.7), it can be seen that Eq. (2.6.7) is very similar to the transpose of Eq. (2.6.2) for the case where the $\psi = 0$. This transpose relationship results from the order in which the rotations are applied. By comparing Eq. (2.6.5) to Eq. (1.2.4) it can be seen that for the aeronautical Euler-angle formulation given in Eq.

(1.2.4), the elevation angle is applied before the bank angle, whereas in Eq. (2.6.5), the dihedral angle is applied before the mounting angle.

The equivalent quaternion formulation can be found by equating the rotation matrix given in Eq. (2.6.7) to the quaternion rotation matrix given in Eq. (2.6.2)

$$\begin{bmatrix} C_{\alpha_0} & 0 & S_{\alpha_0} \\ -S_{\alpha_0}S_{\delta\Gamma} & C_{\delta\Gamma} & C_{\alpha_0}S_{\delta\Gamma} \\ -S_{\alpha_0}C_{\delta\Gamma} & -S_{\delta\Gamma} & C_{\alpha_0}C_{\delta\Gamma} \end{bmatrix} = \begin{bmatrix} e_x^2+e_0^2-e_y^2-e_z^2 & 2(e_xe_y-e_ze_0) & 2(e_xe_z+e_ye_0) \\ 2(e_xe_y+e_ze_0) & e_y^2+e_0^2-e_x^2-e_z^2 & 2(e_ye_z-e_xe_0) \\ 2(e_xe_z-e_ye_0) & 2(e_ye_z+e_xe_0) & e_z^2+e_0^2-e_x^2-e_y^2 \end{bmatrix} \tag{2.6.8}$$

Applying the solution process used by Phillips [2] in developing Eq. (1.6.2), it can be shown that the equivalent quaternion components of the rotation can be computed from

$$\begin{bmatrix} e_0 \\ e_x \\ e_y \\ e_z \end{bmatrix} = \begin{bmatrix} C_{\delta\Gamma/2}C_{\alpha_0/2} \\ -S_{\delta\Gamma/2}C_{\alpha_0/2} \\ C_{\delta\Gamma/2}S_{\alpha_0/2} \\ -S_{\delta\Gamma/2}S_{\alpha_0/2} \end{bmatrix} \tag{2.6.9}$$

Component Position

The position of each component is most commonly defined relative to the vehicle coordinate system. Here we will denote the position of the origin of each component relative to the origin of the vehicle coordinate system as $\mathbf{r}_o$, and the position of the CG of the each component relative to the vehicle coordinate system as $\mathbf{r}_c$. For simple components (such as a rectangular cuboid, cylinder, sphere, or rotor) of constant density, the component CG is equivalent to the origin in the component coordinate system, which yields

$$\mathbf{r}_c = \mathbf{r}_o \tag{2.6.10}$$

For more complex components, such as a wing segment, where the CG of the component is not coincident with the origin of the component, the CG position of the component relative to the vehicle coordinate system can be computed from

$$\mathbf{r}_c = \mathbf{r}_o + [\mathbf{R}] \begin{bmatrix} \bar{x} \\ \bar{y} \\ \bar{z} \end{bmatrix} \tag{2.6.11}$$

where $[\mathbf{R}]$ is the rotation matrix of the component, and $[\bar{x}, \bar{y}, \bar{z}]$ is the location of the center of gravity of the component relative to the origin of the component in the component coordinate system.

Vehicle Mass, Center of Gravity, and Inertia

The total mass of the vehicle $\mathcal{M}$ is simply the summation of all component masses. This can be written as

$$\mathcal{M} = \sum_{i=1}^{N} \mathcal{M}_{c_i} \tag{2.6.12}$$

where $\mathcal{M}_{c_i}$ is the mass of the ith component. The location of the center of gravity of the complete vehicle relative to the vehicle origin will be denoted as $\mathbf{r}_b$ and can be found from the expression

$$\mathbf{r}_b = \frac{\sum_{i=1}^{N} \mathcal{M}_{c_i} \mathbf{r}_{c_i}}{\mathcal{M}} \tag{2.6.13}$$

where $\mathbf{r}_{c_i}$ is the location of the center of gravity of the ith component relative to the vehicle origin in the vehicle coordinate system.

The inertia tensor of the complete vehicle is obtained by summing the inertial influence of each component. However, in general, the CG of each component is not coincident with the vehicle CG, and the ordinates of the component coordinate system are not aligned with the vehicle coordinate system. Hence, the inertia tensor of each component must be rotated and translated before it can be summed to yield the total inertia tensor of the vehicle. This process requires the following steps:

1. Rotate the inertia tensor of each component from the component coordinate system to the vehicle coordinate system
2. Translate the inertia tensor of each component from the component CG location to the vehicle CG location
3. Sum the resulting inertia tensors to obtain the complete vehicle inertia tensor

Each step is outlined here.

Rotation: The inertia tensor of a component $[\mathbf{I}]_{cc}$ about the component CG in the component coordinate system can be expressed as an inertia tensor in the vehicle coordinate frame using the expression

$$[\mathbf{I}]_A = [\mathbf{R}][\mathbf{I}]_{cc}[\mathbf{R}]^{\mathrm{T}} \tag{2.6.14}$$

The subscript A is used here to denote the component's inertia tensor after rotation into the vehicle coordinate system, but prior to translation to the vehicle CG.

Translation: Once the inertia tensor of each component has been rotated to the vehicle coordinate frame, the inertia tensor must be translated such that it represents inertia about the vehicle center of gravity. Let $\mathbf{s}$ be a vector in the vehicle coordinate system that extends from the CG location of the component $\mathbf{r}_c$ to the vehicle CG location $\mathbf{r}_b$. This gives

$$\mathbf{s} = \mathbf{r}_b - \mathbf{r}_c \tag{2.6.15}$$

The inertia tensor of the component evaluated at the vehicle CG location can be found using the parallel axis theorem

$$[\mathbf{I}]_c = [\mathbf{I}]_A + \mathcal{M}_c\left[(\mathbf{s}\cdot\mathbf{s})[\mathbf{E}] - \mathbf{s}\mathbf{s}^{\mathrm{T}}\right] \tag{2.6.16}$$

where $\mathcal{M}_c$ is the mass of the component, and $[\mathbf{E}]$ is the 3×3 identity matrix.

Summation: The total inertia of the vehicle about the vehicle CG is simply the sum of all inertia tensors, one for each component, which have been expressed in the body-fixed coordinate system and translated to be about the vehicle CG. This gives

$$[\mathbf{I}]_{CG} = \sum_{i=1}^{N} [\mathbf{I}]_{c_i} \tag{2.6.17}$$

where $[\mathbf{I}]_{c_i}$ is the inertia tensor of the ith component in the body-fixed coordinate system as computed from Eq. (2.6.16).

The inertia tensor given in Eq. (2.6.17) is about the vehicle CG. Sometimes it can be helpful to know the inertia tensor of the vehicle about the origin instead of the center of gravity. This can also be computed using the parallel axis theorem

$$[\mathbf{I}]_v = [\mathbf{I}]_{CG} + \mathcal{M}\left[(\mathbf{s} \cdot \mathbf{s})[\mathbf{E}] - \mathbf{s}\mathbf{s}^{\mathrm{T}}\right] \tag{2.6.18}$$

where $\mathbf{s}$ is a vector extending from the vehicle center of gravity to the vehicle origin, i.e. $\mathbf{s} = -\mathbf{r}_b$.

EXAMPLE 2.6.1 The Boeing Insitu RQ-21 Blackjack can be modeled as the summation of nine components as given in Appendix C and shown in Fig. 2.6.2. From the information about each component, compute the total vehicle mass, CG location relative to the vehicle origin, inertia tensor about the vehicle CG location, and inertia tensor about the vehicle origin.

Solution: From the information in the appendix, the weight of each component is known, along with the component location and geometric properties. From this information and the relations given in this section, the volume, mass, and CG location of each component is given in Table 2.6.1. Note that the properties for each wing and boom are only given for the right side. These are mirrored across the vehicle plane of symmetry for the left side components. Using the results in Eq. (2.6.12) gives the total mass of the vehicle

$$\mathcal{M} = 3.108095017157 \text{ slug}$$

Figure 2.6.2. Representation of the RQ-21 Blackjack aircraft as a combination of simple components.

Table 2.6.1. Geometric and mass properties of the RQ-21 Blackjack components.

Component	Volume [ft^3]	Mass [slug]	$\mathbf{r}_c$ [ft]
Main Wing	0.4621024	0.1554048	$[-0.07376, 3.40192, 0.00914]$
Winglet	0.0167160	0.0310810	$[-0.69470, 7.78042, -0.45608]$
Horiz. Stab.	0.0443934	0.0621619	$[-3.60226, 0.75, -1.2]$
Vert. Stab.	0.0977751	0.0621619	$[-3.28867, 1.5, -0.24753]$
Fuselage	4.24	1.8648570	$[0.1, 0.0, 0.43]$
Camera	0.2680826	0.4040524	$[3.0, 0.0, 0.5]$
Boom	0.0235619	0.0310810	$[-1.7, 1.5, 0.0]$
Propeller	0.0057312	0.1554048	$[-2.7, 0.0, 0.27]$

The CG location of the vehicle can be computed from Eq. (2.6.13) to yield

$$\mathbf{r}_b = \begin{bmatrix} -0.01590778743561 \\ 0.0 \\ 0.2703908345495 \end{bmatrix} \text{ ft}$$

The inertia of the vehicle about the vehicle CG can be computed by summing the influence of each component. Using the closed-form relations given in this chapter, the inertia tensor for each component about the local component CG $[\mathbf{I}]_{cc}$ are as follows:

Main Wing (right side)

$$[\mathbf{I}]_{cc} = \begin{bmatrix} 0.7430072711916 & 0.06006298313428 & 0.0 \\ 0.06006298313428 & 0.01147173530388 & 0.0 \\ 0.0 & 0.0 & 0.7542834084737 \end{bmatrix} \text{ slug}\cdot\text{ft}^2$$

Winglet (right)

$$[\mathbf{I}]_{cc} = \begin{bmatrix} 0.00256160528553 & 0.00143230363965 & 0.0 \\ 0.00143230363965 & 0.00116243071775 & 0.0 \\ 0.0 & 0.0 & 0.00371339532641 \end{bmatrix} \text{ slug}\cdot\text{ft}^2$$

Horizontal Stabilizer (right):

$$[\mathbf{I}]_{cc} = \begin{bmatrix} 0.01167389266720 & 0.0 & 0.0 \\ 0.0 & 0.001253916209943 & 0.0 \\ 0.0 & 0.0 & 0.01289073617141 \end{bmatrix} \text{ slug}\cdot\text{ft}^2$$

Vertical Stabilizer (right):

$$[\mathbf{I}]_{cc} = \begin{bmatrix} 0.01215599548441 & 0.003763783592799 & 0.0 \\ 0.003763783592799 & 0.004042297113620 & 0.0 \\ 0.0 & 0.0 & 0.01611335509472 \end{bmatrix} \text{ slug}\cdot\text{ft}^2$$

Fuselage:

$$[\mathbf{I}]_{cc} = \begin{bmatrix} 0.2548637914069 & 0.0 & 0.0 \\ 0.0 & 4.520724202454 & 0.0 \\ 0.0 & 0.0 & 4.464778492146 \end{bmatrix} \text{slug}\cdot\text{ft}^2$$

Camera:

$$[\mathbf{I}]_{cc} = \begin{bmatrix} 0.02585935054274 & 0.0 & 0.0 \\ 0.0 & 0.02585935054274 & 0.0 \\ 0.0 & 0.0 & 0.02585935054274 \end{bmatrix} \text{slug}\cdot\text{ft}^2$$

Boom (right):

$$[\mathbf{I}]_{cc} = \begin{bmatrix} 3.885118771446\text{E}{-05} & 0.0 & 0.0 \\ 0.0 & 0.02333013822253 & 0.0 \\ 0.0 & 0.0 & 0.0233301382225 \end{bmatrix} \text{slug}\cdot\text{ft}^2$$

Propeller:

$$[\mathbf{I}]_{cc} = \begin{bmatrix} 0.01439946925865 & 0.0 & 0.0 \\ 0.0 & 0.007270808235622 & 0.0 \\ 0.0 & 0.0 & 0.00727080823562 \end{bmatrix} \text{slug}\cdot\text{ft}^2$$

Using these results in Eq. (2.6.16) yields the inertia of each component computed at the vehicle CG location as follows:

Main Wing (right side)

$$[\mathbf{I}]_{c} = \begin{bmatrix} 2.5522099582153 & 0.0904273331066 & -0.0018920693516 \\ 0.0904273331066 & 0.0225639915613 & 0.1393651384383 \\ -0.0018920693516 & 0.1393651384383 & 2.5532628823544 \end{bmatrix} \text{slug}\cdot\text{ft}^2$$

Winglet (right)

$$[\mathbf{I}]_{c} = \begin{bmatrix} 1.9004476809086 & 0.1643971395854 & -0.0167373374417 \\ 0.1643971395854 & 0.0343607016074 & 0.1761132220646 \\ -0.0167373374417 & 0.1761132220646 & 1.8970431882373 \end{bmatrix} \text{slug}\cdot\text{ft}^2$$

Horizontal Stabilizer (right):

$$[\mathbf{I}]_{c} = \begin{bmatrix} 0.1810370489109 & 0.1672009113971 & -0.3278009168622 \\ 0.1672009113971 & 0.9351731007069 & 0.0685517163920 \\ -0.3278009168622 & 0.0685517163920 & 0.8473789023107 \end{bmatrix} \text{slug}\cdot\text{ft}^2$$

Vertical Stabilizer (right):

$$[\mathbf{I}]_{c} = \begin{bmatrix} 0.1686949370435 & 0.3051620173036 & -0.1091310336666 \\ 0.3051620173036 & 0.6986039076818 & 0.0482927536391 \\ -0.1091310336666 & 0.0482927536391 & 0.8097224596857 \end{bmatrix} \text{slug}\cdot\text{ft}^2$$

Fuselage:

$$[\mathbf{I}]_c = \begin{bmatrix} 0.3023711835545 & 0.0 & -0.0344997525369 \\ 0.0 & 4.5932852309165 & 0.0 \\ -0.0344997525369 & 0.0 & 4.4898321284600 \end{bmatrix} \text{slug}\cdot\text{ft}^2$$

Camera:

$$[\mathbf{I}]_c = \begin{bmatrix} 0.0471611395906 & 0.0 & -0.2797982012162 \\ 0.0 & 3.7223000318162 & 0.0 \\ -0.2797982012162 & 0.0 & 3.7009982427683 \end{bmatrix} \text{slug}\cdot\text{ft}^2$$

Boom (right):

$$[\mathbf{I}]_c = \begin{bmatrix} 0.0722433547439 & 0.0785147792146 & -0.0141531177842 \\ 0.0785147792146 & 0.1137532560570 & 0.0126060060832 \\ -0.0141531177842 & 0.0126060060832 & 0.1814130282729 \end{bmatrix} \text{slug}\cdot\text{ft}^2$$

Propeller:

$$[\mathbf{I}]_c = \begin{bmatrix} 0.0143994929970 & 0.0 & -0.0001630251737 \\ 0.0 & 1.1268612050837 & 0.0 \\ -0.0001630251737 & 0.0 & 1.1268611813453 \end{bmatrix} \text{slug}\cdot\text{ft}^2$$

Summing these results as shown in Eq. (2.6.17) yields

$$[\mathbf{I}]_{CG} = \begin{bmatrix} 10.11319777579 & 0.0 & -1.253889929139 \\ 0.0 & 13.05135638304 & 0.0 \\ -1.253889929139 & 0.0 & 21.89533247430 \end{bmatrix} \text{slug}\cdot\text{ft}^2$$

The inertia of the vehicle about the vehicle origin can be computed from Eq. (2.6.18) and yields

$$[\mathbf{I}]_v = \begin{bmatrix} 10.34043434280 & 0.0 & -1.240521018127 \\ 0.0 & 13.27937947744 & 0.0 \\ -1.240521018127 & 0.0 & 21.89611900168 \end{bmatrix} \text{slug}\cdot\text{ft}^2$$

Complex Structures

From the development above, we see that the mass properties of the entire vehicle can be obtained by summing the contributions of the individual components. Because of this, it is quite straightforward to use the geometric components included in this chapter for much more complex geometries. For example, although the wing was treated as a single object of constant density throughout, additional internal complexity of the wing can be modeled by adding and subtracting other internal geometries. Empty regions between spars and ribs can be modeled by adding cuboids within the wing, which have negative density equal in magnitude to the wing density. A fuel tank can be modeled by adding a cuboid with density in addition to the wing density. A wing skin can be modeled with two co-located wings, the inner slightly smaller than the outer and having the negative density of the outer wing. In general, internal hollow regions can be added within any component by simply setting the density of an internal component equal to the negative of the external component density. This method can be used to create cavities in which other internal components or structures can be added, such as actuators, turbine blades, payloads, hydraulic systems, and support structures. Hence, the geometric shapes included in this chapter can be used to represent much more complex vehicle structures and mass distributions than may first appear obvious.

2.7 Angular Momentum of Spinning Components

In the absence of rigid-body vehicle rotation, the angular momentum of the vehicle can still be non-zero if vehicle components are rotating. Common rotating components include rotors such as propellers and jet engines. Gyroscopes are also often included onboard vehicles for sensor stabilization. The angular momentum vector due to rotating components will be referred to here as the *gyroscopic vector*. The influence of this gyroscopic vector on the motion of the complete vehicle is discussed in Chapter 5.

The angular momentum of an individual component about the component coordinate system can be expressed as a vector

$$\mathbf{h}_c \equiv \begin{bmatrix} h_{x_c} \\ h_{y_c} \\ h_{z_c} \end{bmatrix} \tag{2.7.1}$$

The gyroscopic vector for the entire vehicle can be obtained by rotating the angular momentum vector of each component into the vehicle coordinate system, and summing the results. This yields the vehicle gyroscopic vector

$$\mathbf{h}_v = \mathbf{h}_b \equiv \begin{bmatrix} h_{x_b} \\ h_{y_b} \\ h_{z_b} \end{bmatrix} = \sum_{i=1}^{N} [\mathbf{R}]_i \mathbf{h}_{c_i} \tag{2.7.2}$$

where N is the number of rotating components, $[\mathbf{R}]_i$ is the rotation matrix of the ith component, and $\mathbf{h}_{c_i}$ is the angular momentum vector of the ith component in the component coordinate system. Note that the components of the gyroscopic vector are the same in the vehicle coordinate system as they are in the body-fixed coordinate system, i.e. $\mathbf{h}_v = \mathbf{h}_b$. When the gyroscopic effects are applied to the vehicle equations of motion, the components of the vehicle gyroscopic vector $\mathbf{h}$ are often written in a tensor form $[\mathbf{h}]$ as defined by [14]

$$[\mathbf{h}] \equiv \begin{bmatrix} 0 & -h_{z_b} & h_{y_b} \\ h_{z_b} & 0 & -h_{x_b} \\ -h_{y_b} & h_{x_b} & 0 \end{bmatrix} \tag{2.7.3}$$

This will be referred to as the *vehicle gyroscopic tensor* and will be used in the equations of motion presented in Chapter 5.

2.8 Problems

Theory Problems

2.T1 Show that Eq. (2.4.1) is correct.

2.T2 Show that Eq. (2.4.2) is correct.

2.T3 Show that Eq. (2.4.3) is correct.

2.T4 Beginning with the definitions of υ_0, υ_1, υ_2, and υ_3 given by the integral fractions in Eqs. (2.4.8)–(2.4.11), show that these simplify to the integrals given on the right-hand side of each equation.

2.T5 Using the definition of the thickness distribution of a diamond airfoil given in Eq. (2.4.12), show that the solutions for υ_1, υ_2, and υ_3 given in Eq. (2.4.14) are correct.

2.T6 Using the definition of the thickness distribution from the NACA 4-digit airfoil series given in Eq. (2.4.15), show that the analytic solutions for υ_0, υ_1, υ_2, and υ_3 are those given in Eqs. (2.4.16)–(2.4.19).

2.T7 Use the thickness distribution given in Eq. (2.4.15) to estimate the parameters υ_0, υ_1, υ_2, and υ_3 for a diamond airfoil with a maximum thickness at $\hat{x}_{\text{max}} = 0.25$. Compare the results to the exact solution obtained from Eqs. (2.4.13) and (2.4.14).

2.T8 Select an airfoil from the UIUC airfoil database [12]. Repeat the steps outlined in Example 2.4.1 to compute an estimate for the camber distribution, thickness distribution, and values for υ_0, υ_1, υ_2, and υ_3 for the selected airfoil.

2.T9 Show that using $i = j = k = 0$ in the volume integral given in Eq. (2.4.20) yields the solution given in Eq. (2.4.21).

2.T10 Beginning with the integral given in Eq. (2.4.23), derive the solution given in Eq. (2.4.24).

2.T11 Beginning with the integral given in Eq. (2.2.6), derive the solution given in Eq. (2.4.25).

2.T12 Show that using Eq. (2.4.50) in Eq. (2.2.21) results in Eq. (2.4.51).

Coding Problems

2.C1 Write a parent class for an arbitrary component. This class will contain variable and function definitions inherent to all components and will be inherited by children classes for individual components. The class constructor function should do the following:

1. Accept a dictionary or JSON object containing the component information
2. Initialize component properties including the object name, type, location on the vehicle, and orientation relative to the vehicle coordinate system
3. Initialize member variables for use in the children classes such as object volume, density, mass, weight, and inertia tensor

The class should include two functions that do the following:

1. Compute the density, mass, and weight given any one of the other properties in addition to the object volume
2. Print properties inherent to all components to the screen, such as type, name, location, orientation, CG location, volume, density, mass, weight, and inertia tensor

2.C2 Write a child class for a cuboid that inherits properties from the parent component class created in Problem 2.C1. The constructor function should do the following:

1. Initialize component properties
2. Compute the component volume
3. Compute the component weight, mass, and density given any one of these items (this should be a call to a function in the parent class)
4. Compute the component inertia tensor

The class should also include a function to print the properties of the cuboid to the screen. Test your code using the results given in Example 2.3.1.

2.C3 Repeat Problem 2.C2 for the case of a cylinder component. Test your code using the results given in Example 2.3.2.

2.C4 Repeat Problem 2.C2 for the case of a sphere component. Test your code using the results given in Example 2.3.3.

2.C5 Write an airfoil class that can be used to compute the airfoil integral parameters υ_0, υ_1, υ_2, and υ_3 for a diamond airfoil or an airfoil with a thickness distribution given by Eq. (2.4.15). The constructor should accept a dictionary or JSON object containing either the maximum thickness location $\hat{x}_{max}$ for a diamond airfoil, or the five coefficients a_0, a_1, a_2, a_3, a_4 for an airfoil with a thickness distribution given by Eq. (2.4.15). Test your code using the results given in Example 2.4.1 and Table 2.4.2.

2.C6 Repeat Problem 2.C2 for the case of a wing-segment component. Be sure to include a computation to set the center of gravity of the wing relative to the wing coordinate system. **Note:** It is helpful to include the airfoil class created in Problem 2.C5 as an include statement at the top of the component parent class, and move the include statement for your helper code to the top of the airfoil class. This will allow these to be used by the wing-segment child class, as well as the rotor child class. Test your code using the results given in Example 2.4.2.

2.C7 Repeat Problem 2.C2 for the case of a rotor component. Include a function to compute the angular momentum of the rotor based on a given rotation rate. Test your code using the results given in Examples 2.5.1 and 2.5.2. Code for the required γ factors can be found on the book website.

2.C8 Write a class for a vehicle that contains global vehicle properties, as well as an array of components that make up the vehicle. The class should include a function to initialize the components from a dictionary or JSON object containing the vehicle components, a function to print the properties of each component, a function to compute the vehicle mass properties including the vehicle CG and inertia tensor based on all the component mass properties, and a function to print the vehicle global mass properties to the screen. Test your code using the results given in Example 2.6.1.

2.C9 Add a function to the vehicle class to compute the vehicle angular momentum in the vehicle coordinate system based on the angular momentum of all spinning rotors.

Chapter 3

Aerodynamic Modeling

3.1 Introduction

As an object moves through a fluid, forces and moments are exerted on the object due to its interaction with the fluid. These forces and moments can be divided into two categories: static and dynamic. Static forces and moments are present when an object is submerged in a quiescent fluid without any motion of the object relative to the fluid. Dynamic forces and moments are those resulting from the motion of the object relative to the fluid. In the field of flight dynamics when the vehicle is traversing through the atmosphere, both static and dynamic forces and moments are present, and are referred to as *aerodynamic forces and moments*. If the vehicle is submerged in water, these forces and moments are called *hydrodynamic forces and moments*. In this textbook, we will refer to these forces and moments as *aerodynamic*. However, the principles can be nearly equivalently applied to the case of hydrodynamics.

The aerodynamic forces and moments on an object originate from the interaction of the fluid with the vehicle external geometry. These forces and moments come from two main distributions of fluid characteristics on the surface of the object. These include the pressure distribution and shear distribution. The pressure acts normal to the object's surface, and the shear acts tangential to the object's surface. The total force on the object can be found by integrating the pressure and shear distributions on the surface of the object.

The aerodynamic forces and moments exerted on an object moving through a fluid depend on the fluid properties immediately surrounding the object. Therefore, to predict the dynamics of a vehicle, we must be able to predict the atmospheric properties, including the pressure, density, temperature, and viscosity. These can be modeled or measured many different ways. Here, we will consider a common method used to model atmospheric conditions in the Earth's atmosphere.

3.2 The Standard Atmosphere

The most common method for modeling Earth's atmospheric properties is to assume that they vary with altitude according to what is called the *standard atmosphere*. Other atmospheric models also exist, such as the *desert atmosphere*, the *polar atmosphere*, and the *10-percent hot day*. In the present text, only the definition of the standard atmosphere is included.

The terms used to define the standard atmosphere are typically defined in SI units. Therefore, the development here will be given in SI units, and it is recommended that computer codes developed to predict the atmospheric properties also be developed in SI units. If properties are needed in English units, the results of an atmospheric code can easily be converted to English units. Unit conversion factors used in this book are included in Appendix A. It is highly recommended that students use the conversion factors exactly as they are given in the appendix during code development. This allows code to be debugged more easily, since solutions should match those given in this text out to the precision shown.

According to Newton's law of gravitation, **gravity** is a function of altitude and can be computed from

$$g = g_{\text{ssl}} \left(\frac{R_{E_Z}}{R_{E_Z} + H} \right)^2 \tag{3.2.1}$$

where $g_{\text{ssl}} = 9.80665\ \text{m/s}^2$ is the standard gravitational acceleration at sea level, $R_{E_Z} = 6,356,766$ m is the radius of the Earth suggested for use in the U.S. Standard Atmosphere document published in 1976 [15], and H is the geometric altitude above sea level. From this definition, a **geopotential altitude** can be defined as

$$Z \equiv \frac{R_{E_Z} H}{R_{E_Z} + H} \tag{3.2.2}$$

At low altitudes, the geopotential altitude Z is nearly the same as the geometric altitude H. However, at higher altitudes the difference between the two can be significant. Note that the radius of the Earth R_{E_Z} suggested for use in computing the geopotential altitude in the U.S. is not the same as the mean radius of the Earth at sea level R_E, which will be used later in this text. Hence, values for R_{E_Z} and R_E differ. This distinction is shown in the constants listed in Table A.1.1.

The standard atmosphere is defined in terms of a series of geopotential altitude ranges, each with a defined temperature $\mathcal{T}_i$ at the range base, and a defined temperature gradient $\mathcal{T}_i'$ within the range. The ranges and temperature properties are defined in SI units and given in Table 3.2.1. The **temperature at any geopotential altitude** can be found from

$$\mathcal{T}(Z) = \mathcal{T}_i + \mathcal{T}_i'(Z - Z_i) \tag{3.2.3}$$

where Z falls in the range

$$Z_i \leq Z < Z_{i+1} \tag{3.2.4}$$

At typical altitudes of interest for most aircraft, air can be treated as an ideal gas. Hence, the pressure $\mathcal{P}$ is related to the density ρ and absolute temperature $\mathcal{T}$ according to

$$\mathcal{P} = \rho \mathcal{R} \mathcal{T} \tag{3.2.5}$$

where $\mathcal{R}$ is a gas constant. From these definitions and the assumption of a static fluid, the relationship for the pressure variation with altitude for the standard atmosphere is

$$\frac{d\mathcal{P}}{dZ} = -\frac{g_{\text{ssl}} \mathcal{P}}{\mathcal{R} \mathcal{T}} \tag{3.2.6}$$

Table 3.2.1. Definition of temperature variation with geopotential altitude for the standard atmosphere.

Z_i [m]	Z_{i+1} [m]	$\mathcal{T}_i$ [K]	$\mathcal{T}_i'$ [K/km]
0	11,000	288.150	−6.5
11,000	20,000	216.650	0.0
20,000	32,000	216.650	1.0
32,000	47,000	228.650	2.8
47,000	52,000	270.650	0.0
52,000	61,000	270.650	−2.0
61,000	79,000	252.650	−4.0
79,000	~ 90,000	180.650	0.0

where $\mathcal{R} = 287.0528$ J/(kg·K) is the gas constant for dry air. The **pressure at any geopotential altitude** can be obtained from the integral of Eq. (3.2.6)

$$\mathcal{P}(Z) = \begin{cases} \mathcal{P}_i \exp\left[-\frac{g_{\text{ssl}}(Z-Z_i)}{\mathcal{R}\mathcal{T}_i}\right], & \mathcal{T}_i' = 0 \\ \mathcal{P}_i \left[\frac{\mathcal{T}_i+\mathcal{T}_i'(Z-Z_i)}{\mathcal{T}_i}\right]^{\frac{-g_{\text{ssl}}}{\mathcal{R}\mathcal{T}_i'}}, & \mathcal{T}_i' \neq 0 \end{cases} \tag{3.2.7}$$

where $\mathcal{P}_i$ is the pressure at the base of range i. The pressure at the base of the first range is the pressure at sea level defined as 101,325 Pa for the standard atmosphere.

For any given geometric altitude H, the associated geopotential altitude can be computed from Eq. (3.2.2), after which Eqs. (3.2.3) and (3.2.7) can be used in conjunction with the data given in Table 3.2.1 to compute the pressure and temperature. Note that, since the pressure at any altitude depends on the pressure at the base of the geopotential altitude range, and because pressures are not specifically defined for the standard atmosphere, Eq. (3.2.7) must be used to compute the pressures at any lower ranges before the pressure in the range of interest can be obtained.

The computational cost to compute the pressure at any altitude can be significantly decreased by pre-computing and storing the pressure at the base of each geopotential altitude range. Doing so gives the results shown in Table 3.2.2.

Once the pressure and temperature have been obtained, the **density** can be estimated from the ideal gas relation

$$\rho = \frac{\mathcal{P}}{\mathcal{R}\mathcal{T}} \tag{3.2.8}$$

The **speed of sound** can be computed from

$$a = \sqrt{\gamma \mathcal{R}\mathcal{T}} \tag{3.2.9}$$

where the specific heat ratio for dry air is $\gamma = 1.4$.

Table 3.2.2. Pressure at the base of each geopotential altitude range for the standard atmosphere.

Z_i [m]	$\mathcal{P}_i$ [Pa]
0	1.0132500000000000E+05
11,000	2.2632031822221168E+04
20,000	5.4748735282708267E+03
32,000	8.6801476908672271E+02
47,000	1.1090558898922531E+02
52,000	5.9000524278924367E+01
61,000	1.8209924905017658E+01
79,000	1.0377004548920223E+00

The **dynamic viscosity of atmospheric air** can be estimated using Sutherland's Law according to the relation

$$\mu = \mu_S \left(\frac{\mathcal{T}_S + K_S}{\mathcal{T} + K_S} \right) \left(\frac{\mathcal{T}}{\mathcal{T}_S} \right)^{3/2} \tag{3.2.10}$$

The appropriate values of the constants $\mathcal{T}_S$, μ_S, and K_S for air vary slightly among authors and use cases in the literature. For the example solutions given in this textbook, the values of $\mathcal{T}_S = 273.15$ K, $\mu_S = 1.716\text{E}{-}05$ kg/(m·s), and $K_S = 110.4$ are used. Note that because $\mathcal{T}_S \neq \mathcal{T}_{\text{ssl}}$, μ_S is not the viscosity at sea level in the standard atmosphere.

The *molecular mean free path* is the average distance a fluid molecule travels before colliding with another molecule, and is an important parameter related to whether a fluid can be modeled as a continuum. The **molecular mean free path** λ can be computed from

$$\lambda = \frac{\mu}{\rho a} \sqrt{\frac{\pi \gamma}{2}} \tag{3.2.11}$$

The relation between the molecular mean free path and a fluid continuum will be treated in the following section.

Tables of atmospheric properties as defined by the standard atmosphere using the equations presented in this section are included in Appendix B. Table B.1.2 is in SI units, and Table B.1.3 is in English units.

EXAMPLE 3.2.1 From the definition of the standard atmosphere, compute the gravity, geopotential altitude, temperature, pressure, density, speed of sound, dynamic viscosity, and mean free path at an altitude of 25,000 m in SI units.

Solution: The gravity can be computed from Eq. (3.2.1) with $g_{\text{ssl}} = 9.80665$ m/s^2 and $R_{E_Z} = 6,356,766$ m to yield

$$g = 9.80665 \left(\frac{6356766}{6356766 + 25000} \right)^2 = 9.729967138 \text{ m/s}^2$$

The geopotential altitude can be computed from Eq. (3.2.2) to yield

$$Z = \frac{(6356766)(25000)}{6356766 + 25000} = 24,902.06473 \text{ m}$$

The geopotential altitude falls in the range of the third row of Table 3.2.1. Using the information in that row in Eq. (3.2.3) gives the temperature

$$\mathcal{T}(Z) = 216.650 + (1.0)\frac{24902.06473 - 20000}{1000} = 221.5520647 \text{ K}$$

The pressure can be computed from Eq. (3.2.7). However, in order to compute the pressure at the current altitude, we must know the pressure $\mathcal{P}_i$ at the beginning of the altitude level $Z_i = 20,000$ m. By definition, the pressure at the level $Z_i = 20,000$ depends on the pressure at the level $Z_i = 11,000$, which in turn depends on the pressure at sea level $Z_i = 0$. The pressure at sea level is defined as 101, 325 Pa. Using this along with the corresponding information from the first row of Table 3.2.1 in Eq. (3.2.7) yields the pressure at $Z = 11,000$ m

$$\begin{aligned}\mathcal{P}(11,000) &= 101325\left[\frac{288.150 - 6.5(11000 - 0)/1000}{288.150}\right]^{\frac{(9.80665)(1000)}{(287.0528)(6.5)}} \\ &= 22,632.03182 \text{ Pa}\end{aligned}$$

Repeating this process for the next altitude level and applying the result $\mathcal{P}_i = 22,632.03182$ Pa gives the pressure at $Z = 20,000$ m

$$\begin{aligned}\mathcal{P}(20,000) &= (22632.03182)\exp\left[-\frac{9.80665(20000 - 11000)}{(287.0528)(216.650)}\right] \\ &= 5,474.873528 \text{ Pa}\end{aligned}$$

Note that the solutions computed here for pressure at $Z = 11,000$ m and $Z = 20,000$ m are the same as those given in Table 3.2.2, which can be pre-computed and stored in a code to reduce computation time of atmospheric properties. We can now compute the pressure at the geopotential altitude of interest by using $\mathcal{P}_i = 5,474.873528$ Pa in Eq. (3.2.7) to yield

$$\begin{aligned}\mathcal{P}(Z) &= 5474.873528\left[\frac{216.650 + \frac{1.0}{1000}(24,902.06473 - 20000)}{216.650}\right]^{\frac{(-9.80665)(1000)}{(287.0528)(1.0)}} \\ &= 2,549.214096 \text{ Pa}\end{aligned}$$

The density can now be computed from Eq. (3.2.8) to yield

$$\rho = \frac{\mathcal{P}}{\mathcal{R}\mathcal{T}} = \frac{2549.214096}{(287.0528)(221.5520647)} = 0.040083785 \text{ kg/m}^3$$

The speed of sound can be computed from Eq. (3.2.9) to yield

$$a = \sqrt{\gamma \mathcal{R} \mathcal{T}} = \sqrt{(1.4)(287.0528)(221.5520647)} = 298.3890024 \text{ m/s}$$

The dynamic viscosity can be computed from Eq. (3.2.10) to yield

$$\begin{aligned}\mu &= \mu_S \left(\frac{\mathcal{T}_S + K_S}{\mathcal{T} + K_S}\right)\left(\frac{\mathcal{T}}{\mathcal{T}_S}\right)^{3/2} \\ &= (1.716\text{E}{-05})\left(\frac{273.15 + 110.4}{221.5520647 + 110.4}\right)\left(\frac{221.5520647}{273.15}\right)^{3/2} \\ &= 1.448357564\text{E}{-05} \text{ kg/(m·s)}\end{aligned}$$

Finally, the mean free path can be computed from Eq. (3.2.11) to yield

$$\begin{aligned}\lambda &= \frac{\mu}{\rho a}\sqrt{\frac{\pi\gamma}{2}} = \frac{1.448357564\text{E}{-05}}{(0.040083785)(298.3890024)}\sqrt{\frac{1.4\pi}{2}} \\ &= 1.795759669\text{E}{-06} \text{ m}\end{aligned}$$

EXAMPLE 3.2.2 From the definition of the standard atmosphere, compute the gravity, geopotential altitude, temperature, pressure, density, speed of sound, dynamic viscosity, and mean free path at an altitude of 150,000 ft in English units.

Solution: Because the standard atmosphere is defined in SI units, not in English units, a conversion to SI units must take place at some point in the computation. If a function has been written that computes the standard atmospheric properties in SI units, it is easiest to simply convert the given altitude to SI units, use the function to compute all the atmospheric properties in SI units, and then convert the final atmospheric properties to English units. This can be done in code by simply writing a wrapper to the atmospheric function in SI units. The steps outlined here will be taken to follow this procedure.

The altitude in English units can be converted to SI units using the conversion table in Appendix A. This gives

$$(150000 \text{ ft})\frac{0.3048 \text{ m}}{1 \text{ ft}} = 45,720.0 \text{ m}$$

The gravity can be computed in SI units from Eq. (3.2.1) with $g_{\text{ssl}} = 9.80665 \text{ m/s}^2$ and $R_{E_Z} = 6,356,766$ m to yield

$$g = 9.80665\left(\frac{6356766}{6356766 + 45720}\right)^2 = 9.667091968 \text{ m/s}^2 = 31.7161809966 \text{ ft/s}^2$$

The geopotential altitude can be computed from Eq. (3.2.2) to yield

$$Z = \frac{(6356766)(45720)}{6356766 + 45720} = 45,393.51457 \text{ m}$$

The geopotential altitude falls in the range of the fourth row of Table 3.2.1. Using the information in that row in Eq. (3.2.3) gives the temperature

$$\mathcal{T}(Z) = 228.650 + (2.8)\frac{45393.51457 - 32000}{1000} = 266.1518408 \text{ K}$$

The pressure can be computed from Eq. (3.2.7). However, in order to compute the pressure at the current altitude, we must know the pressure $\mathcal{P}_i$ at the beginning of the altitude level $Z_i = 32,000$ m. By definition, the pressure at the level $Z_i = 32,000$ depends on the pressure at the level $Z_i = 20,000$, which in turn depends on the pressure at sea level $Z_i = 11,000$, which depends on the pressure at sea level $Z_i = 0$. The pressure at sea level is defined as 101,325 Pa. Using this along with the corresponding information from the first row of Table 3.2.1 in Eq. (3.2.7) yields the pressure at $Z = 11,000$ m

$$\begin{aligned}\mathcal{P}(11,000) &= (101325)\left[\frac{288.150 - 6.5(11000 - 0)/1000}{288.150}\right]^{\frac{(9.80665)(1000)}{(287.0528)(6.5)}} \\ &= 22,632.03182 \text{ Pa}\end{aligned}$$

Repeating this process for the next altitude level and applying the result $\mathcal{P}_i = 22,632.03182$ Pa gives the pressure at $Z = 20,000$ m

$$\begin{aligned}\mathcal{P}(20,000) &= (22632.03182)\exp\left[-\frac{9.80665(20000 - 11000)}{(287.0528)(216.650)}\right] \\ &= 5,474.873528 \text{ Pa}\end{aligned}$$

Repeating this process for the next altitude level and applying the result $\mathcal{P}_i = 5,474.873528$ Pa gives the pressure at $Z = 32,000$ m

$$\begin{aligned}\mathcal{P}(32,000) &= (5474.873528)\left[\frac{216.650 + (1.0)(32000 - 20000)/1000}{216.650}\right]^{\frac{(-9.80665)(1000)}{(287.0528)(1.0)}} \\ &= 868.0147691 \text{ Pa}\end{aligned}$$

We can now compute the pressure at the geopotential altitude of interest by using $\mathcal{P}_i = 868.0147691$ Pa in Eq. (3.2.7) to yield

$$\begin{aligned}\mathcal{P}(Z) &= 868.0147691\left[\frac{228.650 + \frac{2.8}{1000}(45393.51457 - 32000)}{228.650}\right]^{\frac{(-9.80665)(1000)}{(287.0528)(2.8)}} \\ &= 136.0693215 \text{ Pa}\end{aligned}$$

The density can now be computed from Eq. (3.2.8) to yield

$$\rho = \frac{\mathcal{P}}{\mathcal{R}\mathcal{T}} = \frac{136.0693215}{(287.0528)(266.1518408)} = 0.0017810207 \text{ kg/m}^3$$

The speed of sound can be computed from Eq. (3.2.9) to yield

$$a = \sqrt{\gamma \mathcal{R}\mathcal{T}} = \sqrt{(1.4)(287.0528)(266.1518408)} = 327.0466076 \text{ m/s}$$

The dynamic viscosity can be computed from Eq. (3.2.10) to yield

$$\begin{aligned}\mu &= \mu_S \left(\frac{\mathcal{T}_S + K_S}{\mathcal{T} + K_S}\right)\left(\frac{\mathcal{T}}{\mathcal{T}_S}\right)^{3/2} \\ &= (1.716\text{E}{-05})\left(\frac{273.15 + 110.4}{266.1518408 + 110.4}\right)\left(\frac{266.1518408}{273.15}\right)^{3/2} \\ &= 1.681151683\text{E}{-05} \text{ kg/(m}\cdot\text{s)}\end{aligned}$$

Finally, the mean free path can be computed from Eq. (3.2.11) to yield

$$\begin{aligned}\lambda &= \frac{\mu}{\rho a}\sqrt{\frac{\pi\gamma}{2}} = \frac{1.681151683\text{E}{-05}}{(0.0017810207)(327.0466076)}\sqrt{\frac{1.4\pi}{2}} \\ &= 4.280083581\text{E}{-05} \text{ m}\end{aligned}$$

We can now convert each of these properties to English units using the unit conversions in Appendix A. The altitude in English units is

$$(45720 \text{ m})\left(\frac{1 \text{ ft}}{0.3048 \text{ m}}\right) = 150{,}000 \text{ ft}$$

Gravity in English units is

$$\left(9.667091968 \ \frac{\text{m}}{\text{s}^2}\right)\left(\frac{1 \text{ ft}}{0.3048 \text{ m}}\right) = 31.71618100 \text{ ft/s}^2$$

The geopotential altitude in English units is

$$(45393.51457 \text{ m})\left(\frac{1 \text{ ft}}{0.3048 \text{ m}}\right) = 148{,}928.8536 \text{ ft}$$

The temperature in Rankine is

$$(1.8)(266.1518408 \text{ K}) = 479.0733134 \text{ R}$$

The pressure in English units is

$$(136.0693215 \text{ Pa})\left(\frac{1 \text{ N/m}^2}{1 \text{ Pa}}\right)\left(\frac{1 \text{ lbf/ft}^2}{47.880258 \text{ N/m}^2}\right) = 2.841866923 \text{ lbf/ft}^2$$

The density in English units is

$$\left(0.0017810207 \text{ kg/m}^3\right)\left(\frac{1 \text{ slug/ft}^3}{515.379 \text{ kg/m}^3}\right) = 3.455749381\text{E}{-06} \text{ slug/ft}^3$$

The speed of sound in English units is

$$(327.0466076 \text{ m/s})\left(\frac{1 \text{ ft}}{0.3048 \text{ m}}\right) = 1,072.987558 \text{ ft/s}$$

The dynamic viscosity in English units is

$$1.681151683\text{E}{-05} \text{ kg/(m·s)}\left(\frac{1 \text{ slug/(ft·s)}}{47.880258 \text{ kg/(m·s)}}\right)$$
$$= 3.511158363\text{E}{-07} \text{ slug/(ft·s)}$$

The mean free path in English units is

$$(4.280083581\text{E}{-05} \text{ m})\left(\frac{1 \text{ ft}}{0.3048 \text{ m}}\right) = 1.404226897\text{E}{-04} \text{ ft}$$

3.3 Governing Equations of Fluid Motion

Important Dimensionless Definitions

A fluid is comprised of molecules in constant motion. The average distance a molecule travels before colliding with another molecule is called the molecular mean free path. The ratio of the molecular mean free path λ to the length scale of the geometry l_{ref} is an important dimensionless ratio called the *Knudsen number*. It is defined as

$$\text{K} \equiv \frac{\lambda}{l_{\text{ref}}} \tag{3.3.1}$$

If the mean free path is sufficiently small compared to the geometry of interest, the fluid can be treated as a continuum. As a first approximation, Phillips [16] suggests that a fluid can typically be modeled as a continuum if

$$\text{K} < 8 \times 10^{-3}\left[\sqrt{1 + \frac{1}{2\text{M}}} - 1\right]^2 \text{M} \tag{3.3.2}$$

where M is the Mach number defined as the ratio of the vehicle velocity to the speed of sound $\text{M} \equiv V/a$. The approximation of modeling the fluid as a continuum holds for most aircraft at typical altitudes. For vehicles at very high-altitudes, the continuum assumption does not hold, and statistical methods must be used. The following discussion of the governing equations of fluid motion assumes the fluid can be treated as a continuum.

The aerodynamic forces and moments acting on a vehicle depend on many properties of the fluid. For example, the aerodynamics depend on the velocity of the fluid relative to the vehicle, fluid density, fluid viscosity, and speed of sound in the fluid. From the Buckingham Pi theorem, it can be shown that in the absence of thermal effects, the general quasi-steady dependence of the aerodynamics on the fluid properties can be simplified to dependence on two dimensionless parameters. These are the Reynolds number and Mach number.

The **Reynolds number** can be thought of as a ratio of the inertial forces to viscous forces and is defined as

$$\mathrm{R_e} \equiv \frac{\rho V l_{\mathrm{ref}}}{\mu} \tag{3.3.3}$$

where l_{ref} is a characteristic length of the geometry of interest. At very low Reynolds numbers, viscosity dominates the behavior of the flow. At very high Reynolds numbers, viscous effects are less significant, and the flow properties are typically dominated by the dynamic pressure of the fluid. Inviscid solutions from the Euler equations can be thought of as solutions to the full Navier-Stokes equations in the limit as $\mathrm{R_e} \to \infty$. Most manned aircraft operate at speeds and geometric length scales that result in Reynolds numbers greater than 1,000,000. For these cases, inviscid aerodynamic solutions provide relatively accurate predictions for pressure on the vehicle surface, as long as the flow remains attached. For small or low-speed aircraft, viscous effects become much more significant and must be carefully accounted for.

The **Mach number** is a ratio of the velocity of the vehicle V to the speed of sound a in the fluid and is defined as

$$\mathrm{M} \equiv \frac{V}{a} = \frac{V}{\sqrt{\gamma \mathcal{R} \mathcal{T}}} \tag{3.3.4}$$

The Mach number provides a sense of the importance of compressibility of the fluid in the aerodynamic calculations. At Mach numbers less than 0.3, compressibility effects are typically negligible. At higher Mach numbers, fluid compressibility significantly alters the aerodynamic forces and moments experienced by the vehicle. At Mach numbers approaching or exceeding 1.0, shock waves typically form in the flow and influence the aerodynamic behavior.

Because the operating Mach number and Reynolds number of the vehicle significantly impact the aerodynamic forces and moments on the vehicle, aerodynamic modeling of a vehicle must account for the effects of Reynolds number and Mach number. The range of speeds of a vehicle during a mission can vary widely, and therefore a vehicle can experience a wide range of Reynolds numbers and Mach numbers during a flight. However, it is typical for a vehicle to spend a large portion of a mission near a single operating condition or cruise speed and altitude. In these cases, a single baseline aerodynamic model developed for a specific Reynolds number and Mach number can be used for the simulation, and corrections for Reynolds number and Mach number can be applied to the baseline aerodynamic model. In practice, corrections for Reynolds number effects are less common, since the dependence of the aerodynamics on Reynolds number are typically small in the range of operating conditions experienced by most aircraft. However, compressibility effects become increasingly important at speeds above Mach 0.3, and should be accounted for in the aerodynamic modeling. Compressibility effects are discussed in a later section of this chapter.

Pressure Definitions

Pressure plays an important role in the mechanics of fluids. Several pressure definitions are commonly used in fluid mechanics, and the student should be familiar with each. Many textbooks use a lower-case p to denote pressure. Here we will always use an upper case $\mathcal{P}$ to denote pressure, to avoid confusion with the vehicle roll rate. The **local fluid pressure** will be denoted by

$$\text{local fluid pressure} \equiv \mathcal{P} \tag{3.3.5}$$

The **freestream fluid pressure** will be denoted by

$$\text{freestream fluid pressure} \equiv \mathcal{P}_\infty \tag{3.3.6}$$

The **dynamic pressure** is a function of the freestream flow velocity V_∞ and density ρ_∞ and is defined as

$$\text{dynamic pressure} \equiv Q \equiv \frac{1}{2}\rho_\infty V_\infty^2 \tag{3.3.7}$$

Many textbooks use the symbol q to denote the dynamic pressure. Here we will use the full definition or the symbol Q to avoid confusion with the vehicle pitch rate. In SI units, density has units of kg/m^3 and velocity has units of m/s. This yields a dynamic pressure in SI units of N/m^2. In English units, density has units of slug/ft^3 and velocity has units of ft/s. This yields a dynamic pressure in English units of lbf/ft^2.

The **stagnation pressure** is the pressure that would exist if the flow were brought to rest relative to the moving body and will be denoted as

$$\text{stagnation pressure} \equiv \mathcal{P}_0 \tag{3.3.8}$$

The **pressure coefficient** is a dimensionless value of the pressure $\mathcal{P}$ at any point in the flow relative to the freestream fluid pressure normalized by the dynamic pressure. It is defined as

$$\text{pressure coefficient} \equiv C_{\mathcal{P}} \equiv \frac{\mathcal{P} - \mathcal{P}_\infty}{\frac{1}{2}\rho_\infty V_\infty^2} \tag{3.3.9}$$

Governing Equations

The governing equations of fluid motion can be developed by applying three fundamental laws of Newtonian physics to a fluid element. These three laws are the conservation of mass, the conservation of momentum, and the conservation of energy. The **conservation of mass** can be expressed in vector notation as

$$\frac{\partial \rho}{\partial t} + \nabla \cdot (\rho \mathbf{V}) = 0 \tag{3.3.10}$$

This is often called the *continuity equation*.

The **conservation of momentum** equation comes from Newton's second law, and can be expressed in vector notation as

$$\frac{\partial(\rho\mathbf{V})}{\partial t} + \nabla \cdot (\rho\mathbf{V}\mathbf{V}) = \mathbf{f}_b + \mathbf{f}_s \tag{3.3.11}$$

where $\mathbf{f}_b$ are body forces acting on the fluid element, and $\mathbf{f}_s$ are surface forces acting on the fluid element. Applying the continuity equation given in Eq. (3.3.10) and assuming that the only body force on the fluid is due to gravity, the conservation of momentum equation can be written in a more standard form

$$\rho\left[\frac{\partial\mathbf{V}}{\partial t} + (\mathbf{V}\cdot\nabla)\mathbf{V}\right] = \nabla\cdot[\sigma] - \rho g_{\text{ssl}}\nabla Z \tag{3.3.12}$$

where $[\sigma]$ is the fluid stress tensor and Z is the geopotential altitude. The fluid stress tensor is comprised of both pressure and shear forces. Because pressure can only act normal to the fluid element, the pressure forces are contained within the diagonal elements of the fluid stress tensor. Equations (3.3.10) and (3.3.12) are often referred to as the *Navier-Stokes equations.*

In the absence of viscosity, the Navier-Stokes equations reduce to the **Euler Equations**

$$\rho\left[\frac{\partial\mathbf{V}}{\partial t} + (\mathbf{V}\cdot\nabla)\,\mathbf{V}\right] = -\nabla\mathcal{P} - \rho g_{\text{ssl}}\nabla Z \tag{3.3.13}$$

Using the mathematical identity $(\mathbf{V}\cdot\nabla)\mathbf{V} = \nabla[(1/2)V^2] - \mathbf{V}\times(\nabla\times\mathbf{V})$, the Euler equations can be rearranged in the form

$$\frac{\partial\mathbf{V}}{\partial t} + \nabla\left(\frac{1}{2}V^2\right) + \frac{\nabla\mathcal{P}}{\rho} + g_{\text{ssl}}\nabla Z = \mathbf{V}\times\mathbf{\Omega} \tag{3.3.14}$$

where $\mathbf{\Omega} \equiv \nabla\times\mathbf{V}$ is the vorticity vector. For inviscid, irrotational flow, $\mathbf{V}\times\mathbf{\Omega} = 0$, and Eq. (3.3.14) reduces to

$$\frac{\partial\mathbf{V}}{\partial t} + \nabla\left(\frac{1}{2}V^2\right) + \frac{\nabla\mathcal{P}}{\rho} + g_{\text{ssl}}\nabla Z = 0 \tag{3.3.15}$$

The **conservation of energy** equation is developed from the first law of thermodynamics, and can be written as

$$\frac{\partial(\rho e)}{\partial t} + \nabla\cdot(\rho\mathbf{V}e) = -\nabla\cdot\mathbf{q} + \nabla\cdot([\sigma]\cdot\mathbf{V}) + U_e''' \tag{3.3.16}$$

where e is the specific energy, $\mathbf{q}$ is the heat flux vector, and U_e''' includes any other volumetric heating. This equation is very important for high-speed vehicles for which heating in the fluid affects the aerodynamics as well as the surface temperature of the vehicle. For vehicles at speeds less than about 5 times the speed of sound, the fluid heating does not significantly impact the vehicle aerodynamics. However, for vehicles traveling faster than about 5 times the speed of sound, these heating and energy effects should not be neglected.

Isentropic Compressible Flow of an Ideal Gas

For modeling purposes, air in the Earth's atmosphere is typically treated as an ideal gas. Additionally, flow outside the boundary layer can typically be treated as inviscid flow. Finally, because the thermal conductivity of air is low, air is often treated as adiabatic. In the absence of entropy-generating effects such as shocks, an inviscid, adiabatic flow is isentropic. For isentropic flow of an ideal gas, Eq. (3.3.15) can be written as

$$\frac{\partial \mathbf{V}}{\partial t} + \nabla\left(\frac{1}{2}V^2 + \frac{\gamma}{\gamma-1}\frac{\mathcal{P}}{\rho} + g_{\text{ssl}}Z\right) = 0 \tag{3.3.17}$$

For the case when the time derivative is zero, Eq. (3.3.17) reduces to

$$\frac{1}{2}V^2 + \frac{\gamma}{\gamma-1}\frac{\mathcal{P}}{\rho} + g_{\text{ssl}}Z = C \tag{3.3.18}$$

where C is a constant. This equation can be considered Bernoulli's equation for compressible flow. Equation (3.3.18) gives rise to the following useful flow properties

$$\frac{\mathcal{P}}{\mathcal{P}_0} = \left(1 + \frac{\gamma-1}{2}\frac{V^2}{\gamma R T}\right)^{-\gamma/(\gamma-1)} = \left(1 + \frac{\gamma-1}{2}\mathrm{M}^2\right)^{-\gamma/(\gamma-1)} \tag{3.3.19}$$

$$\frac{\rho}{\rho_0} = \left(\frac{\mathcal{P}}{\mathcal{P}_0}\right)^{1/\gamma} = \left(1 + \frac{\gamma-1}{2}\mathrm{M}^2\right)^{-1/(\gamma-1)} \tag{3.3.20}$$

$$\frac{T}{T_0} = \left(\frac{\mathcal{P}}{\mathcal{P}_0}\right)^{(\gamma-1)/\gamma} = \left(1 + \frac{\gamma-1}{2}\mathrm{M}^2\right)^{-1} \tag{3.3.21}$$

where $\mathcal{P}_0$, ρ_0, and T_0 are the stagnation pressure, density, and temperature, respectively. These equations can be used to derive the traditional expressions for the stagnation flow properties in terms of freestream conditions for isentropic compressible flow

$$\mathcal{P}_0 = \mathcal{P}_\infty\left(1 + \frac{\gamma-1}{2}\mathrm{M}_\infty^2\right)^{\gamma/(\gamma-1)} \tag{3.3.22}$$

$$\rho_0 = \rho_\infty\left(1 + \frac{\gamma-1}{2}\mathrm{M}_\infty^2\right)^{1/(\gamma-1)} \tag{3.3.23}$$

$$T_0 = T_\infty\left(1 + \frac{\gamma-1}{2}\mathrm{M}_\infty^2\right) \tag{3.3.24}$$

From these relations, it can be shown that by definition, the pressure coefficient at any point in an isentropic compressible flow is

$$C_{\mathcal{P}} \equiv \frac{\mathcal{P} - \mathcal{P}_\infty}{\frac{1}{2}\rho_\infty V_\infty^2} = \frac{2}{\gamma \mathrm{M}_\infty^2}\left(\frac{\mathcal{P}}{\mathcal{P}_\infty} - 1\right) \tag{3.3.25}$$

At subsonic speeds and linearized, small-perturbation flow, this reduces to the well-known *Prandtl-Glauert compressibility correction*

$$C_{\mathcal{P}} = \frac{C_{\mathcal{P}\mathrm{M}0}}{\sqrt{1 - \mathrm{M}_\infty^2}} \tag{3.3.26}$$

where $C_{\mathcal{P}M0}$ is the pressure coefficient at zero Mach number. Note that Eq. (3.3.26) is invalid as $M_\infty \rightarrow 1$, and is not valid for transonic or supersonic flow.

Incompressible Flow

For the special case of incompressible flow, the continuity equation given in Eq. (3.3.10) reduces to

$$\nabla \cdot \mathbf{V} = 0 \tag{3.3.27}$$

Recognizing that $g\nabla H = g_{\text{ssl}}\nabla Z$, Eq. (3.3.15) for steady-state incompressible flow reduces to what is called *Bernoulli's Equation*

$$\frac{1}{2}V^2 + \frac{\mathcal{P}}{\rho} + gH = C \tag{3.3.28}$$

For incompressible flows, the density is constant throughout the flow $\rho = \rho_\infty$ and the stagnation pressure is the sum of the static pressure and dynamic pressure

$$\mathcal{P}_0 = \mathcal{P}_\infty + \frac{1}{2}\rho V_\infty^2 \tag{3.3.29}$$

From the definition of the pressure coefficient given in Eq. (3.3.9), it can be shown that the pressure coefficient for incompressible flow is related to the local velocity and freestream velocity according to

$$C_{\mathcal{P}} \equiv \frac{\mathcal{P} - \mathcal{P}_\infty}{\frac{1}{2}\rho V_\infty^2} = 1 - \left(\frac{V}{V_\infty}\right)^2 \tag{3.3.30}$$

3.4 Aerodynamic Coordinates and Definitions

The pressure and shear distributions on the surface of the object can be integrated to yield the total aerodynamic force and aerodynamic moment vectors, denoted here as **F** and **M**. These two vectors represent the total aerodynamic influence of the fluid on the object. The aerodynamic forces and moments exerted on an object from the surrounding fluid depend on the translational velocity of the object relative to the fluid **V**, the angular rate! of the object ω, and the control states **U**. Hence, for a given vehicle, these can be expressed as

$$\begin{aligned}\mathbf{F} &= f(\mathbf{V}, \omega, \mathbf{U})\\ \mathbf{M} &= f(\mathbf{V}, \omega, \mathbf{U})\end{aligned} \tag{3.4.1}$$

If high-fidelity computational fluid dynamics (CFD) were fast enough to solve the Navier-Stokes equations in real time, CFD could be used in flight simulation to predict the total aerodynamic force and moment vectors on the vehicle. However, this is not feasible with typical computational resources available today. Therefore, it is common to employ *aerodynamic models*, which are pre-assembled correlations for the aerodynamic properties of an object. These are typically based on pre-computed aerodynamic solutions or measurements from CFD simulations, wind-tunnel testing, or closed-form approximations. Aerodynamic models are typically formulated to be functions of Reynolds number, Mach number, aerodynamic angles, body rotation rates, time rate-of-change of aerodynamic angles, and control-surface deflections. The Reynolds number was defined in Eq. (3.3.3) and the Mach number was defined in Eq. (3.3.4). We will now consider the customary definitions of the aerodynamic angles, body rotation rates, and control-surfaces of typical aircraft.

Translational Velocity and Aerodynamic Angles

The aerodynamic forces and moments acting on a vehicle moving through a fluid depend on the direction the vehicle is moving relative to the surrounding fluid. The velocity of the vehicle relative to the surrounding fluid can be expressed as a velocity vector in body-fixed coordinates. Here we will use the body-fixed coordinate system, since the equations of motion discussed in Chapter 5 are typically written in terms of this coordinate system. The components of this vector are typically denoted using the alphabetic triad u, v, and w. Hence, by definition

$$\mathbf{V} \equiv \begin{bmatrix} u \\ v \\ w \end{bmatrix} = \begin{bmatrix} V_{x_b} \\ V_{y_b} \\ V_{z_b} \end{bmatrix} \tag{3.4.2}$$

where the **velocity magnitude** is

$$V \equiv \|\mathbf{V}\| = \sqrt{u^2 + v^2 + w^2} \tag{3.4.3}$$

The velocity of the vehicle relative to the surrounding fluid can also be expressed in terms of the velocity magnitude V and two aerodynamic angles called the *angle of attack* and *sideslip angle*. The **angle of attack** is defined as

$$\alpha \equiv \tan^{-1}\left(\frac{w}{u}\right) \tag{3.4.4}$$

and the **sideslip angle** is defined as

$$\beta \equiv \sin^{-1}\left(\frac{v}{V}\right) \tag{3.4.5}$$

These angles are used to define the direction of the freestream relative to the body-fixed coordinate system.

Because the dynamics of a vehicle depend on the aerodynamic forces and moments, and because the aerodynamic forces and moments are related to the aerodynamic angles, it is often convenient to consider the forces and moments in two coordinate systems related to the aerodynamic angles. These coordinate systems are the *stability coordinate system* and the *wind coordinate system*. These are shown in Fig. 3.4.1.

The **stability coordinate system** is obtained by rotating the body-fixed coordinate system about the y_b-axis by the angle of attack α, as shown in Fig. 3.4.1. Here we will

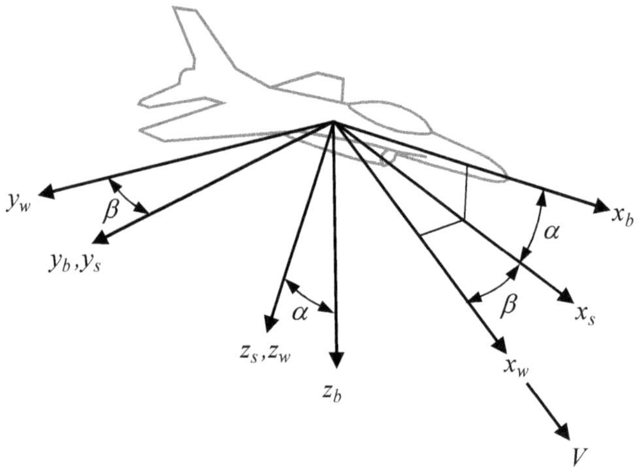

Figure 3.4.1. Body-fixed, stability, and wind coordinate systems.

use a subscript s to denote the stability coordinate system. The transformation of an arbitrary vector in the body-fixed coordinate system to the stability coordinate system is

$$\begin{bmatrix} v_{x_s} \\ v_{y_s} \\ v_{z_s} \end{bmatrix} = \begin{bmatrix} C_\alpha & 0 & S_\alpha \\ 0 & 1 & 0 \\ -S_\alpha & 0 & C_\alpha \end{bmatrix} \begin{bmatrix} v_{x_b} \\ v_{y_b} \\ v_{z_b} \end{bmatrix} \tag{3.4.6}$$

and the inverse transformation is

$$\begin{bmatrix} v_{x_b} \\ v_{y_b} \\ v_{z_b} \end{bmatrix} = \begin{bmatrix} C_\alpha & 0 & -S_\alpha \\ 0 & 1 & 0 \\ S_\alpha & 0 & C_\alpha \end{bmatrix} \begin{bmatrix} v_{x_s} \\ v_{y_s} \\ v_{z_s} \end{bmatrix} \tag{3.4.7}$$

The **wind coordinate system** is obtained by rotating the stability coordinate system about the z_s-axis by the sideslip angle β, as shown in Fig. 3.4.1. Here we will use a subscript w to denote the wind coordinate system. The transformation of an arbitrary vector in the stability coordinate system to the wind coordinate system is

$$\begin{bmatrix} v_{x_w} \\ v_{y_w} \\ v_{z_w} \end{bmatrix} = \begin{bmatrix} C_\beta & S_\beta & 0 \\ -S_\beta & C_\beta & 0 \\ 0 & 0 & 1 \end{bmatrix} \begin{bmatrix} v_{x_s} \\ v_{y_s} \\ v_{z_s} \end{bmatrix} \tag{3.4.8}$$

and the inverse transformation is

$$\begin{bmatrix} v_{x_s} \\ v_{y_s} \\ v_{z_s} \end{bmatrix} = \begin{bmatrix} C_\beta & -S_\beta & 0 \\ S_\beta & C_\beta & 0 \\ 0 & 0 & 1 \end{bmatrix} \begin{bmatrix} v_{x_w} \\ v_{y_w} \\ v_{z_w} \end{bmatrix} \tag{3.4.9}$$

Combining Eqs. (3.4.6) and (3.4.8), the transformation of a vector from the body-fixed coordinate system to the wind coordinate system is

$$\begin{bmatrix} v_{x_w} \\ v_{y_w} \\ v_{z_w} \end{bmatrix} = \begin{bmatrix} C_\beta & S_\beta & 0 \\ -S_\beta & C_\beta & 0 \\ 0 & 0 & 1 \end{bmatrix} \begin{bmatrix} C_\alpha & 0 & S_\alpha \\ 0 & 1 & 0 \\ -S_\alpha & 0 & C_\alpha \end{bmatrix} \begin{bmatrix} v_{x_b} \\ v_{y_b} \\ v_{z_b} \end{bmatrix}$$
$$= \begin{bmatrix} C_\alpha C_\beta & S_\beta & S_\alpha C_\beta \\ -C_\alpha S_\beta & C_\beta & -S_\alpha S_\beta \\ -S_\alpha & 0 & C_\alpha \end{bmatrix} \begin{bmatrix} v_{x_b} \\ v_{y_b} \\ v_{z_b} \end{bmatrix} \tag{3.4.10}$$

Combining Eqs. (3.4.7) and (3.4.9), the transformation of a vector from the wind coordinate system to the body-fixed coordinate system is

$$\begin{bmatrix} v_{x_b} \\ v_{y_b} \\ v_{z_b} \end{bmatrix} = \begin{bmatrix} C_\alpha & 0 & -S_\alpha \\ 0 & 1 & 0 \\ S_\alpha & 0 & C_\alpha \end{bmatrix} \begin{bmatrix} C_\beta & -S_\beta & 0 \\ S_\beta & C_\beta & 0 \\ 0 & 0 & 1 \end{bmatrix} \begin{bmatrix} v_{x_w} \\ v_{y_w} \\ v_{z_w} \end{bmatrix}$$
$$= \begin{bmatrix} C_\alpha C_\beta & -C_\alpha S_\beta & -S_\alpha \\ S_\beta & C_\beta & 0 \\ S_\alpha C_\beta & -S_\alpha S_\beta & C_\alpha \end{bmatrix} \begin{bmatrix} v_{x_w} \\ v_{y_w} \\ v_{z_w} \end{bmatrix} \tag{3.4.11}$$

The velocity vector in the wind coordinate system is simply $(V, 0, 0)$. Equation (3.4.11) can be used to compute the velocity vector in the body-fixed coordinate system from a velocity magnitude in the wind coordinate system and the two aerodynamic angles. This gives

$$\begin{bmatrix} u \\ v \\ w \end{bmatrix} = \begin{bmatrix} V_{x_b} \\ V_{y_b} \\ V_{z_b} \end{bmatrix} = \begin{bmatrix} C_\alpha C_\beta & -C_\alpha S_\beta & -S_\alpha \\ S_\beta & C_\beta & 0 \\ S_\alpha C_\beta & -S_\alpha S_\beta & C_\alpha \end{bmatrix} \begin{bmatrix} V \\ 0 \\ 0 \end{bmatrix} = V \begin{bmatrix} C_\alpha C_\beta \\ S_\beta \\ S_\alpha C_\beta \end{bmatrix} \tag{3.4.12}$$

The sideslip angle defined in Eq. (3.4.5) is not the only method for defining sideslip. The traditional definition given in Eq. (3.4.5) relates the wind coordinate system to the stability coordinate system, and is often used in wind-tunnel testing. Because this traditional definition is the angle between the stability coordinate system and the wind coordinate system, it is not a direct analog to angle of attack, which is an angle relative to the body-fixed coordinate system. The sideslip angle of the freestream velocity vector relative to the body-fixed coordinate system is sometimes known as the *flank angle*, and is defined as

$$\beta_f \equiv \tan^{-1}\left(\frac{v}{u}\right) \tag{3.4.13}$$

Notice that Eq. (3.4.13) is a direct analog to the definition of angle of attack given in Eq. (3.4.4). This is because both the angle of attack and flank angle are measured relative to the body-fixed coordinate system. The flank angle is sometimes the sideslip angle of choice in analytic work or numerical simulations. It is also the angle usually measured by a wind vane in a flight test. Because both the traditional sideslip angle β defined in Eq. (3.4.5) and the flank angle β_f defined in Eq. (3.4.13) are useful definitions in aeronautics, it is important to be clear about the definition used for aerodynamic angles in any application, and to be consistent in implementation. In the present text, the use of either sideslip angle or flank angle will be clearly denoted by using the subscript f to denote the flank angle defined in Eq. (3.4.13), and using no subscript to denote the sideslip angle defined in Eq. (3.4.5).

The relationship between the sideslip angle and the flank angle can be found by applying the velocity components given in Eq. (3.4.12) to Eq. (3.4.13). This gives

$$\tan\beta_f = \frac{\tan\beta}{\cos\alpha} \tag{3.4.14}$$

At small angles of attack, the sideslip angle and the flank angle are very nearly equal, and at an angle of attack of zero, the sideslip and flank angles are identical. However, at large angles of attack, these two angles are significantly different. Both angles are useful in different situations. The key is to be aware of how the angle is defined for any analysis, and to be consistent in application. In any case, if the angle of attack and one definition of sideslip angle is known, the other definition of sideslip angle can be found from Eq. (3.4.14).

The freestream velocity vector given in Eq. (3.4.12) is shown as a function of the angle of attack and traditional sideslip angle. This vector can also be expressed in terms of the angle of attack and flank angle. From the definitions of angle of attack and flank angle given in Eqs. (3.4.4) and (3.4.13), we can write

$$v = u\tan\beta_f \tag{3.4.15}$$

$$w = u\tan\alpha \tag{3.4.16}$$

Using these relations in Eq. (3.4.3), applying trigonometric identities, and applying the result to Eqs. (3.4.15) and (3.4.16) gives [17]

$$\begin{bmatrix} u \\ v \\ w \end{bmatrix} = \frac{V}{\sqrt{1 - S_\alpha^2 S_{\beta_f}^2}} \begin{bmatrix} C_\alpha C_{\beta_f} \\ C_\alpha S_{\beta_f} \\ S_\alpha C_{\beta_f} \end{bmatrix} \tag{3.4.17}$$

It is not uncommon for the aerodynamic properties of a vehicle to depend not only on the aerodynamic angles, but also on the change of those angles with respect to time. For example, the dynamic stall characteristics of a wing can be a function not only of angle of attack, but also of the time rate-of-change of angle of attack. The change in the angle of attack with respect to time is

$$\dot{\alpha} \equiv \frac{\partial\alpha}{\partial t} \tag{3.4.18}$$

and the change in the sideslip angle with respect to time is

$$\dot{\beta} \equiv \frac{\partial\beta}{\partial t} \tag{3.4.19}$$

A dimensionless form of these partial derivatives is commonly defined as

$$\hat{\alpha} \equiv \frac{\dot{\alpha}\bar{c}}{2V} \tag{3.4.20}$$

$$\hat{\beta} \equiv \frac{\dot{\beta}b}{2V} \tag{3.4.21}$$

These dimensionless parameters are often used in aerodynamic models, as will be shown later in this chapter.

Rotation Rate

The rotation rate of a rigid body can be expressed as a single angular rate vector $\omega = (\omega_x, \omega_y, \omega_z)$. The components of this vector can be expressed in the body-fixed coordinate system, the stability coordinate system, or the wind coordinate system. The individual components are typically denoted by the alphabetic triad

$$\begin{bmatrix} p \\ q \\ r \end{bmatrix} \equiv \begin{bmatrix} \omega_x \\ \omega_y \\ \omega_z \end{bmatrix} \tag{3.4.22}$$

where p is the roll rate, q is the pitch rate, and r is the yaw rate. In order to predict the rigid-body dynamics of the vehicle, these components must be known in the body-fixed coordinate system. However, some aerodynamic models or trim algorithms are related to these rotation rates in the stability or wind coordinate systems. Therefore, it is sometimes helpful to transform the components of the angular rate vector between the body-fixed, stability, and wind coordinate systems. Applying the transformations given in Eqs. (3.4.6)–(3.4.11) yields the relations given in Table 3.4.1.

The angular rates are traditionally written in nondimensional form defined as

$$\begin{bmatrix} \bar{p} \\ \bar{q} \\ \bar{r} \end{bmatrix} \equiv \frac{1}{2V} \begin{bmatrix} pb \\ q\bar{c} \\ rb \end{bmatrix} \tag{3.4.23}$$

where b is the wingspan of the main wing, $\bar{c}$ is the mean chord of the main wing, and V is the vehicle velocity magnitude. Because $\bar{q}$ is scaled by the mean chord $\bar{c}$, while $\bar{p}$ and $\bar{r}$ are scaled by the wingspan b, transforming dimensionless rates from one coordinate system to another requires incorporating these differing reference lengths. Table 3.4.2 shows the resulting transformations.

Phillips [18] suggests an alternate dimensionless form for the angular rates, which may be more appropriate for scaling the vehicle dynamics in some situations. He calls these *dynamic angular rates* defined as

$$\begin{bmatrix} \breve{p} \\ \breve{q} \\ \breve{r} \end{bmatrix} \equiv \frac{V}{g} \begin{bmatrix} p \\ q \\ r \end{bmatrix} \tag{3.4.24}$$

Table 3.4.1. Angular rate components in the body-fixed, stability, and wind coordinate systems.

Body-Fixed		From Stability		From Wind
p_b	=	$p_s C_\alpha - r_s S_\alpha$	=	$p_w C_\alpha C_\beta - q_w C_\alpha S_\beta - r_w S_\alpha$
q_b	=	q_s	=	$p_w S_\beta + q_w C_\beta$
r_b	=	$p_s S_\alpha + r_s C_\alpha$	=	$p_w S_\alpha C_\beta - q_w S_\alpha S_\beta + r_w C_\alpha$
Stability		**From Body-Fixed**		**From Wind**
p_s	=	$p_b C_\alpha + r_b S_\alpha$	=	$p_w C_\beta - q_w S_\beta$
q_s	=	q_b	=	$p_w S_\beta + q_w C_\beta$
r_s	=	$r_b C_\alpha - p_b S_\alpha$	=	r_w
Wind		**From Body-Fixed**		**From Stability**
p_w	=	$p_b C_\alpha C_\beta + q_b S_\beta + r_b S_\alpha C_\beta$	=	$p_s C_\beta + q_s S_\beta$
q_w	=	$q_b C_\beta - p_b C_\alpha S_\beta - r_b S_\alpha S_\beta$	=	$q_s C_\beta - p_s S_\beta$
r_w	=	$r_b C_\alpha - p_b S_\alpha$	=	r_s

Table 3.4.2. Dimensionless angular rate components in the body-fixed, stability, and wind coordinate systems.

Body-Fixed		From Stability		From Wind
$\overline{p}_b$	=	$\overline{p}_s C_\alpha - \overline{r}_s S_\alpha$	=	$\overline{p}_w C_\alpha C_\beta - \overline{q}_w (b/\overline{c}) C_\alpha S_\beta - \overline{r}_w S_\alpha$
$\overline{q}_b$	=	$\overline{q}_s$	=	$\overline{p}_w (\overline{c}/b) S_\beta + \overline{q}_w C_\beta$
$\overline{r}_b$	=	$\overline{p}_s S_\alpha + \overline{r}_s C_\alpha$	=	$\overline{p}_w S_\alpha C_\beta - \overline{q}_w (b/\overline{c}) S_\alpha S_\beta + \overline{r}_w C_\alpha$
Stability		**From Body-Fixed**		**From Wind**
$\overline{p}_s$	=	$\overline{p}_b C_\alpha + \overline{r}_b S_\alpha$	=	$\overline{p}_w C_\beta - \overline{q}_w (b/\overline{c}) S_\beta$
$\overline{q}_s$	=	$\overline{q}_b$	=	$\overline{p}_w (\overline{c}/b) S_\beta + \overline{q}_w C_\beta$
$\overline{r}_s$	=	$\overline{r}_b C_\alpha - \overline{p}_b S_\alpha$	=	$\overline{r}_w$
Wind		**From Body-Fixed**		**From Stability**
$\overline{p}_w$	=	$\overline{p}_b C_\alpha C_\beta + \overline{q}_b (b/\overline{c}) S_\beta + \overline{r}_b S_\alpha C_\beta$	=	$\overline{p}_s C_\beta + \overline{q}_s (b/\overline{c}) S_\beta$
$\overline{q}_w$	=	$\overline{q}_b C_\beta - \overline{p}_b (\overline{c}/b) C_\alpha S_\beta - \overline{r}_b (\overline{c}/b) S_\alpha S_\beta$	=	$\overline{q}_s C_\beta - \overline{p}_s (\overline{c}/b) S_\beta$
$\overline{r}_w$	=	$\overline{r}_b C_\alpha - \overline{p}_b S_\alpha$	=	$\overline{r}_s$

where V is the vehicle velocity and g is the gravity. Because the ratio V/g is used for all three angular rates in this dimensionless formulation, the transformation relations given in Table 3.4.1 can be used directly by simply replacing each dimensional angular rate with the dimensionless dynamic angular rate.

Either of the dimensionless formulations given in Eq. (3.4.23) or (3.4.24) can be used in an aerodynamic model, as long as the correct relations are used in the vehicle dynamic simulation to properly convert the components to dimensional form. The key is to be consistent in the generation and use of any dimensionless parameter.

Control Surfaces

The aerodynamic forces and moments are functions of the control inputs **U**, as shown in Eq. (3.4.1). The design and use of control surfaces varies widely across vehicle types and missions. However, the modeling and application of these control surfaces across vehicle types in simulation is very similar. Here we will focus our discussion on the typical control surfaces available on most fixed-wing aircraft. These include the aileron, elevator, and rudder. Each of these is depicted in Fig. 3.4.2.

The **aileron** is typically defined as a control surface located on the outboard section of the main wing. By changing the deflection of the aft portion of this section of the main wing, the aerodynamic rolling moment can be significantly altered. Hence, the aileron is typically used to directly control the aerodynamic rolling moment, ℓ, which directly affects the vehicle rolling rate p. Ailerons are typically asymmetric control surfaces, meaning that the control surface on the left wing moves in the opposite direction as the control surface on the right wing. The deflection angle of the aileron is denoted as δ_a, defined to be positive when the trailing edge of the right-wing aileron is deflected downward.

The **elevator** is typically defined as a control surface located on the horizontal stabilizer. By changing the deflection of the aft portion of the horizontal stabilizer, or by changing the incidence angle of the entire horizontal stabilizer, the aerodynamic pitching moment can be significantly altered. Hence, the elevator is typically used to directly control the aerodynamic pitching moment, m, which directly affects the vehicle pitching rate q. The deflection angle of the elevator is denoted as δ_e, defined to be positive when the trailing edge is deflected downward.

The **rudder** is typically defined as a control surface located on the vertical stabilizer. By changing the deflection of the aft portion of the vertical stabilizer, the aerodynamic

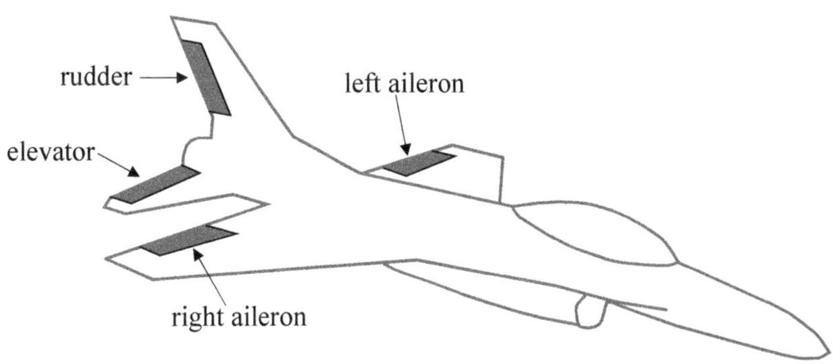

Figure 3.4.2. Typical aircraft control surfaces.

yawing moment can be significantly altered. Hence, the rudder is typically used to directly control the aerodynamic yawing moment, n, which directly affects the vehicle yawing rate r. The deflection angle of the rudder will is denoted as δ_r, defined to be positive when the trailing edge is deflected towards the left of the vehicle.

The aerodynamic forces and moments on a vehicle also depend on the method of propulsion. Here we will refer to the propulsive force as *thrust*, and the control input that controls the thrust as *throttle*. The symbol for throttle used in this text is τ. Propulsion modeling is discussed in Chapter 4.

As stated previously, many other control surfaces are also commonly used on vehicles, including flaps and spoilers. However, their use in an aerodynamic model and application in flight simulation is very similar to the application of the aileron, elevator, and rudder. Hence, once the student is familiar with the application of the aileron, elevator, and rudder, application of any other control surface is rather straightforward. Therefore, in the examples provided in this book, we will focus on application of only the aileron, elevator, rudder, and throttle. The control inputs to the vehicle can be expressed in a single column vector as

$$\mathbf{U} = [\delta_a, \delta_e, \delta_r, \tau]^T \tag{3.4.25}$$

Force and Moment Components

The aerodynamic force on a vehicle can be expressed as a single vector with three components $\mathbf{F} = (F_x, F_y, F_z)$. This vector can be expressed in the body-fixed coordinate system $\mathbf{F}_b = (F_{x_b}, F_{y_b}, F_{z_b})$, the stability coordinate system $\mathbf{F}_s = (F_{x_s}, F_{y_s}, F_{z_s})$, or the wind coordinate system $\mathbf{F}_w = (F_{x_w}, F_{y_w}, F_{z_w})$. Because it is convenient to know the components of the force in each of these coordinate systems, transforming these components from one coordinate system to another is often required in flight simulation. Table 3.4.3 shows the results of applying the transformations given in Eqs. (3.4.6)–(3.4.11).

In the body-fixed coordinate system, the force components are commonly referred to as axial force, side force, and normal force, denoted by the symbols A, Y, and N respectively. In the stability and wind coordinate systems, the force components are commonly referred to as drag, side force, and lift. Because the stability coordinate system is a rotation of the body-fixed coordinate system about the y_b axis, the side force Y is the same in the vehicle and stability coordinate systems. Likewise, because the wind coordinate system is a rotation of the stability coordinate system about the z_s axis, the lift L is the same in the stability and wind coordinate systems. The side force and drag in the stability coordinate system are not equivalent to the side force and drag in the wind coordinate system. Here we denote the side force in the stability and body-fixed coordinate systems as Y, and the side force in the wind coordinate system as S. We also denote the drag in the stability coordinate system as D_s, and the drag in the wind coordinate system as D. Applying these definitions along with the transformations given in Eqs. (3.4.6)–(3.4.11) yields the results in Table 3.4.4.

The aerodynamic moment on a vehicle can also be expressed as a single vector $\mathbf{M} = (M_x, M_y, M_z)$ about a reference point. This vector is most often expressed in

Table 3.4.3. Force components in the body-fixed, stability, and wind coordinate systems.

Body-Fixed		From Stability		From Wind
F_{x_b}	$=$	$F_{x_s}C_\alpha - F_{z_s}S_\alpha$	$=$	$F_{x_w}C_\alpha C_\beta - F_{y_w}C_\alpha S_\beta - F_{z_w}S_\alpha$
F_{y_b}	$=$	F_{y_s}	$=$	$F_{x_w}S_\beta + F_{y_w}C_\beta$
F_{z_b}	$=$	$F_{x_s}S_\alpha + F_{z_s}C_\alpha$	$=$	$F_{x_w}S_\alpha C_\beta - F_{y_w}S_\alpha S_\beta + F_{z_w}C_\alpha$

Stability		From Body-Fixed		From Wind
F_{x_s}	$=$	$F_{x_b}C_\alpha + F_{z_b}S_\alpha$	$=$	$F_{x_w}C_\beta - F_{y_w}S_\beta$
F_{y_s}	$=$	F_{y_b}	$=$	$F_{x_w}S_\beta + F_{y_w}C_\beta$
F_{z_s}	$=$	$-F_{x_b}S_\alpha + F_{z_b}C_\alpha$	$=$	F_{z_w}

Wind		From Body-Fixed		From Stability
F_{x_w}	$=$	$F_{x_b}C_\alpha C_\beta + F_{y_b}S_\beta + F_{z_b}S_\alpha C_\beta$	$=$	$F_{x_s}C_\beta + F_{y_s}S_\beta$
F_{y_w}	$=$	$-F_{x_b}C_\alpha S_\beta + F_{y_b}C_\beta - F_{z_b}S_\alpha S_\beta$	$=$	$-F_{x_s}S_\beta + F_{y_s}C_\beta$
F_{z_w}	$=$	$-F_{x_b}S_\alpha + F_{z_b}C_\alpha$	$=$	F_{z_s}

Table 3.4.4. Force components in the body-fixed, stability, and wind coordinate systems.

Body-Fixed	Symbol	Definition	From Stability	From Wind
Axial	A	$-F_{x_b}$	$D_sC_\alpha - LS_\alpha$	$DC_\alpha C_\beta + SC_\alpha S_\beta - LS_\alpha$
Side	Y	F_{y_b}	Y	$SC_\beta - DS_\beta$
Normal	N	$-F_{z_b}$	$LC_\alpha + D_sS_\alpha$	$DS_\alpha C_\beta + SS_\alpha S_\beta + LC_\alpha$

Stability	Symbol	Definition	From Body-Fixed	From Wind
Drag	D_s	$-F_{x_s}$	$AC_\alpha + NS_\alpha$	$DC_\beta + SS_\beta$
Side	Y	F_{y_s}	Y	$DS_\beta + SC_\beta$
Lift	L	$-F_{z_s}$	$NC_\alpha - AS_\alpha$	L

Wind	Symbol	Definition	From Body-Fixed	From Stability
Drag	D	$-F_{x_w}$	$AC_\alpha C_\beta - YS_\beta + NS_\alpha C_\beta$	$D_sC_\beta - YS_\beta$
Side	S	F_{y_w}	$AC_\alpha S_\beta + YC_\beta + NS_\alpha S_\beta$	$D_sS_\beta + YC_\beta$
Lift	L	$-F_{z_w}$	$NC_\alpha - AS_\alpha$	L

components in the body-fixed coordinate system $\mathbf{M}_b = (M_{x_b}, M_{y_b}, M_{z_b})$ or the stability coordinate system $\mathbf{M}_s = (M_{x_s}, M_{y_s}, M_{z_s})$. However, in some cases it may be useful to consider these components in the wind coordinate system $\mathbf{M}_w = (M_{x_w}, M_{y_w}, M_{z_w})$ as well. In the body-fixed coordinate system, these components are referred to as rolling, pitching, and yawing moments. Here we will use this terminology for all three coordinate systems, and use subscripts to denote to which coordinate system the moment components pertain. Using the transformations given in Eqs. (3.4.6)–(3.4.11), the moments in any of the three coordinate systems can be computed from known moment components in one of the other coordinate systems. This yields the results in Table 3.4.5.

In the absence of Reynolds number and Mach number effects, aerodynamic forces and moments scale with dynamic pressure $Q = (1/2)\rho V^2$. Therefore, forces and moments are typically stored in an aerodynamic model in a dimensionless form by dividing by the dynamic pressure. The units of force can be written as a pressure multiplied by an area. Therefore, to complete the formulation for a dimensionless force, a reference area is needed in addition to the selected dynamic pressure. Typically the area of the main wing S_w is used as the reference area for fixed-wing aircraft. However, any area can be used as the reference area as long as it is used consistently in the model development and application.

The typical definitions for the **dimensionless force components** are

$$
\begin{gathered}
C_x \equiv \frac{F_x}{\frac{1}{2}\rho V^2 S_w}, \quad C_y \equiv \frac{F_y}{\frac{1}{2}\rho V^2 S_w}, \quad C_z \equiv \frac{F_z}{\frac{1}{2}\rho V^2 S_w} \\
C_A \equiv \frac{A}{\frac{1}{2}\rho V^2 S_w}, \quad C_S \equiv \frac{S}{\frac{1}{2}\rho V^2 S_w}, \quad C_N \equiv \frac{N}{\frac{1}{2}\rho V^2 S_w} \\
C_L \equiv \frac{L}{\frac{1}{2}\rho V^2 S_w}, \quad C_D \equiv \frac{D}{\frac{1}{2}\rho V^2 S_w}
\end{gathered}
\tag{3.4.26}
$$

If the components of the dimensionless force vector are known in the body-fixed, stability, or wind coordinate system, they can be evaluated in either of the other coordinate systems by applying the transformations given in Eqs. (3.4.6)–(3.4.11). Because all forces are

Table 3.4.5. Moment components in the body-fixed, stability, and wind coordinate systems.

Body-Fixed	Symbol	Definition	From Stability	From Wind
Roll	ℓ_b	M_{x_b}	$\ell_s C_\alpha - n_s S_\alpha$	$\ell_w C_\alpha C_\beta - m_w C_\alpha S_\beta - n_w S_\alpha$
Pitch	m_b	M_{y_b}	m_s	$\ell_w S_\beta + m_w C_\beta$
Yaw	n_b	M_{z_b}	$\ell_s S_\alpha + n_s C_\alpha$	$\ell_w S_\alpha C_\beta - m_w S_\alpha S_\beta + n_w C_\alpha$

Stability	Symbol	Definition	From Body-Fixed	From Wind
Roll	ℓ_s	M_{x_s}	$\ell_b C_\alpha + n_b S_\alpha$	$\ell_w C_\beta - m_w S_\beta$
Pitch	m_s	M_{y_s}	m_b	$\ell_w S_\beta + m_w C_\beta$
Yaw	n_s	M_{z_s}	$n_b C_\alpha - \ell_b S_\alpha$	n_w

Wind	Symbol	Definition	From Body-Fixed	From Stability
Roll	ℓ_w	M_{x_w}	$\ell_b C_\alpha C_\beta + m_b S_\beta + n_b S_\alpha C_\beta$	$\ell_s C_\beta + m_s S_\beta$
Pitch	m_w	M_{y_w}	$m_b C_\beta - \ell_b C_\alpha S_\beta - n_b S_\alpha S_\beta$	$m_s C_\beta - \ell_s S_\beta$
Yaw	n_w	M_{z_w}	$n_b C_\alpha - \ell_b S_\alpha$	n_s

typically nondimensionalized using the same dynamic pressure and reference area, Tables 3.4.3–3.4.4 can be used directly by simply using the dimensionless forces and moments in the tables rather than the dimensional forces and moments. For example, using Eq. (3.4.26) in Tables 3.4.3 and 3.4.4 at zero sideslip angle gives the **longitudinal relations**

$$\begin{aligned} C_L &= C_N \cos\alpha - C_A \sin\alpha \\ &= C_x \sin\alpha - C_z \cos\alpha \end{aligned} \tag{3.4.27}$$

$$\begin{aligned} C_D &= C_A \cos\alpha + C_N \sin\alpha \\ &= -C_x \cos\alpha - C_z \sin\alpha \end{aligned} \tag{3.4.28}$$

$$C_A = C_D \cos\alpha - C_L \sin\alpha \tag{3.4.29}$$

$$C_N = C_L \cos\alpha + C_D \sin\alpha \tag{3.4.30}$$

$$C_x = C_L \sin\alpha - C_D \cos\alpha \tag{3.4.31}$$

$$C_z = -C_L \cos\alpha - C_D \sin\alpha \tag{3.4.32}$$

The transformation of dimensionless moments is not as straightforward. To complete the formulation for a dimensionless moment, both a reference area and length are needed, in addition to the dynamic pressure. Again, any reference length and area can be used for this purpose, as long as it is used consistently throughout the model development and application. For fixed-wing aircraft, the typical area used in the development of a dimensionless moment is the area of the main wing. The typical length used for the longitudinal moment component M_y is the mean chord of the main wing. The typical length used for the lateral moment components M_x and M_z is the wingspan of the main wing. Hence, the typical definitions for the dimensionless moment components are

$$\begin{aligned} C_\ell &\equiv \frac{\ell}{\frac{1}{2}\rho V^2 S_w b} \\ C_m &\equiv \frac{m}{\frac{1}{2}\rho V^2 S_w \bar{c}} \\ C_n &\equiv \frac{n}{\frac{1}{2}\rho V^2 S_w b} \end{aligned} \tag{3.4.33}$$

Because a different length scale is typically used to formulate dimensionless longitudinal moments than that used to formulate dimensionless lateral moments, the dimensionless components cannot be used directly in the transformations given in Eqs. (3.4.6)–(3.4.11). Rather, these moments must be scaled using ratios of the dimensionless lengths when converting from a longitudinal moment to a lateral moment or vice versa. This process results in the relations given in Table 3.4.6.

The necessary coordinate systems for typical aerodynamic models have now been fully defined, including force and moment components, velocity components, and angular rate components in the body-fixed, stability, and wind coordinate systems. Additionally,

Table 3.4.6. Dimensionless moment components in the body-fixed, stability, and wind coordinate systems.

Body-Fixed		From Stability	From Wind
C_{ℓ_b}	=	$C_{\ell_s}C_\alpha - C_{n_s}S_\alpha$	$C_{\ell_w}C_\alpha C_\beta - C_{m_w}(\bar{c}/b)C_\alpha S_\beta - C_{n_w}S_\alpha$
C_{m_b}	=	C_{m_s}	$C_{\ell_w}(b/\bar{c})S_\beta + C_{m_w}C_\beta$
C_{n_b}	=	$C_{\ell_s}S_\alpha + C_{n_s}C_\alpha$	$C_{\ell_w}S_\alpha C_\beta - C_{m_w}(\bar{c}/b)S_\alpha S_\beta + C_{n_w}C_\alpha$

Stability		From Body-Fixed	From Wind
C_{ℓ_s}	=	$C_{\ell_b}C_\alpha + C_{n_b}S_\alpha$	$C_{\ell_w}C_\beta - C_{m_w}(\bar{c}/b)S_\beta$
C_{m_s}	=	C_{m_b}	$C_{\ell_w}(b/\bar{c})S_\beta + C_{m_w}C_\beta$
C_{n_s}	=	$C_{n_b}C_\alpha - C_{\ell_b}S_\alpha$	C_{n_w}

Wind		From Body-Fixed	From Stability
C_{ℓ_w}	=	$C_{\ell_b}C_\alpha C_\beta + C_{m_b}(\bar{c}/b)S_\beta + C_{n_b}S_\alpha C_\beta$	$C_{\ell_s}C_\beta + C_{m_s}(\bar{c}/b)S_\beta$
C_{m_w}	=	$C_{m_b}C_\beta - (b/\bar{c})(C_{\ell_b}C_\alpha S_\beta + C_{n_b}S_\alpha S_\beta)$	$C_{m_s}C_\beta - C_{\ell_s}(b/\bar{c})S_\beta$
C_{n_w}	=	$C_{n_b}C_\alpha - C_{\ell_b}S_\alpha$	C_{n_s}

dimensionless forms of some components have been defined as well as typical controls on a fixed-wing aircraft. We will now consider an important concept to all fixed-wing aircraft, known as the aerodynamic center.

3.5 The Aerodynamic Center

The *aerodynamic center* is traditionally defined to be the point about which the pitching moment is invariant to small changes in angle of attack. This can be expressed as

$$\frac{\partial C_{m_{ac}}}{\partial \alpha} \equiv 0 \tag{3.5.1}$$

Correctly identifying the location of the aerodynamic center of a lifting body is important in aircraft design and analysis. For example, the location of the aerodynamic center of a complete vehicle relative to the center of gravity is an important measure of static stability. The longitudinal characteristics of a vehicle are dependent on both the axial and normal distances between the aerodynamic center and the center of gravity. The aerodynamic center of a complete vehicle is often referred to as the neutral point and is a function of the aerodynamic centers of all lifting surfaces or wings comprising the vehicle. The distribution of aerodynamic center positions along the span of a single wing depends on the aerodynamic center of the airfoils from which the wing is constructed, as well as the wing sweep, dihedral, and planform. Furthermore, this locus of aerodynamic centers along a wing significantly impacts aeroelastic estimates for flutter and divergence speeds. Correctly predicting the aerodynamic center or neutral point of a complete vehicle during preliminary design depends on the accuracy to which the aerodynamic centers of airfoils and wings can be predicted.

Traditional Relations for the Aerodynamic Center

The traditional estimate for the location of the aerodynamic center is based on approximations that are nearly true for airfoils. However, they should be applied with caution for the complete vehicle. The pitching moment C_m about any point (x, y) in the airfoil plane can be found from a simple transformation of forces and moments about the origin to the point of interest, i.e.,

$$C_m = C_{m_O} + \frac{x}{c} C_N - \frac{y}{c} C_A \tag{3.5.2}$$

where C_{m_O} is the pitching moment about the origin, C_A is the axial-force coefficient, and C_N is the normal-force coefficient. The airfoil can be located anywhere in relation to the origin of the coordinate system. However, for convenience, we will assume that the origin is located at the airfoil leading edge. From Eq. (3.5.2), the pitching moment about the aerodynamic center can be written as

$$C_{m_{ac}} = C_{m_O} + \frac{x_{ac}}{c} C_N - \frac{y_{ac}}{c} C_A \tag{3.5.3}$$

The axial- and normal-force coefficients are related to the lift and drag coefficients through a transformation in angle of attack, as shown in Fig. 1 and expressed as

$$C_A = C_D \cos\alpha - C_L \sin\alpha \tag{3.5.4}$$

$$C_N = C_L \cos\alpha + C_D \sin\alpha \tag{3.5.5}$$

Using Eqs. (3.5.4) and (3.5.5) in Eq. (3.5.3), the pitching moment about the aerodynamic center is

$$C_{m_{ac}} = C_{m_O} + \frac{x_{ac}}{c}(C_L \cos\alpha + C_D \sin\alpha) - \frac{y_{ac}}{c}(C_D \cos\alpha - C_L \sin\alpha) \tag{3.5.6}$$

For a typical airfoil, the vertical offset of the aerodynamic center from the airfoil chord line is small, and the drag is much less than the lift. Additionally, the angle of attack is small for normal flight conditions. Therefore, applying the traditional approximations $C_D \sin\alpha \ll C_L \cos\alpha$, $y_{ac} \sin\alpha \ll x_{ac} \cos\alpha$, $y_{ac} C_D \ll x_{ac} C_L$, and $\cos\alpha \approx 1$ gives

$$C_{m_{ac}} \approx C_{m_O} + \frac{x_{ac}}{c} C_L \tag{3.5.7}$$

Taking the derivative of Eq. (3.5.7) with respect to angle of attack, applying the constraint given by Eq. (3.5.1), and rearranging gives the traditional approximation for the aerodynamic center

$$\frac{x_{ac}}{c} = -\frac{C_{m_O,\alpha}}{C_{L,\alpha}}, \qquad \frac{y_{ac}}{c} = 0 \tag{3.5.8}$$

Note that the y coordinate is traditionally assumed to be zero due to the approximations applied in this traditional development.

The relations shown in Eq. (3.5.8) are widely used today across the aerospace industry and academia. Furthermore, analogous relations have traditionally been used to approximate the location of the neutral point of complete aircraft, and they are often used to evaluate aircraft static stability and compute the aircraft static margin. These traditional approximations suffer from several limiting assumptions and can result in errors on orders that can be significant to vehicle stability.

The assumptions leading to the result given in Eq. (3.5.8) include a constant lift slope and moment slope below stall, and therefore neglect any nonlinearities in the lift and/or pitching moment. Additionally, any influence of drag on the aerodynamic center is entirely neglected. Furthermore, this traditional approach applies the small-angle approximation to reduce the nonlinear trigonometric relations in Eqs. (3.5.4) and (3.5.5) to linear functions of angle of attack. These linearizing approximations significantly hinder our understanding of the effects of nonlinearities associated with the pitch stability of airfoils and aircraft. To improve the accuracy of aerodynamic center predictions, we now relax the linearizing assumptions and consider a more general formulation.

General Relations for the Aerodynamic Center

Phillips et al. [19] presented general relations for the aerodynamic center, which do not include the linearizing approximations used in the traditional approach. Taking the derivative of Eq. (3.5.3) with respect to angle of attack and applying the traditional constraint given in Eq. (3.5.1) yields

$$\frac{x_{ac}}{c}C_{N,\alpha} - \frac{y_{ac}}{c}C_{A,\alpha} = -C_{m_O,\alpha} \tag{3.5.9}$$

Note that this is the equation for a line, not a point. The line given in Eq. (3.5.9) is the neutral axis of the airfoil [19]. All points along this line satisfy the constraint given in Eq. (3.5.1). Therefore, the single constraint given in Eq. (3.5.1) is not sufficient to specify a single point as the aerodynamic center. Phillips et al. [19] suggested a second constraint to isolate the location of the aerodynamic center, namely, that the location of the aerodynamic center must also be invariant to small changes in angle of attack, i.e.,

$$\frac{\partial x_{ac}}{\partial \alpha} \equiv 0 \qquad \frac{\partial y_{ac}}{\partial \alpha} \equiv 0 \tag{3.5.10}$$

Differentiating Eq. (3.5.9) with respect to angle of attack and applying Eq. (3.5.10) gives

$$\frac{x_{ac}}{c}C_{N,\alpha,\alpha} - \frac{y_{ac}}{c}C_{A,\alpha,\alpha} = -C_{m_O,\alpha,\alpha} \tag{3.5.11}$$

The intersection of the two lines specified by Eqs. (3.5.9) and (3.5.11) defines a unique point where both of the constraints are simultaneously satisfied, and therefore defines the location of the aerodynamic center. Solving Eqs. (3.5.9) and (3.5.11) for x_{ac} and y_{ac}, and using the result in Eq. (3.5.3), gives the location of the aerodynamic center and the associated pitching moment [19]

$$\frac{x_{ac}}{c} = \frac{C_{A,\alpha}C_{m_O,\alpha,\alpha} - C_{m_O,\alpha}C_{A,\alpha,\alpha}}{C_{N,\alpha}C_{A,\alpha,\alpha} - C_{A,\alpha}C_{N,\alpha,\alpha}} \tag{3.5.12}$$

$$\frac{y_{ac}}{c} = \frac{C_{N,\alpha} C_{m_O,\alpha,\alpha} - C_{m_O,\alpha} C_{N,\alpha,\alpha}}{C_{N,\alpha} C_{A,\alpha,\alpha} - C_{A,\alpha} C_{N,\alpha,\alpha}} \tag{3.5.13}$$

$$C_{m_{ac}} = C_{m_O} + \frac{x_{ac}}{c} C_N - \frac{y_{ac}}{c} C_A \tag{3.5.14}$$

Equations (3.5.12) and (3.5.13) offer a more accurate description of the location of the aerodynamic center for any lifting body. These expressions allow both the x and y coordinates of the aerodynamic center to be evaluated, unlike the traditional approximations given in Eq. (3.5.8), which always predicts that the aerodynamic center lies on the chord line of an airfoil, or x_b-axis of a complete vehicle. Furthermore, Eqs. (3.5.12) and (3.5.13) correctly allow for the effects of vertical offsets, trigonometric nonlinearities, and aerodynamic nonlinearities such as drag to be included through proper treatment of the first and second aerodynamic derivatives.

EXAMPLE 3.5.1 Consider a tapered wing with a root chord of 2 ft, a tip chord of 1 ft, and a span of 8 ft. The wing has 20 degrees of sweep and is constructed from a symmetric airfoil with a section lift slope of 2π per radian, and a section zero-lift drag coefficient of 0.0055. A numerical lifting-line algorithm [20] was used to generate the aerodynamic data for this wing. This data is shown in Table 3.5.1, with the origin located at the quarter chord of the root airfoil section. Use finite differencing to compute the location of the aerodynamic center for the wing as a function of angle of attack. Compare the resulting locus of aerodynamic centers to the traditional approximation given in Eq. (3.5.8).

Table 3.5.1. Aerodynamic data for the wing used in Example 3.5.1.

α[deg]	C_L	C_D	C_{m_O}
0	0.0000000	0.0055000	0.0000000
2	0.1339588	0.0072888	−0.0664536
4	0.2680906	0.0126613	−0.1328800
6	0.4025653	0.0216352	−0.1992504
8	0.5375477	0.0342398	−0.2655334
10	0.6731936	0.0505157	−0.3316938
12	0.8096484	0.0705143	−0.3976910
14	0.9470438	0.0942973	−0.4634790
16	1.0854954	0.1219365	−0.5290047
18	1.2251006	0.1535228	−0.5942085
20	1.3686288	0.1848091	−0.6590240

Solution: The data in the table can be used along with Eqs. (3.5.4) and (3.5.5) to yield the coefficients in Table 3.5.2. Second-order central differencing can be performed using the algorithms outlined in Appendix D.4. This yields the first and second derivatives given in Table 3.5.3.

Using these results in Eqs. (3.5.12) and (3.5.13), along with the mean chord $\overline{c} = 1.5$ ft yields the dimensional results in Table 3.5.4 and shown in Fig. 3.5.1. Note that due to the symmetry of the wing and airfoil, the aerodynamic data can be mirrored about zero degrees to yield results in the range $-20 \leq \alpha + 20$ deg.

Table 3.5.2. Axial and normal coefficients for the wing used in Example 3.5.1.

α[deg]	C_A	C_N
0	0.0055000	0.0000000
2	0.0026093	0.1341316
4	−0.0060706	0.2683208
6	−0.0205629	0.4026215
8	−0.0409056	0.5370816
10	−0.0671506	0.6717382
12	−0.0993620	0.8066164
14	−0.1376144	0.9417251
16	−0.1819902	1.0770554
18	−0.2325680	1.2125811
20	−0.2944349	1.3492988

Table 3.5.3. First and second aerodynamic derivatives for the wing used in Example 3.5.1.

α[deg]	$C_{A,\alpha}$	$C_{N,\alpha}$	$C_{m_O,\alpha}$	$C_{A,\alpha,\alpha}$	$C_{N,\alpha,\alpha}$	$C_{m_O,\alpha,\alpha}$
0	0.00000	3.84259	−1.90376	−4.74482	0.00000	0.00004
2	−0.16574	3.84341	−1.90337	−4.75115	0.04728	0.02232
4	−0.33192	3.84583	−1.90217	−4.77024	0.09156	0.04596
6	−0.49897	3.84972	−1.90012	−4.80151	0.13077	0.07173
8	−0.66732	3.85481	−1.89711	−4.84398	0.16133	0.10062
10	−0.83733	3.86080	−1.89302	−4.89663	0.18178	0.13394
12	−1.00932	3.86728	−1.88768	−4.95786	0.18925	0.17169
14	−1.18356	3.87375	−1.88093	−5.02552	0.18181	0.21527
16	−1.36011	3.87973	−1.87256	−5.09001	0.16036	0.26418
18	−1.61065	3.89960	−1.86239	−9.26488	0.97836	0.31868

The traditional approximation can be obtained using central differencing on the original dataset to compute values for the lift slope and pitching-moment slope. Here we will use the data at $\alpha = \pm 20$ deg, which yields the solution

$$x_{ac} = -(\overline{c})\frac{C_{m_O,\alpha}}{C_{L,\alpha}} = -(1.5)\frac{-0.6590240 - 0.6590240}{1.3686288 - (-1.3686288)} = 0.722282017 \text{ ft}$$

$$y_{ac} = 0 \text{ ft}$$

Table 3.5.4. Aerodynamic center location for the wing used in Example 3.5.1.

α[deg]	x_{ac}	y_{ac}
0	0.7431542	−0.0000122
2	0.7434652	−0.0144462
4	0.7443898	−0.0287401
6	0.7458987	−0.0427226
8	0.7479175	−0.0560667
10	0.7504165	−0.0688884
12	0.7532367	−0.0806972
14	0.7563275	−0.0916143
16	0.7596630	−0.1017862
18	0.7713289	−0.1330456

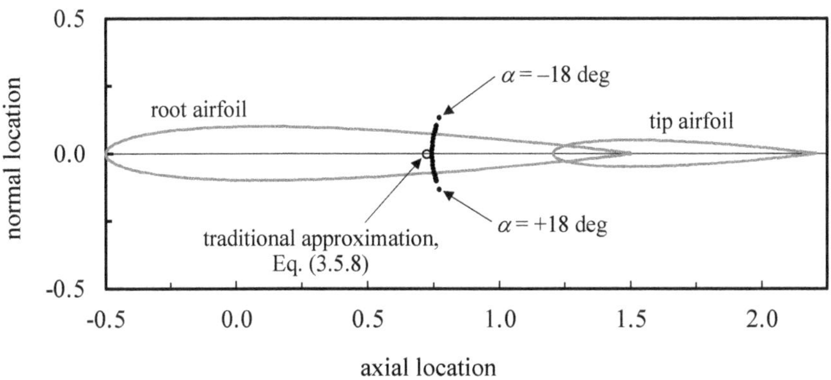

Figure 3.5.1. Locus of aerodynamic centers (filled circles) for a tapered, swept wing. Open circle represents the traditional approximation given in Eq. (3.5.8).

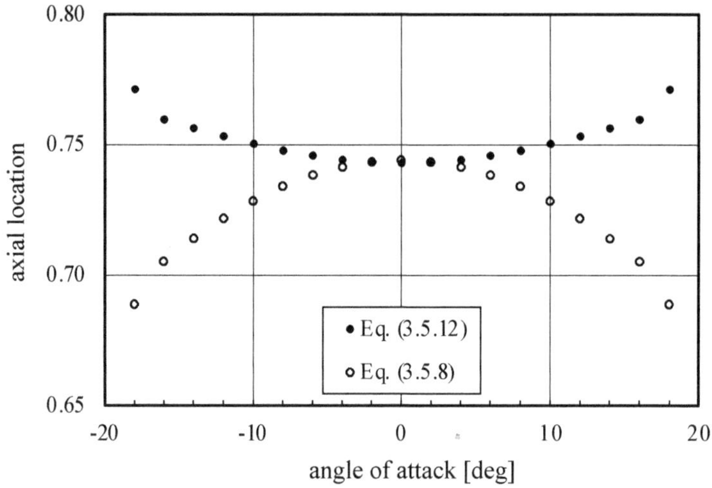

Figure 3.5.2. Axial position of the aerodynamic center as predicted from the traditional approximation given in Eq. (3.5.8) and the more accurate method given in Eq. (3.5.12).

The traditional approximation given in Eq. (3.5.8) could be applied at each angle of attack to capture any change in the local lift slope and pitching-moment slope. The resulting axial position predicted from this traditional equation by computing local derivatives using second-order finite-differencing is shown in Fig. 3.5.2 as a function of angle of attack. Also included in the figure are results from the more accurate solution computed from Eq. (3.5.12). Note that the traditional approximation predicts a forward shift in axial position of the aerodynamic center with increasing angle of attack, whereas the more accurate solution computed from Eq. (3.5.12) predicts an aftward shift with increasing angle of attack. Because the stability of a vehicle is closely tied to the location of the aerodynamic center, care must be taken when estimating the location of the aerodynamic center as a function of angle of attack. As has been shown here, the traditional approximation can result in the prediction of inaccurate location and trends.

If the lift, drag, and pitching moment are known from computational or experimental results at discrete angles of attack, finite differencing methods can be used to evaluate the first and second derivatives needed in Eqs. (3.5.12) and (3.5.13). However, finite differencing methods can be highly sensitive to small errors in data, especially when the data are given at small intervals in angle of attack. Most experimental data and computational results suffer from some amount of scatter, depending on how the experiment or simulation was performed. Therefore, applying finite differencing directly to such datasets can produce significant scatter in the predicted location of the aerodynamic center of an airfoil, wing, or complete vehicle. Hence, rather than using finite differencing methods on the data directly, these data are generally used to fit analytical expressions for the lift, drag, and pitching moment as a function of angle of attack. The analytical derivatives of these expressions can then be used to obtain the required derivatives for Eqs. (3.5.12) and (3.5.13). This method for estimating derivatives reduces the impact of numerical or experimental errors inherent in the dataset. However, such an approach relies on an assumed form of the relationship between the angle of attack, lift, drag, and pitching moment. Below stall, the lift and pitching moment are usually assumed to be linearly related to the angle of attack. This linear relationship is a result of the thin airfoil theory.

Thin-Airfoil-Theory Approximation

Thin airfoil theory was developed by Max Munk during the period between 1914 and 1922 [21, 22], and the development of this theory can be found in many engineering textbooks on aerodynamics. In this classical potential-flow theory, an airfoil is synthesized as the superposition of a uniform flow and a vortex sheet placed along the camber line of the airfoil. Small-camber and small angle-of-attack approximations are applied such that higher-order terms can be neglected. This results in the classical thin airfoil lift and pitching-moment relations

$$C_L = C_{L,\alpha}(\alpha - \alpha_{L0}) \tag{3.5.15}$$

$$C_{m_O} = C_{m_{c/4}} - \frac{C_L}{4} \tag{3.5.16}$$

where $C_{L,\alpha}$ is the lift slope, which from thin airfoil theory is equal to 2π, α_{L0} is the zero-lift angle of attack, and $C_{m_{c/4}}$ is the pitching moment about the quarter-chord. Using these relations in the traditional approximations given in Eqs. (3.5.7) and (3.5.8) yields the aerodynamic center as predicted by thin airfoil theory

$$\frac{x_{ac}}{c} = -\frac{C_{m_O,\alpha}}{C_{L,\alpha}} = \frac{1}{4}, \qquad \frac{y_{ac}}{c} = 0, \qquad C_{m_{ac}} = C_{m_{c/4}} \tag{3.5.17}$$

General Incompressible Inviscid Airfoil Theory

Using conformal mapping, it can be shown that the lift and pitching moment for inviscid airfoils with a single trailing edge are functions of the form [23]

$$C_L = C_{L0,\alpha}(\sin\alpha - \tan\alpha_{L0}\cos\alpha) \tag{3.5.18}$$

$$C_{m_O} = C_{m0,\alpha}\sin(2\alpha) + C_{m,N}C_L\cos\alpha - C_{m,A}C_L\sin\alpha \tag{3.5.19}$$

where $C_{L0,\alpha}$ is the lift slope at zero angle of attack, α_{L0} is the zero-lift angle of attack, and $C_{m0,\alpha}$, $C_{m,N}$, and $C_{m,A}$ are constant coefficients. The forms of Eqs. (3.5.18) and (3.5.19) hold for the potential-flow solution for any arbitrary airfoil with a single trailing edge [23]. These relations were developed without any approximations for airfoil thickness, camber, or angle of attack. They are not constrained under the same limitations that were used in the development of the traditional small-camber and small-angle relations given in Eqs. (3.5.15) and (3.5.16).

In general, the location of the aerodynamic center can be correctly predicted using Eqs. (3.5.12) and (3.5.13). Recall that this definition for the location of the aerodynamic center is a general definition, in that it does not include any linearizing or small-angle approximations. Because inviscid theory does not predict any drag, the inviscid axial and normal coefficients are related to lift according to

$$C_A = -C_L \sin\alpha \tag{3.5.20}$$

$$C_N = C_L \cos\alpha \tag{3.5.21}$$

Using Eqs. (3.5.18) and (3.5.19) in these expressions for the axial and normal coefficients, differentiating twice, and applying the result to Eqs. (3.5.12) and (3.5.13) gives

$$\frac{x_{ac}}{c} = -2\frac{C_{m0,\alpha}}{C_{L0,\alpha}}\cos^2\alpha_{L0} - C_{m,N} \tag{3.5.22}$$

$$\frac{y_{ac}}{c} = \frac{C_{m0,\alpha}}{C_{L0,\alpha}}\sin(2\alpha_{L0}) + C_{m,A} \tag{3.5.23}$$

$$C_{m_{ac}} = C_{m0,\alpha}\sin(2\alpha_{L0}) \tag{3.5.24}$$

Notice that these expressions for the aerodynamic center and associated pitching moment are independent of the angle of attack. Therefore, the location of the aerodynamic center for an arbitrary airfoil in inviscid flow is a single point, independent of the angle of attack. This same conclusion was arrived at by Karamcheti [24], who also used conformal mapping to demonstrate that "there is a point about which the moment remains unaltered as the angle of attack is varied." This point does not, in general, lie on the chord line at the airfoil quarter-chord but is a single deterministic point dependent on airfoil thickness and camber. This solution was developed from the general airfoil theory, which does not make any assumptions for small angles of attack, small camber, or small thickness. It is rather remarkable that, when all geometric and aerodynamic nonlinearities are retained in the lift and pitching-moment equations, along with those in the definition of the aerodynamic center, the relations for the aerodynamic center of an arbitrary airfoil in inviscid flow reduce to such simple expressions, independent of the angle of attack.

Viscous Effects

In the previous two subsections, we neglected the influence of viscosity on the aerodynamic forces and moments created by an airfoil. The largest effect of viscosity on airfoils at angles of attack below stall is the production of friction along the surface of the airfoil. Below stall, this skin friction generally does not significantly impact the lift but does produce drag and alters the pitching moment relative to the inviscid scenario. The section drag coefficient is often modeled as a quadratic function of the section lift coefficient. Here, we use the traditional relation

$$C_D = C_{D_{L0}} + C_{D,L}C_L + C_{D,L^2}C_L^2 \tag{3.5.25}$$

where $C_{D_{L0}}$, $C_{D,L}$, and C_{D,L^2} are constant coefficients for a given drag polar for an airfoil or complete vehicle. Note that the second term on the right-hand side is required to adequately describe drag polars that do not have a minimum at zero lift. Such is the case for most aircraft and non-symmetric airfoils.

Viscosity also affects the pitching moment produced by the airfoil. Recall that Eq. (3.5.19) was developed from the airfoil theory based on conformal mapping, which is built on the assumption of an inviscid flow. This equation for the pitching moment can be extended to account for viscous effects by including the drag component acting in the normal and axial directions

$$C_{m_O} = C_{m0,\alpha}\sin(2\alpha) + C_{m,N}C_N + C_{m,A}C_A \tag{3.5.26}$$

Using Eqs. (3.5.4) and (3.5.5) in Eq. (3.5.26) yields

$$C_{m_O} = C_{m0,\alpha}\sin(2\alpha) + C_{m,N}(C_L\cos\alpha + C_D\sin\alpha) + C_{m,A}(C_D\cos\alpha - C_L\sin\alpha) \tag{3.5.27}$$

The relations for lift and drag given in Eqs. (3.5.18) and (3.5.25) can be used in Eqs. (3.5.4), (3.5.5), and (3.5.26) or (3.5.27), and the results analytically differentiated for use in Eqs. (3.5.12)–(3.5.14). This yields a very complex function. An approximation for the solution can be obtained from a third-order small-angle approximation given by Hunsaker et al. [23]

$$\frac{x_{ac}}{c} \approx -2\frac{C_{m0,\alpha}}{C_{L0,\alpha}}\left[\frac{\kappa_1[3(\alpha\alpha_{L0} - \alpha^2 - \alpha_{L0}^2/2) + 1] - \kappa_2(1 + 3\alpha^2/2) - 1}{\kappa_1(1 + 3\alpha_{L0}^2/2) + 3\kappa_2(\alpha^2/2 - \alpha\alpha_{L0} - 2\kappa_2/3 - 1) - \alpha_{L0}^2 - 1}\right] - C_{m,N} \tag{3.5.28}$$

$$\frac{y_{ac}}{c} \approx -2\frac{C_{m0,\alpha}}{C_{L0,\alpha}}\left[\frac{\kappa_1(3\alpha - 2\alpha_{L0}) + C_{D,L} + 3\alpha\kappa_2 + \alpha_{L0}(1 + \alpha_{L0}^2/3)}{\kappa_1(1 + 3\alpha_{L0}^2/2) + 3\kappa_2(\alpha^2/2 - \alpha\alpha_{L0} - 2\kappa_2/3 - 1) - \alpha_{L0}^2 - 1}\right] - C_{m,A} \tag{3.5.29}$$

$$C_{m_{ac}} \approx 2C_{m0,\alpha}\left[\frac{\alpha_{L0}(\kappa_1 + \kappa_2 - \alpha_{L0}^2/3 - 1) + 6\alpha\kappa_2(\kappa_1 + \kappa_2)}{\kappa_1(1 + 3\alpha_{L0}^2/2) + 3\kappa_2(\alpha^2/2 - \alpha\alpha_{L0} - 2\kappa_2/3 - 1) - \alpha_{L0}^2 - 1}\right] \tag{3.5.30}$$

where

$$\kappa_1 = C_{L0,\alpha}C_{D,L^2}, \qquad \kappa_2 = \frac{C_{D_{L0}}}{2C_{L0,\alpha}} \tag{3.5.31}$$

Although the full expressions for the location of the aerodynamic center including viscous effects are complex, the third-order approximations provide analytical insight by retaining the most important terms. Additionally, note that the inclusion of viscosity results in an aerodynamic center location that varies with angle of attack, whereas the solution for an airfoil in inviscid flow has an aerodynamic center independent of angle of attack. For example solutions for airfoils including viscous effects, see Hunsaker et al. [23].

3.6 Model By Combining Simple Geometries

The aerodynamic forces and moments on the vehicle depend on the external geometry of the vehicle. If CFD or wind-tunnel aerodynamic data are available, they can be used directly for the vehicle simulation. This will be demonstrated later in this chapter. However, during the early design phase of a vehicle, when detailed geometry has not yet been finalized, or before wind-tunnel or CFD data are available, the aerodynamics of the vehicle can be roughly approximated by combining the approximate aerodynamic properties of simple geometric shapes such as rectangular cuboids, cylinders, spheres, and wing segments. Additionally, in the early stages of a flight simulator development, it can be helpful to test the simulation code on simple geometric shapes. This section includes aerodynamic models of simple geometric shapes that can be used for these purposes. Many other aerodynamic models for simplified shapes are available in the literature [25–27], and can be applied in a similar way to those presented here.

In the present development, we shall refer to each simple geometric shape as a component. The aerodynamic properties of each component depend on the velocity of the component relative to the surrounding fluid. We will refer to this velocity as the *component velocity* and denote it as $\mathbf{V}_c$. The component velocity measured at the component origin depends on the velocity vector of the vehicle relative to the surrounding fluid $\mathbf{V}$, which is defined in Eq. (3.4.2), as well as the rotation rate of the vehicle about the vehicle center of gravity

$$\omega_b \equiv \begin{bmatrix} p_b \\ q_b \\ r_b \end{bmatrix} \tag{3.6.1}$$

If both $\mathbf{V}$ and ω_b are known, the component velocity can be found from

$$\mathbf{V}_c = [\mathbf{R}]^{-1}[\mathbf{V} + (\omega_b \times \mathbf{P})] \tag{3.6.2}$$

where $\mathbf{P} \equiv \mathbf{r}_c - \mathbf{r}_b$ is the location vector of the component relative to the vehicle center of gravity in the vehicle coordinate system, and $[\mathbf{R}]$ is the rotation matrix of the component given in Eq. (2.6.2). A unit vector in the direction of the component velocity can be found from

$$\mathbf{u}_c \equiv \frac{\mathbf{V}_c}{V_c} \tag{3.6.3}$$

where V_c is the local velocity magnitude $V_c \equiv \|\mathbf{V}_c\|$.

The component velocity vector can be used to estimate the aerodynamic force $\mathbf{F}_c$ and aerodynamic moment $\mathbf{M}_c$ of each component in the component coordinate system. Estimates for these forces and moments vary by component geometry, and will be presented in the following sections. Once the force and moment for a component has been obtained in the component coordinate system, the force and moment of the component can be rotated to the body-fixed coordinate system to yield

$$\mathbf{F}_b = [\mathbf{R}]\mathbf{F}_c \tag{3.6.4}$$

$$\mathbf{M}_b = [\mathbf{R}]\mathbf{M}_c + \mathbf{P} \times [\mathbf{R}]\mathbf{F}_c = [\mathbf{R}]\mathbf{M}_c + \mathbf{P} \times \mathbf{F}_b \tag{3.6.5}$$

The influence of all components can be summed about the vehicle center of gravity to compute the entire aerodynamic force and moment vectors

$$\mathbf{F}_{b_{\text{total}}} = \sum_{i=1}^{N} \mathbf{F}_{b_i} \tag{3.6.6}$$

$$\mathbf{M}_{b_{\text{total}}} = \sum_{i=1}^{N} \mathbf{M}_{b_i} \tag{3.6.7}$$

where the subscript i represents the properties of the ith component.

A simple model for the aerodynamic forces and moments of rectangular cuboids, cylinders, spheres, and wing segments at subsonic incompressible speeds are considered here. Each depends on the component velocity vector $\mathbf{V}_c$, component velocity magnitude V_c, and fluid density ρ. The examples within each section are designed to help in the development and testing of computer code.

Rectangular Cuboid

The coordinate system and geometric definitions of the rectangular cuboid shown in Fig. 2.3.1 will be used in this section. The aerodynamic forces and moments on a rectangular cuboid can be estimated by treating the cuboid as a bluff body. Because the rectangular cuboid geometry is symmetric about the origin of the cuboid as defined in Fig. 2.3.1, the flow around the cuboid creates an aerodynamic moment about the origin of the cuboid that is approximately zero. This gives

$$\mathbf{M}_c \approx 0 \tag{3.6.8}$$

The aerodynamic force created by the cuboid can be approximated by considering only the drag of the bluff cuboid. Because the drag acts in the direction opposite of the component velocity vector, the aerodynamic force on the cuboid can be estimated as

$$\mathbf{F}_c = -\frac{1}{2}\rho V_c^2 S_{\text{ref}} C_D \mathbf{u}_c \tag{3.6.9}$$

where S_{ref} is the characteristic area of the cuboid, and C_D is a drag coefficient. Here we will assume that the characteristic area is the cross-sectional area of the cuboid normal to the freestream velocity. This area can be computed from the dot product

$$S_{\text{ref}} = \begin{bmatrix} l_y l_z \\ l_x l_z \\ l_x l_y \end{bmatrix} \cdot \begin{bmatrix} |u_{c_x}| \\ |u_{c_y}| \\ |u_{c_z}| \end{bmatrix} \tag{3.6.10}$$

where l_x, l_y, and l_z are the dimensions of the cuboid and $|u_{c_x}|$, $|u_{c_y}|$, and $|u_{c_z}|$ are the absolute values of the components of the velocity unit vector $\mathbf{u}_c$ expressed in the cuboid reference frame. As a conservative estimate, the drag coefficient can be approximated as that of a cube with a face perpendicular to the flow [25]

$$C_D = 1.05 \tag{3.6.11}$$

Using Eqs. (3.6.10) and (3.6.11) in Eq. (3.6.9) gives the total aerodynamic force vector of the cuboid in the coordinate system of the cuboid.

EXAMPLE 3.6.1 A cuboid with $l_x = 5.3$ ft, $l_y = 0.8$ ft, and $l_z = 1.0$ ft is located at $\mathbf{P} = (0.1, 0.0, 0.43)$ ft relative to the vehicle center of gravity and oriented such that the coordinate system of the cuboid is aligned with the coordinate system of the vehicle with $\phi = \theta = \psi = 0$ deg. Consider the case when the vehicle is flying at sea level with a velocity of 100 ft/s, an angle of attack of 5 deg, no sideslip angle, and no rotational velocity ($\omega_b = 0$). Compute the aerodynamic forces and moments caused by the cuboid in both the cuboid coordinate system and the body-fixed coordinate system.

Solution: Using the Euler angles $\phi = \theta = \psi = 0$ in Eq. (2.6.2) gives the rotation matrix and inverse

$$[\mathbf{R}] = [\mathbf{R}]^{-1} = \begin{bmatrix} 1 & 0 & 0 \\ 0 & 1 & 0 \\ 0 & 0 & 1 \end{bmatrix}$$

The components of the translational velocity vector of the vehicle in the body-fixed coordinate system can be computed from Eq. (3.4.12) to yield

$$\mathbf{V} = \begin{bmatrix} 99.61946980917 \\ 0 \\ 8.715574274766 \end{bmatrix} \quad \text{ft/s}$$

Using this along with the position vector $\mathbf{P} = (0.1, 0.0, 0.43)$ ft and $\omega_b = 0$ in Eq. (3.6.2) gives the component velocity vector in the cuboid coordinate system

$$\mathbf{V}_c = \begin{bmatrix} 1 & 0 & 0 \\ 0 & 1 & 0 \\ 0 & 0 & 1 \end{bmatrix} \left(\begin{bmatrix} 99.6195 \\ 0 \\ 8.7156 \end{bmatrix} + \begin{bmatrix} 0 \\ 0 \\ 0 \end{bmatrix} \times \begin{bmatrix} 0.1 \\ 0.0 \\ 0.43 \end{bmatrix} \right) = \begin{bmatrix} 99.6195 \\ 0 \\ 8.7156 \end{bmatrix} \quad \text{ft/s}$$

The component velocity magnitude is

$$V_c = \sqrt{99.6195^2 + 0^2 + 8.7156^2} = 100.0 \quad \text{ft/s}$$

which is the same as the vehicle velocity magnitude because the vehicle is not rotating. The component unit velocity vector in the cuboid coordinate system is

$$\mathbf{u}_c = \frac{\mathbf{V}_c}{V_c} = \begin{bmatrix} 0.9961946980917 \\ 0.0 \\ 0.08715574274766 \end{bmatrix}$$

The characteristic area of the cuboid can be computed from Eq. (3.6.10) to yield

$$S_{\text{ref}} = \begin{bmatrix}(0.8)(1.0)\\(5.3)(1.0)\\(5.3)(0.8)\end{bmatrix} \cdot \begin{bmatrix}0.996195\\0.0\\0.087156\end{bmatrix} = 1.166496107723 \ \text{ft}^2$$

Using these results along with the suggested drag coefficient of $C_D = 1.05$ and atmospheric density at sea level in Eqs. (3.6.9) and (3.6.8) gives the total force and moment vectors in the cuboid coordinate system

$$\mathbf{F}_c = -\frac{1}{2}(0.00237689 \text{ slug/ft}^3)(100.0 \text{ ft/s})^2(1.166496 \text{ ft}^2)(1.05)\begin{bmatrix}0.996195\\0.0\\0.087156\end{bmatrix} = \begin{bmatrix}-14.50094502108\\0.0\\-1.268668299757\end{bmatrix} \text{lbf}$$

$$\mathbf{M}_c = \begin{bmatrix}0.0\\0.0\\0.0\end{bmatrix}$$

Using these results in Eqs. (3.6.4) and (3.6.5) gives the aerodynamic force and moment vectors in the body-fixed coordinate system

$$\mathbf{F}_b = \begin{bmatrix}1&0&0\\0&1&0\\0&0&1\end{bmatrix}\begin{bmatrix}-14.50094502108\\0.0\\-1.268668299757\end{bmatrix} \text{lbf} = \begin{bmatrix}-14.50094502108\\0.0\\-1.268668299757\end{bmatrix} \text{lbf}$$

$$\mathbf{M}_b = \begin{bmatrix}1&0&0\\0&1&0\\0&0&1\end{bmatrix}\begin{bmatrix}0.0\\0.0\\0.0\end{bmatrix} + \begin{bmatrix}0.1\\0.0\\0.43\end{bmatrix} \text{ft} \times \begin{bmatrix}1&0&0\\0&1&0\\0&0&1\end{bmatrix}\begin{bmatrix}-14.50094502108\\0.0\\-1.268668299757\end{bmatrix} \text{lbf} = \begin{bmatrix}0.0\\-6.108539529087\\0.0\end{bmatrix} \text{ft} \cdot \text{lbf}$$

Cylinder

The coordinate system and geometric definitions of the cylinder shown in Fig. 2.3.2 will be used in this section. Because the cylinder is symmetric about its origin, the flow around the cylinder creates an aerodynamic moment about the origin of the cylinder that is approximately zero. This gives

$$\mathbf{M}_c \approx 0 \tag{3.6.12}$$

The aerodynamic force on the cylinder depends on the angle between the component velocity vector and the cylinder axis. Here we will refer to that angle as the angle of attack of the cylinder, α. From the definition of the dot product given in Eq. (D.1.1), and using a unit vector $(1, 0, 0)$ to represent the cylinder axis, the angle of attack can be computed from the x-component of the unit velocity vector $\mathbf{u}_c$

$$\alpha = \cos^{-1}(u_{c_x}) \tag{3.6.13}$$

The total force created by the cylinder can be split into a lift force acting perpendicular to the component velocity vector, and a drag force acting in the opposite direction of the component velocity vector. Both of these forces depend on the velocity magnitude normal to the axis of the cylinder, which can be computed from

$$V_N = V_c \sin(\alpha) = V_c \sqrt{u_{c_y}^2 + u_{c_z}^2} \tag{3.6.14}$$

These forces also depend on the surface area of the cylinder normal to the cylinder axis, which will be referred to as the reference area of the cylinder

$$S_{\text{ref}} = 2 r_2 h_c \tag{3.6.15}$$

Here r_2 is the outer radius of the cylinder and h_c is the height of the cylinder. The lift and drag forces can be computed from

$$L = \frac{1}{2} \rho V_N^2 S_{\text{ref}} C_L \tag{3.6.16}$$

$$D = \frac{1}{2} \rho V_N^2 S_{\text{ref}} C_D \tag{3.6.17}$$

where C_L and C_D are lift and drag coefficients, which depend on the angle of attack. At subcritical Reynolds numbers, C_L and C_D are related to angle of attack according to the *cross-flow principle* [25], and can be approximated as

$$C_L = C_{D0} \sin^2 \alpha \cos \alpha \tag{3.6.18}$$

$$C_D = C_{D0} \sin^3 \alpha + 0.02 \tag{3.6.19}$$

where C_{D0} is a basic drag coefficient. The basic drag coefficient depends on the cross-flow Reynolds number

$$\mathrm{R_e} = \frac{2 \rho V_N r}{\mu} \tag{3.6.20}$$

A host of experimental and analytic studies on cylinder aerodynamics show that the basic drag coefficient can be approximated over a range of Reynolds numbers using the piece-wise function [25, 28]

$$C_{D0} = \begin{cases} 1.18 + \frac{6.8}{\mathrm{R_e^{0.89}}} + \frac{1.96}{\mathrm{R_e^{1/2}}} - \frac{0.0004\mathrm{R_e}}{1+3.64\mathrm{E}{-}07\mathrm{R_e^2}}, & 0 < \mathrm{R_e} \leq 330{,}000 \\ 3.78\mathrm{E}{-}11\mathrm{R_e^2} - 3.56\mathrm{E}{-}05\mathrm{R_e} + 8.7634, & 330{,}000 < \mathrm{R_e} \leq 460{,}000 \\ -5.0\mathrm{E}{-}15\mathrm{R_e^2} + 7.0\mathrm{E}{-}08\mathrm{R_e} + 0.346, & 460{,}000 < \mathrm{R_e} \leq 10{,}000{,}000 \\ 0.55, & \mathrm{R_e} > 10{,}000{,}000 \end{cases} \tag{3.6.21}$$

For the case of $R_e < 0.01$, a computer code can use $C_{D0} = 430.0$ to avoid a singularity at zero velocity. Once the lift and drag forces have been computed, the total force vector can be found from

$$\mathbf{F}_c = D\mathbf{u}_D + L\mathbf{u}_L \tag{3.6.22}$$

where $\mathbf{u}_D$ is the unit vector in the direction of drag

$$\mathbf{u}_D = -\mathbf{u}_c \tag{3.6.23}$$

and $\mathbf{u}_L$ is the unit vector in the direction of lift, which can be found from

$$\mathbf{u}_L = \frac{\mathbf{v}_L}{||\mathbf{v}_L||} \tag{3.6.24}$$

where

$$\mathbf{v}_L = \mathbf{u}_D \times \left(\begin{bmatrix} 1 \\ 0 \\ 0 \end{bmatrix} \times \mathbf{u}_D \right) \tag{3.6.25}$$

EXAMPLE 3.6.2 A vehicle has an external fuel store that can be approximated as a cylinder with a height of $h = 2.5$ ft and a radius of $r = 0.5$ ft. The store is located at $\mathbf{P} = (0.1, 5.0, 1.5)$ ft in the body-fixed coordinate system relative to the vehicle center of gravity. It is mounted on the vehicle relative to the body-fixed coordinate system with the orientation given by the Euler angles $\phi = 0$ deg, $\theta = 10.0$ deg, $\psi = 5$ deg. Consider the case when the vehicle is flying at sea level with a velocity of 200 ft/s, angle of attack of 5 deg, sideslip angle of 5 deg, and roll rate of $p_b = 30$ deg/s. Estimate the aerodynamic forces and moments caused by the store in both the cylinder coordinate system and the body-fixed coordinate system.

Solution: Using the Euler angles $\phi = 0$ deg, $\theta = 10.0$ deg, $\psi = 5$ deg in Eq. (2.6.2) gives the rotation matrix

$$[\mathbf{R}] = \begin{bmatrix} 0.9810602621904 & -0.08715574274766 & 0.1729873939251 \\ 0.08583165117743 & 0.9961946980917 & 0.01513443590134 \\ -0.1736481776669 & 0.0 & 0.9848077530122 \end{bmatrix}$$

The components of the translational velocity vector of the vehicle in the body-fixed coordinate system can be computed from Eq. (3.4.12) to yield

$$\mathbf{V} = \begin{bmatrix} 198.4807753012 \\ 17.43114854953 \\ 17.36481776669 \end{bmatrix} \text{ ft/s}$$

The vehicle rotational velocity vector is

$$\omega_b = \frac{\pi}{180 \text{ deg}} \begin{bmatrix} 30.0 \\ 0.0 \\ 0.0 \end{bmatrix} \text{ deg/s} = \begin{bmatrix} 0.5235987755983 \\ 0.0 \\ 0.0 \end{bmatrix} \text{ rad/s}$$

Using the translational and rotational velocity vectors along with the position vector $\mathbf{P} = (0.1, 5.0, 1.5)$ ft in Eq. (3.6.2) gives the component velocity vector in the cylinder coordinate system

$$\mathbf{V}_c = \begin{bmatrix} 0.981 & -0.087 & 0.173 \\ 0.086 & 0.996 & 0.015 \\ -0.174 & 0.0 & 0.985 \end{bmatrix}^{-1} \left(\begin{bmatrix} 198.481 \\ 17.431 \\ 17.365 \end{bmatrix} + \begin{bmatrix} 0.524 \\ 0.0 \\ 0.0 \end{bmatrix} \times \begin{bmatrix} 0.1 \\ 5.0 \\ 1.5 \end{bmatrix} \right)$$
$$= \begin{bmatrix} 192.6803548707 \\ -0.7163311120834 \\ 54.26582394051 \end{bmatrix} \text{ft/s}$$

The component velocity vector magnitude is

$$V_c = \sqrt{192.680^2 + 0.716^2 + 54.266^2} = 200.1774511061 \text{ ft/s}$$

and the component unit velocity vector in the cylinder coordinate system is

$$\mathbf{u}_c = \frac{\mathbf{V}_c}{V_c} = \begin{bmatrix} 0.9625477485406 \\ -0.003578480533774 \\ 0.2710885948476 \end{bmatrix}$$

The angle of attack of the cylinder can be found from Eq. (3.6.13) to yield

$$\alpha = \cos^{-1}(0.9625477) = 0.2745483319117470 \text{ rad} = 15.73046069090 \text{ deg}$$

The normal velocity magnitude can be found from Eq. (3.6.14) to yield

$$V_N = (200.177451 \text{ ft/s}) \sin(0.274548331912) = 54.27055166666 \text{ ft/s}$$

The surface area of the cylinder normal to the cylinder axis can be found from Eq. (3.6.15) to yield

$$S_{\text{ref}} = (2)(0.5 \text{ ft})(2.5 \text{ ft}) = 2.5 \text{ ft}^2$$

Using atmospheric properties at sea level, the Reynolds number can be found from Eq. (3.6.20) to yield

$$\text{R}_\text{e} = \frac{2(0.00237689 \text{ slug/ft}^3)(54.27055167 \text{ ft/s})(0.5 \text{ ft})}{3.737025865723\text{E}{-}07 \text{ slug/(ft} \cdot \text{s)}} = 345,181.5821132$$

Because the Reynolds number is within the range $330,000 < \text{R}_\text{e} \leq 460,000$, the basic drag coefficient can be computed from the second equation in Eq. (3.6.21) to yield

$$C_{D0} = 3.78\text{E}{-}11\text{R}_\text{e}^2 - 3.56\text{E}{-}05\text{R}_\text{e} + 8.7634 = 0.9788179477910$$

Using this result along with the angle of attack in Eqs. (3.6.18) and (3.6.19) gives the lift and drag coefficients

$$C_L = 0.06925041320041$$
$$C_D = 0.03950514431039$$

The dimensional lift and drag forces can then be computed from Eq. (3.6.16) and (3.6.17) to yield

$$L = (0.5)(0.00238 \text{ slug/ft}^3)(54.271 \text{ ft/s})^2(2.5 \text{ ft}^2)(0.0693) = 0.6059968087488 \text{ lbf}$$
$$D = (0.5)(0.00238 \text{ slug/ft}^3)(54.271 \text{ ft/s})^2(2.5 \text{ ft}^2)(0.0395) = 0.3457017839298 \text{ lbf}$$

The unit vector in the direction of drag can be found from Eq. (3.6.23)

$$\mathbf{u}_D = -\mathbf{u}_c = \begin{bmatrix} -0.9625477485406 \\ 0.003578480533774 \\ -0.2710885948476 \end{bmatrix}$$

The unit vector in the direction of lift can be found from Eqs. (3.6.24) and (3.6.25) to yield

$$\mathbf{u}_L = \begin{bmatrix} 0.2711122125234 \\ 0.01270491782322 \\ -0.9624638971326 \end{bmatrix}$$

Using these results in Eqs. (3.6.22) and (3.6.12) gives the total force and moment vectors in the cylinder coordinate system

$$\mathbf{F}_c = (0.34570178 \text{ lbf}) \begin{bmatrix} -0.96254775 \\ 0.00357848 \\ -0.27108859 \end{bmatrix} + (0.60599681 \text{ lbf}) \begin{bmatrix} 0.27111221 \\ 0.01270492 \\ -0.96246390 \end{bmatrix}$$
$$= \begin{bmatrix} -0.1684613381861 \\ 0.00893622676057 \\ -0.6769658610401 \end{bmatrix} \text{ lbf}$$

$$\mathbf{M}_c = \begin{bmatrix} 0.0 \\ 0.0 \\ 0.0 \end{bmatrix}$$

Using these results in Eqs. (3.6.4) and (3.6.5) gives the aerodynamic force and moment vectors in the body-fixed coordinate system

$$\mathbf{F}_b = \begin{bmatrix} 0.981 & -0.087 & 0.173 \\ 0.086 & 0.996 & 0.015 \\ -0.174 & 0.0 & 0.985 \end{bmatrix} \begin{bmatrix} -0.16846 \\ 0.00894 \\ -0.67696 \end{bmatrix} \text{ lbf} = \begin{bmatrix} -0.2831561281680 \\ -0.01580258952755 \\ -0.6374282240935 \end{bmatrix} \text{ lbf}$$

$$\mathbf{M}_b = \begin{bmatrix} 0.981 & -0.087 & 0.173 \\ 0.086 & 0.996 & 0.015 \\ -0.174 & 0.0 & 0.985 \end{bmatrix} \begin{bmatrix} 0.0 \\ 0.0 \\ 0.0 \end{bmatrix}$$

$$+ \begin{bmatrix} 0.1 \\ 5.0 \\ 1.5 \end{bmatrix} \text{ft} \times \begin{bmatrix} 0.981 & -0.087 & 0.173 \\ 0.086 & 0.996 & 0.015 \\ -0.174 & 0.0 & 0.985 \end{bmatrix} \begin{bmatrix} -0.16846134 \\ 0.00893623 \\ -0.67696586 \end{bmatrix} \text{lbf}$$

$$= \begin{bmatrix} -3.163437236176 \\ -0.3609913698427 \\ 1.414200381887 \end{bmatrix} \text{ft} \cdot \text{lbf}$$

Sphere

The coordinate system and geometric definitions of the sphere shown in Fig. 2.3.3 will be used in this section. A non-spinning sphere exposed to a uniform flow creates no aerodynamic moment about the origin of the sphere. Hence, for a sphere

$$\mathbf{M}_c = 0 \tag{3.6.26}$$

The aerodynamic force created by the sphere is simply equal to the drag of the sphere, which is in the opposite direction of the component velocity vector. This can be computed from

$$\mathbf{F}_c = -\frac{1}{2}\rho V_c^2 \pi r_2^2 C_D \mathbf{u}_c \tag{3.6.27}$$

where r_2 is the outer radius of the sphere, and C_D is a drag coefficient that depends on the Reynolds number

$$\mathrm{R_e} = \frac{2\rho V_c r_2}{\mu} \tag{3.6.28}$$

The drag coefficient can be estimated from a piece-wise function fit to experimental data on **smooth spheres** [25, 28] according to

$$C_D = \begin{cases} \frac{24}{\mathrm{R_e}} + \frac{6}{1+\sqrt{\mathrm{R_e}}} + 0.4, & 0 < \mathrm{R_e} \le 450,000 \\ 1.0\mathrm{E}{+}29\mathrm{R_e}^{-5.211}, & 450,000 < \mathrm{R_e} \le 560,000 \\ -2.0\mathrm{E}{-}23\mathrm{R_e}^3 - 1.0\mathrm{E}{-}16\mathrm{R_e}^2 & \\ \quad + 9.0\mathrm{E}{-}09\mathrm{R_e} + 0.069, & 560,000 < \mathrm{R_e} \le 14,000,000 \\ 0.12, & \mathrm{R_e} > 14,000,000 \end{cases} \tag{3.6.29}$$

For the case of $\mathrm{R_e} < 0.01$, a computer code can use $C_D = 2405.0$ to avoid a singularity at zero velocity.

EXAMPLE 3.6.3 A sphere with a radius of 6 inches is located at $\mathbf{P} = (1.0, 2.0, 3.0)$ ft relative to the vehicle center of gravity and oriented relative to the body-fixed coordinate system by the Euler angles $\phi = 30.0$ deg, $\theta = 45.0$ deg, $\psi = 60.0$ deg. Consider the case when the vehicle is flying at sea level with a velocity of 200 ft/s, an angle of attack of 15 deg, a sideslip angle of 10 deg, and a rotational velocity given by $p_b = 10$ deg/s, $q_b = 5$ deg/s, and $r_b = 3$ deg/s. Compute the aerodynamic forces and moments caused by the sphere in both the sphere coordinate system and the body-fixed coordinate system.

Solution: Using the given Euler angles in Eq. (2.6.2) gives the rotation matrix

$$[\mathbf{R}] = \begin{bmatrix} 0.3535533905933 & -0.5732233047034 & 0.7391989197401 \\ 0.6123724356958 & 0.7391989197401 & 0.2803300858899 \\ -0.7071067811865 & 0.3535533905933 & 0.6123724356958 \end{bmatrix}$$

The components of the translational velocity vector of the vehicle in the body-fixed coordinate system can be computed from Eq. (3.4.12) to yield

$$\mathbf{V} = \begin{bmatrix} 190.2502485128 \\ 34.72963553339 \\ 50.97740044884 \end{bmatrix} \text{ ft/s}$$

The vehicle rotational velocity vector is

$$\omega_b = \frac{\pi}{180 \text{ deg}} \begin{bmatrix} 10.0 \\ 5.0 \\ 3.0 \end{bmatrix} \text{deg/s} = \begin{bmatrix} 0.1745329251994 \\ 0.08726646259972 \\ 0.05235987755983 \end{bmatrix} \text{ rad/s}$$

Using the translational and rotational velocity vectors along with the position vector $\mathbf{P} = (1.0, 2.0, 3.0)$ ft in Eq. (3.6.2) gives the component velocity vector in the sphere coordinate system

$$\mathbf{V}_c = \begin{bmatrix} 0.3536 & -0.5732 & 0.7392 \\ 0.6124 & 0.7392 & 0.2803 \\ -0.7071 & 0.3536 & 0.6124 \end{bmatrix}^{-1} \left(\begin{bmatrix} 190.250 \\ 34.730 \\ 50.977 \end{bmatrix} + \begin{bmatrix} 0.1745 \\ 0.0873 \\ 0.0524 \end{bmatrix} \times \begin{bmatrix} 1.0 \\ 2.0 \\ 3.0 \end{bmatrix} \right)$$

$$= \begin{bmatrix} 52.06646858322 \\ -65.70635526091 \\ 181.7300241541 \end{bmatrix} \text{ ft/s}$$

The component velocity vector magnitude is

$$V_c = \sqrt{52.066^2 + 65.706^2 + 181.730^2} = 200.1350642727 \text{ ft/s}$$

and the component unit velocity vector in the sphere coordinate system is

$$\mathbf{u}_c = \frac{\mathbf{V}_c}{V_c} = \begin{bmatrix} 0.2601566535701 \\ -0.3283100615061 \\ 0.9080369040502 \end{bmatrix}$$

Using the atmospheric properties at sea level, the Reynolds number can be computed from Eq. (3.6.28) to yield

$$\mathrm{R_e} = \frac{2(0.0023769\ \mathrm{slug/ft^3})(200.1351\ \mathrm{ft/s})(0.5\ \mathrm{ft})}{3.737025865723\mathrm{E}{-07}\ \mathrm{slug/(ft\cdot s)}} = 1,272,935.984625$$

From Eq. (3.6.29), the drag coefficient of the sphere at this Reynolds number can be computed as

$$C_D = -2.0\mathrm{E}{-23}\mathrm{R_e^3} - 1.0\mathrm{E}{-16}\mathrm{R_e^2} + 9.0\mathrm{E}{-09}\mathrm{R_e} + 0.069 = 0.08025313481521$$

Using these results in Eqs. (3.6.27) and (3.6.26) gives the total force and moment vectors in the sphere coordinate system

$$\mathbf{F}_c = -\frac{1}{2}(0.0023769\ \mathrm{slug/ft^3})(200.1351\ \mathrm{ft/s})^2\pi(0.5\ \mathrm{ft})^2(0.0803)\begin{bmatrix}0.2602\\-0.3283\\0.9080\end{bmatrix} = \begin{bmatrix}-0.7805714839517\\0.9850583038696\\-2.724465063456\end{bmatrix}\ \mathrm{lbf}$$

$$\mathbf{M}_c = \begin{bmatrix}0.0\\0.0\\0.0\end{bmatrix}$$

Using these results in Eqs. (3.6.4) and (3.6.5) gives the aerodynamic force and moment vectors in the body-fixed coordinate system

$$\mathbf{F}_b = \begin{bmatrix}0.3536 & -0.5732 & 0.7392\\0.6124 & 0.7392 & 0.2803\\-0.7071 & 0.3536 & 0.6124\end{bmatrix}\begin{bmatrix}-0.78057\\0.98506\\-2.72447\end{bmatrix}\ \mathrm{lbf} = \begin{bmatrix}-2.854553702798\\-0.5135959520035\\-0.7681692141086\end{bmatrix}\ \mathrm{lbf}$$

$$\mathbf{M}_b = \begin{bmatrix}0.3536 & -0.5732 & 0.7392\\0.6124 & 0.7392 & 0.2803\\-0.7071 & 0.3536 & 0.6124\end{bmatrix}\begin{bmatrix}0.0\\0.0\\0.0\end{bmatrix} + \begin{bmatrix}1.0\\2.0\\3.0\end{bmatrix}\ \mathrm{ft} \times \begin{bmatrix}0.3536 & -0.5732 & 0.7392\\0.6124 & 0.7392 & 0.2803\\-0.7071 & 0.3536 & 0.6124\end{bmatrix}\begin{bmatrix}-0.78057\\0.98506\\-2.72447\end{bmatrix}\ \mathrm{lbf} = \begin{bmatrix}0.0044494277934\\-7.795491894285\\5.195511453592\end{bmatrix}\ \mathrm{ft\cdot lbf}$$

Note that since a sphere is axially symmetric in all directions, changing the orientation of the sphere will not affect the final forces and moments in the body-fixed coordinate system. The reader may want to test their code using various orientations of the sphere as a check.

Wing Segment

The aerodynamic forces and moment caused by a wing segment can be approximated using the wing-segment geometry information given in Section 2.4 and shown in Fig. 2.4.1, along with a few additional aerodynamic parameters of the wing. The aerodynamics of the wing segment depend on the wing-segment planform area, which will be used for the reference area

$$S_{\text{ref}} = S_w = b\overline{c} \tag{3.6.30}$$

where $\overline{c}$ is the mean chord, which for a simply tapered wing is

$$\overline{c} = \frac{c_r + c_t}{2} \tag{3.6.31}$$

The aspect ratio is defined as

$$R_A \equiv \frac{b^2}{S_w} = \frac{b}{\overline{c}} \tag{3.6.32}$$

The distribution of aerodynamic loading on a finite wing can be replaced by a single force vector and a single moment vector acting at the wing aerodynamic center. The location of the aerodynamic center depends on several factors, including wing planform, taper ratio, and sweep angle. For wings without sweep, the aerodynamic center is typically very near the wing centroidal chord [29]. According to a study by Phillps et al. [30, 31], "wing camber, thickness, and twist have no significant effect on the position of the aerodynamic center of a swept wing." For the purposes of this development, we shall assume that the aerodynamic center of the wing lies at the quarter-chord of the wing centroidal chord. From the geometric definition of the wing, the centroidal chord of the wing lies at the spanwise location

$$y_{ac} = \delta \frac{b}{3} \frac{(c_r + 2c_t)}{(c_r + c_t)} \tag{3.6.33}$$

where δ is defined in Eq. (2.4.29). Accounting for sweep, the corresponding coordinate of the quarter-chord at that spanwise location is

$$x_{ac} = -\frac{b}{3} \frac{(c_r + 2c_t)}{(c_r + c_t)} \tan \Lambda_{c/4} \tag{3.6.34}$$

Hence, the location of the wing aerodynamic center in the wing coordinate system can be estimated as

$$\begin{bmatrix} x_{ac} \\ y_{ac} \\ z_{ac} \end{bmatrix} = \frac{b}{3} \frac{(c_r + 2c_t)}{(c_r + c_t)} \begin{bmatrix} -\tan \Lambda_{c/4} \\ \delta \\ 0 \end{bmatrix} \tag{3.6.35}$$

The location of the wing aerodynamic center in the vehicle coordinate system can be computed from

$$\mathbf{r}_{ac} = \mathbf{r}_o + [\mathbf{R}] \begin{bmatrix} x_{ac} \\ y_{ac} \\ z_{ac} \end{bmatrix} \tag{3.6.36}$$

where $\mathbf{r}_o$ is the vector from the vehicle origin to the origin of the wing section and $[\mathbf{R}]$ is the rotation matrix of the component computed from Eq. (2.6.7).

The component velocity vector at the location we have defined as the aerodynamic center depends on the vehicle translational velocity $\mathbf{V}$, the vehicle rotation rate about the vehicle center of gravity ω_b, and the location of the aerodynamic center of the wing relative to the vehicle center of gravity. This velocity vector can be computed from Eq. (3.6.2) by setting

$$\mathbf{P} = \mathbf{r}_{ac} - \mathbf{r}_b \tag{3.6.37}$$

where $\mathbf{r}_b$ is the vehicle CG location.

Here we will assume that the only components of velocity that affect the forces and moments on the wing are those in the x_c-z_c plane of the wing passing through the aerodynamic center. This velocity vector is

$$\mathbf{V}_w = \begin{bmatrix} V_{c_x} \\ 0 \\ V_{c_z} \end{bmatrix} \tag{3.6.38}$$

The unit vector of the wing velocity vector can be found from

$$\mathbf{u}_w \equiv \frac{\mathbf{V}_w}{V_w} = \begin{bmatrix} u_{w_x} \\ 0 \\ u_{w_z} \end{bmatrix} = \frac{1}{V_w} \begin{bmatrix} V_{c_x} \\ 0 \\ V_{c_z} \end{bmatrix} \tag{3.6.39}$$

where V_w is the wing velocity magnitude

$$V_w = \|\mathbf{V}_w\| = \sqrt{V_{c_x}^2 + V_{c_z}^2} \tag{3.6.40}$$

In this development we will also assume that the resulting forces and moments are only longitudinal within the wing coordinate system. Hence, we will assume that the lift, drag, and pitching moment all act in an x_c-z_c plane of the wing passing through the wing aerodynamic center, and neglect any side force acting in the y_c-direction as well as lateral moments that would act about the x_c or z_c axes of the wing coordinate system located at the aerodynamic center. From these definitions, the unit vector in the direction of drag can be expressed as

$$\mathbf{u}_D = -\mathbf{u}_w = \begin{bmatrix} -u_{w_x} \\ 0 \\ -u_{w_z} \end{bmatrix} \tag{3.6.41}$$

and the unit vector in the direction of lift can be computed from

$$\mathbf{u}_L = \mathbf{u}_D \times \begin{bmatrix} 0 \\ 1 \\ 0 \end{bmatrix} = \begin{bmatrix} u_{w_z} \\ 0 \\ -u_{w_x} \end{bmatrix} \tag{3.6.42}$$

The unit vector in the direction of the pitching moment is simply a vector that aligns with the wing y coordinate

$$\mathbf{u}_m = \begin{bmatrix} 0 \\ 1 \\ 0 \end{bmatrix} \tag{3.6.43}$$

With these unit vectors defined, the force and moment vectors in the wing coordinate system are

$$\mathbf{F}_c = D\mathbf{u}_D + L\mathbf{u}_L \tag{3.6.44}$$

$$\mathbf{M}_c = m\mathbf{u}_m \tag{3.6.45}$$

where L and D are the aerodynamic lift and drag forces produced by the wing, and m is the aerodynamic pitching moment.

The lift, drag, and pitching moment can be computed from

$$L = \frac{1}{2}\rho V_w^2 S_w C_L \tag{3.6.46}$$

$$D = \frac{1}{2}\rho V_w^2 S_w C_D \tag{3.6.47}$$

$$m = \frac{1}{2}\rho V_w^2 S_w \bar{c} C_m \tag{3.6.48}$$

where the lift coefficient, C_L, drag coefficient C_D, and pitching-moment coefficient C_m depend on the angle of attack of the wing, and any control-surface deflection. The angle of attack can be computed from the wing velocity vector in the wing-segment coordinate system

$$\alpha = \tan^{-1}\left(\frac{u_{w_z}}{u_{w_x}}\right) \tag{3.6.49}$$

Note that because the mounting and dihedral angles are accounted for in the rotation matrix $[\mathbf{R}]$, their effects are included in the angle of attack.

Below Stall

At angles of attack below stall, the lift coefficient can be approximated from

$$C_{L_{ss}} = C_{L,\alpha}(\alpha - \alpha_{L0} + \varepsilon_c \delta_c) \tag{3.6.50}$$

where $C_{L,\alpha}$ is the wing lift slope, α_{L0} is the zero-lift angle of attack, ε_c is the control-surface effectiveness, and δ_c is the control-surface deflection. If the wing lift slope is not specified, it can be approximated for a wing below the critical Mach number in a compressible flow from [32]

$$C_{L,\alpha} = \frac{2\pi R_A}{2 + \sqrt{4 + R_A(1 - \mathrm{M}^2 + \tan^2 \Lambda_{c/4})}} \tag{3.6.51}$$

where M is the Mach number and $\Lambda_{c/4}$ is the quarter-chord sweep angle. If the control-surface effectiveness is not specified, it can be approximated from thin-airfoil theory as [11, 33]

$$\varepsilon_c = \eta_f \left(1 - \frac{\theta_f - \sin\theta_f}{\pi}\right) \tag{3.6.52}$$

where

$$\theta_f = \cos^{-1}\left(2\frac{c_f}{c} - 1\right) \tag{3.6.53}$$

In this development, c_f/c is the flap-chord fraction and η_f is a flap efficiency. Typically, $\eta_f = 0.8$ is a reasonable approximation. This approximation for ε_c assumes the control surface extends along the entire span of the wing segment.

Below stall, the drag coefficient can be approximated as a quadratic function of lift as

$$C_{D_{ss}} = C_{D_{L0}} + C_{D,L}C_{L_{ss}} + \frac{C_{L_{ss}}^2}{\pi e_O R_A} \tag{3.6.54}$$

where $C_{D_{L0}}$ and $C_{D,L}$ are coefficients that specify the drag polar, along with the Oswald efficiency e_O. If these properties are not specified, reasonable approximations can be obtained from $0.005 \le C_{D_{L0}} \le 0.01$, $C_{D,L} \approx 0$, and $0.5 \le e_O \le 0.9$ [34–36].

Below stall, the pitching-moment coefficient can be approximated as a linear function of angle of attack α and control-surface deflection δ_c as

$$C_{m_{ss}} = C_{m_0} + C_{m,\alpha}\alpha + C_{m,\delta_c}\delta_c \tag{3.6.55}$$

where C_{m_0} is the pitching moment at zero angle of attack and zero control-surface deflection, $C_{m,\alpha}$ is the pitching-moment slope, and C_{m,δ_c} is the change in pitching moment with respect to control-surface deflection. If C_{m,δ_c} is not specified, it can be approximated from thin-airfoil theory as [11, 33]

$$C_{m,\delta_c} = \eta_f \left(\frac{\sin(2\theta_f) - 2\sin\theta_f}{4}\right) \tag{3.6.56}$$

where θ_f is computed from Eq. (3.6.53). This approximation for C_{m,δ_c} assumes the control surface extends along the entire span of the wing segment.

Above Stall

At angles of attack above stall, the lift, drag, and pitching moment on a wing segment can be approximated from our understanding of flow over a flat plate at high incidence angles. Figure 3.6.1 shows a flat plate inclined at an angle of attack α to the flow direction. As the angle of attack increases, results approach those predicted from Newtonian theory. From this theory, the pressure coefficient on the windward surface inclined to the flow by an angle of attack α can be approximated as

$$C_{\mathcal{P}} = 2\sin^2\alpha \tag{3.6.57}$$

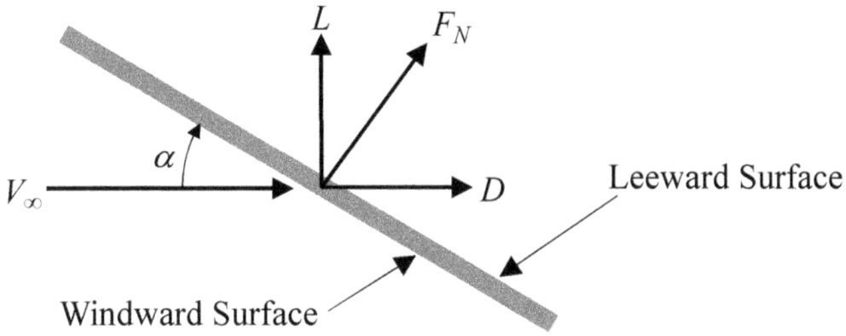

Figure 3.6.1. Newtonian-theory approximations for surface pressure on a flat plate inclined to the incoming flow.

Neglecting any pressure on the leeward side of the surface yields a total force normal to the surface

$$F_N = QS_{\text{ref}}C_{\mathcal{P}} = 2QS_{\text{ref}}\sin^2\alpha \tag{3.6.58}$$

where Q is the dynamic pressure and S is the plate planform area. This normal force can be used to compute the lift and drag coefficients

$$C_{L_N} = 2\text{sign}(\alpha)\sin^2\alpha\cos\alpha \tag{3.6.59}$$

$$C_{D_N} = 2\sin^3(|\alpha|) \tag{3.6.60}$$

where the sign(α) and $|\alpha|$ are used to preserve the correct sign of the solutions.

From Newtonian theory, the center of pressure and aerodynamic center of the flat plate are at the half-chord location. In other words, at the plate half-chord location, Newtonian theory predicts zero pitching moment. In the wing segment development presented in this section, we have chosen to report forces and moments about a location coincident with the wing aerodynamic center below stall. In the sub-stall development, we approximated the aerodynamic center of the wing segment to lie at the quarter-chord of the wing centroidal chord. A first-order approximation for the pitching moment about this location when the wing is stalled can be computed from Newtonian theory by evaluating the pitching moment about the plate quarter-chord location. This yields

$$C_{m_N} = -\frac{F_N}{QS_{\text{ref}}c}\left(\frac{c}{4}\right) = -\frac{1}{2}\text{sign}(\alpha)\sin^2\alpha \tag{3.6.61}$$

where c is the chord length of the flat plate.

Figure 3.6.2 shows the resulting lift, drag, and pitching-moment coefficients as predicted by Newtonian theory on a flat plate at the plate quarter-chord location as computed from Eqs. (3.6.59)–(3.6.61). These can be used to approximate the forces and pitching moment on a wing at angles of attack above stall. Note that some wings are designed to maintain control effectiveness at high angles of attack to assist with stall

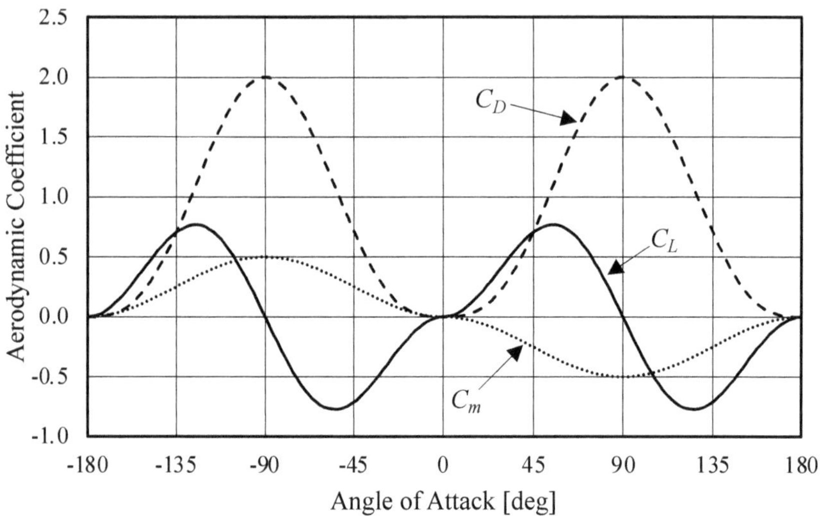

Figure 3.6.2. Newtonian-theory approximations for the aerodynamic forces and moment about the quarter-chord on a flat plate.

recovery and control. For these cases, modifications to this very simple stall model can be made.

The aerodynamic model below stall given in Eqs. (3.6.50), (3.6.54), and (3.6.55) can be blended with the aerodynamic model above stall given in Eqs. (3.6.59), (3.6.60), and (3.6.61) to yield the lift, drag, and pitching moment coefficients required for Eqs. (3.6.46)–(3.6.48) over the entire range of angles of attack. This can be done using the blending function

$$C_L = (1 - \sigma)C_{L_{ss}} + \sigma C_{L_N} \tag{3.6.62}$$

$$C_D = (1 - \sigma)C_{D_{ss}} + \sigma C_{D_N} \tag{3.6.63}$$

$$C_m = (1 - \sigma)C_{m_{ss}} + \sigma C_{m_N} \tag{3.6.64}$$

where σ is a blending parameter that ranges from 0 to 1. The value for σ should be near $\sigma = 0$ in the sub-stall region, and near $\sigma = 1$ once the wing is fully stalled. Beard and McLain [37] suggest a blending function that predicts symmetric stall points at positive and negative angles of attack. This blending function can be relaxed to allow for non-symmetric stall points and expressed as

$$\sigma = \frac{1 + e^{-\lambda_b(\alpha - \alpha_{b0} - \alpha_b)} + e^{\lambda_b(\alpha - \alpha_{b0} + \alpha_b)}}{[1 + e^{-\lambda_b(\alpha - \alpha_{b0} - \alpha_b)}][1 + e^{\lambda_b(\alpha - \alpha_{b0} + \alpha_b)}]} \tag{3.6.65}$$

where α_{b0} is the center or minimum-value angle, α_b is a transition angle, and λ_b is a blending-rate parameter. **Note that values for α, α_{b0}, and α_b must be in radians when evaluating this function.** This sigmoid function produces a value near 0 inside of the range $-\alpha_b < (\alpha - \alpha_{b0}) < \alpha_b$ and a value near 1 outside of this range. At $(\alpha - \alpha_{b0}) \pm \alpha_b$,

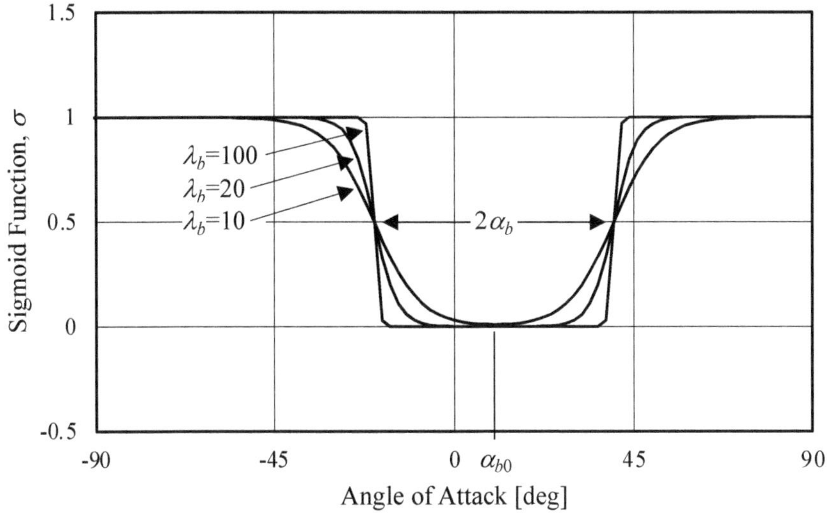

Figure 3.6.3. Sigmoid function with $\alpha_{b0} = 10$ deg, $\alpha_b = 30$ deg and varying values of λ_b.

the function has a value of $\sigma = 0.5$. Typical wings stall in the range of $\alpha = 15 - 25$ degrees, and would have an α_b of 20 to 30 degrees. A blending rate of 20 to 100 usually produces reasonable results, with $\lambda_b = 20$ for a soft stall and $\lambda_b = 100$ for an abrupt stall. Figure 3.6.3 shows example results for this sigmoid function with values of $\alpha_{b0} = 10$ deg, $\alpha_b = 30$ deg and varying values of the blending rate λ_b.

EXAMPLE 3.6.4 Consider a wing segment with a semispan of 5 ft, root chord of 1 ft, tip chord of 0.7 ft, quarter-chord sweep angle of 10 deg, and dihedral angle of 30 deg. The wing segment origin is located relative to the vehicle center of gravity according to the vector $\mathbf{r}_o - \mathbf{r}_b = (0.0, 5.0, 0.5)$ in the vehicle coordinate system. The following aerodynamic properties of the wing are given

$$C_{L,\alpha} = 5.6, \quad \alpha_{L0} = -2.13 \text{ deg},$$
$$C_{D_{L0}} = 0.01, \quad C_{D,L} = -0.004, \quad e_O = 0.8,$$
$$C_{m_0} = -0.14, \quad C_{m,\alpha} = -2.2$$

Consider the case when the vehicle is flying at sea level with a velocity of 200 ft/s, an angle of attack of 15 deg, no sideslip angle, and no rotational velocity. Estimate the aerodynamic force and moment including effects of stall by using the approach given in Eqs. (3.6.59)–(3.6.65) with $\alpha_{b0} = 0$, $\alpha_b = 15$ deg, and $\lambda_b = 40.0$. Compute the aerodynamic forces and moments caused by the wing segment in both the wing coordinate system and the body-fixed coordinate system. Repeat the process for a left wing segment mirrored across the vehicle x-z plane (the left wing is located at $\mathbf{r}_o - \mathbf{r}_b = (0.0, -5.0, 0.5)$).

Solution: Using the given dihedral angle in Eq. (2.6.7) gives the rotation matrix

$$[\mathbf{R}] = \begin{bmatrix} 1 & 0 & 0 \\ 0 & 0.8660254037844 & 0.5 \\ 0 & -0.5 & 0.8660254037844 \end{bmatrix}$$

The location of the wing-segment aerodynamic center in the wing coordinate system can be computed from Eq. (3.6.35) to yield

$$\begin{bmatrix} x_{ac} \\ y_{ac} \\ z_{ac} \end{bmatrix} = \frac{b}{3}\frac{(c_r + 2c_t)}{(c_r + c_t)} \begin{bmatrix} -\tan\Lambda_{c/4} \\ \delta \\ 0 \end{bmatrix} = \begin{bmatrix} -0.4148870134317 \\ 2.352941176471 \\ 0 \end{bmatrix}$$

Using this along with $\mathbf{r}_o - \mathbf{r}_b = (0.0, 5.0, 0.5)$ in Eqs. (3.6.36) and Eq. (3.6.37) gives

$$\mathbf{P} = \begin{bmatrix} -0.4148870134317 \\ 7.037706832434 \\ -0.6764705882353 \end{bmatrix}$$

The components of the translational velocity vector of the vehicle in the body-fixed coordinate system can be computed from Eq. (3.4.12) to yield

$$\mathbf{V} = \begin{bmatrix} 193.1851652578 \\ 0 \\ 51.76380902050 \end{bmatrix} \text{ ft/s}$$

The vehicle rotational velocity vector is $\omega_b = [0, 0, 0]$. Using the translational and rotational velocity vectors along with the position vector $\mathbf{P}$ in Eq. (3.6.2) gives the component velocity vector in the wing coordinate system

$$\mathbf{V}_c = \begin{bmatrix} 1 & 0 & 0 \\ 0 & 0.8660 & 0.5 \\ 0 & -0.5 & 0.8660 \end{bmatrix}^{-1} \left(\begin{bmatrix} 193.1852 \\ 0 \\ 51.7638 \end{bmatrix} + \begin{bmatrix} 0 \\ 0 \\ 0 \end{bmatrix} \times \begin{bmatrix} -0.4149 \\ 7.0377 \\ -0.6765 \end{bmatrix} \right)$$
$$= \begin{bmatrix} 193.1851652578 \\ -25.88190451025 \\ 44.82877360840 \end{bmatrix} \text{ ft/s}$$

Our simplified aerodynamic model considers only the velocity in the wing x-z plane, as shown in Eq. (3.6.38). This yields

$$\mathbf{V}_w = \begin{bmatrix} V_{c_x} \\ 0 \\ V_{c_z} \end{bmatrix} = \begin{bmatrix} 193.1851652578 \\ 0 \\ 44.82877360840 \end{bmatrix}$$

which has a magnitude of

$$V_w = \|\mathbf{V}_w\| = \sqrt{V_{c_x}^2 + V_{c_z}^2} = 198.3182468128$$

The unit vector of the wing velocity vector can then be found from Eq. (3.6.39)

$$\mathbf{u}_w \equiv \frac{\mathbf{V}_w}{V_w} = \begin{bmatrix} 0.9741169476963 \\ 0 \\ 0.2260446243795 \end{bmatrix}$$

From Eqs. (3.6.30)–(3.6.32), the mean chord, area, and aspect ratio are

$$\overline{c} = \frac{c_r + c_t}{2} = 0.85 \text{ ft}$$

$$S_w = b\overline{c} = 4.25 \text{ ft}^2$$

$$R_A = \frac{b^2}{S_w} = 5.882352941176$$

The wing angle of attack can be computed from Eq. (3.6.49) to yield

$$\alpha = \tan^{-1}\left(\frac{u_{w_z}}{u_{w_x}}\right) = 0.228015283857 \text{ rad} = 13.06431342951 \text{ deg}$$

Using these results along with the aerodynamic parameters given in the problem statement in Eqs. (3.6.50)–(3.6.55) yield the following force coefficients below stall

$$\begin{aligned} C_{L_{ss}} &= 1.485068462780 \\ C_{D_{ss}} &= 0.1532367754179 \\ C_{m_{ss}} &= -0.6416336244863 \end{aligned}$$

These results must be blended with the results from the wing stall model. The lift, drag, and moment coefficients for the stall model can be computed from Eqs. (3.6.59)–(3.6.61) to yield

$$C_{L_N} = 2\text{sign}(\alpha)\sin^2\alpha\cos\alpha = 0.09954729462599$$

$$C_{D_N} = 2\sin^3(|\alpha|) = 0.02310003010926$$

$$C_{m_N} = -\frac{1}{2}\text{sign}(\alpha)\sin^2\alpha = -0.02554808610542$$

The blending value σ can be computed from Eq. (3.6.65) to yield

$$\sigma = \frac{1 + e^{-\lambda_b(\alpha-\alpha_{b0}-\alpha_b)} + e^{\lambda_b(\alpha-\alpha_{b0}+\alpha_b)}}{[1 + e^{-\lambda_b(\alpha-\alpha_{b0}-\alpha_b)}][1 + e^{\lambda_b(\alpha-\alpha_{b0}+\alpha_b)}]} = 0.2056474406619$$

Using these results in Eqs. (3.6.62)–(3.6.64) gives

$$C_L = (1-\sigma)C_{L_{ss}} + \sigma C_{L_N} = 1.200139580566$$

$$C_D = (1-\sigma)C_{D_{ss}} + \sigma C_{D_N} = 0.1264744868091$$

$$C_m = (1-\sigma)C_{m_{ss}} + \sigma C_{m_N} = -0.5149372102895$$

These coefficients can now be used in Eqs. (3.6.46)–(3.6.48) to yield

$$L = \frac{1}{2}\rho V_w^2 S_w C_L = 238.4105803447 \text{ lbf}$$

$$D = \frac{1}{2}\rho V_w^2 S_w C_D = 25.12445742748 \text{ lbf}$$

$$m = \frac{1}{2}\rho V_w^2 S_w \bar{c} C_m = -86.94947568105 \text{ ft} \cdot \text{lbf}$$

The unit vectors in the direction of lift, drag, and pitching moment are given in Eqs. (3.6.41)–(3.6.43) and yield

$$\mathbf{u}_L = \begin{bmatrix} u_{w_z} \\ 0 \\ -u_{w_x} \end{bmatrix} = \begin{bmatrix} 0.2260446243795 \\ 0 \\ -0.9741169476963 \end{bmatrix}$$

$$\mathbf{u}_D = \begin{bmatrix} -u_{w_x} \\ 0 \\ -u_{w_z} \end{bmatrix} = \begin{bmatrix} -0.9741169476963 \\ 0 \\ -0.2260446243795 \end{bmatrix}$$

$$\mathbf{u}_m = \begin{bmatrix} 0 \\ 1 \\ 0 \end{bmatrix}$$

The aerodynamic force and moment vectors in the wing coordinate system can then be computed from Eqs. (3.6.44) and (3.6.45) to yield

$$\mathbf{F}_c = D\mathbf{u}_D + L\mathbf{u}_L = \begin{bmatrix} 29.41727030033 \\ 0 \\ -237.9190353658 \end{bmatrix} \text{ lbf}$$

$$\mathbf{M}_c = m\mathbf{u}_m = \begin{bmatrix} 0 \\ -86.94947568105 \\ 0 \end{bmatrix} \text{ ft} \cdot \text{lbf}$$

Using these results in Eqs. (3.6.4) and (3.6.5) gives the aerodynamic force and moment vectors in the body-fixed coordinate system

$$\mathbf{F}_b = \begin{bmatrix} 1 & 0 & 0 \\ 0 & 0.8660 & 0.5 \\ 0 & -0.5 & 0.8660 \end{bmatrix} \begin{bmatrix} 29.417270 \\ 0 \\ -237.919035 \end{bmatrix} \text{ lbf} = \begin{bmatrix} 29.41727030033 \\ -118.9595176829 \\ -206.0439286707 \end{bmatrix} \text{ lbf}$$

$$\mathbf{M}_b = \begin{bmatrix} 1 & 0 & 0 \\ 0 & 0.8660 & 0.5 \\ 0 & -0.5 & 0.8660 \end{bmatrix} \begin{bmatrix} 0 \\ -86.949476 \\ 0 \end{bmatrix}$$

$$+ \begin{bmatrix} -0.414887 \\ 7.037707 \\ -0.676471 \end{bmatrix} \text{ft} \times \begin{bmatrix} 1 & 0 & 0 \\ 0 & 0.8660 & 0.5 \\ 0 & -0.5 & 0.8660 \end{bmatrix} \begin{bmatrix} 29.417270 \\ 0 \\ -237.919035 \end{bmatrix} \text{lbf}$$

$$= \begin{bmatrix} -1530.549379490 \\ -180.6853231318 \\ -114.2006273329 \end{bmatrix} \text{ft} \cdot \text{lbf}$$

Repeating the process for a left wing located at $\mathbf{r}_o - \mathbf{r}_b = (0.0, -5.0, 0.5)$ yields the identical force and moment vectors in the wing coordinate system

$$\mathbf{F}_c = \begin{bmatrix} 29.41727030033 \\ 0 \\ -237.9190353658 \end{bmatrix} \text{lbf}, \qquad \mathbf{M}_c = \begin{bmatrix} 0 \\ -86.94947568105 \\ 0 \end{bmatrix} \text{ft} \cdot \text{lbf}$$

and the following force and moment vectors in the body-fixed coordinate system relative to the vehicle CG

$$\mathbf{F}_b = \begin{bmatrix} 29.41727030033 \\ 118.9595176829 \\ -206.0439286707 \end{bmatrix} \text{lbf}, \qquad \mathbf{M}_b = \begin{bmatrix} 1530.549379490 \\ -180.6853231318 \\ 114.2006273329 \end{bmatrix} \text{ft} \cdot \text{lbf}$$

These detailed results can be used in the development of a computer code for testing and debugging purposes.

Summation of Component Aerodynamics

As stated in the introduction to this section, the aerodynamics of a complete vehicle can be approximated at very low order as the summation of the aerodynamics of the individual components by using Eqs. (3.6.6) and (3.6.7). Extreme caution should be exercised if this approach is taken to model the aerodynamics of a complete vehicle. The following example demonstrates how this can be done.

EXAMPLE 3.6.5 The Boeing Insitu RQ-21 Blackjack can be modeled as the summation of nine components as given in Appendix C.2 and shown in Fig. 2.6.2 with a CG location computed in Example 2.6.1. Consider the case when the vehicle is flying at sea level with a velocity of 100 ft/s, an angle of attack of 5 deg, no sideslip angle, and no angular velocity. Compute the total aerodynamic force and moment vectors in the body-fixed coordinate system by summing the influence of each component.

Solution: Using the methods outlined in this section, the individual forces and moments generated by each component can be computed. Results for these are given in the local component coordinate system in Tables 3.6.1 and 3.6.2. Results for components not included in Table 3.6.2 are exactly zero. Note that the results from the lifting surfaces were generated without any stall model included. The resulting forces and moments in the vehicle body-fixed coordinate system about the center of gravity are given in Tables 3.6.3 and 3.6.4. Summing all the aerodynamic forces and moments about the vehicle center of gravity gives the final aerodynamic force and moment vectors in the body-fixed coordinate system

$$\mathbf{F}_{b_{\text{total}}} = \begin{bmatrix} -10.86807588852 \\ 0.0 \\ -197.0658267795 \end{bmatrix} \text{ lbf}, \quad \mathbf{M}_{b_{\text{total}}} = \begin{bmatrix} 0.0 \\ -96.23836314238 \\ 0.0 \end{bmatrix} \text{ ft-lbf}$$

Table 3.6.1. Aerodynamic forces of the RQ-21 Blackjack components in the local component coordinate system.

Component	Side	F_x [lbf]	F_y [lbf]	F_z [lbf]
Main Wing	Right	10.88117	0.0	–91.72035
	Left	10.88117	0.0	–91.72035
Winglet	Right	–0.13218	0.0	–1.55895
	Left	–0.13218	0.0	–1.55895
Horiz. Stab.	Right	–0.19099	0.0	–5.26373
	Left	–0.19099	0.0	–5.26373
Vert. Stab.	Right	–0.16040	0.0	0.0
	Left	–0.16040	0.0	0.0
Boom	Right	–0.00054	0.0	–0.00026
	Left	–0.00054	0.0	–0.00026
Fuselage	N/A	–14.50095	0.0	–1.26867
Camera	N/A	–1.09054	0.0	–0.09541

Table 3.6.2. Aerodynamic moments of the RQ-21 Blackjack components in the local component coordinate system. Results for components not included are exactly zero.

Component	Side	M_x [ft-lbf]	M_y [ft-lbf]	M_z [ft-lbf]
Main Wing	Right	0.0	–34.64258	0.0
	Left	0.0	–34.64258	0.0
Winglet	Right	0.0	–0.41428	0.0
	Left	0.0	–0.41428	0.0

Table 3.6.3. Aerodynamic forces of the RQ-21 Blackjack components in the vehicle body-fixed coordinate system about the center of gravity.

Component	Side	F_x [lbf]	F_y [lbf]	F_z [lbf]
Main Wing	Right	2.84581	−0.80563	−92.31617
	Left	2.84581	0.80563	−92.31617
Winglet	Right	−0.13218	−1.53527	−0.27071
	Left	−0.13218	1.53527	−0.27071
Horiz. Stab.	Right	−0.19099	0.0	−5.26373
	Left	−0.19099	0.0	−5.26373
Vert. Stab.	Right	−0.16040	0.0	0.0
	Left	−0.16040	0.0	0.0
Boom	Right	−0.00054	0.0	−0.00026
	Left	−0.00054	0.0	−0.00026
Fuselage	N/A	−14.50095	0.0	−1.26867
Camera	N/A	−1.09054	0.0	−0.09541

Table 3.6.4. Aerodynamic moments of the RQ-21 Blackjack components in the vehicle body-fixed coordinate system about the center of gravity.

Component	Side	M_x [ft-lbf]	M_y [ft-lbf]	M_z [ft-lbf]
Main Wing	Right	−334.74094	−28.95107	−10.06634
	Left	334.74094	−28.95107	10.06634
Winglet	Right	−3.25019	−0.13922	2.37653
	Left	3.25019	−0.13922	−2.37653
Horiz. Stab.	Right	−3.94780	−18.05851	0.14324
	Left	3.94780	−18.05851	−0.14324
Vert. Stab.	Right	0.0	0.09495	0.24060
	Left	0.0	0.09495	−0.24060
Boom	Right	−0.00039	−0.00029	0.00081
	Left	0.00039	−0.00029	−0.00081
Fuselage	N/A	0.0	−2.16744	0.0
Camera	N/A	0.0	0.03735	0.0

3.7 Fixed-Wing Aircraft Below Stall

Perhaps the most versatile and powerful method for storing aerodynamic forces and moments of a vehicle as a function of the independent variables α, β, $\overline{p}$, $\overline{q}$, $\overline{r}$, $\hat{\alpha}$, $\hat{\beta}$, δ_a, δ_e, δ_r, $\mathrm{R_e}$, and M is through the use of datasets. Depending on how the datasets are structured and used, they can capture significant nonlinearities and interactions. Such an approach will be discussed in Section 3.11. However, very simple aerodynamic models that capture only the most significant aerodynamic interactions are often sufficient and employed for simulating fixed-wing aircraft. This section will discuss the development and application of a simple aerodynamic model, which can be used for typical fixed-wing aircraft.

In general, an aerodynamic model must provide solutions for the aerodynamic forces and moments as a function of the independent variables. This can be expressed as

$$\begin{aligned}
C_x &= f(\alpha,\beta,\overline{p},\overline{q},\overline{r},\hat{\alpha},\hat{\beta},\delta_a,\delta_e,\delta_r,\mathrm{R_e},\mathrm{M})\\
C_y &= f(\alpha,\beta,\overline{p},\overline{q},\overline{r},\hat{\alpha},\hat{\beta},\delta_a,\delta_e,\delta_r,\mathrm{R_e},\mathrm{M})\\
C_z &= f(\alpha,\beta,\overline{p},\overline{q},\overline{r},\hat{\alpha},\hat{\beta},\delta_a,\delta_e,\delta_r,\mathrm{R_e},\mathrm{M})\\
C_\ell &= f(\alpha,\beta,\overline{p},\overline{q},\overline{r},\hat{\alpha},\hat{\beta},\delta_a,\delta_e,\delta_r,\mathrm{R_e},\mathrm{M})\\
C_m &= f(\alpha,\beta,\overline{p},\overline{q},\overline{r},\hat{\alpha},\hat{\beta},\delta_a,\delta_e,\delta_r,\mathrm{R_e},\mathrm{M})\\
C_n &= f(\alpha,\beta,\overline{p},\overline{q},\overline{r},\hat{\alpha},\hat{\beta},\delta_a,\delta_e,\delta_r,\mathrm{R_e},\mathrm{M})
\end{aligned} \tag{3.7.1}$$

These forces and moments can be in the body-fixed, stability, or wind coordinate system. Once they are known in any one coordinate system, they can be easily transformed to the body-fixed coordinate system using the transformations given in Section 3.4.

For fixed-wing aircraft, the aerodynamic forces of lift and drag are typically developed and analyzed in the wind coordinate system. Therefore, it is common for fixed-wing aerodynamic models to express forces in the wind coordinate system. Hence, an aerodynamic model of an aircraft typically provides the lift, drag, and side force coefficients as a function of the independent variables, and can be expressed as

$$\begin{aligned}
C_L &= f(\alpha,\beta,\overline{p},\overline{q},\overline{r},\delta_a,\delta_e,\delta_r,\mathrm{R_e},\mathrm{M})\\
C_S &= f(\alpha,\beta,\overline{p},\overline{q},\overline{r},\delta_a,\delta_e,\delta_r,\mathrm{R_e},\mathrm{M})\\
C_D &= f(\alpha,\beta,\overline{p},\overline{q},\overline{r},\delta_a,\delta_e,\delta_r,\mathrm{R_e},\mathrm{M})
\end{aligned} \tag{3.7.2}$$

In contrast, it is typical to evaluate the moment coefficients in the body-fixed coordinate system. However, the model for the aerodynamic moments could be expressed in the body-fixed, stability, or wind coordinate system.

Fixed-wing aircraft typically operate at small angles of attack, sideslip angles, rotation rates, translational accelerations, and control-surface deflections. Furthermore, within this common operating region, most of the aerodynamic forces and moments are very nearly linear functions of the independent variables. Because of this, simplified aerodynamic models for fixed-wing aircraft below stall are sometimes developed by retaining only first-order terms in the Taylor-series expansion of the force functions given in Eq. (3.7.2) and the moment functions given in Eq. (3.7.1). This approach neglects all interactions of independent variables. However, some common or known interactions and nonlinearities can be added to the model, as will be demonstrated here.

At a given Reynolds number and Mach number, retaining only the linear terms in the Taylor-series expansion of Eq. (3.7.2) yields the following aerodynamic forces in the **wind coordinate system**

$$\begin{aligned} C_L = C_{L_0} &+ C_{L,\alpha}\alpha + C_{L,\beta}\beta + C_{L,\overline{p}}\overline{p} + C_{L,\overline{q}}\overline{q} + C_{L,\overline{r}}\overline{r} \\ &+ C_{L,\hat{\alpha}}\hat{\alpha} + C_{L,\hat{\beta}}\hat{\beta} + C_{L,\delta_a}\delta_a + C_{L,\delta_e}\delta_e + C_{L,\delta_r}\delta_r \end{aligned} \tag{3.7.3}$$

$$\begin{aligned} C_S = C_{S_0} &+ C_{S,\alpha}\alpha + C_{S,\beta}\beta + C_{S,\overline{p}}\overline{p} + C_{S,\overline{q}}\overline{q} + C_{S,\overline{r}}\overline{r} \\ &+ C_{S,\hat{\alpha}}\hat{\alpha} + C_{S,\hat{\beta}}\hat{\beta} + C_{S,\delta_a}\delta_a + C_{S,\delta_e}\delta_e + C_{S,\delta_r}\delta_r \end{aligned} \tag{3.7.4}$$

$$\begin{aligned} C_D = C_{D_0} &+ C_{D,\alpha}\alpha + C_{D,\beta}\beta + C_{D,\overline{p}}\overline{p} + C_{D,\overline{q}}\overline{q} + C_{D,\overline{r}}\overline{r} \\ &+ C_{D,\hat{\alpha}}\hat{\alpha} + C_{D,\hat{\beta}}\hat{\beta} + C_{D,\delta_a}\delta_a + C_{D,\delta_e}\delta_e + C_{D,\delta_r}\delta_r \end{aligned} \tag{3.7.5}$$

where C_{L_0}, C_{S_0}, and C_{D_0} are the lift, side, and drag coefficients at values of zero for angle of attack, sideslip angle, angular rates, translational acceleration, and control surface deflections. Similarly, at a given Reynolds number and Mach number, retaining only the linear terms in the Taylor-series expansion of the moment coefficients in Eq. (3.7.1) yields the following aerodynamic moments in the **body-fixed coordinate system**

$$\begin{aligned} C_\ell = C_{\ell_0} &+ C_{\ell,\alpha}\alpha + C_{\ell,\beta}\beta + C_{\ell,\overline{p}}\overline{p} + C_{\ell,\overline{q}}\overline{q} + C_{\ell,\overline{r}}\overline{r} \\ &+ C_{\ell,\hat{\alpha}}\hat{\alpha} + C_{\ell,\hat{\beta}}\hat{\beta} + C_{\ell,\delta_a}\delta_a + C_{\ell,\delta_e}\delta_e + C_{\ell,\delta_r}\delta_r \end{aligned} \tag{3.7.6}$$

$$\begin{aligned} C_m = C_{m_0} &+ C_{m,\alpha}\alpha + C_{m,\beta}\beta + C_{m,\overline{p}}\overline{p} + C_{m,\overline{q}}\overline{q} + C_{m,\overline{r}}\overline{r} \\ &+ C_{m,\hat{\alpha}}\hat{\alpha} + C_{m,\hat{\beta}}\hat{\beta} + C_{m,\delta_a}\delta_a + C_{m,\delta_e}\delta_e + C_{m,\delta_r}\delta_r \end{aligned} \tag{3.7.7}$$

$$\begin{aligned} C_n = C_{n_0} &+ C_{n,\alpha}\alpha + C_{n,\beta}\beta + C_{n,\overline{p}}\overline{p} + C_{n,\overline{q}}\overline{q} + C_{n,\overline{r}}\overline{r} \\ &+ C_{n,\hat{\alpha}}\hat{\alpha} + C_{n,\hat{\beta}}\hat{\beta} + C_{n,\delta_a}\delta_a + C_{n,\delta_e}\delta_e + C_{n,\delta_r}\delta_r \end{aligned} \tag{3.7.8}$$

where C_{ℓ_0}, C_{m_0}, and C_{n_0} are the rolling-, pitching-, and yawing-moment coefficients at values of zero for angle of attack, sideslip angle, angular rates, translational acceleration, and control surface deflections.

These equations can be further simplified for many common aircraft designs. For example, most aircraft are very nearly symmetric about the x-z plane. For the case of a **perfectly symmetric aircraft**, the following terms are zero

$$C_{S_0} = C_{\ell_0} = C_{n_0} = 0 \tag{3.7.9}$$

Additionally, for a perfectly symmetric aircraft, the change in lateral aerodynamic forces and moments with respect to the angle of attack and longitudinal accelerations, rotation rates, and elevator deflection are zero

$$\begin{aligned} C_{S,\alpha} &= C_{S,\overline{q}} = C_{S,\hat{\alpha}} = C_{S,\delta_e} = 0 \\ C_{\ell,\alpha} &= C_{\ell,\overline{q}} = C_{\ell,\hat{\alpha}} = C_{\ell,\delta_e} = 0 \\ C_{n,\alpha} &= C_{n,\overline{q}} = C_{n,\hat{\alpha}} = C_{n,\delta_e} = 0 \end{aligned} \tag{3.7.10}$$

Finally, the change in longitudinal aerodynamic forces and moments with respect to the sideslip angle and lateral accelerations, rotation rates, and control surfaces are also zero for a perfectly symmetric aircraft, which yields

$$\begin{aligned} C_{L,\beta} &= C_{L,\overline{p}} = C_{L,\overline{r}} = C_{L,\hat{\beta}} = C_{L,\delta_a} = C_{L,\delta_r} = 0 \\ C_{D,\beta} &= C_{D,\overline{p}} = C_{D,\overline{r}} = C_{D,\hat{\beta}} = C_{D,\delta_a} = C_{D,\delta_r} = 0 \\ C_{m,\beta} &= C_{m,\overline{p}} = C_{m,\overline{r}} = C_{m,\hat{\beta}} = C_{m,\delta_a} = C_{m,\delta_r} = 0 \end{aligned} \tag{3.7.11}$$

The set of terms given in Eq. (3.7.11) are zero due to the fact that changes in the lateral independent variables will have an identical effect on the longitudinal forces and moments whether the independent variables are positive or negative. For example, the change in drag with respect to sideslip angle may be nonzero. However, for a symmetric aircraft, a positive sideslip angle will produce the same change in drag as a negative sideslip angle. Because we are using a linear model relative to the case of no sideslip, the first derivative of the drag with respect to sideslip angle is zero, i.e. $C_{D,\beta} = 0$.

The changes in lateral forces and moments with respect to lateral accelerations $\hat{\beta}$ are typically much less significant for the aerodynamics of a fixed-wing aircraft than the changes in the longitudinal forces and moments with respect to longitudinal acceleration $\hat{\alpha}$. Therefore, it is not uncommon to neglect the following acceleration terms

$$C_{S,\hat{\beta}} \approx C_{\ell,\hat{\beta}} \approx C_{n,\hat{\beta}} \approx 0 \tag{3.7.12}$$

Applying Eqs. (3.7.9)–(3.7.12) to Eqs. (3.7.3)–(3.7.8) gives

$$C_L = C_{L_0} + C_{L,\alpha}\alpha + C_{L,\overline{q}}\overline{q} + C_{L,\hat{\alpha}}\hat{\alpha} + C_{L,\delta_e}\delta_e \tag{3.7.13}$$

$$C_S = C_{S,\beta}\beta + C_{S,\overline{p}}\overline{p} + C_{S,\overline{r}}\overline{r} + C_{S,\delta_a}\delta_a + C_{S,\delta_r}\delta_r \tag{3.7.14}$$

$$C_D = C_{D_0} + C_{D,\alpha}\alpha + C_{D,\overline{q}}\overline{q} + C_{D,\hat{\alpha}}\hat{\alpha} + C_{D,\delta_e}\delta_e \tag{3.7.15}$$

$$C_\ell = C_{\ell,\beta}\beta + C_{\ell,\overline{p}}\overline{p} + C_{\ell,\overline{r}}\overline{r} + C_{\ell,\delta_a}\delta_a + C_{\ell,\delta_r}\delta_r \tag{3.7.16}$$

$$C_m = C_{m_0} + C_{m,\alpha}\alpha + C_{m,\overline{q}}\overline{q} + C_{m,\hat{\alpha}}\hat{\alpha} + C_{m,\delta_e}\delta_e \tag{3.7.17}$$

$$C_n = C_{n,\beta}\beta + C_{n,\overline{p}}\overline{p} + C_{n,\overline{r}}\overline{r} + C_{n,\delta_a}\delta_a + C_{n,\delta_r}\delta_r \tag{3.7.18}$$

The aerodynamic model given in Eqs. (3.7.13)–(3.7.18) is accurate only for very small angles of attack, sideslip angles, and control-surface deflections. To improve this model and capture relevant aerodynamics over a larger range of angles of attack below stall, some nonlinear relationships and coupling should be included. Many of these additional terms come from our understanding of aerodynamic theory.

From lifting-line theory and a host of computational and experimental results, it is well understood that drag below stall can be approximated as a quadratic function of the lift coefficient. Likewise, because the side force on a traditional aircraft is dominated by the lateral force on the vertical stabilizer, the effects of side force on drag can be approximated using a quadratic. Using these quadratic approximations gives

$$C_D \approx C_{D_{L0}} + C_{D,L}C_L + C_{D,L^2}C_L^2 + C_{D,S^2}C_S^2 \tag{3.7.19}$$

where $C_{D_{L0}}$ is the drag at zero lift and zero side force. Notice that we have not included the linear term $C_{D,S}C_S$ due to the assumption that the aircraft is symmetric about the x-z plane. Equation (3.7.19) can be used to develop a drag model superior to that given in Eq. (3.7.15) and account for the changes in drag that can be expected due to angle of attack, pitch rate, normal acceleration, and elevator deflection. To assist in this analysis, we define a *pseudo lift coefficient* as

$$C_{L_1} \equiv C_{L_0} + C_{L,\alpha}\alpha \tag{3.7.20}$$

Applying Eq. (3.7.20) to the relationship for lift coefficient given in Eqs. (3.7.13) and using the result in Eq. (3.7.19) gives

$$\begin{aligned} C_D \approx C_{D_{L0}} &+ C_{D,S^2}C_S^2 \\ &+ C_{D,L}(C_{L_1} + C_{L,\bar{q}}\bar{q} + C_{L,\hat{\alpha}}\hat{\alpha} + C_{L,\delta_e}\delta_e) \\ &+ C_{D,L^2}[C_{L_1}^2 + (C_{L,\bar{q}}\bar{q})^2 + (C_{L,\hat{\alpha}}\hat{\alpha})^2 + (C_{L,\delta_e}\delta_e)^2] \\ &+ 2C_{D,L^2}(C_{L_1}C_{L,\bar{q}}\bar{q} + C_{L_1}C_{L,\hat{\alpha}}\hat{\alpha} + C_{L_1}C_{L,\delta_e}\delta_e) \\ &+ 2C_{D,L^2}(C_{L,\bar{q}}C_{L,\hat{\alpha}}\bar{q}\hat{\alpha} + C_{L,\bar{q}}C_{L,\delta_e}\bar{q}\delta_e + C_{L,\hat{\alpha}}C_{L,\delta_e}\hat{\alpha}\delta_e) \end{aligned} \tag{3.7.21}$$

Since $C_{L,\bar{q}}$, $C_{L,\hat{\alpha}}$, C_{L,δ_e}, $C_{D,L}$, and C_{D,L^2} are constants, combinations of these constants can be renamed. Additionally, we will drop the interaction terms $\bar{q}\hat{\alpha}$, $\bar{q}\delta_e$, and $\hat{\alpha}\delta_e$ and assume the nonlinear terms $\bar{q}^2$ and $\hat{\alpha}^2$ are small and can be neglected in comparison to other terms. Finally, as a first approximation, Phillips [38] suggests that the dependence of drag on $\hat{\alpha}$ can be neglected. Dropping these terms and combining constants gives

$$\begin{aligned} C_D \approx C_{D_{L0}} &+ C_{D,L}C_{L_1} + C_{D,L^2}C_{L_1}^2 + C_{D,S^2}C_S^2 \\ &+ (C_{D,\bar{q}} + C_{D,L\bar{q}}C_{L_1})\bar{q} + (C_{D,\delta_e} + C_{D,L\delta_e}C_{L_1})\delta_e + C_{D,\delta_e^2}\delta_e^2 \end{aligned} \tag{3.7.22}$$

As a final step, it is convenient to replace the C_{L_1} terms outside of the typical drag polar with the expression given in Eq. (3.7.20). Doing so and combining terms gives

$$\begin{aligned} C_D \approx C_{D_{L0}} &+ C_{D,L}C_{L_1} + C_{D,L^2}C_{L_1}^2 + C_{D,S^2}C_S^2 \\ &+ (C_{D,\bar{q}} + C_{D,\alpha\bar{q}}\alpha)\bar{q} + (C_{D,\delta_e} + C_{D,\alpha\delta_e}\alpha)\delta_e + C_{D,\delta_e^2}\delta_e^2 \end{aligned} \tag{3.7.23}$$

This drag model can be significantly more accurate than that given in Eq. (3.7.15). Note that a similar process could be used to further dissect the dependence of drag on the side force coefficients, by using Eq. (3.7.14) in Eq. (3.7.19). However, this level of scrutiny in the drag model is likely not warranted for most conventional aircraft designs.

From lifting-line theory it can be shown that the effects of rolling rate and aileron deflection on the yawing moment can each be approximated as a linear function of angle of attack [29]. Hence, the influence of rolling rate on yawing moment can be approximated as $(C_{n,\overline{p}} + C_{n,\alpha\overline{p}}\alpha)\overline{p}$, and the influence of aileron deflection on yawing moment can be approximated as $(C_{n,\delta_a} + C_{n,\alpha\delta_a}\alpha)\delta_a$. Additionally, an analytic approximation by Phillips [38] shows that the change in rolling moment with respect to yawing rate depends in a linear fashion on the angle of attack. This can be approximated as $(C_{\ell,\overline{r}} + C_{\ell,\alpha\overline{r}}\alpha)\overline{r}$. Finally, the change in side force with respect to rolling rate can also be a function of lift coefficient and can be modeled in the form $(C_{S,\overline{p}} + C_{S,\alpha\overline{p}}\alpha)\overline{p}$. Applying these approximations along with the result given in Eq. (3.7.22) to Eqs. (3.7.13)–(3.7.18) gives the simplified aerodynamic model below stall

$$C_{L_1} = C_{L_0} + C_{L,\alpha}\alpha \tag{3.7.24}$$

$$C_L = C_{L_1} + C_{L,\overline{q}}\overline{q} + C_{L,\hat{\alpha}}\hat{\alpha} + C_{L,\delta_e}\delta_e \tag{3.7.25}$$

$$C_S = C_{S,\beta}\beta + (C_{S,\overline{p}} + C_{S,\alpha\overline{p}}\alpha)\overline{p} + C_{S,\overline{r}}\overline{r} + C_{S,\delta_a}\delta_a + C_{S,\delta_r}\delta_r \tag{3.7.26}$$

$$\begin{aligned} C_D = {} & C_{D_{L0}} + C_{D,L}C_{L_1} + C_{D,L^2}C_{L_1}^2 + C_{D,S^2}C_S^2 \\ & + (C_{D,\overline{q}} + C_{D,\alpha\overline{q}}\alpha)\overline{q} + (C_{D,\delta_e} + C_{D,\alpha\delta_e}\alpha)\delta_e + C_{D,\delta_e^2}\delta_e^2 \end{aligned} \tag{3.7.27}$$

$$C_\ell = C_{\ell,\beta}\beta + C_{\ell,\overline{p}}\overline{p} + (C_{\ell,\overline{r}} + C_{\ell,\alpha\overline{r}}\alpha)\overline{r} + C_{\ell,\delta_a}\delta_a + C_{\ell,\delta_r}\delta_r \tag{3.7.28}$$

$$C_m = C_{m_0} + C_{m,\alpha}\alpha + C_{m,\overline{q}}\overline{q} + C_{m,\hat{\alpha}}\hat{\alpha} + C_{m,\delta_e}\delta_e \tag{3.7.29}$$

$$\begin{aligned} C_n = {} & C_{n,\beta}\beta + (C_{n,\overline{p}} + C_{n,\alpha\overline{p}}\alpha)\overline{p} + C_{n,\overline{r}}\overline{r} \\ & + (C_{n,\delta_a} + C_{n,\alpha\delta_a}\alpha)\delta_a + C_{n,\delta_r}\delta_r \end{aligned} \tag{3.7.30}$$

Equations (3.7.24)–(3.7.30) comprise a reasonable aerodynamic model below stall for traditional fixed-wing aircraft. Note that the coefficients needed in Eqs. (3.7.24)–(3.7.27) are defined in the wind coordinate system, and those needed in Eqs. (3.7.28)–(3.7.30) are often defined in the body-fixed coordinate system.

Estimating Coefficients from Data

Many aerodynamic coefficients needed for Eqs. (3.7.24)–(3.7.30) can be obtained through relatively straight-forward processes from aerodynamic data generated from wind-tunnel testing or computational fluid dynamics. These include aerodynamic coefficients related to angle of attack, sideslip angle, and control-surface deflections. Typically, these equations can be fit to aerodynamic data using common curve-fitting techniques. This will be demonstrated here by example.

EXAMPLE 3.7.1 Using the tabulated data published by Nguyen et al. [39], evaluate appropriate coefficients required to model the lift, drag, and pitching-moment coefficients below stall according to the format given in Eqs. (3.7.24), (3.7.25), (3.7.27), and (3.7.29). A subset of the original data published by Nguyen et al. [39] is included in Appendix C.3, and is sufficient for estimating these parameters.

Solution: The data given in Table C.3.9 contains the C_x and C_z values as a function of angle of attack. These can be used in conjunction with Eqs. (3.4.27) and (3.4.28) to compute the lift and drag coefficients as a function of angle of attack. The results are shown in Table 3.7.1. Note that because the lift includes the effects of angle of attack, the values in the dataset for lift are the values for C_{L_1}.

The first two terms needed in the lift equation are C_{L_0} and $C_{L,\alpha}$. Plotting the computed lift coefficient as a function of angle of attack gives the results in Fig. 3.7.1. Because we are concerned only with developing a model below stall, we will select a subset of data for this analysis. From the figure, it appears that the data from $-15 \leq \alpha \leq 20$ deg is an appropriate range, since stall onset appears to take place outside of that range. In the figure, these selected sub-stall data points are filled in, where the remainder of the data points are shown as open circles. This same range of data will be used for all subsequent analysis for this model of the vehicle below stall. Fitting a line to the subset of data yields the coefficients

$$C_{L_0} \approx 0.0535$$
$$C_{L,\alpha} \approx 3.8459$$

Table 3.7.1. Longitudinal force and moment coefficients computed from Table C.3.9.

α [deg]	C_L	C_D	C_m
−20	−1.0167865	0.4693678	0.0127
−15	−0.9010104	0.3426750	−0.0755
−10	−0.6644347	0.2168727	−0.1025
−5	−0.2790749	0.1031154	−0.0744
0	0.0250000	0.0489000	−0.0598
5	0.3650282	0.0385610	−0.0498
10	0.7471146	0.0819806	−0.0437
15	1.1018549	0.1842595	−0.0407
20	1.3763653	0.3644220	−0.0342
25	1.5575987	0.5828811	−0.0507
30	1.8157790	0.8709785	−0.0459
35	1.8941935	1.1303943	−0.0605
40	1.8831121	1.3775195	−0.0835
45	1.7318459	1.5364016	−0.0923
50	1.5932543	1.6994783	−0.0826
55	1.3899924	1.7759012	−0.0738
60	1.2033331	1.8548341	−0.1414
70	0.8261895	1.9702470	−0.3216
80	0.4288437	1.9592982	−0.4678
90	0.0864000	2.1400000	−0.6184

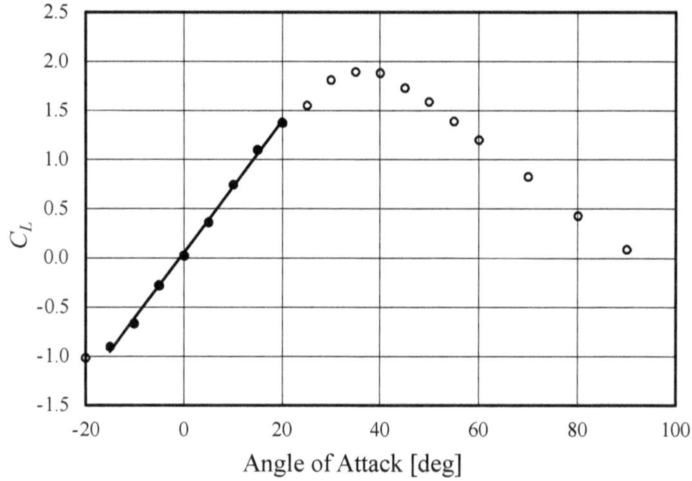

Figure 3.7.1. Lift coefficient computed from [39]. Filled markers represent data used for sub-stall model. Line represents sub-stall model.

Note that the lift slope is given per radian rather than per degree. Repeating the process using the pitching moment data in the same range of angle of attack, gives the first two terms in the pitching-moment equation

$$C_{m_0} \approx -0.0642$$
$$C_{m,\alpha} \approx 0.0955$$

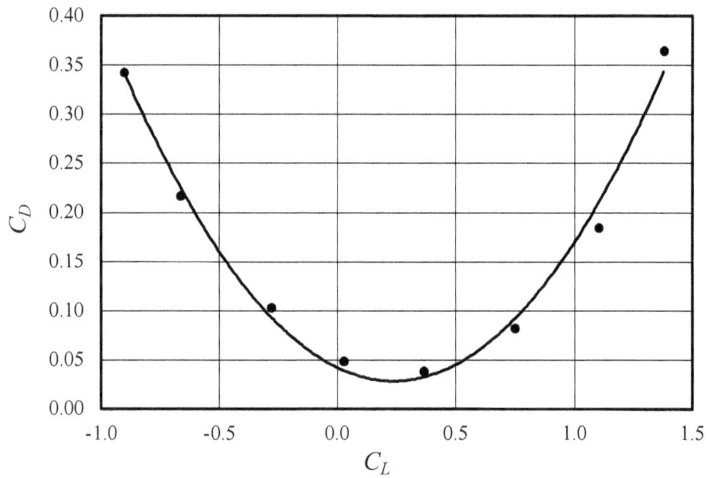

Figure 3.7.2. Drag coefficient as a function of lift coefficient below stall [39]. Curve represents model fit.

The first three terms in the drag equation are $C_{D_{L0}}$, $C_{D,L}$, and C_{D,L^2} and can also be estimated from this subset of data. Plotting the drag coefficient as a function of lift coefficient in the range $-15 \leq \alpha \leq 20$ gives the data in Fig. 3.7.2. Fitting a quadratic to the subset of data yields the coefficients

$$C_{D_{L0}} \approx 0.0425$$
$$C_{D,L} \approx -0.1154$$
$$C_{D,L^2} \approx 0.2427$$

Table C.3.11 contains $C_{x,\bar{q}}$, $C_{z,\bar{q}}$, and $C_{m,\bar{q}}$, as a function of angle of attack. Using Eqs. (3.4.27) and (3.4.28), this data can be used to compute the lift and drag coefficients as a function of angle of attack. The results for data in the range $-15 \leq \alpha \leq 20$ are given in Table 3.7.2 and shown in Fig. 3.7.3.

From Eqs. (3.7.25) and (3.7.29), the model below stall allows only a constant value to be chosen for $C_{L,\bar{q}}$ and $C_{m,\bar{q}}$. In this range, appropriate values for $C_{L,\bar{q}}$ and $C_{m,\bar{q}}$ are approximately

$$C_{L,\bar{q}} \approx 29.5$$
$$C_{m,\bar{q}} \approx -5.5$$

From Eq. (3.7.27), the dependence of drag on pitch rate should be expected to be linear with angle of attack in the region below stall. This is exactly what can be seen in the data in Fig. 3.7.3. Fitting a line through the data yields the coefficients

$$C_{D,\bar{q}} \approx -1.498$$
$$C_{D,\alpha\bar{q}} \approx 22.99$$

Derivatives with respect to the normal acceleration $\hat{\alpha}$ are not given in the dataset published by Nguyen et al. [39]. Therefore, we will neglect all dependence on $\hat{\alpha}$ and set $C_{L,\hat{\alpha}} = C_{m,\hat{\alpha}} = 0.0$.

Table 3.7.2. Longitudinal force and moment derivatives with respect to pitch rate, as computed from Table C.3.11.

α [deg]	$C_{L,\bar{q}}$	$C_{D,\bar{q}}$	$C_{m,\bar{q}}$
−15	22.8389727	−7.1063025	−6.8400
−10	23.3714186	−5.0887132	−6.8400
−5	29.2526522	−4.1151962	−3.4200
0	29.5000000	−1.9000000	−5.4800
5	30.5983414	0.2076112	−5.4500
10	31.3315353	2.5595493	−6.0200
15	29.9284702	4.6028980	−6.7000
20	26.9734612	6.8804063	−5.6900

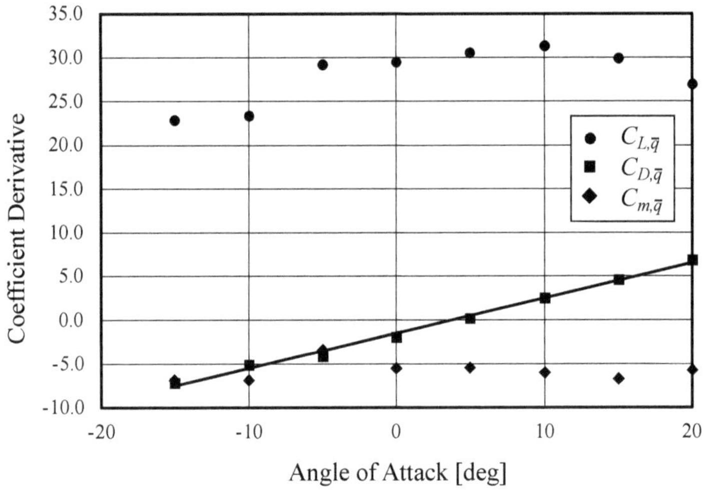

Figure 3.7.3. Longitudinal derivatives with respect to pitch rate as a function of angle of attack below stall [39]. Line represents model fit.

Table 3.7.3. Longitudinal force and moment changes with respect to elevator deflection, as computed from Table C.3.13.

δ_e [deg]	α [deg]	C_L	C_D	C_{L,δ_e}	C_{m,δ_e}	ΔC_D
−10	−15	−0.9922	0.3888	0.5223	−0.6205	0.0462
−10	−10	−0.7827	0.2545	0.6776	−0.6108	0.0376
−10	−5	−0.3469	0.1200	0.3884	−0.4698	0.0169
−10	0	−0.0640	0.0617	0.5099	−0.5890	0.0128
−10	5	0.2844	0.0421	0.4619	−0.5724	0.0036
−10	10	0.6471	0.0736	0.5733	−0.5672	−0.0084
−10	15	0.9732	0.1544	0.7372	−0.6377	−0.0298
−10	20	1.2725	0.3225	0.5954	−0.5821	−0.0420
10	−15	−0.7939	0.3173	0.6135	−0.5523	−0.0254
10	−10	−0.5680	0.2110	0.5526	−0.6463	−0.0058
10	−5	−0.1964	0.1077	0.4739	−0.6555	0.0046
10	0	0.1140	0.0606	0.5099	−0.5798	0.0117
10	5	0.4864	0.0626	0.6954	−0.6348	0.0241
10	10	0.8415	0.1166	0.5410	−0.6366	0.0346
10	15	1.1584	0.2246	0.3237	−0.5987	0.0403
10	20	1.3882	0.4019	0.0681	−0.5283	0.0375

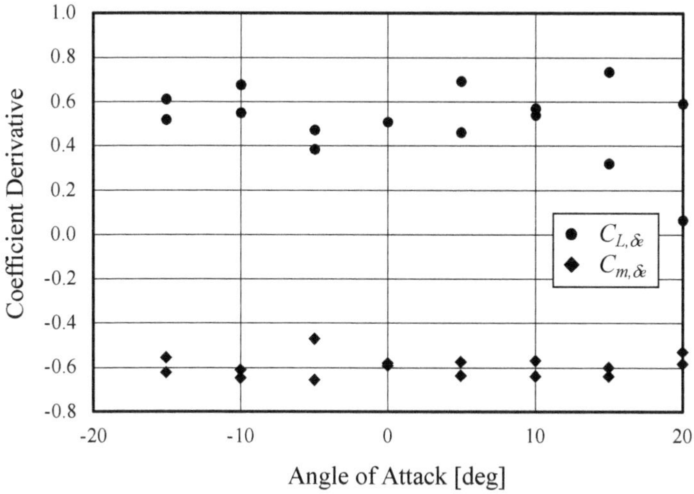

Figure 3.7.4. Longitudinal derivatives with respect to elevator deflection as a function of angle of attack below stall [39].

Let us next consider changes with respect to elevator deflection. The data given in Table C.3.13 can be used to compute the results given in Table 3.7.3, by using finite differencing relative to the value at zero deflection. These results are shown in Fig. 3.7.4. Note that derivatives are given per radian rather than per degree. From the figure it can be seen that C_{L,δ_e} and C_{m,δ_e} are nearly constant and can be approximated as the mean values within the dataset, which yields

$$C_{L,\delta_e} \approx 0.5153$$
$$C_{m,\delta_e} \approx -0.5926$$

Equation (3.7.27) gives the following relationship for the change in drag due to elevator deflection

$$(\Delta C_D)_{\delta_e} = (C_{D,\delta_e} + C_{D,\alpha\delta_e}\alpha)\delta_e + C_{D,\delta_e^2}\delta_e^2$$

The coefficients values for this model that best fit the data given in Table 3.7.3 can be found using a simple solver routine. For example, in Excel^{TM}, best-fit values for these coefficients can be obtained by creating three columns. The first column contains the ΔC_D values from Table 3.7.3. The second column is a model column, with values computed from the $(\Delta C_D)_{\delta_e}$ equation given here. This second column references cells that will hold the values of the three coefficients $C_{D,\alpha\delta_e}$, C_{D,δ_e}, and C_{D,δ_e^2}, which have yet to be determined. These cells can be initialized with values of zero. The final column is an error column, containing the squared error between the values in the first two columns. Finally, a single cell must be created that contains the sum of the error column. Once this is set up, the Solver add-in can be used to minimize the value in the cell containing the sum of the error, by changing the values of the three coefficients. Running the solver produces the results

$$C_{D,\delta_e} = -0.0006$$
$$C_{D,\alpha\delta_e} = 0.7084$$
$$C_{D,\delta_e^2} = 0.3251$$

Figure 3.7.5 shows the original values from the data as well as the model fit. The term C_{D,δ_e^2} is necessary to allow a non-zero value of the model at $\alpha = 0$. Notice that the value for C_{D,δ_e} is very small compared to the other two coefficient values for the elevator deflection. Setting this value to zero has very little affect on the model accuracy for this particular dataset.

The final term required for the drag equation is C_{D,S^2}. This can be estimated from the data given in Table C.3.10. To avoid capturing stall on the vertical tail, we will use only the data in the range $-20 \leq \beta \leq 20$ deg. Using the relations given in Table 3.4.3 yields the data in the wind coordinate system shown in Table 3.7.4. Fitting a parabola to ΔC_D as a function of C_S yields the coefficient

$$C_{D,S^2} = 0.8733$$

Note that in order to obtain a perfectly symmetric parabola, the data was mirrored about $\beta = 0$, since the data is not perfectly symmetric. Figure 3.7.6 shows the results of this fit, along with the original data, including the mirrored data.

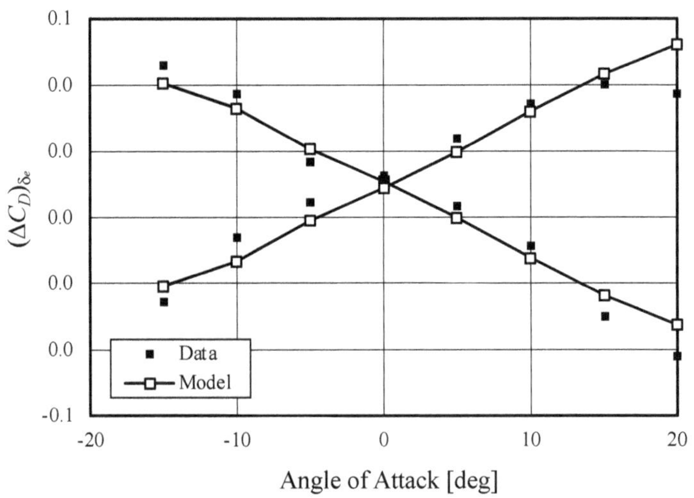

Figure 3.7.5. Change in drag due to elevator deflection as a function of angle of attack below stall [39]. Lines represent model fit.

Table 3.7.4. Drag and side force with respect to sideslip angle, as computed from Table C.3.10.

β [deg]	C_S	C_D	ΔC_D
−20	0.3816	0.1819	0.1330
−15	0.2875	0.1201	0.0712
−10	0.1905	0.0793	0.0304
−8	0.1474	0.0673	0.0184
−6	0.1083	0.0585	0.0096
−4	0.0691	0.0527	0.0038
−2	0.0354	0.0498	0.0009
0	0.0000	0.0489	0.0000
2	−0.0377	0.0496	0.0007
4	−0.0729	0.0527	0.0038
6	−0.1136	0.0589	0.0100
8	−0.1595	0.0678	0.0189
10	−0.2026	0.0797	0.0308
15	−0.2986	0.1213	0.0724
20	−0.3922	0.1839	0.1350

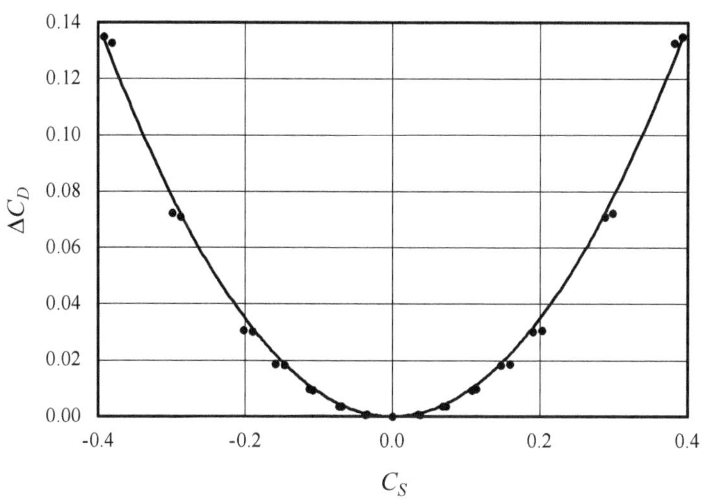

Figure 3.7.6. Change in drag due to side force below stall [39].

The previous example demonstrates how coefficients within the model equations given in Eqs. (3.7.24)–(3.7.30) could be obtained from experimental or other tabulated aerodynamic data. The example considered only the longitudinal equations C_L, C_D, and C_m, but a similar approach can be used to obtain the coefficients required for the lateral equations C_S, C_ℓ, and C_n. This is left as an exercise for the student.

Minimum Required Dataset

At this point, it may be worth considering what the minimum required dataset must contain in order to estimate all the coefficients required in Eqs. (3.7.24)–(3.7.30). For this discussion we will assume that all forces and moments can be measured and reported for any given case, and that second-order central differencing is used to evaluate derivatives. First, a base case is needed, with all independent variables set to zero. This provides data for computing C_{L_0} and C_{m_0}, as well as providing a data point for evaluating other coefficients. Next, two cases for each of the aerodynamic angles α and β are needed in order to compute the derivatives $C_{L,\alpha}$, $C_{m,\alpha}$, $C_{\ell,\beta}$, and $C_{n,\beta}$. These data, along with the base data, can also be used to compute values for the coefficients $C_{D_{L0}}$, $C_{D,L}$, C_{D,L^2}, and C_{D,S^2}. Two cases are needed with varying $\hat{\alpha}$ to compute $C_{L,\hat{\alpha}}$ and $C_{m,\hat{\alpha}}$. In order to compute derivatives with respect to the rotation rates $\overline{p}$, $\overline{q}$, and $\overline{r}$, four cases are needed for each rotational rate. These four cases include positive and negative values of each rotation rate, in combination with positive and negative values of angle of attack. For example, the cases shown in Table 3.7.5 show the cases required to compute the derivatives with respect to pitch rate $C_{D,\alpha\overline{q}}$ and $C_{D,\overline{q}}$. These data are also more than sufficient to then compute the derivatives $C_{L,\overline{q}}$ and $C_{m,\overline{q}}$. A similar combination of four cases is required to compute the derivatives with respect to $\overline{p}$ ($C_{S,\alpha\overline{p}}$, $C_{S,\overline{p}}$, $C_{\ell,\overline{p}}$, $C_{n,\alpha\overline{p}}$, and $C_{n,\overline{p}}$), $\overline{r}$ ($C_{S,\overline{r}}$, $C_{\ell,\alpha\overline{r}}$, $C_{\ell,\overline{r}}$, and $C_{n,\overline{r}}$), as well as the control-surface deflections δ_a (C_{S,δ_a}, C_{ℓ,δ_a}, $C_{n,\alpha\delta_a}$, and C_{n,δ_a}) and δ_e (C_{L,δ_e}, $C_{D,\alpha\delta_e}$, C_{D,δ_e}, C_{D,δ_e^2}, and C_{m,δ_e}). Finally, two cases are needed to compute the derivatives with respect to δ_r (C_{S,δ_r}, C_{ℓ,δ_r}, and C_{n,δ_r}). Table 3.7.6 summarizes the total number of cases required to compute the derivatives required for this aerodynamic model using central differencing. This yields a total of 29 cases. Due to vehicle symmetry, a few of these cases can be omitted while retaining second-order central differencing. This can be done with careful bookkeeping and attention paid to how the independent variables affect the dependent variables.

Table 3.7.5. Minimum set of data required to compute the derivatives in Eqs. (3.7.24)–(3.7.30) with respect to pitch rate.

α	$\overline{q}$
+	+
+	−
−	+
−	−

Table 3.7.6. Minimum set of cases required to compute the derivatives in the aerodynamic model given in Eqs. (3.7.24)–(3.7.30).

Independent Variable	Number of Cases
Base Case	1
α	2
β	2
$\hat{\alpha}$	2
$\overline{p}$	4
$\overline{q}$	4
$\overline{r}$	4
δ_a	4
δ_e	4
δ_r	2
Total Cases	29

Analytic Approximations

If aerodynamic data is not readily available for the vehicle of interest, estimates for some of the parameters can be obtained from basic information about the vehicle geometry. Here we will present approximations published by Phillips [40]. **The approximations given in this section should be used with extreme caution, and only when more accurate aerodynamic information about the vehicle is not available.**

Phillips presents many aerodynamic approximations that can be used in the initial phase of aircraft design. The longitudinal approximations presented here are based on the simple wing and tail model without vertical offsets [41, 42]. Additionally, effects of the engine or propeller have not been accounted for here. Coefficients required for the lift model given in Eqs. (3.7.24) and (3.7.25) can be approximated from

$$C_{L_0} \approx C_{L_w,\alpha}(\alpha_{0_w} - \alpha_{L0_w}) + \frac{S_h}{S_w}\eta_h C_{L_h,\alpha}(\alpha_{0_h} - \alpha_{L0_h} - \varepsilon_{d0}) \tag{3.7.31}$$

$$C_{L,\alpha} \approx C_{L_w,\alpha} + \frac{S_h}{S_w}\eta_h C_{L_h,\alpha}(1 - \varepsilon_{d,\alpha}) \tag{3.7.32}$$

$$C_{L,\overline{q}} \approx 2\frac{l_w}{\overline{c}_w} C_{L_w,\alpha} + 2\frac{S_h l_h}{S_w \overline{c}_w}\eta_h C_{L_h,\alpha} \tag{3.7.33}$$

$$C_{L,\hat{\alpha}} \approx \eta_h \frac{4 S_h l_{wt}}{\pi b^2 \overline{c}} C_{L_w,\alpha} C_{L_h,\alpha} \tag{3.7.34}$$

$$C_{L,\delta_e} \approx \frac{S_h}{S_w}\eta_h C_{L_h,\alpha}\varepsilon_e \tag{3.7.35}$$

where

$$l_{wt} = \begin{cases} 1.1(x_{b\,\text{wingtip}} - x_{bh}), & x_{b\,\text{wingtip}} > x_{bh} \\ 0.0, & x_{bh} > x_{b\,\text{wingtip}} \end{cases} \tag{3.7.36}$$

and $x_{b\text{wingtip}}$ and x_{bh} are the x locations of the main wing tip and aerodynamic center of the horizontal tail respectively, both in the body-fixed coordinate system. The lift coefficient on the main wing and horizontal tail are

$$C_{L_w} = C_{L_w,\alpha}(\alpha + \alpha_{0_w} - \alpha_{L0_w}) \tag{3.7.37}$$

$$C_{L_h} = C_{L_h,\alpha}(\alpha + \alpha_{0_h} - \alpha_{L0_h} - \varepsilon_d + \varepsilon_e \delta_e) \tag{3.7.38}$$

If the spanwise lift distributions on the main wing and horizontal tail are nearly elliptic, and the wings are thin, the lift slope of these lifting surfaces can be approximated as

$$C_{L_w,\alpha} = \frac{2\pi}{1 + 2/R_{A_w}} \tag{3.7.39}$$

$$C_{L_h,\alpha} = \frac{2\pi}{1 + 2/R_{A_h}} \tag{3.7.40}$$

If the spanwise lift distribution on the main wing is elliptic, the downwash on a horizontal tail ε_d sufficiently far downstream from the main wing can be approximated as

$$\varepsilon_d = \varepsilon_{d0} + \varepsilon_{d,\alpha} \approx \frac{16}{\pi^3} \frac{C_{L_w}}{R_{A_w}} \tag{3.7.41}$$

where

$$\varepsilon_{d0} \approx \frac{16}{\pi^3} \frac{C_{L_w,\alpha}(\alpha_{0_w} - \alpha_{L0_w})}{R_{A_w}} \tag{3.7.42}$$

$$\varepsilon_{d,\alpha} \approx \frac{16}{\pi^3} \frac{C_{L_w,\alpha}}{R_{A_w}} \tag{3.7.43}$$

The elevator effectiveness can be approximated from thin-airfoil theory as [33]

$$\varepsilon_e \approx 0.8 \left(1 - \frac{\theta_f - \sin\theta_f}{\pi}\right) \tag{3.7.44}$$

where

$$\theta_f = \cos^{-1}\left(2\frac{c_f}{c} - 1\right) \tag{3.7.45}$$

and c_f/c is the flap-chord fraction of the control surface. This approximation assumes the elevator spans the entire length of the horizontal stabilizer, and can be reduced based on the elevator span.

Coefficients required for the pitching-moment model given in Eq. (3.7.29) can be approximated from

$$C_{m0} \approx C_{m_w} + \frac{S_h \bar{c}_h}{S_w \bar{c}_w} \eta_h C_{m0h} - \frac{l_w}{\bar{c}_w} C_{L_w,\alpha}(\alpha_{0_w} - \alpha_{L0_w}) - \frac{S_h l_h}{S_w \bar{c}_w} \eta_h C_{L_h,\alpha}(\alpha_{0_h} - \alpha_{L0_h} - \varepsilon_{d0}) \tag{3.7.46}$$

$$C_{m,\alpha} \approx -\frac{l_w}{\bar{c}_w} C_{L_w,\alpha} - \frac{S_h l_h}{S_w \bar{c}_w} \eta_h C_{L_h,\alpha}(1 - \varepsilon_{d,\alpha}) \tag{3.7.47}$$

$$C_{m,\bar{q}} \approx -2\frac{l_w^2}{\bar{c}_w^2} C_{L_w,\alpha} - 2\frac{S_h l_h^2}{S_w \bar{c}_w^2} \eta_h C_{L_h,\alpha} \tag{3.7.48}$$

$$C_{m,\hat{\alpha}} \approx \frac{x_{bh}}{\bar{c}_w} C_{L,\hat{\alpha}} \tag{3.7.49}$$

$$C_{m,\delta_e} \approx \frac{S_h \bar{c}_h}{S_w \bar{c}_w} \eta_h C_{m_h,\delta_e} - \frac{S_h l_h}{S_w \bar{c}_w} \eta_h C_{L_h,\alpha} \varepsilon_e \tag{3.7.50}$$

(3.7.51)

The influence of the fuselage or nacelles on the vehicle pitching moment can also be approximately accounted for as suggested by Finck [27, 43] as

$$(\Delta C_{m0})_f = (\Delta C_{m,\alpha})_f \alpha_{0_f} \tag{3.7.52}$$

$$(\Delta C_{m,\alpha})_f \approx -2\frac{S_f l_f}{S_w \bar{c}_w} \left[1 - 1.76 \left(\frac{d_f}{c_f}\right)^{3/2}\right] \tag{3.7.53}$$

where S_f is the maximum cross-sectional area of the fuselage, l_f is the distance aft of the center of gravity to the center of pressure of the fuselage, c_f is the fuselage chord length, and d_f is the equivalent diameter of a circle having the same area as the maximum cross-section of the fuselage.

Likewise, several of the lateral coefficients of fixed-wing aircraft can be approximated from a simplified model of the vehicle, as suggested by Phillips [38, 44], which includes effects of the main-wing dihedral angle, vertical tail, and fuselage. In the following approximations, the effects of main-wing sweep have been neglected, all lifting surfaces are assumed to be thin, and the horizontal stabilizer is assumed to have no dihedral.

The coefficients for the side-force model given in Eq. (3.7.26) can be approximated as

$$C_{S,\beta} \approx -\eta_v \frac{S_v}{S_w} C_{L_v,\alpha}(1 - \varepsilon_{s,\beta}) \tag{3.7.54}$$

$$C_{S,\overline{p}} \approx 0 \tag{3.7.55}$$

$$C_{S,\alpha\overline{p}} \approx 0 \tag{3.7.56}$$

$$C_{S,\overline{r}} \approx -\eta_v \frac{2S_v x_{b_v}}{S_w b_w} C_{L_v,\alpha}(1 - \varepsilon_{s,\beta}) \tag{3.7.57}$$

$$C_{S,\delta_r} \approx \eta_v \varepsilon_r \frac{S_v}{S_w} C_{L_v,\alpha} \tag{3.7.58}$$

Additionally, C_{S,δ_a} is typically small and can be neglected in an approximate model. Hence, $C_{S,\delta_a} \approx 0$.

Coefficients required for the rolling-moment model given in Eq. (3.7.28) can be approximated as

$$C_{\ell,\beta} \approx -\frac{2\sin\Gamma}{3\pi\cos^4\Gamma}\kappa_\Gamma \kappa_\ell C_{L_w,\alpha} - \eta_v \frac{S_v}{S_w}\frac{(h_v \pm 0.08b_h)}{b_w} C_{L_v,\alpha}(1 - \varepsilon_{s,\beta}) \tag{3.7.59}$$

$$C_{\ell,\overline{p}} \approx -\frac{\kappa_{\ell\overline{p}}}{8} C_{L_w,\alpha} \tag{3.7.60}$$

$$C_{\ell,\overline{r}} \approx \left(1 - \frac{3\kappa_{\ell\overline{p}}}{\pi R_{A_w}} C_{L_w,\alpha}\right) \frac{C_{L_w,\alpha}(\alpha_{0_w} - \alpha_{L0_w})}{4} + \eta_v \frac{2S_v x_{b_v} z_{b_v}}{S_w b_w^2} C_{L_v,\alpha}(1 - \varepsilon_{s,\beta}) \tag{3.7.61}$$

$$C_{\ell,\alpha\overline{r}} \approx \left(1 - \frac{3\kappa_{\ell\overline{p}}}{\pi R_{A_w}} C_{L_w,\alpha}\right) \frac{C_{L_w,\alpha}}{4} \tag{3.7.62}$$

$$C_{\ell,\delta_a} \approx -\frac{4\pi}{S_w b_w} \int_{y_1}^{y_2} y_b \tilde{\varepsilon}_a c_w \, dy_b \tag{3.7.63}$$

$$C_{\ell,\delta_r} \approx \eta_v \varepsilon_r C_{L_v,\alpha} \frac{S_v}{S_w} \frac{(h_v \pm 0.08b_h)}{b_w} \tag{3.7.64}$$

where Γ represents the main wing dihedral angle. For a main wing with an elliptic lift distribution,

$$\kappa_\ell \approx 1.0 \tag{3.7.65}$$

$$\kappa_{\ell\overline{p}} \approx \frac{1 + 2/R_{A_w}}{1 + 4/R_{A_w}} \tag{3.7.66}$$

Values for κ_Γ for elliptic wings range from $\kappa_\Gamma \approx 0.75$ for a wing of aspect ratio of 4 to $\kappa_\Gamma \approx 0.92$ for an aspect ratio of 20, with $\kappa_\Gamma \approx 0.84$ for an aspect ratio of 8. The ± sign in the expressions above should be positive if the horizontal stabilizer is mounted above the vertical stabilizer in a "T-tail" configuration, and negative if the horizontal stabilizer is mounted below the vertical stabilizer. To approximate the effects of a high-wing or low-wing configuration on the roll-stability derivative, 2–3 degrees of main-wing dihedral can be added to represent the stabilizing effect of a high-wing configuration, and 3–4 degrees of anhedral can be added to represent the destabilizing effects of a low-wing configuration. The sidewash gradient on the vertical tail $\varepsilon_{s,\beta}$ can be approximated as suggested by Phillips [45]. The rudder effectiveness ε_r can be computed in the same way the elevator effectiveness is, as given in Eq. (3.7.44). The approximation for the roll-control derivative C_{ℓ,δ_a} is based on strip theory, where y_1–y_2 represent the bounds of the ailerons in the vehicle y coordinate system. Phillips states that strip theory is a poor model, and suggests a more accurate model can be obtained from lifting-line theory [46].

The coefficients for the yawing-moment model given in Eq. (3.7.30) can be approximated as

$$C_{n,\beta} \approx \eta_v \frac{S_v l_v}{S_w b_w} C_{L_v,\alpha}(1 - \varepsilon_{s,\beta}) - (\Delta C_{m,\alpha})_f \frac{\bar{c}_w}{b_w} \tag{3.7.67}$$

$$C_{n,\bar{p}} \approx -\left(1 - \frac{3\kappa_{\ell\bar{p}}}{\pi R_{A_w}} C_{L_w,\alpha}\right) \frac{C_{L_w,\alpha}(\alpha_{0_w} - \alpha_{L0_w})}{8} \tag{3.7.68}$$

$$C_{n,\alpha\bar{p}} \approx -\left(1 - \frac{3\kappa_{\ell\bar{p}}}{\pi R_{A_w}} C_{L_w,\alpha}\right) \frac{C_{L_w,\alpha}}{8} \tag{3.7.69}$$

$$C_{n,\bar{r}} \approx -\eta_v \frac{2 S_v x_{b_v}^2}{S_w b_w^2} C_{L_v,\alpha}(1 - \varepsilon_{s,\beta}) \tag{3.7.70}$$

$$C_{n,\delta_a} \approx -\frac{3 C_{\ell,\delta_a}}{\pi R_{A_w}} C_{L_w,\alpha}(\alpha_{0_w} - \alpha_{L0_w}) \tag{3.7.71}$$

$$C_{n,\alpha\delta_a} \approx -\frac{3 C_{\ell,\delta_a}}{\pi R_{A_w}} C_{L_w,\alpha} \tag{3.7.72}$$

$$C_{n,\delta_r} \approx -\eta_v \frac{S_v l_v}{S_w b_w} \left(\varepsilon_r C_{L_v,\alpha} - \frac{\bar{c}_v}{l_v} C_{m_v,\delta_r}\right) \tag{3.7.73}$$

Again, the reader is reminded that the analytic approximations presented here are very rough approximations. **These approximations should be used with extreme caution, and only if better aerodynamic information about the vehicle is not available**.

3.8 Stall

Fixed-wing aircraft stall at high angles of attack. Therefore, any realistic simulator must have some method for approximating the aerodynamic characteristics of stall. The stall characteristics of each vehicle are different, and depend on the vehicle design. For example, the pitching moment of aircraft at high angles of attack can vary widely, depending on many vehicle characteristics such as main wing sweep, horizontal tail size and location, and vertical offset between the lifting surfaces and the center of gravity. Most aircraft are designed to produce a strong negative pitching moment when the aircraft stalls. This helps with stall recovery [47].

Because aircraft do not normally fly for extended periods of time in a stalled condition, the aerodynamics of stall must not be particularly accurate for most flight simulators.

Rather, some representation for the forces and moments in a stalled condition must be included in the aerodynamic model to produce a somewhat realistic simulation. If aerodynamic information at high angles of attack is available, it can be used directly in a dataset as discussed in Section 3.11. In the absence of stall information, a reasonable stall model can be developed using Newtonian approximations for the force and pitching moment on a flat plate.

At high angles of attack, the lift, drag, and pitching moment on a finite wing can be approximated from Newtonian theory, as discussed in Section 3.6. If the geometry of the lifting surfaces of the vehicle is known, along with an approximation for the downwash on the horizontal tail, the method outlined in Section 3.6 can be applied to each lifting surface individually and the results summed to predict the total longitudinal force and moment on the vehicle at high angles of attack. Such an approach can be used to capture the interactions between the main wing and the horizontal stabilizer at high angles of attack. This method of modeling the aerodynamics of each lifting surface individually requires some effort to set up and tune properly to produce accurate stalled characteristics.

As a simple first-order approximation, the lift, drag, and pitching moment of the vehicle at high angles of attack can be modeled from those of a single flat plate with the same area as the main wing, or using the entire planform area of the vehicle. The lift and drag coefficients on a flat plate at high incidence angles as predicted from Newtonian theory are given in Eqs. (3.6.59) and (3.6.60). For this first-order approximation, these can be used directly by using the angle of attack of the vehicle in place of the angle of attack of the flat plate. Use of the results for pitching moment given in Eq. (3.6.61) requires some discussion.

In the development of Eq. (3.6.61), the pitching moment was computed about the quarter-chord of the flat plate. This was done through the use of the quantity $(c/4)$ in Eq. (3.6.61), which resulted in a minimum pitching moment of $-1/2$ at 90 degrees. This was an appropriate approximation for a wing segment, since the moments were reported about the wing-segment aerodynamic center, and the aerodynamic center of a wing segment is typically near the quarter-chord of the wing centroidal chord. However, use of Eq. (3.6.61) directly on a fixed-wing aircraft is not appropriate for at least two reasons. First, the aerodynamic model of the full vehicle may be reported about a different location than the main wing aerodynamic center. For example, the F-16 data included in Appendix C.3 is reported about the 0.35 mean-aerodynamic-chord location, not the quarter-chord of the main wing centroidal chord. Second, in the development of Eq. (3.6.61), the center of pressure on a flat plate was assumed to lie at the half-chord of the plate. Most fixed-wing aircraft include a horizontal stabilizer. This aft lifting surface shifts the center of pressure and aerodynamic center of the vehicle aft from that of the main wing. Both of these assumptions, however, affect only the magnitude of Eq. (3.6.61). They do not change the function form. This yields the Newtonian relation

$$C_{m_N} = C_{m_{\min}} \operatorname{sign}(\alpha) \sin^2 \alpha \tag{3.8.1}$$

where $C_{m_{\min}}$ is the pitching moment at an angle of attack of 90 deg. If pitching-moment data is available above stall for a vehicle, the value of $C_{m_{\min}}$ can be selected empirically to fit the data. This will be demonstrated in the following example.

The sub-stall aerodynamic data or model, such as that given in Eqs. (3.7.25), (3.7.27), and (3.7.29), can be blended with the models from Newtonian theory to yield the final lift, drag, and pitching-moment coefficients over the entire range of angles of attack. The blending between the sub-stall and Newtonian models can be performed using the functions given in Eqs. (3.6.62)–(3.6.64). Recall that the σ value is a blending parameter that ranges from 0 to 1. It can be computed from the sigmoid function given in Eq. (3.6.65) with appropriate values for the parameters α_{b0}, α_b, and λ_b. Note that these parameters may vary for the lift, drag, and pitching-moment blending functions.

EXAMPLE 3.8.1 Using the aerodynamic model and coefficients from Example 3.7.1 for the F-16 below stall, along with the Newtonian model for a flat plate given in Eqs. (3.6.59), (3.6.60), and (3.8.1), blend these models using Eqs. (3.6.62)–(3.6.64) to create a model for the lift, drag, and pitching moment of the aircraft from −180 deg to +180 deg. Use the sigmoid function given in Eq. (3.6.65), and select appropriate values for α_{b0}, α_b, and λ_b that best match the data in Appendix C.3 for the F-16 over the range of data available. Note that different values for the parameters α_{b0}, α_b, and λ_b for the lift, drag, and pitching-moment models may be needed to adequately match the data.

Solution: The sub-stall model given in Eqs. (3.7.24) and (3.7.25), the Newtonian model given in Eq. (3.6.59), and lift-coefficient data computed from the data given in Appendix C.3 were included on a single plot as shown in Fig. 3.8.1. The sigmoid function given in Eq. (3.6.65) was used along with the blending function given in Eq. (3.6.62) to compute a blended model. Values for α_{b0}, α_b, and λ_b were manually altered until the blended model appeared to approximate the data as well as possible. The final values chosen for the lift model were

$$\alpha_{b0} = 0.0 \text{ deg}, \qquad \alpha_b = 43.0 \text{ deg}, \qquad \lambda_b = 7.0$$

Results are shown in Fig. 3.8.1. This process was repeated for the drag coefficient with final values of

$$\alpha_{b0} = 5.0 \text{ deg}, \qquad \alpha_b = 45.0 \text{ deg}, \qquad \lambda_b = 7.0$$

Drag results are shown in Fig. 3.8.2.

The pitching-moment Newtonian model given in Eq. (3.8.1) requires one additional parameter $C_{m_{\min}}$, which is the pitching moment on the vehicle at 90 deg angle of attack. From the pitching-moment data included in the appendix, this value is

$$C_{m_{\min}} = -0.6184$$

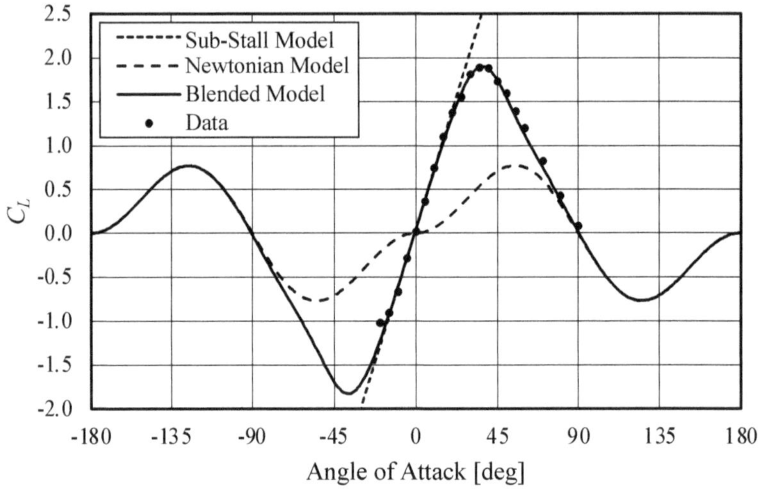

Figure 3.8.1. Models for the lift coefficient over a large range of angles of attack.

The sigmoid-function parameters were then selected to produce a blended model that matched the data reasonably well. The final values chosen were

$$\alpha_{b0} = 20.0 \text{ deg}, \qquad \alpha_b = 40.0 \text{ deg}, \qquad \lambda_b = 6.0$$

Pitching-moment results are shown in Fig. 3.8.3.

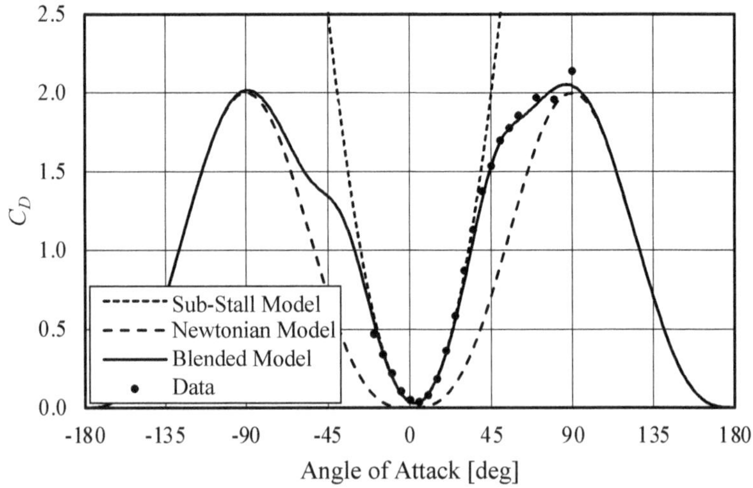

Figure 3.8.2. Models for the drag coefficient over a large range of angles of attack.

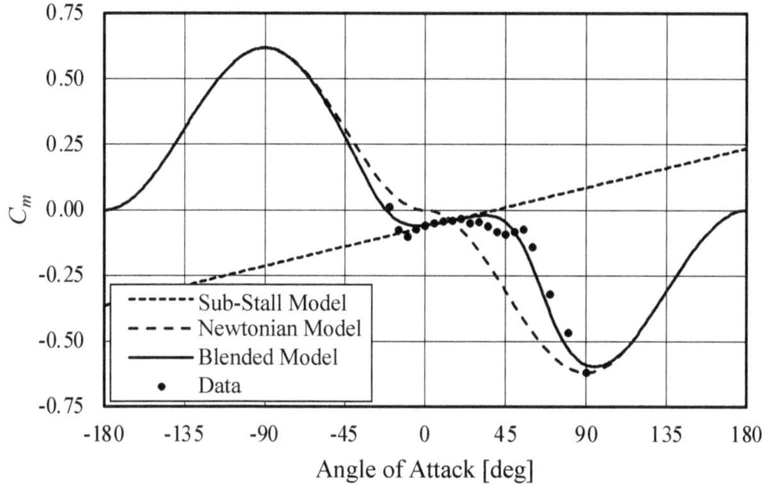

Figure 3.8.3. Models for the pitching-moment coefficient over a large range of angles of attack.

The blended models for lift and drag match the data very well in the range of available data. The blended model for the pitching-moment does not accurately capture the trends seen in the data in the range from $30.0 \le \alpha \le 60.0$ deg. These trends are likely caused by the complex aerodynamics of vortices being shed from the vehicle nose at high angles of attack, as well as the effect of these vortices and downwash on the horizontal tail. The degrees of freedom available in applying the Newtonian model only on the main wing, and using the sigmoid function given in Eq. (3.6.65) are not sufficient to capture these variations. However, the blended model for the pitching moment shown here is likely sufficient for most simulation frameworks that do not require high fidelity modeling of the vehicle at angles of attack above 30 degrees.

The stall model suggested here can be incorporated very easily into most simulators. Typically, once the aerodynamic forces and moments have been computed from a dataset or model, including the effects of rotational rates and control-surface deflections, the final lift, drag, and pitching-moment coefficients can be blended with the Newtonian approximations using Eqs. (3.6.62)–(3.6.64). This method naturally reduces the elevator effectiveness at high angles of attack. Other methods can be implemented to retain control-surface effectiveness at high angles of attack if appropriate. The approach demonstrated in this section represents a simple first-order approximation for the effects of stall. However, using the principles discussed in this section, more sophisticated models can be developed to predict the aerodynamic characteristics of the vehicle at high angles of attack.

If information about the lateral aerodynamics of the aircraft beyond stall are known, modifications to the lateral forces and moments can also be made. These are particularly useful for modeling the effectiveness of control surfaces on the moments at high angles of attack. However, because most fixed-wing aircraft are not designed to spend significant amounts of time in the post-stall region, highly accurate models for lateral forces and moments above stall are usually not as important as reasonable models for the longitudinal forces and moments above stall.

3.9 Compressibility

At Mach numbers below about 0.3, compressibility of the flow can be neglected. Hence, this range is typically called the *incompressible* regime. At Mach numbers greater than 0.3, the compressibility of the air impacts the aerodynamics of the vehicle. Hence, this range is typically called the *compressible* regime. As speeds near Mach 1, a portion of the flow is supersonic, while some of the flow is subsonic. This range is called the *transonic* regime. At speeds sufficiently greater than Mach 1, the flow is fully supersonic. This is called the *supersonic* regime.

Dimensionless aerodynamic coefficients for typical fixed-wing aircraft are nearly independent of velocity at high Reynolds numbers in the incompressible regime. However, in the compressible regime, aerodynamic coefficients of the vehicle vary with Mach number. At speeds below the transonic regime, pressure on the surface of the vehicle scales nearly with the Prandtl-Glauert compressibility correction given in Eq. (3.3.26). As a first approximation at sub-transonic speeds, the aerodynamic coefficients dominated by pressure can simply be scaled using the Prandtl-Glauert compressibility correction. Since the lift and side forces are dominated by the pressure, C_L and C_S can be approximated as scaling with this correction. Additionally, the aerodynamic moments C_ℓ, C_m, and C_n are each typically dominated by the pressure on the vehicle surface, and can each be approximated as scaling with this correction. This yields

$$C_L \approx \frac{C_{L\mathrm{M0}}}{\sqrt{1-\mathrm{M}^2}} \tag{3.9.1}$$

$$C_S \approx \frac{C_{S\mathrm{M0}}}{\sqrt{1-\mathrm{M}^2}} \tag{3.9.2}$$

$$C_\ell \approx \frac{C_{\ell\mathrm{M0}}}{\sqrt{1-\mathrm{M}^2}} \tag{3.9.3}$$

$$C_m \approx \frac{C_{m\mathrm{M0}}}{\sqrt{1-\mathrm{M}^2}} \tag{3.9.4}$$

$$C_n \approx \frac{C_{n\mathrm{M0}}}{\sqrt{1-\mathrm{M}^2}} \tag{3.9.5}$$

where the subscript M0 denotes the value at zero Mach number.

The drag coefficient at low Mach numbers is a function of both inviscid and viscous effects. As the Mach number increases, wave drag becomes increasingly important and the drag on the vehicle increases until supersonic speeds are reached. The dependence of drag on Mach number depends heavily on the airfoil cross sections used for the lifting surfaces, the sweep angle of the wings, and the shape of the vehicle fuselage. A first-order approximation can be developed by considering only the effect of wing sweep. A rough fit to data published by NASA for the drag on swept wings can be expressed as

$$C_D \approx C_{DM0}\left(1 + C_{M_1} M^{C_{M_2}}\right) \tag{3.9.6}$$

where

$$C_{M_1} = \frac{2.13}{(\Lambda + 0.15)^2} \tag{3.9.7}$$

and

$$C_{M_2} = 15.35\Lambda^2 - 19.64\Lambda + 16.86 \tag{3.9.8}$$

where Λ is the wing sweep in radians. Equation (3.9.6) does not predict a maximum drag coefficient in the transonic regime, and continues to increase as Mach number increases. Therefore, it should only be used at speeds below the transonic regime. More complex functions could be developed to model drag more accurately as a function of Mach number.

A significant amount of data showing how aerodynamic parameters vary with Mach number is available in the literature. For some examples, see [48–50]. Many corrections have been suggested or could be obtained by fitting published data. The compressibility corrections given in this section are simply examples of how compressibility can be modeled, and should be used with extreme caution. When possible, wind tunnel or CFD data specific to the vehicle should be used. However, in the absence of such data, the relations given in this section can be used to provide reasonable trends.

3.10 Ground Effect

When a wing comes in the presence of the ground, the downwash is reduced across the wing. This produces an increase in lift and a decrease in induced drag. The *lift influence ratio* is the ratio of the lift in ground effect to that of the lift out of ground effect, both computed at the same angle of attack

$$\text{lift influence ratio} = R_L \equiv \frac{[C_L(\alpha)]_g}{[C_L(\alpha)]_\infty} \tag{3.10.1}$$

Here we will use the symbol g to denote the values in ground effect with the wing at a height h above the ground. Since the induced-drag coefficient for a wing without twist is proportional to the lift coefficient squared, a common measure of the influence of the ground on induced drag is computed by the ratio of the induced drag divided by the square of the lift coefficient evaluated in ground effect to that of the same ratio out of ground effect. This is called the *induced-drag influence ratio* and is defined as

$$\text{induced-drag influence ratio} = R_D \equiv \frac{(C_{Di}/C_L^2)_g}{(C_{Di}/C_L^2)_\infty} \tag{3.10.2}$$

Historical Approximations

Many approximations have been developed for the induced-drag influence ratio R_D of a wing. Most of these models are strong functions of the ratio of the wing height above ground h to the wing span b. Here we will present four models that have historically been used, as well as a more accurate model that captures the effects of planform and aspect ratio. Example results for each of these models are shown in Fig. 3.10.1.

A model published by Hoerner [51] in 1975 can be written as

$$R_D = \frac{33(h/b)^{1.5}}{1 + 33(h/b)^{1.5}} \tag{3.10.3}$$

Raymer [52] also suggests the use of this model.

In the first edition of his textbook, McCormick [53] suggests the relation

$$R_D = \frac{(16h/b)^2}{1 + (16h/b)^2} \tag{3.10.4}$$

This model has appeared in other textbooks by Anderson [54, 55] and Phillips [56], however was most likely a typographical error, and significantly overpredicts R_D in the range $0.05 < h/b < 1.0$. This model should never be used.

In a later printing of his textbook, McCormick [53] revised the relation to

$$R_D = \frac{[16h/(\pi b)]^2}{1 + [16h/(\pi b)]^2} \tag{3.10.5}$$

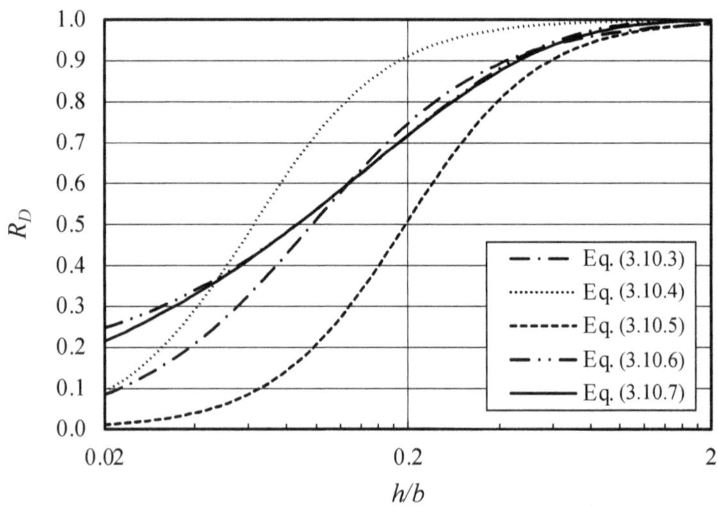

Figure 3.10.1. Example solutions for the induced-drag influence ratio computed from various models.

This revised relation significantly underpredicts R_D in the range $h/b < 1.0$.

In 1982, Torenbeek [57] published a relation for R_D that includes dependence on lift coefficient and aspect ratio in the form

$$R_D = \frac{1 - \exp[-2.48(2h/b)^{0.768}]}{1 - \kappa_t C_L/(4\pi R_A h/b)} \tag{3.10.6}$$

where $\kappa_t = \sqrt{1 + (2h/b)^2} - 2h/b$. The results from Eq. (3.10.6) shown in Fig. 3.10.1 were computed using $C_L = 0.5$ and $R_A = 8$. It should be noted that the relation suggested by Torenbeek gives unreasonable results at very low values of h/b. Divergence from the trends shown in Fig. 3.10.1 typically occur at values of h/b lower than those shown in the figure.

In 2013, Phillips and Hunsaker [58] suggested a simplified form of Torenbeek's relation that was fit to a large dataset of ground-effect solutions for wings of various planforms and aspect ratios generated from the numerical lifting-line algorithm [20]. This relation can be written as

$$R_D = 1 - \exp[-4.01(h/b)^{0.717}] \tag{3.10.7}$$

This is perhaps the simplest and most accurate model that can be applied in the early stages of design, when the planform and aspect ratio of the main wing have not yet been finalized, and is also included in Fig. 3.10.1.

Although several models for R_D exist, very few exist for R_L. Additionally, although many historical models neglect the dependence of wing aerodynamics in ground effect on wing planform, the lift and drag of a wing in ground effect are functions of aspect ratio and taper ratio. Hence, in later stages of design, wing planform should be included to produce a more accurate model of the wing aerodynamics in ground effect. A model published by Phillips and Hunsaker [58] will be presented here, since it considers both the lift and induced-drag influence ratios R_L and R_D, and includes the effects of planform and aspect ratio.

Phillips and Hunsaker Model

Phillips and Hunsaker [58] used a numerical lifting-line algorithm [20] to conduct a large computational study on ground effect for wings without twist over a range of planform, aspect ratio, and taper ratio. Closed-form relations were fit to the computational results, and can be used as an approximation for computing both lift and induced-drag influence ratios in the presence of the ground. The resulting lift and induced-drag influence ratios can be computed from

$$R_L = \frac{1 + 288\delta_L (h/b)^{0.787} \exp[-9.14(h/b)^{0.327}]/R_A^{0.882}}{\beta_L} \tag{3.10.8}$$

$$R_D = \{1 - \delta_D \exp[-4.74(h/b)^{0.814}] - (h/b)^2 \exp[-3.88(h/b)^{0.758}]\}\beta_D \tag{3.10.9}$$

where

$$\beta_L = 1 + \frac{0.269 C_{L_g}^{1.45}}{R_A^{3.18}(h/b)^{1.12}} \tag{3.10.10}$$

$$\beta_D = 1 + \frac{0.0361 C_{L_g}^{1.21}}{R_A^{1.19}(h/b)^{1.51}} \tag{3.10.11}$$

and C_{L_g} is the lift coefficient in ground effect. For an elliptic wing, $\delta_L = \delta_D = 1.0$. For linearly tapered wings,

$$\delta_L = 1 - 2.25(R_T^{0.00273} - 0.997)(R_A^{0.717} + 13.6) \tag{3.10.12}$$

$$\delta_D = 1 - 0.157(R_T^{0.775} - 0.373)(R_A^{0.417} - 1.27) \tag{3.10.13}$$

Note that **this model is only good for positive values of lift coefficient.** If C_{L_g} is negative, Eqs. (3.10.10) and (3.10.11) yield complex numbers, which are non-physical. Additionally, it is important to note that this model diverges at very low values for h/b.

Because C_{L_g} is the lift coefficient in ground effect, these relations are rather straightforward to apply for the case when the lift of the aircraft is supporting the weight of the aircraft in a trim condition near the ground. In this case, the lift coefficient is equal to the weight coefficient, and can be easily found from the weight, velocity, and altitude of the aircraft. However, these equations are not as straightforward to implement in a simulation environment, where the lift coefficient in ground effect is an unknown. In fact, the purpose of applying these relations to a simulation environment is to solve for the lift in ground effect as a function of the state of the aircraft and control inputs.

For the case of flight simulation, the aerodynamic model of the aircraft can be used to obtain the lift coefficient at a given angle of attack in the absence of ground effect C_L. This is then used in Eq. (3.10.8) to obtain the lift coefficient in ground effect C_{L_g}. However, since the lift coefficient in ground effect is needed in Eq. (3.10.10) to solve for β_L, solving for the lift coefficient in ground effect requires an iterative process. Using the lift in the absence of ground effect as the initial guess for C_{L_g}, Eqs. (3.10.10) and (3.10.8) can be solved. This gives a new estimate for the lift in ground effect, and the process can be repeated using this updated value for C_{L_g}. This iterative process converges very rapidly and typically reaches double-precision convergence within about 10 iterations.

It is often helpful to minimize computational time required to predict the aerodynamics of the aircraft in a flight simulator. In this case, one may choose not to perform any iterations to solve for the lift in ground effect, and rather use the lift out of ground effect as an approximation for C_{L_g} in Eq. (3.10.10). This method typically produces errors in R_L of less than 0.5%. Since most aerodynamic models are not accurate to within 1%, this method can be used to speed up computation time with little loss in total simulation accuracy. Once the lift in ground effect C_{L_g} has been obtained, the results can be applied to Eqs. (3.10.11) and (3.10.9) to obtain the change in induced-drag due to ground effect.

Application to an Aerodynamic Model

Once R_L and R_D have been obtained, approximations for ground effect can be added to the aerodynamic model given in Eqs. (3.7.24)–(3.7.30) by making a few assumptions. First we assume that the lift on the aircraft is dominated by the lift on the main wing, and that in ground effect, that lift is dominated by the angle of attack. Therefore, we will use the ratio R_L to alter C_{L_1} as

$$C_{L_g} = R_L C_{L_1} \tag{3.10.14}$$

The quantity C_{L_g} is then used in Eqs. (3.7.24)–(3.7.30) for every occurrence of the term C_{L_1}. Additionally, we will assume that the induced drag is the dominant effect in the term C_{D,L^2} which appears in Eq. (3.7.27). Hence, we can use the ratio R_D in Eq. (3.7.27) to yield an estimate for the total drag in ground effect

$$\begin{aligned} C_D = {} & C_{D_{L0}} + C_{D,L}C_{L_g} + R_D C_{D,L^2}C_{L_g}^2 + C_{D,S^2}C_S^2 \\ & + (C_{D,\alpha\overline{q}}\alpha + C_{D,\overline{q}})\overline{q} + (C_{D,\alpha\delta_e}\alpha + C_{D,\delta_e})\delta_e + C_{D,\delta_e^2}\delta_e^2 \end{aligned} \tag{3.10.15}$$

This approach can be used to alter the lift and induced drag in ground effect based on any relations for the ratios R_L and R_D.

EXAMPLE 3.10.1 Compute the lift coefficient and total drag coefficient for the F-16 aircraft in ground effect when the aircraft is at an angle of attack of 10 deg and the main wing is 20 ft above the ground. Assume $\overline{q} = \hat{\alpha} = \delta_e = 0$. Use the aerodynamic model for C_L and C_D given in Eqs. (3.7.25) and (3.7.27), along with the aerodynamic coefficients computed from Example 3.7.1. Additional information about the F-16 can be found in Appendix C.3.

Solution: Let us begin by computing the lift coefficient out of ground effect. This can be computed by using the coefficients from Example 3.7.1 in Eq. (3.7.25), which yields

$$C_L = C_{L_1} = (0.0535)(3.8459)(10)(\pi/180) = 0.724736177024$$

From the vehicle information available in the appendix, the aspect ratio of the main wing is 3 and the taper ratio is 0.2275. Using this information in Eqs. (3.10.12) and (3.10.13) yields

$$\delta_L = 1.036751070889$$
$$\delta_D = 1.002713745554$$

In order to compute β_L, we must know the lift coefficient in ground effect, C_{L_g}. As an initial guess, we will use the lift coefficient out of ground effect. Using $C_{L_g} \approx C_{L_1}$ in Eq. (3.10.10) gives

$$\beta_L = 1.008072160979$$

Using the result in Eq. (3.10.8) gives

$$R_L = 1.019258331138$$

Table 3.10.1. Example iterative solutions for the lift ground effect ratio.

Iteration	Initial C_{L_g}	β_L	R_L	Final C_{L_g}
1	0.724736177024	1.00807216	1.01925833	0.738693386310
2	0.738693386310	1.00829855	1.01902949	0.738527533529
3	0.738527533529	1.00829584	1.01903222	0.738529512254
4	0.738529512254	1.00829588	1.01903218	0.738529488648
5	0.738529488648	1.00829588	1.01903218	0.738529488929
6	0.738529488929	1.00829588	1.01903218	0.738529488926
7	0.738529488926	1.00829588	1.01903218	0.738529488926

Finally, using this in Eq. (3.10.14) gives the first estimate for the lift coefficient in ground effect,

$$C_{L_g} \approx (1.019258331138)(0.724736177024) = 0.7386933863097$$

This is an initial guess, which can be improved by repeating the process and using the newly computed C_{L_g} in the calculation for β_L of the next iteration. Doing so yields the results given in Table 3.10.1. The solution for the lift in ground effect stops changing out to machine precision after 7 iterations, and yields an increase in lift coefficient of about 1.9% relative to the lift coefficient out of ground effect. Because the final few digits of machine precision can vary between computers, the entire machine precision is not shown in the table. However, if we take the solution for the 7th iteration as the correct solution, the error in the solution for each iteration can be computed. The error in the initial guess, or first iteration, is about 0.022%, and decreases by two orders of magnitude with every iteration. Because aerodynamic models are not typically accurate to within less than about 1%, iterating on this solution during a flight simulation may not be necessary.

Before we compute the drag coefficient in ground effect, it is insightful to compute the drag coefficient out of ground effect. Using the coefficients from Example 3.7.1, along with the lift coefficient out of ground effect in Eq. (3.7.27) yields

$$C_D = 0.0425 - (0.1154)C_L + (0.2427)C_L^2 = 0.08634180630149$$

We will now compute the drag coefficient in ground effect. Using the results generated above in Eqs. (3.10.11), (3.10.9), and (3.10.15) yields the solutions in Table 3.10.2. After 7 iterations, the model predicts a decrease in the induced-drag coefficient of about 3.4% relative to the case out of ground effect. Again, there can be error in this solution relative to the ground effect model if a value of C_{L_g} is used that is not fully converged. If we again take the solution for the 7th iteration as the correct solution, the error in the solution for each iteration can be computed. The error in the initial guess, or first iteration, is about 0.045%, and decreases by two orders of

Table 3.10.2. Example iterative solutions for the lift ground effect ratio.

Iteration	C_{L_g}	β_D	R_D	C_D
1	0.738693386310	1.012487411	0.952917314	0.0834530574102
2	0.738527533529	1.012484019	0.952914122	0.0834151120073
3	0.738529512254	1.012484059	0.952914160	0.0834155646417
4	0.738529488648	1.012484059	0.952914159	0.0834155592417
5	0.738529488929	1.012484059	0.952914159	0.0834155593061
6	0.738529488926	1.012484059	0.952914159	0.0834155593054
7	0.738529488926	1.012484059	0.952914159	0.0834155593054

magnitude with every iteration. Again, this points to the fact that due to the inaccuracies inherent in most aerodynamic models, iterating on the ground-effect solution during a flight simulation may not be necessary.

3.11 Aerodynamic Datasets

A common approach for evaluating aerodynamic properties of a vehicle in flight simulation is to use a pre-determined dataset of aerodynamic properties of the vehicle. This approach is typically employed by many of the most accurate simulators in the industry. In this approach, aerodynamic forces and moments on the vehicle are stored in a dataset as a function of vehicle operating conditions. The vehicle operating conditions can include aerodynamic angles, vehicle rotation rates, translational accelerations, control-surface deflections, Reynolds number, Mach number, and others. The aerodynamic information contained in the dataset is generated a priori from CFD simulations, wind-tunnel tests, panel codes, flight-test results, or other aerodynamic prediction or measurement techniques. In the present section, we will consider what a typical aerodynamic dataset might consist of, and how it can be structured and used during simulation.

Aerodynamic forces and moments for a vehicle can be stored in the body-fixed coordinate system, the stability coordinate system, or the wind coordinate system. However, since the forces and moments on the vehicle are typically needed in the body-fixed coordinate system, storing the information in the body-fixed coordinate system can reduce the computational cost during simulation. In the present development, the force and moment vectors will not be sub-scripted to denote the coordinate system, since the same approach can be used independent of the coordinate system of choice. If the data is stored in a coordinate system other than the body-fixed coordinate system, the transformations given in Section 3.4 can be used to compute the forces and moments in the body-fixed coordinate system.

At any point in the simulation, a way to evaluate the three components of force $\mathbf{F} = (F_x, F_y, F_z)$ and the three components of moment $\mathbf{M} = (M_x, M_y, M_z)$ is needed. Since the aerodynamic forces and moments scale with dynamic pressure, the dimensionless forces and moments defined in Eqs. (3.4.26) and (3.4.33) are typically stored in the dataset, rather than the dimensional values. Hence, we will refer to C_x, C_y, C_z, C_ℓ, C_m, and C_n as the *dependent variables*. The dimensionless aerodynamic forces and moments of a vehicle depend on the aerodynamic angles, vehicle angular rates, translational accelerations, control-surface deflections, Reynolds number, Mach number, and other variables. We will refer to these as the *independent variables*.

A dataset is a method for storing information about the dependent variables at a discrete number of intervals and combinations of independent variables. Interpolation techniques are then used to evaluate the dependent variables at any combination of independent-variable values. In general, the dataset and interpolation schemes allow for a function call for a typical fixed-wing aircraft of the form

$$\begin{aligned}
C_x &= C_x(\alpha, \beta, \overline{p}, \overline{q}, \overline{r}, \hat{\alpha}, \hat{\beta}, \delta_a, \delta_e, \delta_r, \mathrm{R_e}, \mathrm{M}) \\
C_y &= C_y(\alpha, \beta, \overline{p}, \overline{q}, \overline{r}, \hat{\alpha}, \hat{\beta}, \delta_a, \delta_e, \delta_r, \mathrm{R_e}, \mathrm{M}) \\
C_z &= C_z(\alpha, \beta, \overline{p}, \overline{q}, \overline{r}, \hat{\alpha}, \hat{\beta}, \delta_a, \delta_e, \delta_r, \mathrm{R_e}, \mathrm{M}) \\
C_\ell &= C_\ell(\alpha, \beta, \overline{p}, \overline{q}, \overline{r}, \hat{\alpha}, \hat{\beta}, \delta_a, \delta_e, \delta_r, \mathrm{R_e}, \mathrm{M}) \\
C_m &= C_m(\alpha, \beta, \overline{p}, \overline{q}, \overline{r}, \hat{\alpha}, \hat{\beta}, \delta_a, \delta_e, \delta_r, \mathrm{R_e}, \mathrm{M}) \\
C_n &= C_n(\alpha, \beta, \overline{p}, \overline{q}, \overline{r}, \hat{\alpha}, \hat{\beta}, \delta_a, \delta_e, \delta_r, \mathrm{R_e}, \mathrm{M})
\end{aligned} \tag{3.11.1}$$

Note that it is customary to store the data as a function of rotation rates p, q, and r in terms of the dimensionless rotation rates $\overline{p}$, $\overline{q}$, and $\overline{r}$ defined in Eq. (3.4.23). However, this is not absolutely necessary, and other definitions for dimensionless rates could also be used, such as that given in Eq. (3.4.24). Likewise, data can be stored as a function of the dimensional translational accelerations $\dot{\alpha}$ and $\dot{\beta}$, or the dimensionless translational accelerations $\hat{\alpha}$ and $\hat{\beta}$. Here we will consider various methods for structuring a dataset, as well as methods for dataset interpolation.

Datasets can be formatted many different ways. The purpose of this section is not to give an exhaustive list of options, but to present three practical approaches. These include rectilinear datasets, unstructured datasets, and recursive datasets.

Rectilinear Dataset

The simplest dataset structure for storage and interpolation is a *rectilinear dataset*. Such a dataset consists of a dimension for each independent variable, and an equal number of entries for all other independent variables at each value of the independent variable of interest. For example, if the only independent variables were angle of attack and sideslip angle, a rectilinear dataset could be created by storing the information at N_α values in angle of attack, and N_β values in sideslip angle. This would create a rectilinear dataset with $N_\alpha \times N_\beta$ entries. Note that N_α need not be equal to N_β, and that the incremental spacing in angle of attack need not be equal to that for the sideslip angle. The only requirement is that the spacing is consistent in angle of attack for each sideslip angle, and vice versa.

Table 3.11.1. Example rectilinear dataset format. The dependent variable is aerodynamic C_x data from Nguyen et al. [39].

Angle of Attack	Sideslip Angle, β [deg]				
α [deg]	−6.0	−2.0	0.0	2.0	6.0
−20.0	−0.18720	−0.18600	−0.18680	−0.18990	−0.19000
−15.0	−0.18520	−0.18770	−0.18750	−0.18980	−0.18680
−10.0	−0.17070	−0.17720	−0.17870	−0.17690	−0.17110
−5.0	−0.14200	−0.14370	−0.14320	−0.14250	−0.14100
0.0	−0.11220	−0.11300	−0.11320	−0.11290	−0.11100
5.0	−0.06680	−0.06900	−0.06930	−0.06860	−0.06640
10.0	−0.00780	−0.01160	−0.01200	−0.01230	−0.00880
15.0	0.05350	0.05380	0.05370	0.05330	0.05270
20.0	0.09240	0.09480	0.09510	0.09750	0.09130
35.0	0.14780	0.16230	0.16630	0.16670	0.16370
40.0	0.16010	0.17260	0.17390	0.17110	0.16550
60.0	0.17040	0.17190	0.17180	0.17280	0.17340
70.0	0.17880	0.17380	0.16950	0.17100	0.17300
80.0	0.15350	0.15660	0.15980	0.15730	0.15860
90.0	0.16860	0.16690	0.16600	0.16720	0.16640

Table 3.11.1 shows an example of a rectilinear dataset with the angle of attack and sideslip angles as the independent variables, and C_x as the dependent variable. Figure 3.11.1 shows the independent-variable locations in graphical form for the same data. These data were taken from an aerodynamic dataset published by Nguyen et al. [39]. Some data have been purposefully deleted here from the original dataset for demonstration purposes. Note that the data are not equally spaced in either independent variable. For example, the step size between $\beta = 0.0$ deg and $\beta = 2.0$ deg is $\Delta\beta = 2.0$ deg, whereas the step size between $\beta = 2.0$ deg and $\beta = 6.0$ deg is $\Delta\beta = 4.0$ deg. Likewise, α increments by 5.0 deg for $-20.0 \leq \alpha \leq 20.0$ deg, and then by various intervals for $20.0 \leq \alpha \leq 90.0$ deg. However, the dataset is still considered rectilinear because there is a full column and row of data for each specified value of the independent variables α and β.

Rectilinear datasets are attractive for two main reasons. First, they are simple to visualize and store. The multi-dimensional arrays needed to store rectilinear dataset information are straightforward to create in most programming languages. Second, because the relation between data points in the dataset is structured and well-defined, interpolation is straightforward. Linear, quadratic, cubic, or higher-order interpolation can easily be performed in such a structured dataset. There are many open-source tools available for interpolating on rectilinear datasets.

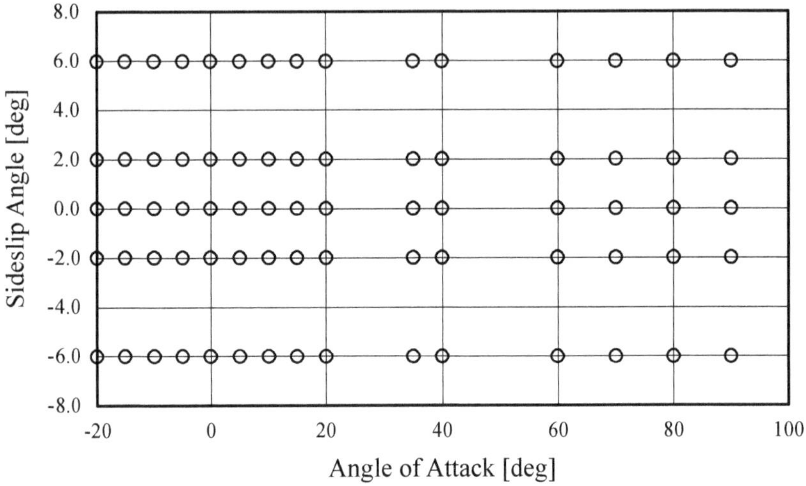

Figure 3.11.1. Example independent-variable values in a rectilinear dataset.

Although rectilinear datasets are attractive, the main challenge of rectilinear datasets is that they become extremely large and impractical as the number of independent variables increases. For example, in order to accurately model the aerodynamics of a fixed-wing aircraft at a given Reynolds number and Mach number, the typical independent variables include two aerodynamic angles (angle of attack and sideslip angle), three control surfaces (e.g. aileron, elevator, rudder), and three angular rates (p,q,r) for a total of eight independent variables. If just 10 points are stored in each dimension, this results in 10^8 = 100 million cases. If the independent variables of Mach and Reynolds number are added, this results in 10^{10} = 10 billion cases. Rectilinear datasets are ideal for situations in which data can be rapidly generated. However, these sizes of datasets are nearly impossible to produce from high-fidelity CFD or wind-tunnel tests using today's technology. Therefore, **rectilinear datasets including all independent variables are typically not used in flight simulation**. Rather, smaller datasets are typically generated, and then combined in an additive approach, as will be discussed later in this section.

Unstructured Dataset

Another attractive dataset format is what will be referred to here as an *unstructured dataset*. In this format, individual, uncorrelated, and unstructured points are known in the multi-dimensional space. The number and location of data points are not related in any of the dimensions. Sophisticated methods are then used to interpolate within the unstructured format of the data.

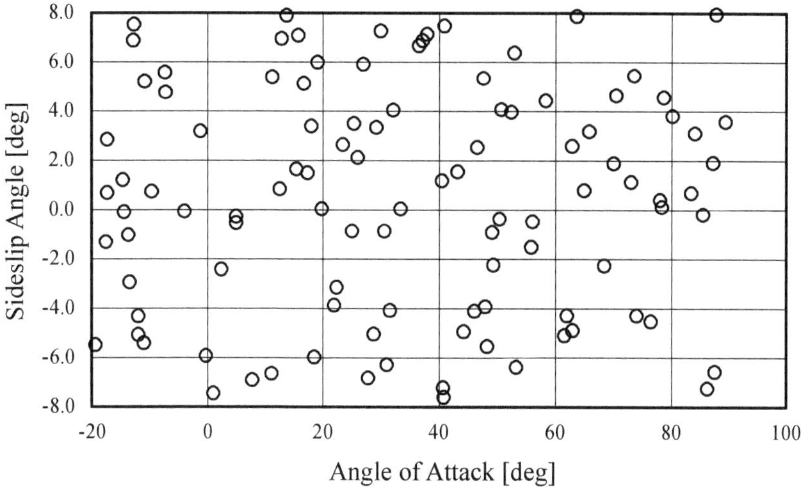

Figure 3.11.2. Example independent-variable values in an unstructured dataset.

Figure 3.11.2 shows locations of how data could be located for two independent variables α and β of an example unstructured dataset. Note that there is no structure to the values of the independent variables. Table 3.11.2 shows how data could be formatted for an unstructured dataset with three independent variables.

The main advantage of an unstructured dataset is the fact that data can be easily added or removed. Because data can be stored in arrays in a text file or within the software, adding or removing data is very straightforward, and can be performed without concern for how the data is related to the dataset as a whole. Additionally, more data points can be used in certain areas of the domain to locally improve fidelity of the model without significantly increasing the size of the dataset.

Table 3.11.2. Sample data format for an unstructured dataset.

Independent Variables			Dependent Variable
Reynolds Number	β [deg]	α [deg]	C_x
1,000,000	−10.0	−5.0	−0.0727
1,500,000	−20.0	0.0	−0.0450
1,300,000	0.0	0.0	−0.0489
1,100,000	5.0	5.0	−0.0278
1,600,000	10.0	−10.0	0.0156

The main challenge with an unstructured dataset lies in interpolating the data. Sophisticated algorithms have been developed for interpolation within unstructured datasets, such as methods based on Delaunay triangulation [59]. However, both triangulation and interpolation within an unstructured dataset can be computationally expensive and time consuming compared to that within a structured dataset, especially for higher-dimensional datasets. Furthermore, high-order interpolation such as cubic interpolation is significantly more challenging in an unstructured dataset. Libraries for handling these types of datasets can be found for some programming languages.

Recursive Dataset

A very useful dataset structure for flight simulation is what we will refer to as a *recursive dataset*. This type of dataset is sometimes called a *hierarchical* or *nested* dataset. Here we will refer to it as a recursive dataset, partially due to the fact that recursive algorithms can be used for interpolation within the dataset. This will be demonstrated in the example at the end of this section.

A recursive dataset is a dataset in which data is organized recursively into subsets of data for each independent variable. The data in a lower-dimension independent variable is available at only a subset of values of a higher-dimension independent variable. For example, Fig. 3.11.3 shows possible values of the independent variables α and β for a recursive dataset, with α as the next lower dimension from β. Note that all of the α values lie along horizontal lines of only a subset of β values, including $\beta = -6.0, -2.0, 0.0, 2.0$, or 6.0 deg. However, for any given value of β, any number of data points in α can be specified, and these can be completely uncorrelated in number or value from those specified at another value of β.

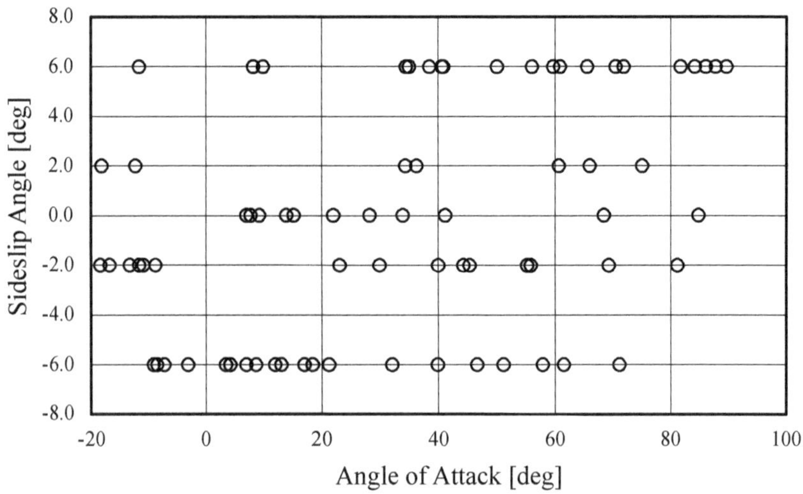

Figure 3.11.3. Example independent-variable values in a recursive dataset.

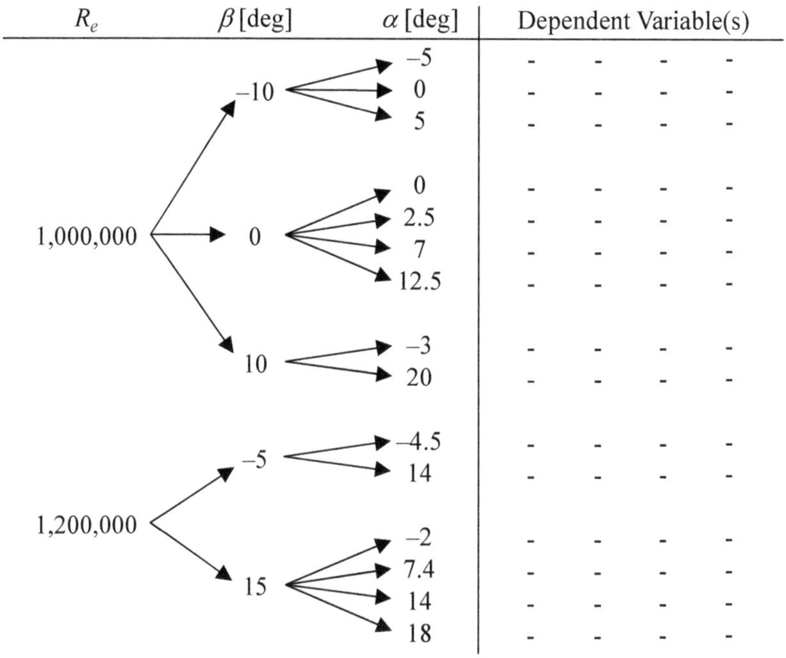

Figure 3.11.4. Example recursive dataset structure with independent-variable relations.

An alternate way of looking at a recursive dataset is in a tree-like format shown in Fig. 3.11.4 where the highest-dimension independent variable is on the left, and the dimensions decrease to the right. The same data is shown in table format in Table 3.11.3. In this example, notice that the data is defined at only two values of Reynolds number. For each Reynolds number, there is a subset of data defined at a series of sideslip angles. For each sideslip angle, there is a subset of data defined at a series of angle-of-attack values. Dependent variable information is associated with the lowest-dimension independent variable.

Note that the size of a series within a subset does not need to be equal to the size of a series within another subset. This offers significant flexibility in generating data for a recursive dataset, and data can easily be added to refine the solution in important regions of the dataset. It may be interesting to note that the rectilinear dataset is a special case of a recursive dataset where the data in a higher dimension is only available at a subset of values of the lower dimension.

Whereas the order of dimensions in a rectilinear dataset is not significantly important, the organization of dimensions in a recursive dataset is a key element in producing a useful dataset. The data is typically organized such that lower dimensions have more discrete points than higher dimensions. For example, in aircraft simulation, higher-resolution data is typically needed in angle of attack than Reynolds number. Data as a

Table 3.11.3. Sample data format for a recursive dataset.

Independent Variables			Dependent Variables		
Reynolds Number	β [deg]	α [deg]	C_x	C_y	C_z
1,000,000	−10.0	−5.0	−0.0727	0.1833	0.2880
1,000,000	−10.0	0.0	−0.0450	0.2014	−0.0360
1,000,000	−10.0	5.0	−0.0036	0.2028	−0.3590
1,000,000	0.0	0.0	−0.0489	0.0000	−0.0250
1,000,000	0.0	2.5	−0.0278	0.0000	−0.1960
1,000,000	0.0	7.0	0.0156	0.0000	−0.5202
1,000,000	0.0	12.5	0.0781	0.0000	−0.9310
1,000,000	10.0	−5.0	−0.0703	−0.2072	0.2600
1,000,000	10.0	−2.5	−0.0568	−0.2103	0.1130
1,000,000	10.0	0.0	−0.0433	−0.2134	−0.0340
1,000,000	10.0	10.0	0.0503	−0.2171	−0.7350
1,000,000	10.0	20.0	0.1259	−0.1729	−1.3820
1,200,000	−5.0	−10.0	−0.0913	0.0597	0.6915
1,200,000	−5.0	−5.5	−0.0899	0.0623	0.6712
1,200,000	−5.0	5.5	0.0004	−0.0184	−0.4024
1,200,000	−5.0	10.0	0.0504	0.0940	−0.7390
1,200,000	15.0	0.0	−0.0399	−0.3198	−0.0480
1,200,000	15.0	5.0	0.0013	−0.3257	−0.3770
1,200,000	15.0	8.3	0.0328	−0.3222	−0.6067

function of angle of attack may be recorded at more than 20 data points, while data as a function of Reynolds number may be recorded at only a few points. Therefore, it is typical to organize the data such that Reynolds number is in a higher dimension than angle of attack.

Data formatted in tables such as Table 3.11.3 are typically organized with the highest dimension in the first column and decreasing dimensions to the right. Note that if the data is stored in tabulated form as shown in Table 3.11.3, the independent variables in each dimension must be properly sorted in either ascending or descending order. The data shown in this table is in ascending order. This sorting process is easily done in most programming languages, and must be completed before recursive interpolation can be performed.

Interpolation within a recursive dataset is straightforward. Interpolation is performed by considering only a single dimension at a time. To interpolate the value at any combination of independent variables, the bounding points in the highest dimension are first found. The subset of data at these bounding points in the next lower dimension are then considered. The bounding points in the next lower dimension within that data are then found, and this process is repeated recursively until the lowest dimension is reached. Interpolation is then performed on the dependent variables associated with the independent variables in the lowest dimension. The interpolated solutions for the dependent variables are then passed to the next higher dimension. Interpolation is performed in each dimension and the solutions passed recursively up to the next higher dimension until the highest dimension is reached. Once interpolation has been performed in the highest dimension, the solution is complete. The entire process of interpolation within the dataset is a recursive process. This is perhaps best understood by example.

EXAMPLE 3.11.1 Using the data given in Table 3.11.3, use recursion and linear interpolation to estimate values for the three dependent variables C_x, C_y, and C_z for the case of $\alpha = 5.0$ deg and $\beta = 5.0$ deg at a Reynolds number of 1,100,000.

Solution: We begin by looking at the highest dimension and finding the bounding values for a Reynolds number of 1,100,000. In the highest dimension, data is available at only two values, which bound the value of interest. These are $\mathrm{R_e} = 1,000,000$ and $\mathrm{R_e} = 1,200,000$. We will define these as

$$\mathrm{R_e^-} = 1,000,000$$
$$\mathrm{R_e^+} = 1,200,000$$

Within those sets of data, we find the bounding values in the next hightest dimension, β, for $\beta = 5$ deg. For $\mathrm{R_e} = 1,000,000$, these are $\beta = 0.0$ deg and $\beta = 10.0$ deg. For $\mathrm{R_e} = 1,200,000$, these are $\beta = -5.0$ deg and $\beta = 15.0$ deg. These are boxed in Table 3.11.4 and yield

$$\beta^-|_{\mathrm{R_e^-}} = 0.0 \text{ deg}$$
$$\beta^+|_{\mathrm{R_e^-}} = 10.0 \text{ deg}$$

$$\beta^-|_{\mathrm{R_e^+}} = -5.0 \text{ deg}$$
$$\beta^+|_{\mathrm{R_e^+}} = 15.0 \text{ deg}$$

Within those sets of data, we find the bounding values in the lowest dimension, α, for $\alpha = 5$ deg. These are boxed in Table 3.11.4 and yield

$$\alpha^-|_{\mathrm{R_e^-},\beta^-} = 2.5 \text{ deg}, \qquad \alpha^-|_{\mathrm{R_e^-},\beta^+} = 0.0 \text{ deg}$$
$$\alpha^+|_{\mathrm{R_e^-},\beta^-} = 7.0 \text{ deg}, \qquad \alpha^+|_{\mathrm{R_e^-},\beta^+} = 10.0 \text{ deg}$$

$$\alpha^-|_{\mathrm{R_e^+},\beta^-} = -5.5 \text{ deg}, \qquad \alpha^-|_{\mathrm{R_e^+},\beta^+} = 5.0 \text{ deg}$$
$$\alpha^+|_{\mathrm{R_e^+},\beta^-} = \ \ 5.5 \text{ deg}, \qquad \alpha^+|_{\mathrm{R_e^+},\beta^+} = 5.0 \text{ deg}$$

Since we have now reached the lowest dimension, the interpolation is performed on the associated dependent variables. For example, from the dependent data, C_x at a value of $\alpha = 2.5$ deg is

$$C_x|_{\mathrm{R_e^-},\beta^-,\alpha^-} = -0.0278$$

and C_x at a value of $\alpha = 7.0$ deg is

$$C_x|_{\mathrm{R_e^-},\beta^-,\alpha^+} = 0.0156$$

Linearly interpolating to $\alpha = 5$ deg gives

$$C_x|_{\mathrm{R_e^-},\beta^-} = \left(\frac{0.0156 - (-0.0278)}{7.0 - 2.5}\right)(5.0 - 2.5) + (-0.0278) = -0.00368888889$$

Table 3.11.4. Example interpolation within a recursive dataset.

Reynolds Number	β [deg]	α [deg]	C_x	C_y	C_z
1,000,000	−10.0	−5.0	−0.0727	0.1833	0.2880
1,000,000	−10.0	0.0	−0.0450	0.2014	−0.0360
1,000,000	−10.0	5.0	−0.0036	0.2028	−0.3590
1,000,000	0.0	0.0	−0.0489	0.0000	−0.0250
1,000,000	0.0	2.5	−0.0278	0.0000	−0.1960
1,000,000	0.0	7.0	0.0156	0.0000	−0.5202
1,000,000	0.0	12.5	0.0781	0.0000	−0.9310
1,000,000	10.0	−5.0	−0.0703	−0.2072	0.2600
1,000,000	10.0	−2.5	−0.0568	−0.2103	0.1130
1,000,000	10.0	0.0	−0.0433	−0.2134	−0.0340
1,000,000	10.0	10.0	0.0503	−0.2171	−0.7350
1,000,000	10.0	20.0	0.1259	−0.1729	−1.3820
1,200,000	−5.0	−10.0	−0.0913	0.0597	0.6915
1,200,000	−5.0	−5.5	−0.0899	0.0623	0.6712
1,200,000	−5.0	5.5	0.0004	−0.0184	−0.4024
1,200,000	−5.0	10.0	0.0504	0.0940	−0.7390
1,200,000	15.0	0.0	−0.0399	−0.3198	−0.0480
1,200,000	15.0	5.0	0.0013	−0.3257	−0.3770
1,200,000	15.0	8.3	0.0328	−0.3222	−0.6067

Repeating this process for the other three interpolations in the lowest dimension yields

$$C_x|_{R_e^-,\beta^+} = 0.0035$$
$$C_x|_{R_e^+,\beta^-} = -0.00370454545$$
$$C_x|_{R_e^+,\beta^+} = 0.0013$$

These four results can now be passed up one dimension and interpolated for the case of $\beta = 5$ deg. This yields

$$C_x|_{R_e^-} = \left(\frac{0.0035 - (-0.003689)}{10.0 - 0.0}\right)(5.0 - 0.0) + (-0.003689) = -0.00009444444$$
$$C_x|_{R_e^+} = \left(\frac{0.0013 - (-0.0037)}{15.0 - (-5)}\right)(5.0 - (-5)) + (-0.0037) = -0.0012022727272$$

The final interpolation can now be performed in the highest dimension to yield

$$C_x = \left(\frac{-0.0012 - (-0.000094)}{1200000 - 1000000}\right)(1100000 - 1000000) + (-0.000094)$$
$$= -0.000648358586$$

Repeating this process for the other dependent variables gives

$$C_y = -0.138920454545$$
$$C_z = -0.372802777778$$

The recursive dataset is attractive because it does not require uniform sizes in any dimension. Therefore, additional data can be generated in certain areas of the dataset where higher resolution information is needed. Additionally, interpolation within the dataset is straightforward due to the structured nature of the dataset. Interpolation in each dimension is performed recursively, such that interpolation only happens in a single dimension at a time. Therefore, linear interpolation and higher-order interpolation such as cubic interpolation is straightforward. Finally, due to the nature of sweeps that are typically performed in CFD computations and wind-tunnel experiments, data is often naturally generated in a format conducive to a recursive dataset.

Dataset Usage in Flight Simulation

Datasets are an excellent method for storing aerodynamic information that can be rapidly interpolated during flight simulation to approximate the aerodynamics of the vehicle at a given state. In general, a dataset approach can be employed to yield the solution to Eq. (3.11.1). In theory, this could be accomplished with a single dataset with the independent

Table 3.11.5. Table expression for a single general dataset at a set Reynolds number and Mach number.

Dependent Variables						Independent Variables									
C_x	C_y	C_z	C_ℓ	C_m	C_n	α	β	$\overline{p}$	$\overline{q}$	$\overline{r}$	δ_a	δ_e	δ_r	$\hat{\alpha}$	$\hat{\beta}$
×	×	×	×	×	×	×	×	×	×	×	×	×	×	×	×

variables α, β, $\overline{p}$, $\overline{q}$, $\overline{r}$, δ_a, δ_e, δ_r, $\hat{\alpha}$, $\hat{\beta}$, $\mathrm{R_e}$, and M, and the dependent variables C_x, C_y, C_z, C_ℓ, C_m, and C_n. For brevity, the total dependent variable force and moment coefficients will be denoted as $C_{(\mathbf{F},\mathbf{M})}$ in the following development. In the following development, a dataset denoting this dependence will be expressed as

$$C_{(\mathbf{F},\mathbf{M})}(\alpha, \beta, \overline{p}, \overline{q}, \overline{r}, \delta_a, \delta_e, \delta_r, \hat{\alpha}, \hat{\beta}, \mathrm{R_e}, \mathrm{M}) \tag{3.11.2}$$

which indicates that the force and moment coefficients are stored in a dataset as a function of the dependent variables listed. In mathematical terms,

$$C_{(\mathbf{F},\mathbf{M})} = f(\alpha, \beta, \overline{p}, \overline{q}, \overline{r}, \delta_a, \delta_e, \delta_r, \hat{\alpha}, \hat{\beta}, \mathrm{R_e}, \mathrm{M}) \tag{3.11.3}$$

It is sometimes helpful to express the dataset components in terms of a table. For example, the dataset described by Eq. (3.11.2), at a set Reynolds number and Mach number can be written as shown in Table 3.11.5.

Additive Δ Datasets

A single general dataset such as that given in Eq. (3.11.2) or shown in Table 3.11.5 could be generated for a vehicle. However, such an approach is not typically practical due to the size of the single dataset needed as well as the complexity of interpolation in multiple dimensions. Therefore, it is more common to use a combination of smaller datasets and employ an additive approach for combining the interpolated solutions from these datasets to approximate the aerodynamics of the vehicle at a given state.

For example, at a given Reynolds number and Mach number, a base dataset can be generated as a function of angle of attack and sideslip of the form

$$C_{(\mathbf{F},\mathbf{M})_0}(\alpha, \beta) \tag{3.11.4}$$

The subscript 0 denotes that all other independent variables were held at a value of zero while generating the dataset. Deviations to this base dataset due to rotation rates, translational accelerations, and control-surface deflections can then be stored in separate datasets. If these deviations are also a function of angle of attack and sideslip angle, the datasets can be expressed as

$$\Delta C_{(\mathbf{F},\mathbf{M})}(\alpha,\beta,\overline{p}) \tag{3.11.5}$$
$$\Delta C_{(\mathbf{F},\mathbf{M})}(\alpha,\beta,\overline{q}) \tag{3.11.6}$$
$$\Delta C_{(\mathbf{F},\mathbf{M})}(\alpha,\beta,\overline{r}) \tag{3.11.7}$$
$$\Delta C_{(\mathbf{F},\mathbf{M})}(\alpha,\beta,\delta_a) \tag{3.11.8}$$
$$\Delta C_{(\mathbf{F},\mathbf{M})}(\alpha,\beta,\delta_e) \tag{3.11.9}$$
$$\Delta C_{(\mathbf{F},\mathbf{M})}(\alpha,\beta,\delta_r) \tag{3.11.10}$$
$$\Delta C_{(\mathbf{F},\mathbf{M})}(\alpha,\beta,\hat{\alpha}) \tag{3.11.11}$$
$$\Delta C_{(\mathbf{F},\mathbf{M})}(\alpha,\beta,\hat{\beta}) \tag{3.11.12}$$

The deviation datasets are generated by obtaining the aerodynamic information at a given angle of attack, sideslip angle, and third independent variable, and then subtracting off the solution at the same angle of attack and sideslip angle with a value of zero for all independent variables. For dependence on the roll rate, this can be expressed as

$$\Delta C_{(\mathbf{F},\mathbf{M})}(\alpha,\beta,\overline{p}) \equiv C_{(\mathbf{F},\mathbf{M})}(\alpha,\beta,\overline{p}) - C_{(\mathbf{F},\mathbf{M})_0}(\alpha,\beta) \tag{3.11.13}$$

and definitions for Eqs. (3.11.6)–(3.11.12) can be similarly expressed. Such an approach yields a solution in the form of a sum of multiple datasets expressed as

$$\begin{aligned} C_{(\mathbf{F},\mathbf{M})} = C_{(\mathbf{F},\mathbf{M})_0}(\alpha,\beta) &+ \Delta C_{(\mathbf{F},\mathbf{M})}(\alpha,\beta,\overline{p}) + \Delta C_{(\mathbf{F},\mathbf{M})}(\alpha,\beta,\overline{q}) \\ &+ \Delta C_{(\mathbf{F},\mathbf{M})}(\alpha,\beta,\overline{r}) + \Delta C_{(\mathbf{F},\mathbf{M})}(\alpha,\beta,\delta_a) \\ &+ \Delta C_{(\mathbf{F},\mathbf{M})}(\alpha,\beta,\delta_e) + \Delta C_{(\mathbf{F},\mathbf{M})}(\alpha,\beta,\delta_r) \\ &+ \Delta C_{(\mathbf{F},\mathbf{M})}(\alpha,\beta,\hat{\alpha}) + \Delta C_{(\mathbf{F},\mathbf{M})}(\alpha,\beta,\hat{\beta}) \end{aligned} \tag{3.11.14}$$

The datasets used in Eq. (3.11.14) can be written in table form as shown in Table 3.11.6, where each row of the table represents a different dataset.

The approximation of the vehicle aerodynamics from Eq. (3.11.14) neglects interactions of the rotational rates, control-surface deflections, and translational accelerations, which could be captured in a more general dataset such as that given in Eq. (3.11.2). However, the approximate solution given in Eq. (3.11.14) requires significantly less data to generate, store, and interpolate within.

For example, consider the size of a rectilinear dataset required to store all six components of the force and moment coefficients at a given Mach number and Reynolds number. If these are stored at 10 values for each of the 10 remaining independent variables listed in Eq. (3.11.2), this rectilinear dataset would require generating $10^{10} = 10$ billion cases and storing $(6)10^{10} = 60$ billion entries. On the other hand, the base table in Eq. (3.11.4) would require generating 100 cases and storing $(6)100 = 600$ entries, and the datasets listed in Eqs. (3.11.5)–(3.11.12) would require generating $1,000$ cases and storing $(6)1,000 = 6,000$ entries each. Hence, the solution given in Eq. (3.11.14)

Table 3.11.6. Table expression for a the datasets given in Eq. (3.11.14).

Dependent Variables						Independent Variables									
C_x	C_y	C_z	C_ℓ	C_m	C_n	α	β	$\overline{p}$	$\overline{q}$	$\overline{r}$	δ_a	δ_e	δ_r	$\hat{\alpha}$	$\hat{\beta}$
×	×	×	×	×	×	×	×								
Δ	Δ	Δ	Δ	Δ	Δ	×	×	×							
Δ	Δ	Δ	Δ	Δ	Δ	×	×		×						
Δ	Δ	Δ	Δ	Δ	Δ	×	×			×					
Δ	Δ	Δ	Δ	Δ	Δ	×	×				×				
Δ	Δ	Δ	Δ	Δ	Δ	×	×					×			
Δ	Δ	Δ	Δ	Δ	Δ	×	×						×		
Δ	Δ	Δ	Δ	Δ	Δ	×	×							×	
Δ	Δ	Δ	Δ	Δ	Δ	×	×								×

for this case requires generating 8, 100 cases and storing 48, 600 entries. This is less than 0.0001% of the number of cases required for the general dataset.

The amount of data required for the approximation given in Eq. (3.11.14) can be reduced even further, by recognizing that dependence on β is not always significant in the range of sideslip angles typically experienced during flight, and that some longitudinal and lateral coupling is weak and can be neglected. For example, Nguyen et al. [39] published datasets to simulate the aerodynamics of a fixed-wing fighter aircraft that included the following dependencies [60]

$$\begin{aligned} C_{(\mathbf{F},\mathbf{M})} &= C_{(x,y,z,\ell,m,n)_0}(\alpha,\beta,\delta_e) + \Delta C_{(y,\ell,n)}(\alpha,\overline{p}) \\ &+ \Delta C_{(x,z,m)}(\alpha,\overline{q}) + \Delta C_{(y,\ell,n)}(\alpha,\overline{r}) \\ &+ \Delta C_{(y,\ell,n)}(\alpha,\beta,\delta_a) + \Delta C_{(y,\ell,n)}(\alpha,\beta,\delta_r) \end{aligned} \tag{3.11.15}$$

Note that the actual aerodynamic model published by Nguyen et al. looks significantly different than that shown in Eq. (3.11.15), and that the published model included additional independent variables for leading-edge flaps and speed brakes. However, Eq. (3.11.15) demonstrates all dependencies included in the published model that would apply to a typical fixed-wing aircraft with the control surfaces included in Eq. (3.11.2).

The datasets used in Eq. (3.11.15) can be written in table form as shown in Table 3.11.7, where each row of the table represents a different dataset. Note that dependence on the translational terms $\hat{\alpha}$ and $\hat{\beta}$ were neglected. If 10 values are used for each independent variable and rectilinear datasets are generated for each term, this approach requires generating only 3, 300 cases and storing only 12, 900 entries, which is significant savings relative to those from Eqs. (3.11.2) and (3.11.14).

Table 3.11.7. Table expression for a the datasets given in Eq. (3.11.15).

Dependent Variables						Independent Variables							
C_x	C_y	C_z	C_ℓ	C_m	C_n	α	β	$\overline{p}$	$\overline{q}$	$\overline{r}$	δ_a	δ_e	δ_r
×	×	×	×	×	×	×	×					×	
	Δ		Δ		Δ	×		×					
Δ		Δ		Δ		×			×				
	Δ		Δ		Δ	×				×			
	Δ		Δ		Δ	×	×				×		
	Δ		Δ		Δ	×	×						×

Another useful form for datasets can be obtained by considering the simple aerodynamic model for a symmetric fixed-wing aircraft given in Eqs. (3.7.24)–(3.7.30). The interactions in this model can be captured from datasets defined as

$$\begin{aligned} C_{(\mathbf{F},\mathbf{M})} = {} & C_{(L,D,m)_0}(\alpha) + \Delta C_{(S,D,\ell,n)}(\beta) \\ & + \Delta C_{(L,D,m)}(\alpha,\overline{q}) + \Delta C_{(L,D,m)}(\alpha,\delta_e) + \Delta C_{(L,m)}(\hat{\alpha}) \\ & + \Delta C_{(S,\ell,n)}(\alpha,\overline{p}) + \Delta C_{(S,\ell,n)}(\alpha,\overline{r}) \\ & + \Delta C_{(S,\ell,n)}(\alpha,\delta_a) + \Delta C_{(S,\ell,n)}(\delta_r) \end{aligned} \tag{3.11.16}$$

Note that because this is a model for a symmetric vehicle, it is assumed that at zero sideslip angle $C_S = C_\ell = C_n = 0$. The datasets used in Eq. (3.11.16) can be written in table form as shown in Table 3.11.8, where each row represents a different dataset.

Many other dataset formulations could be developed to capture the relevant aerodynamic data for various vehicles. The reader is reminded that the simplified models shown here should be used with caution. Care must be taken to ensure that the complete aerodynamic dataset captures all significant aerodynamic interactions. The specific vehicle interactions that must be included can vary with vehicle design.

Table 3.11.8. Table expression for a the datasets given in Eq. (3.11.16).

Dependent Variables						Independent Variables								
C_L	C_S	C_D	C_ℓ	C_m	C_n	α	β	$\overline{p}$	$\overline{q}$	$\overline{r}$	δ_a	δ_e	δ_r	$\hat{\alpha}$
×		×		×		×								
	Δ	Δ	Δ		Δ		×							
Δ		Δ		Δ		×			×					
Δ		Δ		Δ		×						×		
Δ				Δ										×
	Δ		Δ		Δ	×		×						
	Δ		Δ		Δ	×				×				
	Δ		Δ		Δ	×					×			
	Δ		Δ		Δ								×	

Additive Derivative Datasets

The approach to modeling deviations in forces and moments relative to a base aerodynamic dataset by using deviation datasets like those shown in Eqs. (3.11.5)–(3.11.12) is not the only way to capture the dependence of the aerodynamics on other independent variables relative to a base aerodynamic dataset. If the change in the aerodynamic properties of the vehicle is nearly linear with the change in an independent variable, dataset sizes can be significantly reduced by storing the first derivative of the force and moment components with respect to the independent variable, rather than the force and moment components themselves. For example, if the force and moment components are very nearly linear functions of roll rate, the deviation from the base dataset can be approximated as

$$\Delta C_{(\mathbf{F},\mathbf{M})}(\alpha,\beta,\overline{p}) \approx C_{(\mathbf{F},\mathbf{M}),\overline{p}}(\alpha,\beta)\overline{p} \tag{3.11.17}$$

where $C_{(\mathbf{F},\mathbf{M}),\overline{p}}(\alpha,\beta)$ represents a dataset of the first derivative of the force and moment coefficients with respect to $\overline{p}$ stored as a function of α and β. Because fixed-wing aircraft typically operate in small regions of angle of attack, sideslip angle, rotation rates, and control-surface deflections, the aerodynamic deviations due to the independent variables are often approximated as being linear. This is equivalent to employing a Taylor-series expansion of the aerodynamics, and retaining only first-order terms. Although this is not appropriate for all independent variables, this approach can be used with caution to further decrease the dataset sizes from those given above.

For example, Nguyen et al. [39, 60] employed the following approximations

$$\begin{aligned} C_{(\mathbf{F},\mathbf{M})} &= C_{(x,y,z,\ell,m,n)_0}(\alpha,\beta,\delta_e) + C_{(x,z,m),\overline{q}}(\alpha)\overline{q} \\ &\quad + C_{(y,\ell,n),\delta_a}(\alpha,\beta)\delta_a + C_{(y,\ell,n),\delta_r}(\alpha,\beta)\delta_r \\ &\quad + C_{(y,\ell,n),\overline{r}}(\alpha)\overline{r} + C_{(y,\ell,n),\overline{p}}(\alpha)\overline{p} \end{aligned} \tag{3.11.18}$$

If each dataset were rectilinear and the same resolution of 10 points were used for each independent variable, and if central differencing were used to evaluate derivatives, this dataset would require generating 1,460 cases and storing 7,380 entries. This is less than half of the cases required to generate the data for Eq. (3.11.15).

The interactions in the simple aerodynamic model for a symmetric fixed-wing aircraft given in Eqs. (3.7.24)–(3.7.30) can be captured in the datasets given in Eq. (3.11.16). However, the model equations given in Eqs. (3.7.24)–(3.7.30) assume linear relationships below stall. Therefore, a complete dataset that captures these same linear relationships can be expressed as

$$\begin{aligned} C_{(\mathbf{F},\mathbf{M})} &= C_{(L,D,m)_0}(\alpha) + \Delta C_{(S,D,\ell,n)}(\beta) \\ &\quad + C_{(L,D,m),\overline{q}}(\alpha)\overline{q} + \Delta C_{(L,D,m)}(\alpha,\delta_e) + C_{(L,m),\hat{\alpha}}\hat{\alpha} \\ &\quad + C_{(S,\ell,n),\overline{p}}(\alpha)\overline{p} + C_{(S,\ell,n),\overline{r}}(\alpha)\overline{r} \\ &\quad + C_{(S,\ell,n),\delta_a}(\alpha)\delta_a + C_{(S,\ell,n),\delta_r}\delta_r \end{aligned} \tag{3.11.19}$$

Note that some dataset dependencies were retained to capture the nonlinear terms in the drag relations given in Eq. (3.7.27).

The reader is reminded that approximations given in this section for simplifying the size and complexity of the datasets to be generated and stored must be used with extreme caution. The number of independent variables needed to accurately model the aerodynamics of vehicles varies, depending on the vehicle geometry and control-surface design. Hence, the approximations given in this section should be used as a reference, but not as a definitive solution for every vehicle. Accurately modeling the aerodynamics of some vehicles may require significantly different approaches to dataset generation, storage, and interpolation. For example, a fixed-wing aircraft with a rotating empennage studied by Bolander et al. [61, 62] required capturing more complex interactions than the approximate solutions given in this section. Care must be taken in designing and storing aerodynamic datasets for flight simulation to ensure that the datasets are small enough so that interpolation can be performed rapidly while preserving all significant aerodynamic relationships.

EXAMPLE 3.11.2 Consider the F-16 aircraft specified in Appendix C.3 with the aerodynamic model F16-AD given in Table C.3.15. Compute the aerodynamic force and moment coefficients for the cases given in the following table.

Independent Variable	Units	Case A	B	C	D
α	deg	5.0	25.0	45.0	-15.0
β	deg	5.0	-5.0	-20.0	20.0
$\overline{p}$	none	0.01	-0.01	0.01	0.01
$\overline{q}$	none	0.01	0.01	-0.01	0.01
$\overline{r}$	none	0.01	0.01	0.01	-0.01
δ_a	deg	5.0	-5.0	20.0	-20.0
δ_e	deg	5.0	10.0	-25.0	25.0
δ_r	deg	5.0	10.0	-20.0	-20.0
δ_{sb}	deg	5.0	10.0	60.0	0.0
δ_{lef}	deg	5.0	25.0	0.0	10.0

Solution: Linearly interpolating within the rectilinear datasets given in Table C.3.8 for Case A yields

$$C_{(x,y,z,\ell,m,n)_0}(\delta_e,\beta,\alpha) \rightarrow \begin{bmatrix} C_{x_0} \\ C_{y_0} \\ C_{z_0} \\ C_{\ell_0} \\ C_{m_0} \\ C_{n_0} \end{bmatrix} = \begin{bmatrix} -0.00860 \\ -0.09870 \\ -0.49000 \\ -0.01292 \\ -0.10585 \\ 0.01856 \end{bmatrix}$$

$$\Delta C_{(m)}(\beta,\alpha) \rightarrow \Delta C_m = 0.0567$$

$$\Delta C_{(m)}(\delta_e,\alpha) \rightarrow \Delta C_m = 0.0$$

$$C_{(\ell,n),\beta}(\alpha)\beta \rightarrow \begin{bmatrix}\Delta C_\ell\\ \Delta C_n\end{bmatrix} = \begin{bmatrix}0.0\\0.0\end{bmatrix}\frac{5\pi}{180.0} = \begin{bmatrix}0.0\\0.0\end{bmatrix}$$

$$C_{(x,y,z,\ell,m,n),\delta_{lef}}(\beta,\alpha)\delta_{lef} \rightarrow \begin{bmatrix}\Delta C_x\\ \Delta C_y\\ \Delta C_z\\ \Delta C_\ell\\ \Delta C_m\\ \Delta C_n\end{bmatrix} = \begin{bmatrix}-8.36518\text{E}-03\\ -8.93814\text{E}-03\\ 1.26051\text{E}-01\\ 4.81285\text{E}-03\\ -8.64020\text{E}-02\\ 1.14592\text{E}-04\end{bmatrix}\frac{5\pi}{180.0} = \begin{bmatrix}-7.3\text{E}-04\\ -7.8\text{E}-04\\ 1.1\text{E}-02\\ 4.2\text{E}-04\\ -7.54\text{E}-03\\ 1.0\text{E}-05\end{bmatrix}$$

$$C_{(x,z,m),\overline{q}}(\alpha)\overline{q} \rightarrow \begin{bmatrix}\Delta C_x\\ \Delta C_z\\ \Delta C_m\end{bmatrix} = \begin{bmatrix}0.88\\ -31.8\\ -5.18\end{bmatrix}(0.01) = \begin{bmatrix}0.0088\\ -0.318\\ -0.0518\end{bmatrix}$$

$$C_{(x,z,m),\overline{q}\delta_{lef}}(\alpha)\overline{q}\delta_{lef} \rightarrow \begin{bmatrix}\Delta C_x\\ \Delta C_z\\ \Delta C_m\end{bmatrix} = \begin{bmatrix}3.62109\\ 2.97938\\ -0.61879\end{bmatrix}(0.01)\frac{5\pi}{180.0} = \begin{bmatrix}0.00316\\ 0.0026\\ -0.00054\end{bmatrix}$$

$$C_{(x,z,m),\delta_{sb}}(\alpha)\delta_{sb} \rightarrow \begin{bmatrix}\Delta C_x\\ \Delta C_z\\ \Delta C_m\end{bmatrix} = \begin{bmatrix}-3.419\text{E}-02\\ -2.564\text{E}-01\\ 2.760\text{E}-02\end{bmatrix}\frac{5\pi}{180.0} = \begin{bmatrix}-2.98333\text{E}-03\\ -2.23750\text{E}-02\\ 2.40833\text{E}-03\end{bmatrix}$$

$$C_{(y,\ell,n),\overline{p}}(\alpha)\overline{p} \rightarrow \begin{bmatrix}\Delta C_y\\ \Delta C_\ell\\ \Delta C_n\end{bmatrix} = \begin{bmatrix}0.12800\\ -0.414\\ -0.0085\end{bmatrix}(0.01) = \begin{bmatrix}0.00128\\ -0.00414\\ -0.000085\end{bmatrix}$$

$$C_{(y,\ell,n),\overline{r}}(\alpha)\overline{r} \rightarrow \begin{bmatrix}\Delta C_y\\ \Delta C_\ell\\ \Delta C_n\end{bmatrix} = \begin{bmatrix}0.966\\ 0.124\\ -0.381\end{bmatrix}(0.01) = \begin{bmatrix}0.00966\\ 0.00124\\ -0.00381\end{bmatrix}$$

$$C_{(y,\ell,n),\delta_a}(\beta,\alpha)\delta_a \rightarrow \begin{bmatrix}\Delta C_y\\ \Delta C_\ell\\ \Delta C_n\end{bmatrix} = \begin{bmatrix}7.43413\text{E}-02\\ -1.46964\text{E}-01\\ -2.70723\text{E}-02\end{bmatrix}\frac{5\pi}{180.0} = \begin{bmatrix}6.4875\text{E}-03\\ -1.2825\text{E}-02\\ -2.3625\text{E}-03\end{bmatrix}$$

$$C_{(y,\ell,n),\delta_r}(\beta,\alpha)\delta_r \rightarrow \begin{bmatrix}\Delta C_y\\ \Delta C_\ell\\ \Delta C_n\end{bmatrix} = \begin{bmatrix}1.894\text{E}-01\\ 2.702\text{E}-02\\ -8.690\text{E}-02\end{bmatrix}\frac{5\pi}{180.0} = \begin{bmatrix}1.6525\text{E}-02\\ 2.35833\text{E}-03\\ -7.58333\text{E}-03\end{bmatrix}$$

$$C_{(y,\ell,n),\overline{p}\delta_{lef}}(\alpha)\overline{p}\delta_{lef} \rightarrow \begin{bmatrix}\Delta C_y\\ \Delta C_\ell\\ \Delta C_n\end{bmatrix} = \begin{bmatrix}-1.37739\text{E}-01\\ -4.58366\text{E}-02\\ -2.95646\text{E}-02\end{bmatrix}\frac{0.05\pi}{180.0} = \begin{bmatrix}-1.202\text{E}-04\\ -4.0\text{E}-05\\ -2.58\text{E}-05\end{bmatrix}$$

$$C_{(y,\ell,n),\overline{r}\delta_{lef}}(\alpha)\overline{r}\delta_{lef} \rightarrow \begin{bmatrix}\Delta C_y\\ \Delta C_\ell\\ \Delta C_n\end{bmatrix} = \begin{bmatrix}-6.18794\text{E}-02\\ -8.25059\text{E}-02\\ -3.66693\text{E}-02\end{bmatrix}\frac{0.05\pi}{180.0} = \begin{bmatrix}-5.4\text{E}-05\\ -7.2\text{E}-05\\ -3.2\text{E}-05\end{bmatrix}$$

$$C_{(y,\ell,n),\delta_a\delta_{lef}}(\beta,\alpha)\delta_a\delta_{lef} \rightarrow \begin{bmatrix}\Delta C_y\\ \Delta C_\ell\\ \Delta C_n\end{bmatrix} = \begin{bmatrix}7.550\text{E}-02\\ 1.346\text{E}-02\\ -1.871\text{E}-02\end{bmatrix}\left(\frac{5\pi}{180.0}\right)^2 = \begin{bmatrix}5.75\text{E}-04\\ 1.025\text{E}-04\\ -1.425\text{E}-04\end{bmatrix}$$

Summing the results as outlined in Eq. (C.3.1) yields the final aerodynamic coefficients

$$C_x = -3.5333333333\text{E}{-04}, \quad C_y = -6.5126700000\text{E}{-02}, \quad C_z = -8.167750000\text{E}{-01}$$
$$C_\ell = -2.5876166667\text{E}{-02}, \quad C_m = -1.0662166667\text{E}{-01}, \quad C_n = 4.5288666667\text{E}{-03}$$

The results are shown out to as high of precision as room allows in order to facilitate code debugging. Repeating the process for the remaining cases yields the results shown in Table 3.11.9.

Table 3.11.9. Results for the cases in Example 3.11.2.

	Case			
	A	B	C	D
C_x	-0.0003533333	0.0900666667	-0.1096000000	-0.0768300000
C_y	-0.0651267000	0.1165141667	0.0008500000	-0.4117516667
C_z	-0.8167750000	-2.0746500000	-1.4092000000	0.3714000000
C_ℓ	-0.0258761667	0.0398088201	0.0193966667	0.0070986667
C_m	-0.1066216667	-0.1591166667	0.1729000000	-0.2365920000
C_n	0.0045288667	-0.0361026868	0.0340000000	0.1134510000

3.12 Problems

Theory Problems

3.T1 From the definition of the body-fixed, stability, and wind coordinate systems, develop the relationships given in Table 3.4.1.

3.T2 From the definition of the body-fixed, stability, and wind coordinate systems, along with the definitions of dimensionless rotational rates given in Eq. (3.4.23), develop the relationships given in Table 3.4.2.

3.T3 From the definition of the body-fixed, stability, and wind coordinate systems, develop the relationships given in Table 3.4.3.

3.T4 From the definition of the body-fixed, stability, and wind coordinate systems, develop the relationships given in Table 3.4.4.

3.T5 From the definition of the body-fixed, stability, and wind coordinate systems, develop the relationships given in Table 3.4.5.

3.T6 From the definition of the body-fixed, stability, and wind coordinate systems, along with the definition of dimensionless moments given in Eq. (3.4.33), develop the relationships given in Table 3.4.6.

3.T7 Search the literature to find an aerodynamic model of a cuboid, or data from which a model can be constructed. If data is available, fit the data to an equation to develop a model as a function of the important independent parameters. Compare the model to that given in Eqs. (3.6.8)–(3.6.11).

3.T8 Search the literature to find an aerodynamic model of a cylinder, or data from which a model can be constructed. If data is available, fit the data to an equation to develop a model as a function of the important independent parameters. Compare the model to that given in Eqs. (3.6.12)–(3.6.22).

3.T9 Search the literature to find an aerodynamic model of a sphere, or data from which a model can be constructed. If data is available, fit the data to an equation to develop a model as a function of the important independent parameters. Compare the model to that given in Eqs. (3.6.26)–(3.6.29).

3.T10 Search the literature to find an aerodynamic model of a wing segment below stall, or data from which a model can be constructed. If data is available, fit the data to an equation to develop a model as a function of the important independent parameters. Compare the model to that given in Eqs. (3.6.30)–(3.6.55).

3.T11 Search the literature to find an aerodynamic model of a wing segment above stall, or data from which a model can be constructed. If data is available, fit the data to an equation to develop a model as a function of the important independent parameters. Compare the model to that given in Eqs. (3.6.59)–(3.6.61).

3.T12 Using the experimental data published by NASA in 1979 [39], estimate coefficients that can be used to model the aerodynamics of the F-16 below stall in a model of the form given in Eqs. (3.7.24)–(3.7.30). Appendix C.3 contains sufficient tabulated data for this problem. Compare your results for the longitudinal coefficients to those given in Example 3.7.1, and your results for the lateral coefficients to those given in Appendix C.3.

3.T13 Based on the geometry of the F-16 given in Appendix C.3, use the analytic approximations given in Section 3.7 to approximate as many of the derivatives needed for the model given in Eqs. (3.7.24)–(3.7.30) as possible. Compare the results to the results from Problem 3.T12, which were obtained from experimental data.

3.T14 Search the literature to find data for a fixed-wing aircraft at high angles of attack above stall. Fit the model given in Eqs. (3.6.59), (3.6.60), and (3.8.1) and select values for α_b and λ_b that gives a model that best fits the data.

3.T15 Search the literature to find experimental or CFD data that shows how aerodynamic parameters of a vehicle vary with Mach number. Fit a curve through the data to be used as a correction model for your vehicle aerodynamic parameters as a function of Mach number. You may wish to combine multiple sources of data to develop your correction model. Compare your model to those given in Eqs. (3.9.1)–(3.9.6).

3.T16 Plot the ground effect models for R_D given in Eqs. (3.10.3)–(3.10.7) and compare the results to those obtained from Eq. (3.10.9) for a wing with an aspect ratio of 8, and a taper ratio of 0.5. Comment on the results.

Coding Problems

3.C1 Write a function to compute gravity as a function of altitude from Eq. (3.2.1). The input to the function call should be the geometric altitude in meters, and the function should return gravity in SI units of m/s^2. Test your code against the solutions given in Table B.1.1.

3.C2 Write a function to compute gravity as a function of geometric altitude in English units. The input should be the geometric altitude in feet, and the function should return gravity in English units of ft/s^2. Test your code against the solutions given in Table B.1.1.

3.C3 Write a function to compute standard-atmospheric properties in SI units as a function of geometric altitude. Given a geometric altitude in meters, your code must return the geopotential altitude [m], temperature [K], pressure [N/m^2], density [kg/m^3], dynamic viscosity [kg/(m·s)], and speed of sound [m/s]. Use your code to compute the atmospheric properties at altitudes varying from 0 to 90,000 m in increments of 5,000 m. Compare your results to the solutions given in Appendix B. Higher-precision solutions can be found on the book website.

3.C4 Write a function to compute standard-atmospheric properties in English units as a function of geometric altitude. Given a geometric altitude in feet, your code must return the geopotential altitude [ft], temperature [°R], pressure [lbf/ft^2], density [slug/ft^3], dynamic viscosity [slug/(ft·s)], and speed of sound [ft/s]. The easiest way to do this is to create a wrapper to the function written for Problem 3.C3. The wrapper would accept the altitude in feet, convert the altitude to meters, call the function in SI units, and convert the results to English units. Use your code to compute the atmospheric properties at altitudes varying from 0 to 200,000 ft in increments of 10,000 ft. Compare your results to the solutions given in Appendix B. Higher-precision solutions can be found on the book website.

3.C5 Modify your code written for Problems 3.C3 and 3.C4 to run as quickly as possible. This can be tested by generating a wrapper code that calls the functions for thousands of different altitudes. You can set up timing procedures to compute the processor time required for the wrapper code to run. Use this setup to test the speed of your code and find faster ways of computing correct atmospheric properties. An example wrapper code to perform this speed test can be found on the book website.

3.C6 Write a function that accepts the components of a vector given in the body-fixed coordinate system and returns the vector in the stability coordinate system.

3.C7 Write a function that accepts the components of a vector given in the body-fixed coordinate system and returns the vector in the wind coordinate system.

3.C8 Write a function that accepts the components of a vector given in the stability coordinate system and returns the vector in the body-fixed coordinate system.

3.C9 Write a function that accepts the components of a vector given in the stability coordinate system and returns the vector in the wind coordinate system.

3.C10 Write a function that accepts the components of a vector given in the wind coordinate system and returns the vector in the body-fixed coordinate system.

3.C11 Write a function that accepts the components of a vector given in the wind coordinate system and returns the vector in the stability coordinate system.

3.C12 Add a function to the parent class written for Problem 2.C1 that computes the local velocity in the component coordinate system for an arbitrary component, which depends on the position and orientation of the component relative to the vehicle center of gravity. The input to the function will be the global vehicle properties of velocity, angle of attack, sideslip angle, and rotation rates.

3.C13 Add a function to the parent class written for Problem 2.C1 that transforms the aerodynamic forces and moments from the component coordinate system to the body-fixed coordinate system about the vehicle center of gravity.

3.C14 Add a function to the cuboid child class written for Problem 2.C2 that computes the aerodynamics of a cuboid based on the local velocity vector of the cuboid computed from the function developed for Problem 3.C12. The function should include the ability to print the properties of the cuboid to the screen. Test your code using the results given in Example 3.6.1.

3.C15 Repeat Problem 3.C14 for the case of a cylinder component. Test your code using the results given in Example 3.6.2.

3.C16 Repeat Problem 3.C14 for the case of a sphere component. Test your code using the results given in Example 3.6.3.

3.C17 Repeat Problem 3.C14 for the case of a wing-segment component. Test your code using the results given in Example 3.6.4.

3.C18 Add a function to the vehicle class that sums all aerodynamic influences of the components and returns the global aerodynamic force and moment vector in the body-fixed coordinate system. Test your code using the results given in Example 3.6.5.

3.C19 Write a vehicle-class function that reads all coefficients for the aerodynamic model given in Eqs. (3.7.24)–(3.7.30) from a JSON object formatted like that given in Appendix C.3.

3.C20 Write a function that computes the aerodynamic forces and moments of a vehicle in the body-fixed coordinate system, based on the model given in Eqs. (3.7.24)–(3.7.30). This can be a function within a component class, or a vehicle class, in which all the aerodynamic coefficients needed for the model are known. These coefficient values can be read in from a JSON file during initialization.

3.C21 Modify the function written for Problem 3.C20 to include the effects of stall. Test your code using the results given in Example 3.8.1.

3.C22 Modify the function written for Problem 3.C20 to approximate the effects of compressibility.

3.C23 Modify the function written for Problem 3.C20 so that the influence of ground effect is included for the lift and drag coefficients. Test your code using the results given in Example 3.10.1.

3.C24 Write a dataset class that can be used to read and interpolate dataset information structured as a rectilinear dataset or a recursive dataset. The class constructor should do the following:

1. Accept the name of a file that contains the tabulated data
2. Initialize the dataset properties and parameters based on the type of dataset

Write a function within the class that accepts an array of independent variables and returns the interpolated dependent variables. Test your code using the results given in Examples 3.11.1 and 3.11.2.

Chapter 4

Propulsion Modeling

4.1 Introduction

Many methods of aircraft propulsion have been developed in the past century, including propellers, turboprops, turbojets, turbofans, ramjets, scramjets, and rocket engines. Methods of powering these systems range from traditional fuel-based engines to electric, hydrogen fuel cell, and hybrid systems. Accurately modeling these systems can be very complex, since they often involve a combination of electrical, mechanical, fluid, and combustion systems and dynamics.

For practical simulation purposes, propulsion modeling can take on many different forms ranging from simple or ideal engine models to complex data tables. The most accurate models typically come from what are termed *engine decks*. An engine deck is a computer code or tabulated data from which engine performance can be obtained as a function of engine input parameters and operating conditions. An engine deck can properly account for many intricacies, including fuel-flow rates, ram pressure, operating temperatures, air bleed, power extraction, and other parameters. An engine deck is commonly provided by the engine manufacturer to the customer when an engine is purchased. However, engine-deck information is often considered proprietary. Therefore, detailed propulsion models or performance data as a function of input parameters and operating conditions for specific engines are not always available in the open literature.

Methods for predicting engine performance can be obtained from models used in the design and analysis of propulsion systems. For example, ideal engine models for turbojets, turbofans, ramjets, and scramjets can be obtained from textbooks on the subject [63, 64]. For flight simulation focused on vehicle dynamics, detailed modeling of a propulsion system is not always necessary. Hence, rather than presenting detailed modeling methods, the present chapter will focus on simple models that can be used for a variety of propulsion systems. As will be shown, reasonable first-order approximations for very complex systems can be developed quite easily based on very limited data. A more detailed model for propellers and electric systems is also included in this chapter, since propellers are commonly used for aerospace vehicles.

The term *throttle* is sometimes used to denote a single control input to a propulsion system. In general, there are many different inputs to propulsion systems. For example, inputs to a traditional piston-engine with a variable-pitch propeller could include both fuel rate and propeller pitch angle. To simplify the work presented in this section, the term throttle will always be a number between zero and one, and represent a percentage relative to the maximum thrust or power available. Dynamic lag in the propulsion system can be modeled in the same way actuation lag is modeled, as described in Chapter 9.

For simulation purposes, a propulsion system will be treated as a component attached to a main vehicle. The local velocity of a component attached to the main vehicle is related to the vehicle velocity, vehicle angular rate, and location of the component relative to the vehicle CG. This relationship is given in Eq. (3.6.2), and the resulting unit vector in the direction of the local velocity relative to the propulsion component is given in Eq. (3.6.3). These same relationships can be used to denote the local freestream velocity relative to the propulsion system.

The model for a propulsion system must be able to predict the resulting forces and moments as a function of throttle setting, operating altitude, and velocity. Constructing such models for a range of propulsion systems is the focus of this chapter. The resulting forces and moments from the propulsion system will be denoted as $\mathbf{F}_c$ and $\mathbf{M}_c$ in the component coordinate system. These can be summed and translated to the vehicle center of gravity as was shown in Eqs. (3.6.4)–(3.6.7).

4.2 A General Propulsion Model

A first-order general model for a propulsion system can be expressed as

$$T = \tau \left(\frac{\rho}{\rho_{\text{ssl}}} \right)^{T_a} (T_0 + T_1 V + T_2 V^2) \tag{4.2.1}$$

where ρ is the air density at the operating condition, ρ_{ssl} is the air density at sea level, and T_a, T_0, T_1, and T_2 are model constants. Note that T_0 represents the maximum static thrust at sea level. This model can be used to fit data from a wide range of propulsion systems.

EXAMPLE 4.2.1 The Pratt and Whitney PW4056 engine is a dual-spool, axial-flow, high-bypass-ratio turbofan engine used on the Boeing 747. McCormick [65] published maximum cruise thrust for this engine as a function of altitude and Mach number for the case of 100% ram recovery, no airbleed, and no power extraction on a standard day. Fit the data in the plots published by McCormick to the model given in Eq. (4.2.1) to obtain model coefficients that can be used to represent the thrust of this engine as a function of altitude, velocity, and throttle setting.

Solution: The original performance curves published by McCormick include maximum cruise thrust as a function of altitude and Mach number. Data along these curves were selected and digitized. The resulting data are shown in Fig. 4.2.1. Using the standard-day atmospheric model from Chapter 3, the density and speed of sound at each specified altitude was obtained. The speed of sound was used to convert the Mach number to velocity in units of ft/s. The resulting data was fit to Eq. (4.2.1) using the Solver function in Excel® to minimize the error between the data and model. This yields the following coefficients

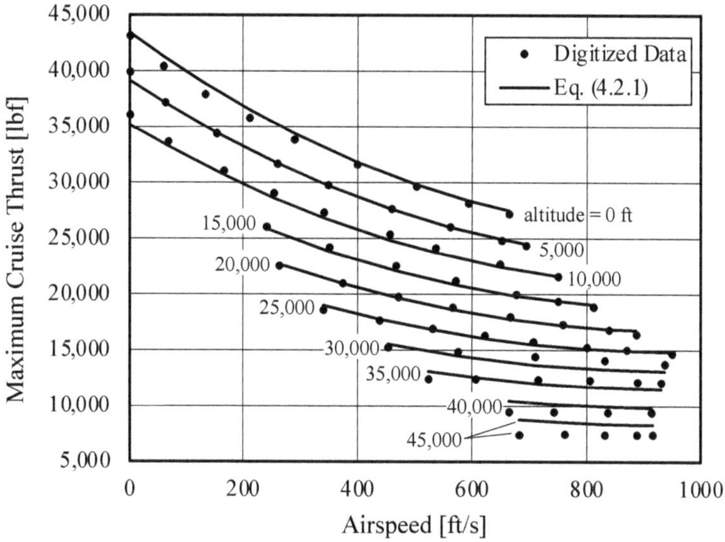

Figure 4.2.1. Maximum cruise thrust for the PW4056 turbofan engine digitized from plots published by McCormick [65], and a fit to the data using Eq. (4.2.1).

$$T_a = 0.6941, \quad T_0 = 43,434 \text{ lbf},$$
$$T_1 = -36.753 \text{ lbf}\cdot\text{s/ft}, \quad T_2 = 0.01927 \text{ lbf}\cdot(\text{s/ft})^2$$

Results for this model are shown as the solid line in Fig. 4.2.1. Note that the model is not as accurate at higher altitudes, most likely due to the fact that fewer points were used to represent the data at higher altitudes than at lower altitudes. The accuracy at higher altitudes could be improved by adding more data points at higher altitudes, or by weighting the error at higher altitudes more heavily when computing the error.

4.3 Thrust Independent of Airspeed

A simple first-order model of some propulsion systems can be generated by approximating the system as one that produces a maximum thrust at a given altitude, independent of airspeed V. Applying this approximation to Eq. (4.2.1) yields the simple model

$$T = \tau T_0 \left(\frac{\rho}{\rho_{\text{ssl}}}\right)^{T_a} \tag{4.3.1}$$

where T_0 is the maximum thrust at sea level. This approximation is sometimes appropriate for modeling a turbojet engine. This model is particularly useful, since maximum thrust values at sea level are sometimes publicly available for specific engines, even if variation of that thrust with altitude is not specifically known. In such a case, a reasonable model can be obtained by simply using the maximum thrust value T_0 along with a value of $T_a = 1$, yielding the model

$$T = \tau T_0 \left(\frac{\rho}{\rho_{\text{ssl}}} \right) \tag{4.3.2}$$

This will be demonstrated in the following example.

EXAMPLE 4.3.1 The Pratt and Whitney JT4A engine is an axial-flow turbojet engine used to power some Boeing 707 and Douglas DC-8 aircraft. McCormick [65] published maximum cruise thrust for a version of this engine as a function of altitude and airspeed for the case of 100% ram recovery on a standard day. Figure 4.3.1 shows data extracted from the plots published by McCormick. Fit the data in the plot to the models given in Eqs. (4.3.1) and (4.3.2) to obtain model coefficients that can be used to represent the thrust of the engine as a function of throttle setting and altitude, but independent of airspeed. Compare the results of the models.

Solution: Points along the performance curves published by McCormick were selected and digitized as a function of velocity. At each altitude, the maximum, minimum, and average were computed from the digitized points. These are shown in Fig. 4.3.2. Using the standard-day atmospheric model from Chapter 3, the density at each specified altitude was obtained. The data was fit to Eqs. (4.3.1) and (4.3.2) using the Solver function in Excel® to minimize the error between each model and the data points. The resulting coefficients for the model from Eq. (4.3.1) are

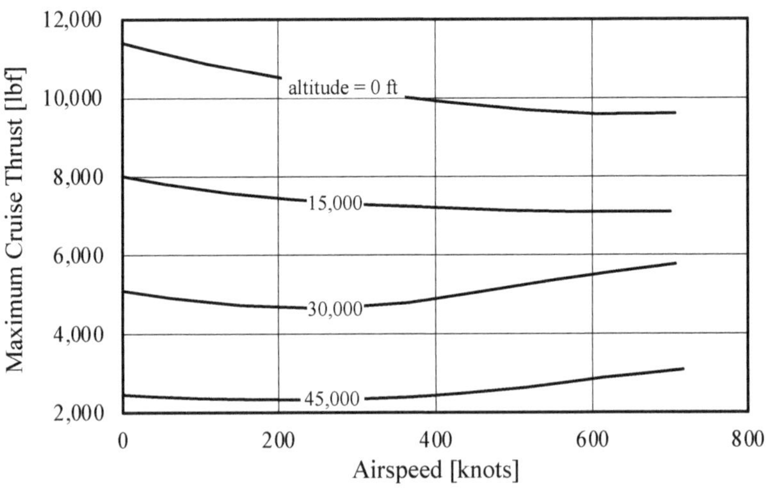

Figure 4.3.1. Digitized data for the maximum cruise thrust of the JT4A turbojet engine as a function of altitude and airspeed, published by McCormick [65].

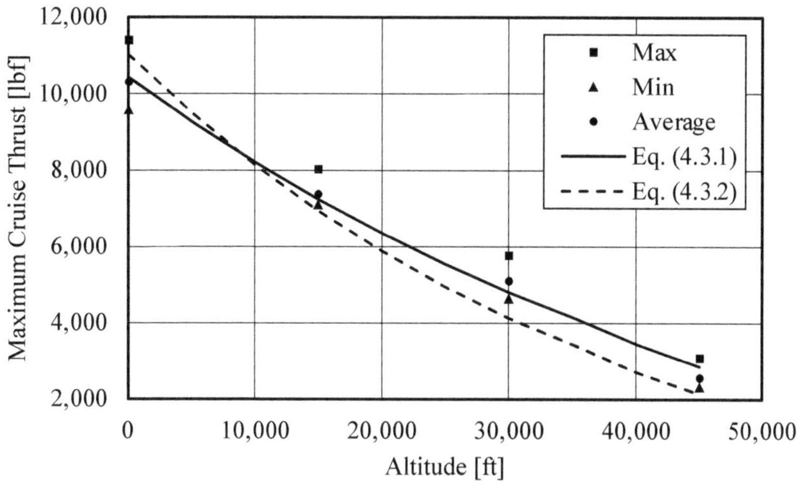

Figure 4.3.2. Maximum cruise thrust for the JT4A turbojet engine including data digitized from McCormick [65], along with two model fits to the data.

$$T_a = 0.79, \quad T_0 = 10,438 \text{ lbf}$$

and results for this model are shown as the solid line in Fig. 4.3.2. The resulting coefficient for the model from Eq. (4.3.2) is

$$T_0 = 11,031 \text{ lbf}$$

and results for this model are shown as the dashed line in Fig. 4.3.2.

As a first-order approximation, results from Eq. (4.3.1) fit the data better than the results from Eq. (4.3.2). Hence, Eq. (4.3.1) should be used for engines of this type when maximum thrust is known as a function of altitude. However, in cases when only the maximum thrust at sea level is known, Eq. (4.3.2) can be used as an estimate for the thrust available from engines of this type.

4.4 Power Independent of Airspeed

A first-order model of some propulsion systems can be generated by approximating the system as one that produces available power that is nearly independent of airspeed V and proportional to the density ratio $(\rho/\rho_{ssl})^{T_a}$ and throttle setting τ. This yields

$$P_a = \tau P_{max} \left(\frac{\rho}{\rho_{ssl}} \right)^{T_a} \tag{4.4.1}$$

where $P_{\max}$ is the maximum power available at sea level. Because available power is equal to thrust multiplied by the true airspeed of the incoming flow, this can alternately be written in the form

$$TV = \tau P_{\max} \left(\frac{\rho}{\rho_{\text{ssl}}} \right)^{T_a} \tag{4.4.2}$$

which yields a relationship for thrust in the form

$$T = \tau \frac{P_{\max}}{V} \left(\frac{\rho}{\rho_{\text{ssl}}} \right)^{T_a} \tag{4.4.3}$$

This model can be appropriate for piston engines connected to a variable-pitch propeller, which can be adjusted such that the maximum power available is nearly independent of velocity.

EXAMPLE 4.4.1 The Wright H-3 piston engine produced power as a function of altitude as published by Raffaelli [66]. Fit the model given by Eq. (4.4.1) to this data.

Solution: Points along the performance curve published by Raffaelli were selected and digitized. The resulting data is shown in Fig. 4.4.1. Raffaelli also included density ratios at each altitude, which were used here instead of the atmospheric model given in Chapter 3. The data was fit to Eq. (4.4.1) using the Solver function in Excel® to minimize the error between the model and the digitized data. This yields the following coefficients

$$T_a = 1.19, \quad P_{\max} = 316.4 \text{ hp}$$

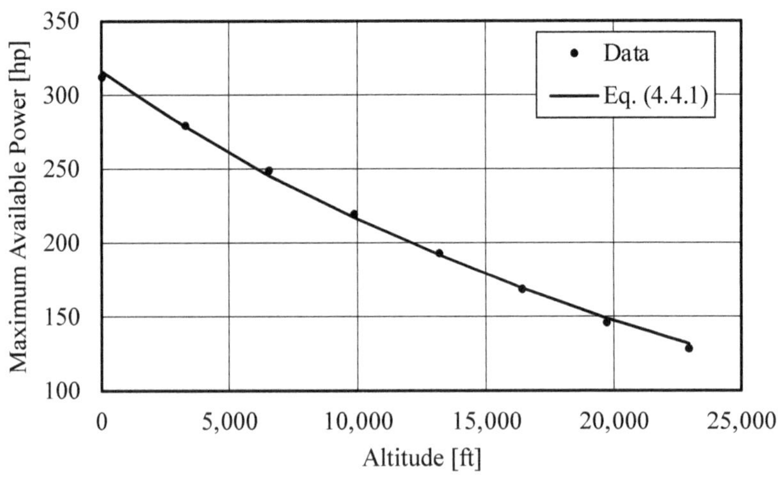

Figure 4.4.1. Maximum power available for the Wright H-3 piston engine as digitized from Raffaelli [66], along with a model fit to the data.

Results for this model are shown as the solid line in Fig. 4.4.1. In the event that only the sea-level maximum power $P_{\max}$ for a piston engine is known, rather than the power as a function of altitude, Eq. (4.4.1) can be used with $T_a = 1$ as a reasonable estimate.

4.5 Rotors

General Propulsive Properties

Rotors are used on many air and water vehicles for propulsion and control. The terminology and rotor model developed in this section applies most directly to fixed-wing aircraft. However, similar models can be developed for rotary aircraft such as helicopters and multi-rotor designs.

The main propulsive property of interest for a rotor is the thrust, T. However, a rotor also produces other forces and moments. For example, the torque required to turn the rotor must be reacted by the vehicle. Torque will be given the symbol ℓ_r. When a rotor is perfectly aligned with the incoming flow, the dominant force is the thrust, and the dominant moment is the torque. When a rotor is not aligned with the incoming velocity, the rotor produces an additional force called the *normal force* N, and an additional moment called the *yawing moment* n. Both the normal force and yawing moment are caused by the fact that the aerodynamic loading on the blades are not axisymmetric when the incoming flow is not aligned with the rotor axis. The normal force is caused by the asymmetric drag on the rotor blades, and the yawing moment is caused by the asymmetric lift on the rotor blades. Hence, in general, the four main propulsive properties of a rotor include the thrust T, the torque ℓ_r, the normal force N, and the yawing moment n.

For modeling purposes, the propulsive properties are typically expressed in dimensionless form. The thrust is related to the dimensionless thrust coefficient C_T according to

$$T = \rho(\omega_r/2\pi)^2 d_r^4 C_T \tag{4.5.1}$$

where ω_r is the rotor angular velocity in radians per second and d_r is the rotor diameter. The torque is typically expressed in terms of the power that must be provided by the engine to turn the rotor. This is called the *brake power*, and is defined as the product of the torque ℓ_r and rotor angular rate

$$P_b \equiv \ell_r \omega_r \tag{4.5.2}$$

The brake power is related to the dimensionless brake-power coefficient C_{P_b} according to

$$P_b = \rho(\omega_r/2\pi)^3 d_r^5 C_{P_b} \tag{4.5.3}$$

where the term $(\omega_r/2\pi)$ has the units of rotations per second. The normal force and yawing moment are directly proportional to the angle of attack. The angle of attack will be defined here as the angle between the rotor velocity vector and rotor axis, which can be found from

$$\alpha = \cos^{-1}(u_{c_x}) \tag{4.5.4}$$

where $\mathbf{u}_c$ is defined in Eq. (3.6.3). The normal force is related to the dimensionless normal-force coefficient $C_{N,\alpha}$ according to

$$N = \rho(\omega_r/2\pi)^2 d_r^4 C_{N,\alpha}\alpha \tag{4.5.5}$$

The yawing moment is related to the dimensionless yawing-moment coefficient $C_{n,\alpha}$ according to

$$n = \rho(\omega_r/2\pi)^2 d_r^5 C_{n,\alpha}\alpha \tag{4.5.6}$$

Hence, in order to model the main propulsive properties of a rotor, we must have expressions for the dimensionless parameters C_T, C_{P_b}, $C_{N,\alpha}$, and $C_{n,\alpha}$.

Dimensionless Parameters

In general, the propulsive properties of a rotor depend on the rotor geometry, number of blades, rotor rotation rate, and rotor velocity vector. The definitions for the rotor geometry used in Chapter 2 were used to estimate the mass properties of a rotor, and are quite accurate for those purposes. Accurately predicting the propulsive properties that result from a spinning rotor requires some additional information. For example, the geometric definition in Chapter 2 did not include the blade twist distribution (pitch) or any chord distribution other than a simply tapered blade. However, the aerodynamics of a rotor depend heavily on the chord and twist distributions. If the full geometry of the propeller is known, a method such as blade theory [67] or CFD simulations can be used to estimate these coefficients. These coefficients can also be obtained from wind-tunnel testing.

The propulsive properties of a rotor are typically a very strong function of a dimensionless number called the *advance ratio*. The advance ratio is a ratio of the distance that the propeller advances in the flow in one revolution divided by the propeller diameter, and is mathematically defined as

$$J \equiv \frac{2\pi V_c}{\omega_r d_r} \tag{4.5.7}$$

The rotation rate of the rotor in radians per second is denoted as ω_r, and in rotations per minute (rpm) as N_r. These are related according to

$$\omega_r = N_r \frac{\pi}{30} \tag{4.5.8}$$

Note that the total freestream velocity V_c is applied in the definition of advance ratio used in this book, which is typical for a fixed-wing aircraft. For rotorcraft, more accurate propeller models can be developed by considering the propulsive properties of the rotor in terms of both the axial-velocity component and a normal- or side-velocity component.

The dimensionless thrust, brake power, normal-force, and yawing-moment coefficients of a rotor can be modeled as polynomial functions of advance ratio of the form

$$C_T = C_{T0} + C_{T1}J + C_{T2}J^2 \tag{4.5.9}$$

$$C_{P_b} = C_{P0} + C_{P1}J + C_{P2}J^2 \tag{4.5.10}$$

$$C_{N,\alpha} = C_{N1}J + C_{N2}J^2 + C_{N3}J^3 \tag{4.5.11}$$

$$C_{n,\alpha} = C_{n1}J + C_{n2}J^2 + C_{n3}J^3 \tag{4.5.12}$$

Because the normal force and yawing moment are zero when the freestream velocity is zero, the parabolic functions for the normal force and yawing moment do not have a constant term. As mentioned above, the coefficients in these functions depend on the rotor pitch, diameter, and airfoil properties. For modeling purposes, the dimensionless coefficients C_{T0}, C_{T1}, C_{T2}, C_{P0}, C_{P1}, C_{P2}, C_{N1}, C_{N2}, C_{N3}, C_{n1}, C_{n2}, and C_{n3} can be accepted as input to the simulation code by the user.

Example Values

Phillips [67] provides an example for how to use propeller blade theory to estimate these coefficients. In his work, he considers a family of fixed-pitch propellers, each with two blades, and each employing an elliptic chord distribution. The properties of the propellers are considered as a function of the pitch-to-diameter ratio K_c. The following relationships were obtained by fitting the polynomials given in Eqs. (4.5.9)–(4.5.12) to digitized plots published by Phillips for this family of propellers. The coefficients required for C_T can be approximated as

$$C_{T0} = -0.119K_c^2 + 0.238K_c - 0.0194 \tag{4.5.13}$$

$$C_{T1} = 0.146K_c^2 - 0.0816K_c - 0.0612 \tag{4.5.14}$$

$$C_{T2} = 0.175K_c^3 - 0.496K_c^2 + 0.441K_c - 0.211 \tag{4.5.15}$$

The coefficients required for C_{P_b} can be approximated as

$$C_{P0} = -0.953K_c^4 + 2.5K_c^3 - 2.243K_c^2 + 0.885K_c - 0.115 \tag{4.5.16}$$

$$C_{P1} = 2.17K_c^4 - 5.55K_c^3 + 4.9K_c^2 - 1.75K_c + 0.225 \tag{4.5.17}$$

$$C_{P2} = -0.991K_c^4 + 2.34K_c^3 - 1.81K_c^2 + 0.541K_c - 0.132 \tag{4.5.18}$$

The coefficients required for $C_{N,\alpha}$ can be approximated as

$$C_{N1} = 0.0147K_c^2 + 0.0116K_c - 0.0034 \tag{4.5.19}$$

$$C_{N2} = -0.0311K_c^2 + 0.0279K_c + 0.00984 \tag{4.5.20}$$

$$C_{N3} = 0.0171K_c^2 - 0.0139K_c + 0.0176 \tag{4.5.21}$$

The coefficients required for $C_{n,\alpha}$ can be approximated as

$$C_{n1} = 0.0222K_c^2 - 0.0454K_c + 0.0034 \tag{4.5.22}$$

$$C_{n2} = -0.037K_c^2 + 0.0384K_c - 0.0065 \tag{4.5.23}$$

$$C_{n3} = 0.0206K_c^2 - 0.0266K_c + 0.0123 \tag{4.5.24}$$

Similar methods can be used to obtain coefficients for other rotor geometries.

Computational Method

The rotor model included here can be used in different ways for a simulation code. For example, this model can be used to compute the resulting propulsive properties of thrust, brake power, normal force, and yawing moment from a specified rotor rotation rate. In some instances, it is helpful to be able to compute the rotor rotation rate from a specified thrust or torque. In either case, once the angular rate is known, the advance ratio can be computed from Eq. (4.5.7). Solutions to Eqs. (4.5.9)–(4.5.12) can then be found and used in Eqs. (4.5.1), (4.5.3), (4.5.5), and (4.5.6) to solve for the thrust, brake power, normal force, and yawing moment.

Scenario 1: Computing Thrust for a specified Rotation Rate: If the rotor rotation rate is specified, the thrust can be found by computing the advance ratio from Eq. (4.5.7) and using the result in Eq. (4.5.9) to find the thrust coefficient. The thrust coefficient can then be used in Eq. (4.5.1) to compute the dimensional thrust.

Scenario 2: Computing Rotation Rate for specified Thrust: It is sometimes useful to specify a prescribed thrust for a given operating condition. If the thrust is specified, the corresponding rotor rotation rate must be found. **If the thrust coefficient is a quadratic function of advance ratio** as shown in Eq. (4.5.9), the rotation rate can be found analytically. Using Eqs. (4.5.7) and (4.5.9) in Eq. (4.5.1) yields the thrust as a quadratic function of angular rate, which can be solved using the quadratic equation to yield

$$\omega_r = \frac{\pi}{d_r C_{T0}} \left[-V_c C_{T1} \pm \sqrt{V_c^2 C_{T1}^2 - 4C_{T0}\left(V_c^2 C_{T2} - \frac{T}{\rho d_r^2}\right)} \right] \tag{4.5.25}$$

The positive solution is usually the correct solution.

If the thrust coefficient is given in a database format or is not a quadratic function of advance ratio, iterative methods such as Newton's Method can be used to rapidly converge to the correct solution. In any case, once the angular rate is known, the advance ratio can be found from Eq. (4.5.7), and the solution process follows that described above.

Scenario 3: Computing Rotation Rate for specified Torque: It is sometimes useful to specify a prescribed torque for a given operating condition. If the torque is specified, the corresponding rotor rotation rate must be found. **If the torque coefficient is a quadratic function of advance ratio** as shown in Eq. (4.5.10), the rotation rate can be found analytically. Using Eqs. (4.5.7), (4.5.10), and (4.5.3), in Eq. (4.5.2) and rearranging yields the torque as a quadratic function of angular rate, which can be solved using the quadratic equation to yield

$$\omega_r = \frac{\pi}{d_r C_{P0}}\left[-V_c C_{P1} \pm \sqrt{V_c^2 C_{P1}^2 - 4C_{P0}\left(V_c^2 C_{P2} - \frac{2\pi \ell_r}{\rho d_r^3}\right)}\right] \tag{4.5.26}$$

The positive solution is usually the correct solution.

If the brake-power coefficient is given in a database format or is not a quadratic function of advance ratio, iterative methods such as Newton's Method can be used to rapidly converge to the correct solution. In any case, once the angular rate is known, the advance ratio can be found from Eq. (4.5.7), and the solution process follows that described above.

Total Force and Moment Vectors

The total force resulting from the rotor thrust and normal force can be expressed in vector form in the rotor coordinate system as

$$\mathbf{F}_c = \begin{bmatrix} T \\ 0 \\ 0 \end{bmatrix} + N\mathbf{u}_N \tag{4.5.27}$$

where

$$\mathbf{u}_N = \frac{\mathbf{v}_N}{|\mathbf{v}_N|} \tag{4.5.28}$$

$$\mathbf{v}_N = -\begin{bmatrix} 0 \\ u_{c_y} \\ u_{c_z} \end{bmatrix} \tag{4.5.29}$$

The rotor torque can be found by rearranging Eq. (4.5.2) to give

$$\ell_r = \frac{P_b}{\omega_r} \tag{4.5.30}$$

The total moment resulting from the rotor torque and yawing moment can be expressed in vector form in the rotor coordinate system as

$$\mathbf{M}_c = \begin{bmatrix} -\delta \ell_r \\ 0 \\ 0 \end{bmatrix} - \delta n \mathbf{u}_N \tag{4.5.31}$$

where δ depends on the direction of the rotor and takes on the values

$$\delta \equiv \begin{cases} 1, & \text{right-handed rotor} \\ -1, & \text{left-handed rotor} \end{cases} \tag{4.5.32}$$

Efficiency

The power that gets imparted to the surrounding air is called the *propulsive power* and is defined as the product of the thrust and local velocity magnitude

$$P_a \equiv T V_c \tag{4.5.33}$$

The ratio of the propulsive power to the brake power is called the *propulsive efficiency*

$$\eta_P \equiv \frac{P_a}{P_b} = \frac{T V_c}{\ell_r \omega_r} = \frac{C_T J}{C_{P_b}} \tag{4.5.34}$$

EXAMPLE 4.5.1 Consider a rotor with a diameter of $d_r = 2$ ft and a pitch-to-diameter ratio of $K_c = 0.8$ that has the properties given in Eqs. (4.5.13)–(4.5.24). Compute the coefficients C_T, C_{P_b}, $C_{N,\alpha}$, $C_{n,\alpha}$, and the propulsive efficiency as a function of advance ratio in the range of positive thrust. Predict the total force and moment generated by the rotor in the rotor coordinate system when the rotor is operating at sea level with a rotation rate of $N_r = 5,000$ rpm, a velocity of 100 ft/s, and an angle of attack of 5 deg.

Solution: The pitch-to-diameter ratio $K_c = 0.8$ can be used in Eqs. (4.5.13)–(4.5.24) to yield the following coefficients

$$C_{T0} = 0.09484, \quad C_{T1} = -0.03304, \quad C_{T2} = -0.08604,$$

$$C_{P0} = 0.0471312, \quad C_{P1} = 0.008232, \quad C_{P2} = -0.0654336,$$

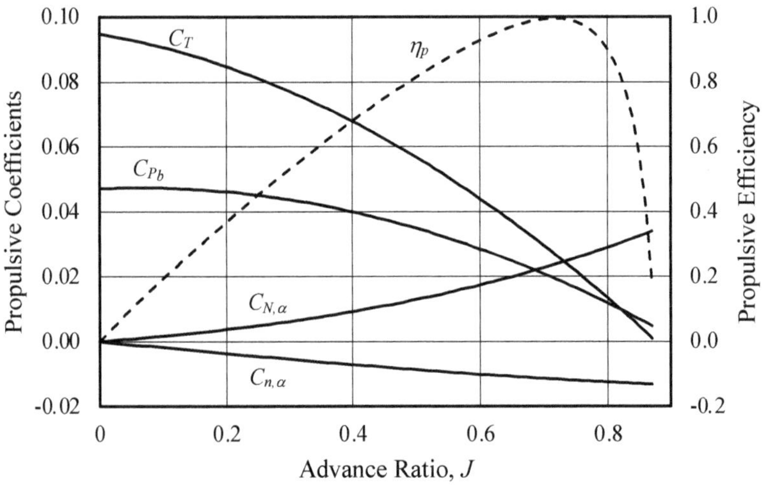

Figure 4.5.1. Propulsive properties of the propeller in Example 4.5.1.

$$C_{N1} = 0.015288, \quad C_{N2} = 0.012256, \quad C_{N3} = 0.017424,$$

$$C_{n1} = -0.018712, \quad C_{n2} = 0.00054, \quad C_{n3} = 0.004204$$

Using these coefficients in Eqs. (4.5.9)–(4.5.12) yields the results shown in Fig. 4.5.1. At the operating condition specified in the problem, the rotation rate and advance ratio can be computed as

$$\omega_r = \left(\frac{5000 \text{ rotations}}{\text{minute}}\right)\left(\frac{1 \text{ minute}}{60 \text{ second}}\right)\left(\frac{2\pi \text{ radians}}{\text{rotation}}\right) = 523.59877559830 \text{ rad/s}$$

$$J = \frac{2\pi V_c}{\omega_r d_r} = \frac{2\pi(100 \text{ ft/s})}{(523.5988 \text{ rad/s})(2 \text{ ft})} = 0.6$$

Using the advance ratio along with the coefficients computed above in Eqs. (4.5.9)–(4.5.12) yields

$$C_T = 0.0440416, \quad C_{P_b} = 0.028514304,$$

$$C_{N,\alpha} = 0.017348544, \quad C_{n,\alpha} = -0.010124736$$

These results can be used in Eqs. (4.5.1)–(4.5.6) along with the density at sea level to yield

$$T = \rho(\omega_r/2\pi)^2 d_r^4 C_T = 11.63134831186 \text{ lbf}$$

$$P_b = \rho(\omega_r/2\pi)^3 d_r^5 C_{P_b} = 1255.100487169 \text{ ft·lbf/s}$$

$$N = \rho(\omega_r/2\pi)^2 d_r^4 C_{N,\alpha}\alpha = 0.3998318412727 \text{ lbf}$$

$$n = \rho(\omega_r/2\pi)^2 d_r^5 C_{n,\alpha}\alpha = -0.4666895201442 \text{ ft·lbf}$$

The torque can be computed from Eq. (4.5.30) to yield

$$\ell_r = \frac{P_b}{\omega_r} = 2.397065359321 \text{ ft·lbf}$$

At the angle of attack of $\alpha = 5$ deg,

$$\mathbf{u}_c = \begin{bmatrix} \cos(\alpha) \\ 0 \\ \sin(\alpha) \end{bmatrix} = \begin{bmatrix} 0.996194698091746 \\ 0 \\ 0.0871557427476582 \end{bmatrix}$$

$$\mathbf{v}_N = -\begin{bmatrix} 0 \\ u_{c_y} \\ u_{c_z} \end{bmatrix} = \begin{bmatrix} 0 \\ 0 \\ -0.0871557427476582 \end{bmatrix}$$

$$\mathbf{u}_N = \frac{\mathbf{v}_N}{|\mathbf{v}_N|} = \begin{bmatrix} 0 \\ 0 \\ -1 \end{bmatrix}$$

Using these results in Eq. (4.5.27) yields

$$\mathbf{F}_c = \begin{bmatrix} T \\ 0 \\ 0 \end{bmatrix} + N\mathbf{u}_N = \begin{bmatrix} 11.63134831186 \\ 0 \\ -0.3998318412727 \end{bmatrix} \text{ lbf}$$

For this right-hand propeller, $\delta = 1$ and Eq. (4.5.31) yields

$$\mathbf{M}_c = \begin{bmatrix} -\delta\ell_r \\ 0 \\ 0 \end{bmatrix} - \delta n\mathbf{u}_N = \begin{bmatrix} -2.397065359321 \\ 0 \\ -0.4666895201442 \end{bmatrix} \text{ ft·lbf}$$

Note that to match these values to the precision shown here, the atmospheric density model from Chapter 3 must be used.

EXAMPLE 4.5.2 Consider the rotor used for the Simple Quadrotor model given in Appendix C.4 with properties specified in Tables C.4.1 and C.4.2. Predict the total force and moment generated by each rotor in the rotor coordinate system and the vehicle body-fixed coordinate system when the vehicle is hovering at 10,000 ft with a rotor rotation rate of $N_r = 7{,}000$ rpm. Also compute the rotor angular momentum in the rotor coordinate system and the vehicle body-fixed coordinate system.

Solution: At the operating condition specified in the problem, the rotation rate is

$$\omega_r = \left(\frac{7000 \text{ rotations}}{\text{minute}}\right)\left(\frac{1 \text{ minute}}{60 \text{ second}}\right)\left(\frac{2\pi \text{ radians}}{\text{rotation}}\right) = 733.038285838 \text{ rad/s}$$

When the vehicle is hovering, the airspeed is zero and the advance ratio is $J = 0$. From Eqs. (C.4.1)–(C.4.4), the dimensionless thrust, brake power, normal-force, and yawing-moment coefficients are

$$C_T = 0.0560, \quad C_{P_b} = 0.0137, \quad C_N = C_n = 0.0$$

These results can be used in Eqs. (4.5.1)–(4.5.6) and (4.5.30) along with the density at an altitude of 10,000 ft to yield

$$T = \rho(\omega_r/2\pi)^2 d_r^4 C_T = 3.266891482031 \text{ lbf}$$

$$\ell_r = \rho(\omega_r/2\pi)^3 d_r^5 C_{P_b}/\omega_r = 0.159000098490 \text{ ft·lbf}$$

$$N = \rho(\omega_r/2\pi)^2 d_r^4 C_{N,\alpha}\alpha = 0.0 \text{ lbf}$$

$$n = \rho(\omega_r/2\pi)^2 d_r^5 C_{n,\alpha}\alpha = 0.0 \text{ ft·lbf}$$

Because $C_N = C_n = 0$ when the advance ratio is zero, using these results in Eq. (4.5.27) yields

$$\mathbf{F}_c = \begin{bmatrix} T \\ 0 \\ 0 \end{bmatrix} + N\mathbf{u}_N = \begin{bmatrix} 3.266891482031 \\ 0 \\ 0 \end{bmatrix} \text{ lbf}$$

Table 4.5.1. Example 4.5.2 forces and moments in the rotor coordinate system.

Rotor	Rotation	$\mathbf{F}_c$ [lbf]	$\mathbf{M}_c$ [ft·lbf]
1	RH	[3.266891482031, 0, 0]	[-0.159000098490, 0, 0]
2	LH	[3.266891482031, 0, 0]	[0.159000098490, 0, 0]
3	RH	[3.266891482031, 0, 0]	[-0.159000098490, 0, 0]
4	LH	[3.266891482031, 0, 0]	[0.159000098490, 0, 0]

For a right-hand propeller, $\delta = 1$ and Eq. (4.5.31) yields

$$\mathbf{M}_c = \begin{bmatrix} -\delta\ell_r \\ 0 \\ 0 \end{bmatrix} - \delta n \mathbf{u}_N = \begin{bmatrix} -0.159000098490 \\ 0 \\ 0 \end{bmatrix} \text{ ft·lbf} \tag{4.5.35}$$

Table 4.5.1 shows the forces and moments of each rotor in the rotor coordinate system. Using this information along with the location and orientation information of the rotors in Eqs. (3.6.4) and (3.6.5) yields the forces and moments of each rotor in the body-fixed coordinate system shown in Table 4.5.2.

The rotor angular momentum in the rotor coordinate system can be computed from the rotor angular rate ω_r and rotor inertia about the x axis using Eq. (2.5.39). For a right-hand propeller, this yields

$$\mathbf{h}_c = \begin{bmatrix} I_{xx}\omega_x \\ 0 \\ 0 \end{bmatrix} = \begin{bmatrix} (0.000095231)(733.038285838) \\ 0 \\ 0 \end{bmatrix} = \begin{bmatrix} 0.0698079689986 \\ 0 \\ 0 \end{bmatrix} \frac{\text{slug·ft}^2}{\text{s}}$$

Table 4.5.3 shows the angular momentum of each rotor in the rotor coordinate system. These results can be rotated to the body-fixed coordinate system, which yields the angular momentum of each rotor in the body-fixed coordinate system also shown in Table 4.5.3. Note that to match these values to the precision shown here, the atmospheric density model from Chapter 3 must be used.

Table 4.5.2. Example 4.5.2 forces and moments in the body-fixed coordinate system.

Rotor	$\mathbf{F}_b$ [lbf]	$\mathbf{M}_b$ [ft·lbf]		
1	[0, 0, -3.26689148]	[-3.26689148,	3.26689148,	0.1590000985]
2	[0, 0, -3.26689148]	[-3.26689148,	-3.26689148,	-0.1590000985]
3	[0, 0, -3.26689148]	[3.26689148,	-3.26689148,	0.1590000985]
4	[0, 0, -3.26689148]	[3.26689148,	3.26689148,	-0.1590000985]

Table 4.5.3. Example 4.5.2 rotor angular momentum in the rotor and body-fixed coordinate systems.

Rotor	Rotation	$\mathbf{h}_c$ [slug·ft²/s]	$\mathbf{h}_b$ [slug·ft²/s]
1	RH	[0.0698079689986, 0, 0]	[0, 0, -0.0698079689986]
2	LH	[-0.0698079689986, 0, 0]	[0, 0, 0.0698079689986]
3	RH	[0.0698079689986, 0, 0]	[0, 0, -0.0698079689986]
4	LH	[-0.0698079689986, 0, 0]	[0, 0, 0.0698079689986]

EXAMPLE 4.5.3 Repeat Example 4.5.2 for the case when the vehicle is in forward flight at an airspeed of 10 ft/s and an angle of attack and elevation angle of $\alpha = \theta = -5$ deg.

Solution: At the operating condition specified in the problem, the rotation rate is

$$\omega_r = \left(\frac{7000 \text{ rotations}}{\text{minute}}\right)\left(\frac{1 \text{ minute}}{60 \text{ second}}\right)\left(\frac{2\pi \text{ radians}}{\text{rotation}}\right) = 733.038285838 \text{ rad/s}$$

From Eq. (4.5.7), the advance ratio is

$$J \equiv \frac{2\pi V_c}{\omega_r d_r} = 0.0685714285714$$

From Eq. (4.5.4), the rotor angle of attack is

$$\alpha = \cos^{-1}(u_{c_x}) = 1.483529864195 \text{ rad} = 85.0 \text{ deg}$$

From Eqs. (C.4.1)–(C.4.4), the dimensionless thrust, brake power, normal-force, and yawing-moment coefficients are

$$C_T = 0.0506896718367, \quad C_{P_b} = 0.0142877551020,$$
$$C_{N,\alpha} = 0.0003268616956, \quad C_{n,\alpha} = -0.0007527519534$$

These results can be used in Eqs. (4.5.1)–(4.5.6) and (4.5.30) along with the density at an altitude of 10,000 ft to yield

$$T = \rho(\omega_r/2\pi)^2 d_r^4 C_T = 2.957101020542 \text{ lbf}$$

$$\ell_r = \rho(\omega_r/2\pi)^3 d_r^5 C_{P_b}/\omega_r = 0.1658214940460 \text{ ft·lbf}$$
$$N = \rho(\omega_r/2\pi)^2 d_r^4 C_{N,\alpha}\alpha = 0.02828831010055 \text{ lbf}$$
$$n = \rho(\omega_r/2\pi)^2 d_r^5 C_{n,\alpha}\alpha = -0.08143383336950 \text{ ft·lbf}$$

Table 4.5.4. Example 4.5.3 forces and moments in the rotor coordinate system.

Rotor	Rotation	$\mathbf{F}_c$ [lbf]	$\mathbf{M}_c$ [ft·lbf]
1	RH	[2.9571010, 0, -0.0282883]	[-0.1658215, 0, -0.0814338]
2	LH	[2.9571010, 0, -0.0282883]	[0.1658215, 0, 0.0814338]
3	RH	[2.9571010, 0, -0.0282883]	[-0.1658215, 0, -0.0814338]
4	LH	[2.9571010, 0, -0.0282883]	[0.1658215, 0, 0.0814338]

Table 4.5.5. Example 4.5.3 forces and moments in the body-fixed coordinate system.

Rotor	$\mathbf{F}_b$ [lbf]	$\mathbf{M}_b$ [ft·lbf]		
1	[-0.0282883, 0, -2.9571010]	[-3.0385349,	2.9641731,	0.1941098]
2	[-0.0282883, 0, -2.9571010]	[-2.8756672,	-2.9500289,	-0.1375332]
3	[-0.0282883, 0, -2.9571010]	[2.8756672,	-2.9500289,	0.1375332]
4	[-0.0282883, 0, -2.9571010]	[3.0385349,	2.9641731,	-0.1941098]

Using these results in Eq. (4.5.27) and Eq. (4.5.31) yields the results shown in Table 4.5.4. Using this information along with the location and orientation information of the rotors in Eqs. (3.6.4) and (3.6.5) yields the forces and moments of each rotor in the body-fixed coordinate system shown in Table 4.5.5.

Because the rotor angular rate is identical to that in Example 4.5.2, the rotor angular momentum results are identical to those given in Table 4.5.3. Note that to match these values to the precision shown here, the atmospheric density model from Chapter 3 must be used.

EXAMPLE 4.5.4 For the Simple Quadrotor model given in Appendix C.4, altitude can be maintained if the vehicle is level and each rotor produces thrust equivalent to a quarter of the weight. Compute the rotor angular rate, torque, normal force, and yawing moment for the case when the Simple Quadrotor is hovering at sea level.

Solution: Because the quadrotor weighs 5 lbf at sea level, each rotor must produce 1.25 lbf. Using the thrust coefficient expression given in Eq. (C.4.1), the specified thrust of 1.25 lbf, the density at sea level, and a component velocity of $V_c = 0$ for hover in Eq. (4.5.25) yields the rotation rate

$$\begin{aligned}\omega_r &= \frac{\pi}{d_r C_{T0}}\left[-V_c C_{T1} \pm \sqrt{V_c^2 C_{T1}^2 - 4C_{T0}\left(V_c^2 C_{T2} - \frac{T}{\rho d_r^2}\right)}\right] \\ &= \frac{\pi}{d_r C_{T0}}\sqrt{4C_{T0}\frac{T}{\rho d_r^2}} = 389.6871580612 \quad \text{rad/s}\end{aligned}$$

From Eq. (4.5.7), the advance ratio is

$$J \equiv \frac{2\pi V_c}{\omega_r d_r} = 0.0$$

From Eqs. (C.4.1)–(C.4.4), the dimensionless thrust, brake power, normal-force, and yawing-moment coefficients are

$$C_T = 0.0560, \quad C_{P_b} = 0.0137, \quad C_N = C_n = 0.0$$

These results can be used in Eqs. (4.5.1)–(4.5.6) and (4.5.30) along with sea-level density to yield

$$T = \rho(\omega_r/2\pi)^2 d_r^4 C_T = 1.25 \text{ lbf}$$
$$\ell_r = \rho(\omega_r/2\pi)^3 d_r^5 C_{P_b}/\omega_r = 0.06083768751002 \text{ ft·lbf}$$
$$N = \rho(\omega_r/2\pi)^2 d_r^4 C_{N,\alpha}\alpha = 0.0 \text{ lbf}$$
$$n = \rho(\omega_r/2\pi)^2 d_r^5 C_{n,\alpha}\alpha = 0.0 \text{ ft·lbf}$$

Note that to match these values to the precision shown here, the atmospheric density model from Chapter 3 must be used.

EXAMPLE 4.5.5 For a single rotor from the Simple Quadrotor model given in Appendix C.4, compute the rotor angular rate, torque, normal force, and yawing moment for the case when the rotor is operating at an altitude of 10,000 ft, has zero airspeed, and a rotor torque specified to be $\ell_r = 0.05$ ft·lbf.

Solution: Using the brake-power coefficient expression given in Eq. (C.4.2), the density at the specified altitude, and a component velocity of $V_c = 0$ for zero airspeed in Eq. (4.5.26) yields the rotation rate

$$\omega_r = \frac{\pi}{d_r C_{P0}}\left[-V_c C_{P1} \pm \sqrt{V_c^2 C_{P1}^2 - 4C_{P0}\left(V_c^2 C_{P2} - \frac{2\pi\ell_r}{\rho d_r^3}\right)}\right]$$
$$= \frac{\pi}{d_r C_{P0}}\left[\sqrt{4C_{P0}\frac{2\pi\ell_r}{\rho d_r^3}}\right] = 411.0673312662 \text{ rad/s}$$

The advance ratio and the dimensionless thrust, brake power, normal-force, and yawing-moment coefficients are identical to those in Example 4.5.4. These results can be used in Eqs. (4.5.1)–(4.5.6) and (4.5.30) along with the density at an altitude of 10,000 ft to yield

$$T = \rho(\omega_r/2\pi)^2 d_r^4 C_T = 1.027323729057 \text{ lbf}$$
$$\ell_r = \rho(\omega_r/2\pi)^3 d_r^5 C_{P_b}/\omega_r = 0.05 \text{ ft·lbf/s}$$

$$N = \rho(\omega_r/2\pi)^2 d_r^4 C_{N,\alpha}\alpha = 0.0 \text{ lbf}$$

$$n = \rho(\omega_r/2\pi)^2 d_r^5 C_{n,\alpha}\alpha = 0.0 \text{ ft·lbf}$$

Note that to match these values to the precision shown here, the atmospheric density model from Chapter 3 must be used.

4.6 Electric Propulsion

An electric propulsion system typically consists of a battery, an electronic speed controller (ESC), an electric motor, and a rotor or propeller. The battery supplies DC power to the ESC. The ESC accepts throttle commands from the pilot and controls the power to the electric motor. The electric motor converts the electric energy to mechanical shaft power, and the rotor converts the rotational shaft power to thrust.

Electric motors produce torque that is nearly proportional to the current. Because the ESC can rapidly adjust the current, the system is capable of rapidly adjusting the total thrust due to changes in throttle setting. This makes electric propulsion attractive for systems that must be able to rapidly alter and precisely control thrust, such as multirotor vehicles. From a practical standpoint, electric propulsion systems are often easier to maintain and operate than internal combustion engines. For these reasons, many aircraft in development today use electric propulsion, including most hobby aircraft, small drones, and even some larger vehicles and manned aircraft.

Although electric propulsion offers several advantages over internal combustion engines, it also presents two significant disadvantages. First, the energy density of current batteries is significantly less than fuel used for internal combustion engines. Second, the vehicle does not lose weight as energy is consumed. Whereas internal combustion engines burn fuel throughout the flight, which reduces the weight of the vehicle with time, the weight of a battery must be carried throughout the duration of the flight. Despite these drawbacks, electric propulsion has become a common propulsion method for many types of aircraft. Understanding how to correctly model these systems will continue to be important in flight-simulation frameworks.

Electric propulsion systems can be modeled at many varying levels of fidelity, and many models have been proposed and used in the literature. The present section outlines a simple low-fidelity model that should be used with caution.

Battery Properties

Batteries are typically characterized by the properties shown in Table 4.6.1. The battery capacity Q_{b_0} is a total charge of the battery, which is often expressed in terms of current multiplied by time. For example, a 3,000 mAh battery is rated to supply 3,000 milliamps of current for 1 hour. The capacity reduces with time as current is drawn from the battery. The capacity at any point in time can be estimated as the original capacity less the integral of the current with time, i.e.

$$Q_b = Q_{b_0} - \int_0^t I_b dt \tag{4.6.1}$$

The voltage when the battery is at full capacity and under no load will be denoted as $E_{b_{0F}}$. The no-load voltage at any point in time E_{b_0} is sometimes approximated as a constant independent of time or charge remaining, i.e. $E_{b_0} \approx E_{b_{0F}}$. However, more accurate models include the degradation of the no-load voltage with time, as a function of charge remaining in the battery, Q_b. A simple model for the no-load voltage of a battery was suggested by Maoquan and Haixin [68]

$$E_{b_0} = E_{b_{0F}} - E_A \frac{Q_{b_0}}{Q_b} + E_B \exp\left(-E_C \int_0^t I_b dt\right) \tag{4.6.2}$$

where E_A, E_B, and E_C are model constants, which can be obtained from fitting Eq. (4.6.2) to experimental time-history data for the battery [68]. The internal resistance of the battery R_b is an effective resistance that causes the battery terminal voltage to drop when current is drawn. The terminal battery voltage under load is related to the battery current draw and resistance according to

$$E_b = E_{b_0} - I_b R_b \tag{4.6.3}$$

The C-rating C_b of a battery is a measure of the maximum rate at which charge can be safely discharged from the battery. The maximum current is a function of the C-rating and battery capacity and can be computed from

$$I_{b_{\max}} = C_b Q_b \tag{4.6.4}$$

For example, if a 3000 mAh (3 Ah) battery has a C-rating of $C_b = 20$ hour^{-1}, the fastest discharge rate that can be safely used is $I_{b_{\max}} = (20)(3) = 60$ amps.

Table 4.6.1. Properties that characterize a battery.

Property	Symbol	Typical Units
Full-Charge Capacity	Q_{b_0}	milliamp hours (mAh)
No-Load Voltage at Full Capacity	$E_{b_{0F}}$	volts
Internal Resistance	R_b	ohms
C-Rating	C_b	hours^{-1} or "C's"
Voltage Constant in Eq. (4.6.2)	E_A	volts
Voltage Constant in Eq. (4.6.2)	E_B	volts
Voltage Constant in Eq. (4.6.2)	E_C	(amp hours)$^{-1}$

Electric Motor Properties

Electric motors are typically characterized by the properties shown in Table 4.6.2. The speed constant K_v provides the rate at which the motor will spin under no load per volt applied. For example, if a motor is rated at $K_v = 1000$ rpm/volt and 12 volts are applied across the motor, it will spin at $N_m = 12,000$ rpm under no load. Note that here we will denote the motor rotation rate in terms of rpm as N_m, and in terms of rad/s as ω_m. A rotation rate in rpm is related to the rotation rate in rad/s according to

$$N_m = \frac{30}{\pi}\omega_m \tag{4.6.5}$$

Electric motors are also characterized by a no-load current I_{m_0} and armature resistance R_m. The motor rotation rate N_m is related to the speed constant K_v, voltage supplied to the motor E_m, motor current I_m, and motor resistance R_m according to

$$N_m = K_v(E_m - I_m R_m) \tag{4.6.6}$$

The motor torque is related to the motor current according to

$$\ell_m = \frac{C_I}{K_v}(I_m - I_{m_0}) \tag{4.6.7}$$

where C_I is a conversion factor. If the torque has units of N·m, the conversion factor is

$$C_I = \left(\frac{1 \text{ rotation}}{2\pi \text{ rad}}\right)\left(\frac{60 \text{ sec}}{1 \text{ min}}\right) = \frac{30}{\pi} \tag{4.6.8}$$

If the torque has units of ft·lbf, the conversion factor is

$$C_I = \left(\frac{1 \text{ lbf}}{4.4482216152605 \text{ N}}\right)\left(\frac{1 \text{ ft}}{0.3048 \text{ m}}\right)\left(\frac{1 \text{ rotation}}{2\pi \text{ rad}}\right)\left(\frac{60 \text{ sec}}{1 \text{ min}}\right) = 7.04319971369755 \tag{4.6.9}$$

Note that the no-load current I_{m_0} is sometimes considered a constant, but is more correctly a function of the motor voltage. If this relationship is known, $I_{m_0} = f(E_m)$ can easily be computed and used in Eq. (4.6.7).

Table 4.6.2. Properties that characterize an electric motor.

Property	Symbol	Typical Units
Speed Constant	K_v	rpm/volt
No-Load Current	I_{m_0}	amps
Armature Resistance	R_m	ohms
Gear Ratio	G_m	None
Gearbox Efficiency	η_g	None

The motor gear ratio G_m is a ratio of the rotation rate of the motor to that of the output shaft. For example, if a motor has a gear ratio of $G_m = 4$, the motor spins 4 times faster than the output shaft. Hence, the motor rotation rate is related to the shaft rotation rate according to

$$G_m = \frac{N_m}{N_s} = \frac{\omega_m}{\omega_s} \tag{4.6.10}$$

The power output from the shaft is related to the power produced by the motor

$$\ell_s \omega_s = \eta_g \ell_m \omega_m \tag{4.6.11}$$

where η_g is the efficiency of the gearbox. Using Eqs. (4.6.10) and (4.6.11), it can be shown that the output shaft torque ℓ_s is related to the motor torque and gear ratio according to

$$\ell_s = \eta_g G_m \ell_m \tag{4.6.12}$$

The electrical power $E_m I_m$ supplied to the motor is converted by the motor to mechanical power $\ell_m \omega_m$ and resistive losses $I_m^2 R_m$ according to

$$E_m I_m = \ell_m \omega_m + I_m^2 R_m \tag{4.6.13}$$

Electronic Speed Controller Properties

A speed controller is a power converter that converts DC voltage from the battery into rapidly-switching three-phase voltages applied to the motor. The ESC receives a control signal from the receiver, which is typically encoded using pulse-width-modulation (PWM). Here we will call this signal the throttle setting and use the symbol τ, which can take on values $0 \leq \tau \leq 1$, with $\tau = 0$ equivalent to no throttle, and $\tau = 1$ denoting full throttle. Electronic speed controllers are typically characterized by the properties shown in Table 4.6.3. The resistance of the speed controller R_c characterizes both the resistance and impedance in the ESC as current moves through the controller. The maximum current $I_{c_{\max}}$ denotes the maximum current the electronics of the speed controller are designed to handle.

The power conversion from battery to motor must follow energy conservation. Therefore,

$$E_m I_m = \eta_c E_b I_b \tag{4.6.14}$$

where η_c represents the efficiency of the conversion. The speed controller efficiency typically reduces with a reduction in throttle setting. For example, a simple linear model

Table 4.6.3. Properties that characterize an electronic speed controller (ESC).

Property	Symbol	Typical Units
Resistance	R_c	Ohms
Maximum Current	$I_{c_{\max}}$	Amps

in which the efficiency is 100% at full throttle and decreases by 10% to 90% at zero throttle can be expressed as

$$\eta_c = 1 - 0.1(1 - \tau) \tag{4.6.15}$$

Note that this efficiency approximation is only a model, and not a law. The ESC controls the effective voltage applied to the motor using pulse-width modulation based on throttle setting. Hence, the motor voltage is related to the battery voltage, the motor current, and the resistance of the ESC according to

$$E_m = \tau E_b - I_m R_c \tag{4.6.16}$$

A Simple Electric Propulsion Model

Batteries, motors, and speed controllers can be combined and used in various ways for electric propulsion systems. Here we will consider a simple model in which a single battery, ESC, and motor are connected as shown in Fig. 4.6.1 to drive a single propeller. The reader is reminded that this is a simplified model, and should be applied with caution.

The characteristics of the system depend on the operating condition, which includes a throttle setting and airspeed. The throttle setting controls the amount of power available to the motor, and the airspeed controls the characteristics of the propeller or rotor. For this given operating condition, the torque and rotation rate of the propeller must match the torque and rotation rate of the motor output shaft. The process for solving the state of the system typically requires an iterative process in which the rotation rate of the shaft and rotor are varied until the torque of the shaft and rotor match.

The solution process outlined here begins with an initial guess for the rotor rotation rate. Given this rotation rate and rotor airspeed, the rotor torque can be computed. The following process can be used to compute the shaft rotation rate that will produce a shaft

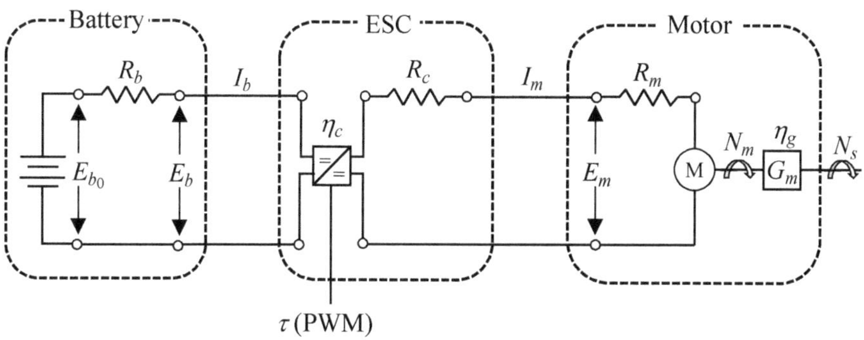

Figure 4.6.1. Schematic of a typical battery, ESC, and motor combination.

torque equal to the rotor torque. First, the shaft torque is set to be equal to the rotor torque. The motor torque can then be computed by rearranging Eq. (4.6.12) to yield

$$\ell_m = \frac{\ell_s}{\eta_g G_m} \tag{4.6.17}$$

The motor current can be computed by rearranging Eq. (4.6.7) to yield

$$I_m = \frac{\ell_m K_v}{C_I} + I_{m0} \tag{4.6.18}$$

Eq. (4.6.14) can be rearranged to yield

$$I_b = \frac{E_m I_m}{\eta_c E_b} \tag{4.6.19}$$

Using this in Eq. (4.6.3) yields

$$E_b = E_{b0} - \frac{E_m I_m}{\eta_c E_b} R_b \tag{4.6.20}$$

which can be rearranged as a quadratic in battery voltage E_b in the form

$$E_b^2 - E_{b0} E_b + \frac{E_m I_m}{\eta_c} R_b = 0 \tag{4.6.21}$$

This expression can be solved to yield the solution

$$E_b = \frac{1}{2}\left(E_{b0} + \sqrt{E_{b0}^2 - 4\frac{E_m I_m}{\eta_c} R_b}\right) \tag{4.6.22}$$

Using Eq. (4.6.22) in Eq. (4.6.16) and rearranging yields a quadratic in E_m, which can be solved to yield

$$E_m = \frac{1}{2}\left(-B + \sqrt{B^2 - 4C}\right) \tag{4.6.23}$$

where

$$B = 2I_m R_c - \tau E_{b0} + \frac{\tau^2}{\eta_c} I_m R_b \tag{4.6.24}$$

$$C = I_m^2 R_c^2 - \tau E_{b0} I_m R_c \tag{4.6.25}$$

This result is used in Eq. (4.6.6) to yield the motor rotation rate. The resulting shaft rotation rate is computed from the gear ratio according to

$$N_s = \frac{N_m}{G_m} \tag{4.6.26}$$

The error in the solution is the difference between the shaft rotation rate and the rotor rotation rate. A new guess for the correct rotor rotation rate can be selected using

$$N_r = N_r + \Gamma(N_s - N_r) \tag{4.6.27}$$

where Γ is a relaxation factor. This process is repeated until the difference between the shaft rotation rate and the rotor rotation rate is sufficiently small. Once the solution has converged, the battery voltage and current can be computed. Equation (4.6.6) can be rearranged to yield

$$E_m = \frac{N_m}{K_v} + I_m R_m \tag{4.6.28}$$

Equating this to Eq. (4.6.16) and rearranging yields the battery voltage

$$E_b = \frac{N_m/K_v + I_m(R_m + R_c)}{\tau} \tag{4.6.29}$$

The battery current can then be computed from Eq. (4.6.19) or by rearranging Eq. (4.6.3) to yield

$$I_b = \frac{E_{b_0} - E_b}{R_b} \tag{4.6.30}$$

The complete solution process for using this method to find the correct rotation rate of the rotor and shaft follows the steps shown in Algorithm 4.6.1.

Algorithm 4.6.1 Iterative method for solving the simple electrical propulsion model.

Input
- τ — (real) throttle setting
- V_c — (real) rotor airspeed
- Γ — (real) relaxation factor
- ε_t — (real) tolerance for convergence

$N_r = 1000$ — initial guess (can be stored from a previous solution)
$\varepsilon = 1.0$ — initialize error to large value
while $\text{abs}(\varepsilon) > \varepsilon_t$ **do**
    Compute rotor torque for given V_c and N_r
    $\ell_s = \ell_r$ — set shaft torque equal to rotor torque
    $\ell_m =$ Eq. (4.6.17) — motor torque
    $I_m =$ Eq. (4.6.18) — motor current
    $\eta_c =$ Eq. (4.6.15) — ESC efficiency
    $B =$ Eq. (4.6.24)
    $C =$ Eq. (4.6.25)
    $E_m =$ Eq. (4.6.23) — motor voltage
    $N_m =$ Eq. (4.6.6) — motor rotation rate
    $N_s =$ Eq. (4.6.26) — shaft rotation rate
    $\varepsilon = N_s - N_r$ — rotation-rate error
    $N_r =$ Eq. (4.6.27) new estimate for rotor rotation rate
end while
Other useful quantities to compute:
$E_b =$ Eq. (4.6.29) — battery voltage
$I_b =$ Eq. (4.6.19) or Eq. (4.6.30) — battery current

EXAMPLE 4.6.1 For a single rotor from the Simple Quadrotor model given in Appendix C.4, compute the rotor rotation rate, rotor thrust, motor voltage, motor current, battery voltage, and battery current when the vehicle is stationary at an altitude of 10,000 ft and at a throttle setting of 46%. Assume the rotor is connected to a single battery, ESC, and motor as shown in Fig. 4.6.1. The battery has the properties

$$E_{b_0} = 22.2 \text{ volts}, \quad R_b = 0.025 \text{ ohms}$$

The ESC has the properties

$$R_c = 0.01 \text{ ohms}$$

$$\eta_c = 1 - 0.078(1 - \tau)$$

The motor has the properties

$$K_v = 450 \text{ rpm/volt}, \quad R_m = 0.12 \text{ ohms}, \quad I_{m_0} = 0.3 \text{ amps}, \quad G_m = 1.0$$

Solution: Using Algorithm 4.6.1 with an initial guess of $N_r = 1,000$ rpm and a relaxation factor of $\Gamma = 0.9$ yields the following results for the first iteration.

$$\ell_r = 0.0032448999692 \text{ ft·lbf}$$

$$\ell_s = 0.0032448999692 \text{ ft·lbf}$$

$$\ell_m = 0.0032448999692 \text{ ft·lbf}$$

$$I_m = 0.507321252484 \text{ amps}$$

$$\eta_c = 0.95788$$

$$B = -10.199051836292$$

$$C = -0.051781908818$$

$$E_m = 10.204126441073 \text{ volts}$$

$$N_m = 4564.4615508487 \text{ rpm}$$

$$N_s = 4564.4615508487 \text{ rpm}$$

$$\varepsilon = -3564.4615508487 \text{ rpm}$$

$$N_r = 4208.0153957638 \text{ rpm}$$

Table 4.6.4 shows the results for several iterations until the solution converges to double-precision accuracy.

Table 4.6.4. Example 4.6.1 RPM iteration convergence history.

Iteration	N_r [rpm]	N_s [rpm]	ε [rpm]
1	1000.00000000	4564.46155085	-3.564461551E+03
2	4208.01539576	4353.25906902	-1.452436733E+02
3	4338.73470170	4339.13802868	-4.033269784E−01
4	4339.09769598	4339.09821494	-5.189609928E−04
5	4339.09816304	4339.09816371	-6.656036931E−07
6	4339.09816364	4339.09816364	-8.521965356E−10
7	4339.09816364	4339.09816364	-2.728484105E−12
8	4339.09816364	4339.09816364	0.0000000000000

The motor voltage and current for the converged solution are

$$E_m = 10.146848058089 \text{ volts}$$

$$I_m = 4.203397453657 \text{ amps}$$

and the battery voltage and current for the converged solution are

$$E_b = 22.149743549185 \text{ volts}$$

$$I_b = 2.010258032581 \text{ amps}$$

Note that in this example, the motor voltage is significantly lower than the battery voltage, and the motor current is significantly higher than the battery current. Also note the battery voltage sag due to load from the nominal voltage of 22.2 volts.

4.7 Problems

Theory Problems

4.T1 Search the literature to find maximum thrust data for a turbojet engine as a function of altitude and velocity. Fit the data to the model given in Eq. (4.2.1). Comment on the accuracy of the model fit.

4.T2 For the engine used in Problem 4.T1, fit the maximum thrust data to the model given in Eqs. (4.3.1) and (4.3.2). Evaluate the accuracy of the simplified models relative to the original data.

4.T3 Repeat Problems 4.T1 and 4.T2 for a turbofan engine of your choice.

4.T4 Using the data published by Nguyen [39], evaluate coefficients that can be used in the propulsion model given by Eq. (4.3.1) to predict the thrust as a function of throttle setting τ and density ρ.

4.T5 Beginning with Eqs. (4.5.1), (4.5.7), and (4.5.9), show that Eq. (4.5.25) yields the solution for the angular rate for a specified thrust if the thrust coefficient is a quadratic function of advance ratio.

4.T6 Beginning with Eqs. (4.5.3), (4.5.2), (4.5.7), and (4.5.10), show that Eq. (4.5.26) yields the solution for the angular rate for a specified torque if the brake-power coefficient is a quadratic function of advance ratio.

Coding Problems

4.C1 Write a child class for a generic propulsion component that inherits properties from the parent component class created in Problem 2.C1. The class should accept input values for T_a, T_0, T_1 and T_2. If a value for one of these parameters is not included in the input, the class should set it to a value of zero. The class should also include a function to print the properties of the propulsive component to the screen. The class should have a function that computes the thrust based on the current operating conditions. Test your code using the results given in Examples 4.2.1, 4.3.1, and 4.4.1.

4.C2 Extend your code written for Problem 4.C1 so that the thrust as a function of velocity can be entered as a polynomial with an arbitrary number of coefficients. For an input file in JSON format, this can take on the form

```
"engine_name" : {
    "type" : "T=f(V)",
    "T_coeff" : [43434.0, -36.753, 0.01927],
    "T_a" : 0.6941
}
```

where type specifies that the thrust is a function of velocity. The `T_coeff` key is used to store an array of arbitrary length, where the index of the value in the array specifies the power of V to be multiplied. Hence, the values shown here would represent a thrust model of the form

$$T = \tau \left(\frac{\rho}{\rho_{\text{ssl}}}\right)^{0.6941} (43434.0 - 36.753V + 0.01927V^2)$$

4.C3 Modify the child class written for Problem 2.C7 to compute the thrust, torque, normal force, and yawing moment produced by a rotor. The class should accept input array values that specify arbitrary polynomials for the thrust, brake power, normal force, and yawing moment coefficients C_T, C_{P_b}, $C_{N,\alpha}$, and $C_{n,\alpha}$. The index of the value in each array specifies the power of advance ratio J to be multiplied with the corresponding coefficient. For example, the rotor used in Example 4.5.1 could be specified with the following JSON input

```
"prop_name" : {
    "CT" : [0.09484, -0.03304, -0.08604],
```

```
        "CPb" : [0.0471312, 0.008232, -0.0654336],
        "CN,alpha[1/rad]" : [0.0, 0.015288, 0.012256, 0.017424],
        "Cn,alpha[1/rad]" : [0.0, -0.018712, 0.00054, 0.004204]
    }
```

The class should also include a function to print the propulsive properties of the rotor component to the screen. The class should have a function that computes the total component forces and moments based on the current operating conditions, including a specified rotation rate in RPM. Test your code using the results given in Examples 4.5.1, 4.5.2, and 4.5.3.

4.C4 Modify the code written for Problem 4.C3 to allow the user to specify the rotor thrust rather than the rotor rotation rate. Test your code using the results given in Example 4.5.4.

4.C5 Modify the code written for Problem 4.C3 to allow the user to specify the rotor torque rather than the rotor rotation rate. Test your code using the results given in Example 4.5.5.

4.C6 Modify the code written for Problem 4.C3 to allow the user to specify battery, motor, and ESC properties associated with the rotor. The code should solve for the rotor rotation rate and resulting forces and moments given a throttle setting. Test your code using the results given in Example 4.6.1.

Chapter 5

Equations of Motion

5.1 Newton's Second Law

Newton's second law is the foundation of rigid-body mechanics. This law states that the time rate of change in translational momentum is equal to the force exerted on the object, and that the time rate of change in rotational momentum is equal to the moment or torque exerted on the object. For more information on Newton's second law, the reader is referred to any of several dynamics textbooks. In the present development, we will apply Newton's second law to the vehicle in the body-fixed coordinate system.

Translational Momentum

In the body-fixed coordinate system, Newton's second law for translational momentum of a rigid-body vehicle can be expressed as

$$\mathbf{F}_s + \mathbf{W} = \frac{d}{dt}(\mathcal{M}\mathbf{V}) + \omega \times (\mathcal{M}\mathbf{V}) \tag{5.1.1}$$

where $\mathbf{F}_s$ is the net surface force vector on the vehicle, $\mathbf{W}$ is the force of gravity, $\mathcal{M}$ is the mass of the vehicle, and $\mathbf{V}$ and ω are the translational and rotational velocities, respectively, of the body-fixed coordinate frame relative to an inertial coordinate frame.

Because mass is a scalar, Eq. (5.1.1) can be rearranged as

$$\frac{d}{dt}(\mathcal{M}\mathbf{V}) = \mathbf{F}_s + \mathbf{W} - \mathcal{M}\omega \times \mathbf{V} \tag{5.1.2}$$

The term $\omega \times \mathbf{V}$ is often referred to as the *Coriolis acceleration*. When this term is multiplied by the vehicle mass $\mathcal{M}$, it results in what is called the *Coriolis force*.

The mass of a vehicle typically changes with time. For example, the mass decreases as fuel is burned or as a payload is released from the vehicle. The mass of a vehicle can also increase during flight, as in the case of aerial refueling. As mass enters or exits the vehicle, it carries with it momentum. This momentum can be expressed as $\dot{\mathcal{M}}_e \mathbf{V}_e$ where $\dot{\mathcal{M}}_e$ is the mass flow rate entering or exiting the vehicle and $\mathbf{V}_e$ is the velocity of $\dot{\mathcal{M}}_e$ relative to the inertial coordinate system. Here we will define $\dot{\mathcal{M}}_e$ as positive if the mass is exiting the vehicle, and negative if the mass is entering the vehicle.

Using the product rule for the term on the left-hand side of Eq. (5.1.2), and adding in the momentum term from any mass entering or exiting the vehicle yields

$$\mathcal{M}\frac{d\mathbf{V}}{dt} + \mathbf{V}\frac{d\mathcal{M}}{dt} + \dot{\mathcal{M}}_e\mathbf{V}_e = \mathbf{F}_s + \mathbf{W} - \mathcal{M}\omega \times \mathbf{V} \tag{5.1.3}$$

The conservation of mass requires that the mass entering or exiting the vehicle must be equal to the change in mass of the vehicle. This can be written as

$$\frac{d\mathcal{M}}{dt} = -\dot{\mathcal{M}}_e \tag{5.1.4}$$

Using this in Eq. (5.1.3) yields

$$\mathcal{M}\dot{\mathbf{V}} = \mathbf{F}_s - \dot{\mathcal{M}}_e(\mathbf{V}_e - \mathbf{V}) + \mathbf{W} - \mathcal{M}\omega \times \mathbf{V} \tag{5.1.5}$$

The term $-\dot{\mathcal{M}}_e(\mathbf{V}_e - \mathbf{V})$ represents the force on the vehicle due to mass entering or exiting the vehicle, and is typically included as a thrust force. In the present development, we will combine this force with the surface force vector $\mathbf{F}_s$ to yield a *pseudo aerodynamic force* $\overline{\mathbf{F}}$ defined as

$$\overline{\mathbf{F}} \equiv \mathbf{F}_s - \dot{\mathcal{M}}_e(\mathbf{V}_e - \mathbf{V}) \tag{5.1.6}$$

which includes both the aerodynamic force on the surface of the vehicle, as well as the thrust force. Hence, Newton's second law for translational momentum can be written as

$$\mathcal{M}\dot{\mathbf{V}} = \overline{\mathbf{F}} + \mathbf{W} - \mathcal{M}\omega \times \mathbf{V} \tag{5.1.7}$$

In terms of the body-fixed components given in Eqs. (3.4.2) and (3.4.22), this can be written as

$$\mathcal{M}\begin{bmatrix}\dot{u}\\ \dot{v}\\ \dot{w}\end{bmatrix} = \begin{bmatrix}\overline{F}_{x_b}\\ \overline{F}_{y_b}\\ \overline{F}_{z_b}\end{bmatrix} + \begin{bmatrix}W_{x_b}\\ W_{y_b}\\ W_{z_b}\end{bmatrix} - \mathcal{M}\begin{vmatrix}\mathbf{i} & \mathbf{j} & \mathbf{k}\\ p & q & r\\ u & v & w\end{vmatrix}$$
$$= \begin{bmatrix}\overline{F}_{x_b} + W_{x_b} + \mathcal{M}(rv - qw)\\ \overline{F}_{y_b} + W_{y_b} + \mathcal{M}(pw - ru)\\ \overline{F}_{z_b} + W_{z_b} + \mathcal{M}(qu - pv)\end{bmatrix} \tag{5.1.8}$$

where u, v, and w are the body-fixed translational velocity components of the vehicle relative to an inertial coordinate system, and p, q, and r are the body-fixed angular velocity components of the vehicle relative to an inertial coordinate system. Recall from Section 1.1 that the atmosphere is assumed to be aligned with the Earth-fixed coordinate system and translating relative to the Earth-fixed coordinate system at a constant velocity. Therefore, the atmospheric coordinate system can be considered an inertial coordinate system for the vehicle, and the velocity components u, v, and w represent the body-fixed translational velocity of the vehicle relative to the atmosphere.

For purposes of numerical integration, it is convenient to rearrange Eq. (5.1.8) as

$$\begin{bmatrix}\dot{u}\\ \dot{v}\\ \dot{w}\end{bmatrix} = \begin{bmatrix}\overline{F}_{x_b}/\mathcal{M} + W_{x_b}/\mathcal{M} + (rv - qw)\\ \overline{F}_{y_b}/\mathcal{M} + W_{y_b}/\mathcal{M} + (pw - ru)\\ \overline{F}_{z_b}/\mathcal{M} + W_{z_b}/\mathcal{M} + (qu - pv)\end{bmatrix} \tag{5.1.9}$$

In this form, if the translational and rotational velocity of the vehicle are known in the body-fixed coordinate system, along with the vehicle mass, pseudo aerodynamic force vector, and weight vector, the change in the translational velocity of the vehicle can be computed.

It is important to note that the translational relation given in Eqs. (5.1.7), (5.1.8), and (5.1.9) is not based on the assumption of constant mass, and can be applied to a vehicle with mass changing very rapidly with respect to time. However, any momentum due to the changing mass must be accounted for in the pseudo aerodynamic force, which includes the effects of thrust. The pseudo aerodynamic force must account for the difference between the velocity of the vehicle and the velocity of the mass entering or exiting the vehicle.

Rotational Momentum

Newton's second law for angular momentum can be expressed as

$$\mathbf{M}_s = \frac{d}{dt}\left([\mathbf{I}]\omega + \mathbf{h}\right) + \omega \times \left([\mathbf{I}]\omega + \mathbf{h}\right) \tag{5.1.10}$$

where $\mathbf{M}_s$ is the net surface moment vector on the vehicle about the vehicle center of gravity, $[\mathbf{I}]$ is the inertia tensor of the vehicle, and $\mathbf{h}$ is the angular momentum of any objects on the vehicle relative to the body-fixed coordinate system. Equation (5.1.10) is the angular momentum equation for rigid bodies, with one exception. **In this expression, we have retained the influence of spinning components, such as spinning rotors, engines, propellers, or other gyroscopic effects common to aircraft.** However, we have neglected any motion to elastic members or other non-rigid-body motion other than gyroscopic effects. Therefore, we will continue to refer to this expression as that for rigid-body vehicles. The inertia tensor is defined as

$$[\mathbf{I}] \equiv \begin{bmatrix} I_{xx_b} & -I_{xy_b} & -I_{xz_b} \\ -I_{yx_b} & I_{yy_b} & -I_{yz_b} \\ -I_{zx_b} & -I_{zy_b} & I_{zz_b} \end{bmatrix} \tag{5.1.11}$$

where

$$I_{xx_b} = \iiint\limits_{\mathcal{M}} (y_b^2 + z_b^2)\, d\mathcal{M} \tag{5.1.12}$$

$$I_{yy_b} = \iiint\limits_{\mathcal{M}} (x_b^2 + z_b^2)\, d\mathcal{M} \tag{5.1.13}$$

$$I_{zz_b} = \iiint\limits_{\mathcal{M}} (x_b^2 + y_b^2)\, d\mathcal{M} \tag{5.1.14}$$

$$I_{xy_b} = I_{yx_b} = \iiint\limits_{\mathcal{M}} x_b y_b\, d\mathcal{M} \tag{5.1.15}$$

$$I_{yz_b} = I_{zy_b} = \iiint_{\mathcal{M}} y_b z_b d\mathcal{M} \tag{5.1.16}$$

$$I_{xz_b} = I_{zx_b} = \iiint_{\mathcal{M}} x_b z_b d\mathcal{M} \tag{5.1.17}$$

Equation (5.1.10) can be rearranged to yield

$$\frac{d}{dt}\left([\mathbf{I}]\omega + \mathbf{h}\right) = \mathbf{M}_s - \omega \times \left([\mathbf{I}]\omega + \mathbf{h}\right) \tag{5.1.18}$$

Just as the mass of a vehicle changes with time, the inertia of the vehicle also changes. As mass enters or exits the vehicle, it can carry with it angular momentum. Applying the conservation of mass and including the momentum effects due to mass entering and exiting the vehicle in Eq. (5.1.18), as was done in the development of the translational momentum equation, and rearranging yields

$$[\mathbf{I}]\dot{\omega} = \overline{\mathbf{M}} - \omega \times \left([\mathbf{I}]\omega + \mathbf{h}\right) - \dot{\mathbf{h}} \tag{5.1.19}$$

where $\overline{\mathbf{M}}$ is a *pseudo aerodynamic moment*. This pseudo aerodynamic moments includes both the aerodynamic moment due to pressure and shear distributions on the surface of the vehicle, as well as thrust effects due to mass entering and exiting the vehicle. Note that for a spinning rotor at constant speed, $\dot{\mathbf{h}} = 0$. However, for variable-speed rotors or other spinning components that change speed with time, $\dot{\mathbf{h}}$ is non-zero and must be retained. Equation (5.1.19) can be written in terms of body-fixed components as

$$\begin{bmatrix} I_{xx_b} & -I_{xy_b} & -I_{xz_b} \\ -I_{xy_b} & I_{yy_b} & -I_{yz_b} \\ -I_{xz_b} & -I_{yz_b} & I_{zz_b} \end{bmatrix} \begin{bmatrix} \dot{p} \\ \dot{q} \\ \dot{r} \end{bmatrix} = \begin{bmatrix} \overline{M}_{x_b} \\ \overline{M}_{y_b} \\ \overline{M}_{z_b} \end{bmatrix} + \begin{bmatrix} 0 & -h_{z_b} & h_{y_b} \\ h_{z_b} & 0 & -h_{x_b} \\ -h_{y_b} & h_{x_b} & 0 \end{bmatrix} \begin{bmatrix} p \\ q \\ r \end{bmatrix}$$
$$+ \begin{bmatrix} (I_{yy_b} - I_{zz_b})qr + I_{yz_b}(q^2 - r^2) + I_{xz_b}pq - I_{xy_b}pr \\ (I_{zz_b} - I_{xx_b})pr + I_{xz_b}(r^2 - p^2) + I_{xy_b}qr - I_{yz_b}pq \\ (I_{xx_b} - I_{yy_b})pq + I_{xy_b}(p^2 - q^2) + I_{yz_b}pr - I_{xz_b}qr \end{bmatrix} - \begin{bmatrix} \dot{h}_{x_b} \\ \dot{h}_{y_b} \\ \dot{h}_{z_b} \end{bmatrix} \tag{5.1.20}$$

Just as Eqs. (5.1.7) and (5.1.8) do not require constant mass, Eqs. (5.1.19) and (5.1.20) do not require constant inertia.

For purposes of numerical integration, it is convenient to rearrange Eq. (5.1.20) as

$$\begin{bmatrix} \dot{p} \\ \dot{q} \\ \dot{r} \end{bmatrix} = \begin{bmatrix} I_{xx_b} & -I_{xy_b} & -I_{xz_b} \\ -I_{xy_b} & I_{yy_b} & -I_{yz_b} \\ -I_{xz_b} & -I_{yz_b} & I_{zz_b} \end{bmatrix}^{-1} \left\{ \begin{bmatrix} \overline{M}_{x_b} \\ \overline{M}_{y_b} \\ \overline{M}_{z_b} \end{bmatrix} + \begin{bmatrix} 0 & -h_{z_b} & h_{y_b} \\ h_{z_b} & 0 & -h_{x_b} \\ -h_{y_b} & h_{x_b} & 0 \end{bmatrix} \begin{bmatrix} p \\ q \\ r \end{bmatrix} \right.$$
$$\left. + \begin{bmatrix} (I_{yy_b} - I_{zz_b})qr + I_{yz_b}(q^2 - r^2) + I_{xz_b}pq - I_{xy_b}pr \\ (I_{zz_b} - I_{xx_b})pr + I_{xz_b}(r^2 - p^2) + I_{xy_b}qr - I_{yz_b}pq \\ (I_{xx_b} - I_{yy_b})pq + I_{xy_b}(p^2 - q^2) + I_{yz_b}pr - I_{xz_b}qr \end{bmatrix} - \begin{bmatrix} \dot{h}_{x_b} \\ \dot{h}_{y_b} \\ \dot{h}_{z_b} \end{bmatrix} \right\} \tag{5.1.21}$$

In this form, if the rotational velocity of the vehicle is known in the body-fixed coordinate system, along with the vehicle inertia, pseudo aerodynamic moment vector, and gyroscopic vectors **h** and $\dot{\mathbf{h}}$, the change in the rotational velocity of the vehicle can be computed.

For a vehicle that is symmetric about the x_b-z_b plane, and neglecting any gyroscopic effects, Eq. (5.1.20) reduces to

$$\begin{bmatrix} I_{xx_b} & 0 & -I_{xz_b} \\ 0 & I_{yy_b} & 0 \\ -I_{xz_b} & 0 & I_{zz_b} \end{bmatrix} \begin{bmatrix} \dot{p} \\ \dot{q} \\ \dot{r} \end{bmatrix} = \begin{bmatrix} \overline{M}_{x_b} + (I_{yy_b} - I_{zz_b})qr + I_{xz_b}pq \\ \overline{M}_{y_b} + (I_{zz_b} - I_{xx_b})pr + I_{xz_b}(r^2 - p^2) \\ \overline{M}_{z_b} + (I_{xx_b} - I_{yy_b})pq - I_{xz_b}qr \end{bmatrix} \tag{5.1.22}$$

and Eq. (5.1.21) reduces to

$$\begin{bmatrix} \dot{p} \\ \dot{q} \\ \dot{r} \end{bmatrix} = \begin{bmatrix} I_{xx_b} & 0 & -I_{xz_b} \\ 0 & I_{yy_b} & 0 \\ -I_{xz_b} & 0 & I_{zz_b} \end{bmatrix}^{-1} \begin{bmatrix} \overline{M}_{x_b} + (I_{yy_b} - I_{zz_b})qr + I_{xz_b}pq \\ \overline{M}_{y_b} + (I_{zz_b} - I_{xx_b})pr + I_{xz_b}(r^2 - p^2) \\ \overline{M}_{z_b} + (I_{xx_b} - I_{yy_b})pq - I_{xz_b}qr \end{bmatrix} \tag{5.1.23}$$

5.2 Pseudo Aerodynamic Forces and Moments

The pseudo aerodynamic force and moment vectors $\overline{\mathbf{F}}$ and $\overline{\mathbf{M}}$ include the aerodynamic force and moment, as well as the force and moment vector from thrust. This can be expressed as

$$\overline{\mathbf{F}} = \mathbf{F} + \mathbf{F}_T \tag{5.2.1}$$

$$\overline{\mathbf{M}} = \mathbf{M} + \mathbf{M}_T \tag{5.2.2}$$

where the subscript T denotes the effects due to thrust. For many vehicles, these two components of the pseudo aerodynamic force and moment vectors are very distinct. For example, on a rocket, the aerodynamic forces and moments **F** and **M** can be obtained by integrating the pressure and shear distributions on the surface of the rocket, and those due to thrust, $\mathbf{F}_T$ and $\mathbf{M}_T$, can be obtained from analysis of the rocket motor. A similar approach can be taken for many aircraft that use a propeller or jet engine for propulsion. In these cases, the aerodynamic force and moment vectors **F** and **M** can be computed from the methods presented in Chapter 3, and the thrust force and moment vectors $\mathbf{F}_T$ and $\mathbf{M}_T$ can be computed from the methods presented in Chapter 4.

However, the aerodynamic and thrust components that comprise the pseudo aerodynamic force and moment vectors are not always completely distinct. For example, the design of a hypersonic vehicle typically includes designing the lower surface of the vehicle to act as an inlet for the engine. Hence, the engine performance of hypersonic vehicles often includes integrating the pressure and shear over a substantial portion of the vehicle surface. In such cases, the distinction between aerodynamic and thrust effects on the total vehicle forces and moments is not as clear. In any case, **care must be taken to ensure that the aerodynamic and thrust characteristics of a vehicle are not counted twice during simulation**.

The pseudo aerodynamic force and moment vectors $\overline{\mathbf{F}}$ and $\overline{\mathbf{M}}$ required in Eqs. (5.1.9) and (5.1.21) must be in the body-fixed coordinate system. If the aerodynamic or thrust force and moment vectors are given in the wind or stability coordinate systems, the relations in Section 3.4 can be used to transform them to the body-fixed coordinate system. For fixed-wing aircraft, the thrust force and moment vectors are typically known in the body-fixed coordinate system, but the aerodynamic force and moment vectors may be known in either the stability or wind coordinate system. Hence, common relations for the pseudo aerodynamic force and moment vectors on a fixed-wing aircraft can be expressed as

$$\begin{aligned}\begin{bmatrix}\overline{F}_{x_b}\\ \overline{F}_{y_b}\\ \overline{F}_{z_b}\end{bmatrix} &= \begin{bmatrix}F_{T_x}\\ F_{T_y}\\ F_{T_z}\end{bmatrix} + \frac{1}{2}\rho V^2 S_w \begin{bmatrix}-C_A\\ C_Y\\ -C_N\end{bmatrix}\\ &= \begin{bmatrix}F_{T_x}\\ F_{T_y}\\ F_{T_z}\end{bmatrix} + \frac{1}{2}\rho V^2 S_w \begin{bmatrix}C_L S_\alpha - C_{D_s} C_\alpha\\ C_Y\\ -C_L C_\alpha - C_{D_s} S_\alpha\end{bmatrix}\\ &= \begin{bmatrix}F_{T_x}\\ F_{T_y}\\ F_{T_z}\end{bmatrix} + \frac{1}{2}\rho V^2 S_w \begin{bmatrix}C_L S_\alpha - C_S C_\alpha S_\beta - C_D C_\alpha C_\beta\\ C_S C_\beta - C_D S_\beta\\ -C_L C_\alpha - C_S S_\alpha S_\beta - C_D S_\alpha C_\beta\end{bmatrix}\end{aligned} \tag{5.2.3}$$

$$\begin{aligned}\begin{bmatrix}\overline{M}_{x_b}\\ \overline{M}_{y_b}\\ \overline{M}_{z_b}\end{bmatrix} &= \begin{bmatrix}M_{T_x}\\ M_{T_y}\\ M_{T_z}\end{bmatrix} + \frac{1}{2}\rho V^2 S_w \begin{bmatrix}b_w C_\ell\\ \overline{c}_w C_m\\ b_w C_n\end{bmatrix}\\ &= \begin{bmatrix}M_{T_x}\\ M_{T_y}\\ M_{T_z}\end{bmatrix} + \frac{1}{2}\rho V^2 S_w \begin{bmatrix}b_w (C_{\ell_s} C_\alpha - C_{n_s} S_\alpha)\\ \overline{c}_w C_{m_s}\\ b_w (C_{\ell_s} S_\alpha + C_{n_s} C_\alpha)\end{bmatrix}\\ &= \begin{bmatrix}M_{T_x}\\ M_{T_y}\\ M_{T_z}\end{bmatrix} + \frac{1}{2}\rho V^2 S_w \begin{bmatrix}b_w (C_{\ell_w} C_\alpha C_\beta - C_{n_w} S_\alpha) - \overline{c}_w C_{m_w} C_\alpha S_\beta\\ b_w C_{\ell_w} S_\beta + \overline{c}_w C_{m_w} C_\beta\\ b_w (C_{\ell_w} S_\alpha C_\beta + C_{n_w} C_\alpha) - \overline{c}_w C_{m_w} S_\alpha S_\beta\end{bmatrix}\end{aligned} \tag{5.2.4}$$

The aerodynamic forces and moments on a vehicle depend on the external geometry of the vehicle, as well as the operating condition and control settings as discussed in Chapter 3. Models for various propulsion systems were discussed in Chapter 4. The total pseudo aerodynamic force and moment vectors are the sum of the aerodynamic and thrust force and moment vectors.

EXAMPLE 5.2.1 The Boeing Insitu RQ-21 Blackjack can be modeled as the summation of nine components as given in Appendix C.2 and shown in Fig. 2.6.2. Consider the case when the vehicle is flying at sea level with a velocity of 100 ft/s, an angle of attack of 5 deg, a propeller RPM of 5,000, no sideslip angle, and no angular velocity ($\omega = 0$). Compute the total pseudo aerodynamic force and moment vectors in the body-fixed coordinate system by summing the influence of each component.

Solution: Using the methods outlined in Chapters 3 and 4, the individual forces and moments generated by each component can be computed. Results for the aerodynamic forces and moments from each component are given in the local component coordinate system in Tables 3.6.1 and 3.6.2 from Example 3.6.5. The resulting aerodynamic forces and moments in the vehicle body-fixed coordinate system about the center of gravity are given in Tables 3.6.3 and 3.6.4. The forces and moments resulting from the propeller in the component coordinate system were computed in Example 4.5.1 and yield

$$\mathbf{F}_{c_{\text{prop}}} = \begin{bmatrix} 11.63134831186 \\ 0.0 \\ -0.3998318412727 \end{bmatrix} \text{lbf}, \quad \mathbf{M}_{c_{\text{prop}}} = \begin{bmatrix} -2.397065359321 \\ 0.0 \\ -0.4666895201442 \end{bmatrix} \text{ft-lbf}$$

The forces and moments resulting from the propeller in the body-fixed coordinate system about the center of gravity are

$$\mathbf{F}_{b_{\text{prop}}} = \begin{bmatrix} 11.63134831186 \\ 0.0 \\ -0.3998318412727 \end{bmatrix} \text{lbf}, \quad \mathbf{M}_{b_{\text{prop}}} = \begin{bmatrix} -2.397065359321 \\ -1.077731464273 \\ -0.4666895201442 \end{bmatrix} \text{ft-lbf}$$

Summing the aerodynamic and propulsive forces and moments yields the pseudo forces and moments about the vehicle center of gravity in the body-fixed coordinate system

$$\overline{\mathbf{F}}_b = \begin{bmatrix} 0.7632724233459 \\ 0.0 \\ -197.4656586208 \end{bmatrix} \text{lbf}, \quad \overline{\mathbf{M}}_b = \begin{bmatrix} -2.397065359321 \\ -97.31609460665 \\ -0.4666895201442 \end{bmatrix} \text{ft-lbf}$$

Reference Point Movement

The body-fixed, stability, and wind coordinate systems shown in Fig. 3.4.1 share an origin at the center of gravity of the vehicle. This origin was chosen for convenience, since the pseudo aerodynamic forces and moments about the center of gravity are needed in the equations of motion of the vehicle. However, the aerodynamic moments measured in a wind tunnel test or those computed from CFD may have been computed about an arbitrary point that does not coincide with the vehicle center of gravity. Furthermore, due to fuel consumption and other passenger or cargo movement during a flight, it is not uncommon for the center of gravity of a vehicle to shift during flight. Hence, in order to accurately model the forces and moments on a vehicle in flight, we must be able to transfer the moments computed about one location to another location in the vehicle coordinate system.

In the development of a computer code, it is good practice to compute the pseudo aerodynamic force and moment vectors first about the vehicle origin. These will be denoted here as $\overline{\mathbf{F}}_v$ and $\overline{\mathbf{M}}_v$. Once these are known at the origin, the pseudo aerodynamic force and moment vectors $\overline{\mathbf{F}}_b$ and $\overline{\mathbf{M}}_b$ about the vehicle CG can be obtained. Since the body-fixed coordinate system is aligned with the vehicle coordinate system, the pseudo aerodynamic force vector is identical in both coordinate systems, i.e. $\overline{\mathbf{F}}_b = \overline{\mathbf{F}}_v$.

In general, a moment vector about any location $\mathbf{P}_1 = (x_1, y_1, z_1)$ relative to an origin is related to the moment vector at the origin $\mathbf{M}_O$ and the force vector $\mathbf{F}$ according to

$$\mathbf{M}_1 = \mathbf{M}_O - \mathbf{P}_1 \times \mathbf{F} \tag{5.2.5}$$

Hence, for the case when the origin is the vehicle origin, and the point of interest is the vehicle center of gravity, this yields

$$\overline{\mathbf{M}}_b = \overline{\mathbf{M}}_v - \mathbf{r}_b \times \overline{\mathbf{F}}_v \tag{5.2.6}$$

where $\mathbf{r}_b$ is the vector from the vehicle origin to the vehicle center of gravity.

The reference location of the aerodynamic model is sometimes known and specified relative to the vehicle center of gravity. For example, some versions of the aerodynamic models given in Table C.3.15 include reference locations relative to the vehicle center of gravity. In these cases, the aerodynamic moment vector about the center of gravity can be computed from

$$\mathbf{M}_b = \mathbf{M}_{\text{ref}} + \mathbf{P}_{\text{ref}} \times \mathbf{F}_{\text{ref}} \tag{5.2.7}$$

where $\mathbf{M}_{\text{ref}}$ is the reference aerodynamic moment vector, $\mathbf{F}_{\text{ref}}$ is the reference aerodynamic force vector, and $\mathbf{P}_{\text{ref}}$ is the aerodynamic reference location relative to the vehicle center of gravity in the body-fixed coordinate system

EXAMPLE 5.2.2 The aerodynamic information for the F-16 Fighting Falcon given in Appendix C.3 was taken about the point located at 35% of the mean aerodynamic chord. This point is considered the nominal CG location for the study published by NASA [39]. Consider the case when the vehicle is flying at sea level with a velocity of 220 ft/s, an angle of attack of $\alpha = 5.0$ deg, with all other states and controls set to zero. Using the aerodynamic coefficients in Table C.3.4, compute the aerodynamic force and moment vectors in the body-fixed coordinate system about the original model reference location. Then compute the aerodynamic force and moment vectors in the body-fixed coordinate system for the case if the aerodynamic reference point was located 1 ft aft of the center of gravity.

Solution: Using the coefficients from Table C.3.4 in the aerodynamic-model equations given in Eqs. (3.7.24)–(3.7.30) at this operating condition yields the total force and moment vectors about the original aerodynamic reference point

$$\overline{\mathbf{F}}_v = \begin{bmatrix} -5.161618753591 \\ 0.0 \\ -6,740.814755613 \end{bmatrix} \text{ lbf}, \quad \overline{\mathbf{M}}_v = \begin{bmatrix} 0.0 \\ -10,912.90859785 \\ 0.0 \end{bmatrix} \text{ ft-lbf}$$

For the case when the center of gravity is 1 ft forward of the reference point, Eq. (5.2.7) can be used with

$$\mathbf{P}_{\text{ref}} = \begin{bmatrix} -1.0 \\ 0.0 \\ 0.0 \end{bmatrix} \text{ ft}$$

Using these results in Eq. (5.2.7) yields the total force and moment vectors in the body-fixed coordinate system about the vehicle center of gravity

$$\overline{\mathbf{F}}_b = \begin{bmatrix} -5.161618753591 \\ 0.0 \\ -6,740.814755613 \end{bmatrix} \text{ lbf}, \quad \overline{\mathbf{M}}_b = \begin{bmatrix} 0.0 \\ -17,653.72335347 \\ 0.0 \end{bmatrix} \text{ ft-lbf}$$

The aerodynamic forces and moments about the aerodynamic reference point include a nose-down pitching moment in addition to a lift force of roughly 6,700 lbf. The aerodynamic forces about the CG, which is 1 ft forward of the reference location, are identical to those about the reference point. However, the lift force generates additional nose-down pitching moment about the CG, resulting in a total nose-down pitching moment greater than that about the reference point.

5.3 Flat-Earth Kinematic Transformation Equations

Newton's second law, as discussed in the previous section, gives a relation for the translational and rotational motion of the vehicle in the body-fixed coordinate system. To complete our formulation, we must find relations that describe the **position** and **orientation** of the body-fixed coordinate system **relative to an inertial coordinate system**. Relations that provide the motion of a coordinate system relative to another coordinate system are called *kinematic transformation equations*. In this section, we will present the kinematic transformation equations for the case when the Earth-fixed coordinate system is used as the inertial coordinate system.

The Earth-fixed coordinate system is a flat-Earth coordinate system fixed to the surface of the Earth, with its origin at sea level. The x axis points to the north, the y axis points to the east, and the z axis points toward the center of the Earth. The orientation of the vehicle relative to the Earth-fixed coordinate system can be expressed in Euler angles, as discussed in Section 1.2, or in terms of the quaternion formulation discussed in Section 1.5. Here we will consider each formulation.

Euler-Angle Formulation

The typical Euler angles used in aircraft dynamics are the bank angle ϕ, elevation angle θ, and azimuth angle ψ, with limits given in Table 1.2.1. A vector in the Earth-fixed coordinate system can be expressed in body-fixed coordinates using Euler angles by applying the relation given in Eq. (1.2.5), which yields

$$\begin{bmatrix} v_{x_b} \\ v_{y_b} \\ v_{z_b} \end{bmatrix} = \begin{bmatrix} C_\theta C_\psi & C_\theta S_\psi & -S_\theta \\ S_\phi S_\theta C_\psi - C_\phi S_\psi & S_\phi S_\theta S_\psi + C_\phi C_\psi & S_\phi C_\theta \\ C_\phi S_\theta C_\psi + S_\phi S_\psi & C_\phi S_\theta S_\psi - S_\phi C_\psi & C_\phi C_\theta \end{bmatrix} \begin{bmatrix} v_{x_f} \\ v_{y_f} \\ v_{z_f} \end{bmatrix} \tag{5.3.1}$$

Likewise, a vector in the body-fixed coordinate system can be expressed in Earth-fixed coordinates using Euler angles by applying the relation given in Eq. (1.2.6), which yields

$$\begin{bmatrix} v_{x_f} \\ v_{y_f} \\ v_{z_f} \end{bmatrix} = \begin{bmatrix} C_\theta C_\psi & S_\phi S_\theta C_\psi - C_\phi S_\psi & C_\phi S_\theta C_\psi + S_\phi S_\psi \\ C_\theta S_\psi & S_\phi S_\theta S_\psi + C_\phi C_\psi & C_\phi S_\theta S_\psi - S_\phi C_\psi \\ -S_\theta & S_\phi C_\theta & C_\phi C_\theta \end{bmatrix} \begin{bmatrix} v_{x_b} \\ v_{y_b} \\ v_{z_b} \end{bmatrix} \tag{5.3.2}$$

For example, the weight vector in Earth-fixed coordinates is $\mathbf{W}_f = (0, 0, W)$. Using this in Eq. (5.3.1) gives the weight vector in the body-fixed coordinate system

$$\begin{bmatrix} W_{x_b} \\ W_{y_b} \\ W_{z_b} \end{bmatrix} = W \begin{bmatrix} -S_\theta \\ S_\phi C_\theta \\ C_\phi C_\theta \end{bmatrix} \tag{5.3.3}$$

The velocity of the vehicle relative to the surrounding atmosphere is expressed in body-fixed coordinates as $\mathbf{V}_b = (u, v, w)$. Because the atmosphere coordinate system, as defined in Section 1.1, has the same orientation as the Earth-fixed coordinate system, the velocity in Earth-fixed coordinates, $\mathbf{V}_f = (\dot{x}_f, \dot{y}_f, \dot{z}_f)$, can be found from Eq. (5.3.2) as

$$\begin{bmatrix} \dot{x}_f \\ \dot{y}_f \\ \dot{z}_f \end{bmatrix} = \begin{bmatrix} C_\theta C_\psi & S_\phi S_\theta C_\psi - C_\phi S_\psi & C_\phi S_\theta C_\psi + S_\phi S_\psi \\ C_\theta S_\psi & S_\phi S_\theta S_\psi + C_\phi C_\psi & C_\phi S_\theta S_\psi - S_\phi C_\psi \\ -S_\theta & S_\phi C_\theta & C_\phi C_\theta \end{bmatrix} \begin{bmatrix} u \\ v \\ w \end{bmatrix} + \begin{bmatrix} V_{wx_f} \\ V_{wy_f} \\ V_{wz_f} \end{bmatrix} \tag{5.3.4}$$

where the velocity of the wind relative to the Earth-fixed coordinate system is $\mathbf{V}_{w_f} = (V_{wx_f}, V_{wy_f}, V_{wz_f})$. Equation (5.3.4) is an important kinematic transformation equation, which relates the translational velocity of the vehicle in the body-fixed coordinate system to the translational velocity of the vehicle in the Earth-fixed coordinate system. Given an initial position of the vehicle, the position as a function of time can be obtained from Eq. (5.3.4) by integration.

Because the vehicle can rotate, the Euler angles are a function of time. The change in the Euler angles, which describe the orientation of the vehicle relative to the Earth-fixed coordinate system, are related to the rotation rate of the body-fixed coordinate system according to Eq. (1.2.7). Applying the rotation rate of the body-fixed coordinate system $\omega = (p, q, r)$ to Eq. (1.2.7) yields

$$\begin{bmatrix} \dot{\phi} \\ \dot{\theta} \\ \dot{\psi} \end{bmatrix} = \begin{bmatrix} 1 & S_\phi S_\theta / C_\theta & C_\phi S_\theta / C_\theta \\ 0 & C_\phi & -S_\phi \\ 0 & S_\phi / C_\theta & C_\phi / C_\theta \end{bmatrix} \begin{bmatrix} p \\ q \\ r \end{bmatrix} \tag{5.3.5}$$

Equation (5.3.5) is an important kinematic transformation equation, which relates the rotational velocity of the vehicle in the body-fixed coordinate system to the change in orientation of the vehicle relative to the Earth-fixed coordinate system. Given an initial orientation of the vehicle, the orientation as a function of time can be obtained from Eq. (5.3.5) by integration.

Quaternion Formulation

The orientation of the body-fixed coordinate system relative to the Earth can also be expressed in terms of a quaternion, as discussed in Section 1.5. A vector in the Earth-fixed coordinate system can be expressed in body-fixed coordinates using a quaternion representation by applying the relation given in Eq. (1.5.4), which yields

$$\begin{bmatrix} v_{x_b} \\ v_{y_b} \\ v_{z_b} \end{bmatrix} = \begin{bmatrix} e_0 \\ -e_x \\ -e_y \\ -e_z \end{bmatrix} \otimes \left(\begin{bmatrix} 0 \\ v_{x_f} \\ v_{y_f} \\ v_{z_f} \end{bmatrix} \otimes \begin{bmatrix} e_0 \\ e_x \\ e_y \\ e_z \end{bmatrix} \right) \tag{5.3.6}$$

Likewise, a vector in the body-fixed coordinate system can be expressed in Earth-fixed coordinates using a quaternion representation by applying the relation given in Eq. (1.5.5), which yields

$$\begin{bmatrix} v_{x_f} \\ v_{y_f} \\ v_{z_f} \end{bmatrix} = \begin{bmatrix} e_0 \\ e_x \\ e_y \\ e_z \end{bmatrix} \otimes \left(\begin{bmatrix} 0 \\ v_{x_b} \\ v_{y_b} \\ v_{z_b} \end{bmatrix} \otimes \begin{bmatrix} e_0 \\ -e_x \\ -e_y \\ -e_z \end{bmatrix} \right) \tag{5.3.7}$$

Efficient algorithms for computing the transformations given in Eqs. (5.3.6) and (5.3.7) are given in Section 1.5.

The weight vector in Earth-fixed coordinates is $\mathbf{W}_f = (0, 0, W)$. Using this in Eq. (5.3.6) gives the weight vector in the body-fixed coordinate system

$$\begin{bmatrix} W_{x_b} \\ W_{y_b} \\ W_{z_b} \end{bmatrix} = W \begin{bmatrix} 2(e_x e_z - e_y e_0) \\ 2(e_y e_z + e_x e_0) \\ e_z^2 + e_0^2 - e_x^2 - e_y^2 \end{bmatrix} \tag{5.3.8}$$

The velocity of the vehicle relative to the surrounding atmosphere is expressed in body-fixed coordinates as $\mathbf{V}_b = (u, v, w)$. Because the atmosphere coordinate system, as defined in Section 1.1, has the same orientation as the Earth-fixed coordinate system, the velocity in Earth-fixed coordinates, $\mathbf{V}_f = (\dot{x}_f, \dot{y}_f, \dot{z}_f)$, can be found from Eq. (5.3.7) as

$$\begin{bmatrix} \dot{x}_f \\ \dot{y}_f \\ \dot{z}_f \end{bmatrix} = \begin{bmatrix} e_0 \\ e_x \\ e_y \\ e_z \end{bmatrix} \otimes \left(\begin{bmatrix} 0 \\ u \\ v \\ w \end{bmatrix} \otimes \begin{bmatrix} e_0 \\ -e_x \\ -e_y \\ -e_z \end{bmatrix} \right) + \begin{bmatrix} V_{wx_f} \\ V_{wy_f} \\ V_{wz_f} \end{bmatrix} \tag{5.3.9}$$

where the velocity of the wind relative to the Earth-fixed coordinate system is $\mathbf{V}_{w_f} = (V_{wx_f}, V_{wy_f}, V_{wz_f})$. Equation (5.3.4) is an important kinematic transformation equation, which relates the translational velocity of the vehicle in the body-fixed coordinate system to the translational velocity of the vehicle in the Earth-fixed coordinate system. Given an initial position of the vehicle, the position as a function of time can be obtained from Eq. (5.3.9) by integration.

Because the vehicle can rotate, the components of the quaternion are a function of time. The change in the quaternion is related to the rotation rate of the body-fixed coordinate system according to Eq. (1.5.9). Applying the rotation rate of the body-fixed coordinate system $\omega = (p, q, r)$ to Eq. (1.5.9) yields

$$\begin{bmatrix} \dot{e}_0 \\ \dot{e}_x \\ \dot{e}_y \\ \dot{e}_z \end{bmatrix} = \frac{1}{2} \begin{bmatrix} -e_x & -e_y & -e_z \\ e_0 & -e_z & e_y \\ e_z & e_0 & -e_x \\ -e_y & e_x & e_0 \end{bmatrix} \begin{bmatrix} p \\ q \\ r \end{bmatrix} \tag{5.3.10}$$

Equation (5.3.10) is an important kinematic transformation equation, which relates the rotational velocity of the vehicle in the body-fixed coordinate system to the change in orientation of the vehicle relative to the Earth-fixed coordinate system. Given an initial orientation of the vehicle, the orientation as a function of time can be obtained from Eq. (5.3.10) by integration.

5.4 Flat-Earth Rigid-Body Equations of Motion

The relations for Newton's second law, given in Section 5.1, can be combined with the kinematic transformation equations given in Section 5.3 to yield the flat-Earth rigid-body equations of motion. The reader is reminded that rigid-body assumptions were used in the development of Newton's second law in Section 5.1 with one exception. Gyroscopic effects due to rotating propellers, engines, or other components were included in the development. However, no aeroelastic effects were included for flexible structures or other moving parts. Additionally, the reader is reminded that the relations for Newton's second law were developed without assuming constant mass or inertia. Therefore, the following relations can be used for vehicles with changing mass and inertia, as long as the pseudo aerodynamic force and moment vectors include the effects of mass entering or exiting the vehicle, such as thrust from an engine.

Euler-Angle Formulation

Combining Eqs. (5.1.9), (5.1.21), (5.3.3), (5.3.4) and (5.3.5) yields the **flat-Earth rigid-body equations of motion using the Euler-angle formulation**

$$\begin{bmatrix}\dot{u}\\ \dot{v}\\ \dot{w}\end{bmatrix} = \frac{1}{\mathcal{M}}\begin{bmatrix}\overline{F}_{x_b}\\ \overline{F}_{y_b}\\ \overline{F}_{z_b}\end{bmatrix} + g\begin{bmatrix}-S_\theta\\ S_\phi C_\theta\\ C_\phi C_\theta\end{bmatrix} + \begin{bmatrix}rv - qw\\ pw - ru\\ qu - pv\end{bmatrix} \tag{5.4.1}$$

$$\begin{aligned}\begin{bmatrix}\dot{p}\\ \dot{q}\\ \dot{r}\end{bmatrix} = &\begin{bmatrix}I_{xx_b} & -I_{xy_b} & -I_{xz_b}\\ -I_{xy_b} & I_{yy_b} & -I_{yz_b}\\ -I_{xz_b} & -I_{yz_b} & I_{zz_b}\end{bmatrix}^{-1}\left\{\begin{bmatrix}\overline{M}_{x_b}\\ \overline{M}_{y_b}\\ \overline{M}_{z_b}\end{bmatrix} + \begin{bmatrix}0 & -h_{z_b} & h_{y_b}\\ h_{z_b} & 0 & -h_{x_b}\\ -h_{y_b} & h_{x_b} & 0\end{bmatrix}\begin{bmatrix}p\\ q\\ r\end{bmatrix}\right.\\ &\left. + \begin{bmatrix}(I_{yy_b} - I_{zz_b})qr + I_{yz_b}(q^2 - r^2) + I_{xz_b}pq - I_{xy_b}pr\\ (I_{zz_b} - I_{xx_b})pr + I_{xz_b}(r^2 - p^2) + I_{xy_b}qr - I_{yz_b}pq\\ (I_{xx_b} - I_{yy_b})pq + I_{xy_b}(p^2 - q^2) + I_{yz_b}pr - I_{xz_b}qr\end{bmatrix} - \begin{bmatrix}\dot{h}_{x_b}\\ \dot{h}_{y_b}\\ \dot{h}_{z_b}\end{bmatrix}\right\}\end{aligned} \tag{5.4.2}$$

$$\begin{bmatrix}\dot{x}_f\\ \dot{y}_f\\ \dot{z}_f\end{bmatrix} = \begin{bmatrix}C_\theta C_\psi & S_\phi S_\theta C_\psi - C_\phi S_\psi & C_\phi S_\theta C_\psi + S_\phi S_\psi\\ C_\theta S_\psi & S_\phi S_\theta S_\psi + C_\phi C_\psi & C_\phi S_\theta S_\psi - S_\phi C_\psi\\ -S_\theta & S_\phi C_\theta & C_\phi C_\theta\end{bmatrix}\begin{bmatrix}u\\ v\\ w\end{bmatrix} + \begin{bmatrix}V_{wx_f}\\ V_{wy_f}\\ V_{wz_f}\end{bmatrix} \tag{5.4.3}$$

$$\begin{bmatrix}\dot{\phi}\\ \dot{\theta}\\ \dot{\psi}\end{bmatrix} = \begin{bmatrix}1 & S_\phi S_\theta / C_\theta & C_\phi S_\theta / C_\theta\\ 0 & C_\phi & -S_\phi\\ 0 & S_\phi / C_\theta & C_\phi / C_\theta\end{bmatrix}\begin{bmatrix}p\\ q\\ r\end{bmatrix} \tag{5.4.4}$$

Equations (5.4.1) and (5.4.2) represent Newton's second law for a rigid-body vehicle. The pseudo aerodynamic forces and moments $\overline{\mathbf{F}}_b$ and $\overline{\mathbf{M}}_b$ in these two equations include the aerodynamic forces and moments of the vehicle and thrust. The aerodynamic forces and moments are functions of the velocity and aerodynamic angles, as well as pilot control inputs, as discussed in Chapter 3. Equations (5.4.3) and (5.4.4) are kinematic transformation equations, which relate the translational and rotational velocity components of the body-fixed coordinate system to the Earth-fixed coordinate system. The 12 time-dependent variables on the left-hand side of these equations, u, v, w, p, q, r, x_f, y_f, z_f, ϕ, θ, and ψ, are called the *state variables*. The left-hand side of these equations represents the change of the states with respect to time. If the state of the vehicle at a point in time is known, along with required controls for the aerodynamic model, Eqs. (5.4.1)–(5.4.4) can be integrated forward to find the new state of the aircraft.

The Euler-angle formulation of the flat-Earth rigid-body equations of motion is not typically used in flight simulation due to two main reasons. First, this formulation suffers from singularities in Eq. (5.4.4) known as gimbal lock, which is discussed in Chapter 1. Second, computing the requisite sine and cosine functions required within Eqs. (5.4.1), (5.4.3), and (5.4.4) is relatively computationally expensive compared to the quaternion formulation. Therefore, although Euler angles are relatively simple to visualize, most modern simulators use alternate formulations, such as the quaternion formulation.

Quaternion Formulation

Combining Eqs. (5.1.9), (5.1.21), (5.3.8), (5.3.9) and (5.3.10) yields the **flat-Earth rigid-body equations of motion using the quaternion formulation**

$$\begin{bmatrix}\dot{u}\\ \dot{v}\\ \dot{w}\end{bmatrix} = \frac{1}{\mathcal{M}}\begin{bmatrix}\overline{F}_{xb}\\ \overline{F}_{yb}\\ \overline{F}_{zb}\end{bmatrix} + g\begin{bmatrix}2(e_x e_z - e_y e_0)\\ 2(e_y e_z + e_x e_0)\\ e_z^2 + e_0^2 - e_x^2 - e_y^2\end{bmatrix} + \begin{bmatrix}rv - qw\\ pw - ru\\ qu - pv\end{bmatrix} \tag{5.4.5}$$

$$\begin{aligned}\begin{bmatrix}\dot{p}\\ \dot{q}\\ \dot{r}\end{bmatrix} = \begin{bmatrix}I_{xx_b} & -I_{xy_b} & -I_{xz_b}\\ -I_{xy_b} & I_{yy_b} & -I_{yz_b}\\ -I_{xz_b} & -I_{yz_b} & I_{zz_b}\end{bmatrix}^{-1} &\left\{\begin{bmatrix}\overline{M}_{xb}\\ \overline{M}_{yb}\\ \overline{M}_{zb}\end{bmatrix} + \begin{bmatrix}0 & -h_{z_b} & h_{y_b}\\ h_{z_b} & 0 & -h_{x_b}\\ -h_{y_b} & h_{x_b} & 0\end{bmatrix}\begin{bmatrix}p\\ q\\ r\end{bmatrix}\right.\\ &\left. + \begin{bmatrix}(I_{yy_b} - I_{zz_b})qr + I_{yz_b}(q^2 - r^2) + I_{xz_b}pq - I_{xy_b}pr\\ (I_{zz_b} - I_{xx_b})pr + I_{xz_b}(r^2 - p^2) + I_{xy_b}qr - I_{yz_b}pq\\ (I_{xx_b} - I_{yy_b})pq + I_{xy_b}(p^2 - q^2) + I_{yz_b}pr - I_{xz_b}qr\end{bmatrix} - \begin{bmatrix}\dot{h}_{x_b}\\ \dot{h}_{y_b}\\ \dot{h}_{z_b}\end{bmatrix}\right\}\end{aligned} \tag{5.4.6}$$

$$\begin{bmatrix}\dot{x}_f\\ \dot{y}_f\\ \dot{z}_f\end{bmatrix} = \begin{bmatrix}e_0\\ e_x\\ e_y\\ e_z\end{bmatrix} \otimes \left(\begin{bmatrix}0\\ u\\ v\\ w\end{bmatrix} \otimes \begin{bmatrix}e_0\\ -e_x\\ -e_y\\ -e_z\end{bmatrix}\right) + \begin{bmatrix}V_{wx_f}\\ V_{wy_f}\\ V_{wz_f}\end{bmatrix} \tag{5.4.7}$$

$$\begin{bmatrix}\dot{e}_0\\ \dot{e}_x\\ \dot{e}_y\\ \dot{e}_z\end{bmatrix} = \frac{1}{2}\begin{bmatrix}-e_x & -e_y & -e_z\\ e_0 & -e_z & e_y\\ e_z & e_0 & -e_x\\ -e_y & e_x & e_0\end{bmatrix}\begin{bmatrix}p\\ q\\ r\end{bmatrix} \tag{5.4.8}$$

Equations (5.4.5) and (5.4.6) represent Newton's second law for a rigid-body vehicle. The pseudo aerodynamic forces and moments $\overline{\mathbf{F}}_b$ and $\overline{\mathbf{M}}_b$ in these two equations include the aerodynamic forces and moments of the vehicle and thrust. The aerodynamic forces and moments are functions of the velocity and aerodynamic angles, as well as pilot control inputs, as discussed in Chapter 3. Equations (5.4.7) and (5.4.8) are kinematic transformation equations, which relate the translational and rotational velocity components of the body-fixed coordinate system to the Earth-fixed coordinate system. The 13 time-dependent variables on the left-hand side of these equations, u, v, w, p, q, r, x_f, y_f, z_f, e_0, e_x, e_y, and e_z, are called the *state variables*. The left-hand side of these equations represents the change of the states with respect to time. If the state of the vehicle at a point in time is known, along with required controls for the aerodynamic model, Eqs. (5.4.5)–(5.4.8) can be integrated forward to find the new state of the aircraft.

Notice that the quaternion formulation has 13 state variables, whereas the Euler-angle formulation has only 12. This is because the quaternion, containing four components, is used to express the orientation of the vehicle, which has only three degrees of freedom. As discussed in Section 1.5, the length of the quaternion must be unity. Hence, the constraint given in Eq. (1.5.11) is the final relation that closes this formulation.

5.5 Gravity Relief Due to Earth's Curvature

The flat-Earth equations of motion presented in the previous section neglect the effects of Earth's curvature on the dynamics of the vehicle. One effect of curvature is the so-called *gravity relief*, which can be significant for vehicles traveling at high speeds. As a vehicle travels around the Earth, it experiences a fictitious outward force (centrifugal force) relative to the center of the Earth, due to the fact that the vehicle is traveling a curved path about Earth's center, as shown in Fig. 5.5.1. This force results from acceleration experienced by the vehicle as it travels along a curved path. This acceleration is related to the tangential velocity about Earth's center and the altitude of the vehicle according to

$$a_c = \frac{V_t^2}{R_E + H} \tag{5.5.1}$$

where R_E is the radius of the Earth, H is the altitude of the vehicle, and V_t is the horizontal or tangential velocity, which can be computed from the flat Earth-fixed velocity components

$$V_t = \sqrt{\dot{x}_f^2 + \dot{y}_f^2} \tag{5.5.2}$$

Because the centrifugal acceleration is in the opposite direction of gravity, this effect is often called *gravity relief.*

Gravity relief can be accounted for in the equations of motion developed in the previous section, by replacing the gravity constant g in Eqs. (5.4.1) and (5.4.5) with the term $g - a_c$ where

$$a_c = \frac{V_t^2}{R_E + H} = \frac{\dot{x}_f^2 + \dot{y}_f^2}{R_E + H} \tag{5.5.3}$$

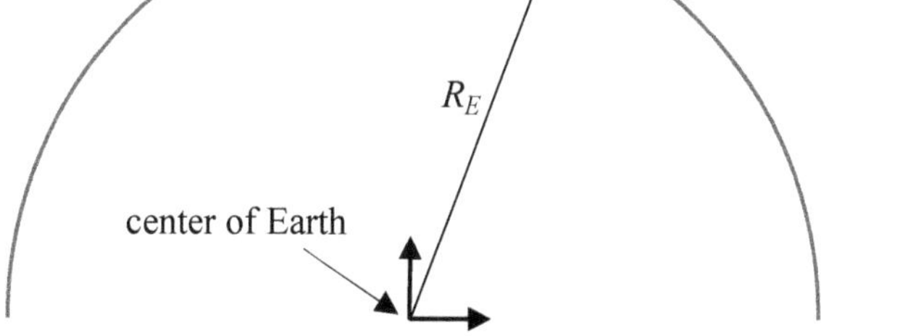

Figure 5.5.1. Acceleration of a vehicle due to gravity and curvature of the Earth.

For example, including gravity relief in Eq. (5.4.1) yields

$$\begin{bmatrix} \dot{u} \\ \dot{v} \\ \dot{w} \end{bmatrix} = \frac{1}{\mathcal{M}} \begin{bmatrix} \overline{F}_{x_b} \\ \overline{F}_{y_b} \\ \overline{F}_{z_b} \end{bmatrix} + (g - a_c) \begin{bmatrix} -S_\theta \\ S_\phi C_\theta \\ C_\phi C_\theta \end{bmatrix} + \begin{bmatrix} rv - qw \\ pw - ru \\ qu - pv \end{bmatrix} \tag{5.5.4}$$

and including gravity relief in Eq. (5.4.5) yields

$$\begin{bmatrix} \dot{u} \\ \dot{v} \\ \dot{w} \end{bmatrix} = \frac{1}{\mathcal{M}} \begin{bmatrix} \overline{F}_{x_b} \\ \overline{F}_{y_b} \\ \overline{F}_{z_b} \end{bmatrix} + (g - a_c) \begin{bmatrix} 2(e_x e_z - e_y e_0) \\ 2(e_y e_z + e_x e_0) \\ e_z^2 + e_0^2 - e_x^2 - e_y^2 \end{bmatrix} + \begin{bmatrix} rv - qw \\ pw - ru \\ qu - pv \end{bmatrix} \tag{5.5.5}$$

Because the radius of the Earth R_E is so large, the acceleration a_c is typically very small compared to the acceleration of gravity for most aircraft. However, for aircraft at very high speeds, this term can be significant. In order to account for gravity relief, Eqs. (5.5.4) and (5.5.5) should be used in place of Eqs. (5.4.1) and (5.4.5).

EXAMPLE 5.5.1 Compute the acceleration from the centrifugal force of a hypersonic vehicle flying at Mach 5 and an altitude of 70, 000 ft. Compare this to the acceleration of gravity at the same altitude.

Solution: We will first compute the gravity in English units. The altitude in English units can be converted to SI units using the conversion table in Appendix A. This yields

$$(70000\ \text{ft})\frac{0.3048\ \text{m}}{1\ \text{ft}} = 21,336.0\ \text{m}$$

The gravity can be computed in SI units from Eq. (3.2.1) with $g_{ssl} = 9.80665\ \text{m/s}^2$ and $R_{E_Z} = 6,356,766$ m to yield

$$g = 9.80665 \left(\frac{6356766}{6356766 + 21336} \right)^2 = 9.741149411\ \text{m/s}^2 = 31.95915161084\ \text{ft/s}^2$$

Using the process outlined in Chapter 3, the speed of sound at an altitude of 70, 000 ft is 970.89696 ft/s. If the vehicle is traveling Mach 5, it is traveling at 5 times the speed of sound. The acceleration from the centrifugal force can then be computed from Eq. (5.5.1) to yield

$$a_c = \frac{V_t^2}{R_E + H} = \frac{(5 \times 970.89696)^2}{6,366,707.01949371/0.3048 + 70,000} = 1.124432583\ \text{ft/s}^2$$

The ratio a_c/g yields

$$\frac{a_c}{g} = \frac{1.124432583}{31.95915161084} = 0.03518343$$

which is a gravity relief of about 3.52%.

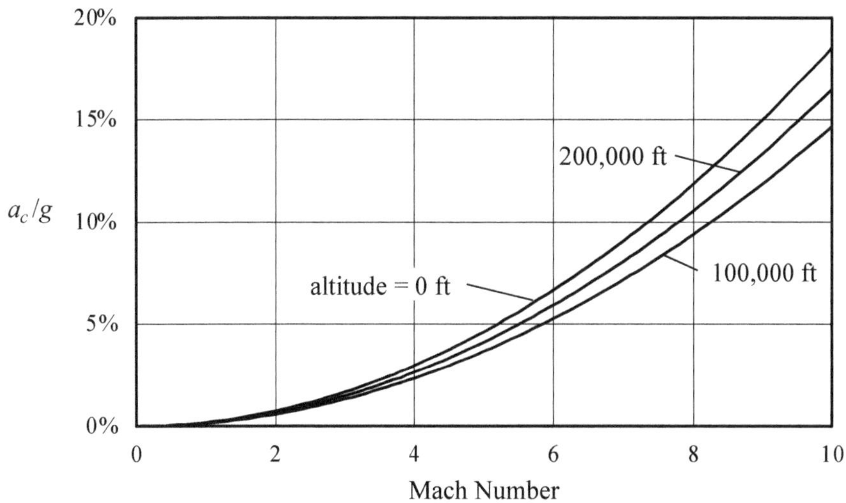

Figure 5.5.2. Gravity relief as a percent of gravity over a range of Mach and altitude.

The process used in the above example can be easily replicated over a range of altitudes and Mach numbers. For example, Fig. 5.5.2 shows results for the ratio a_c/g at three different altitudes using a standard-day atmosphere up to a Mach number of 10. Note that the results for 100,000 ft are lower than those at both sea level and 200,000 ft. This is due to the fact that for the standard-day atmospheric profile, the speed of sound decreases with altitude at low altitudes, but increases with altitude at higher altitudes.

Note that each curve in Fig. 5.5.2 is a perfect quadratic in velocity. Hence, these results can easily be extrapolated to higher Mach numbers. The high percentage of gravity relief at large Mach numbers shows that gravity relief should not be neglected when simulating the physics of high-speed vehicles. When the gravity relief is equal to the value of gravity, i.e. $a_c/g = 1$, the net effective gravity is zero. In the absence of aerodynamic forces and thrust, this condition corresponds to a circular orbit at that altitude.

Because gravity relief is significant for high-speed vehicles, and because it is simple to model for all vehicles, gravity relief is included in the equations of motion used for most of the examples in the remainder of this textbook. Note that the gravity relief discussed above is due to the vehicle's tangential speed, and is distinct from the centrifugal effect associated with Earth's rotation, which is typically insignificant for most aircraft. For example, the centrifugal acceleration due to Earth's rotation on the surface of the Earth at the equator is on the order of 0.35% of gravity. This effect decreases with latitude until it approaches zero at the poles. Acceleration due to Earth's rotation will be neglected in this textbook.

5.6 Summary of Quaternion Formulation

The quaternion formulation of the flat-Earth rigid-body equations of motion given in Eqs. (5.4.5)–(5.4.8) can be adjusted to include gravity relief as explained in the previous section. Applying the gravity-relief term a_c to Eqs. (5.4.5)–(5.4.8) yields the final **quaternion formulation of the rigid-body equations of motion including gravity relief**

$$\begin{bmatrix} \dot{u} \\ \dot{v} \\ \dot{w} \end{bmatrix} = \frac{1}{\mathcal{M}} \begin{bmatrix} \overline{F}_{x_b} \\ \overline{F}_{y_b} \\ \overline{F}_{z_b} \end{bmatrix} + (g - a_c) \begin{bmatrix} 2(e_x e_z - e_y e_0) \\ 2(e_y e_z + e_x e_0) \\ e_z^2 + e_0^2 - e_x^2 - e_y^2 \end{bmatrix} + \begin{bmatrix} rv - qw \\ pw - ru \\ qu - pv \end{bmatrix} \tag{5.6.1}$$

$$\begin{aligned} \begin{bmatrix} \dot{p} \\ \dot{q} \\ \dot{r} \end{bmatrix} = \begin{bmatrix} I_{xx_b} & -I_{xy_b} & -I_{xz_b} \\ -I_{xy_b} & I_{yy_b} & -I_{yz_b} \\ -I_{xz_b} & -I_{yz_b} & I_{zz_b} \end{bmatrix}^{-1} & \left\{ \begin{bmatrix} \overline{M}_{x_b} \\ \overline{M}_{y_b} \\ \overline{M}_{z_b} \end{bmatrix} + \begin{bmatrix} 0 & -h_{z_b} & h_{y_b} \\ h_{z_b} & 0 & -h_{x_b} \\ -h_{y_b} & h_{x_b} & 0 \end{bmatrix} \begin{bmatrix} p \\ q \\ r \end{bmatrix} \right. \\ & \left. + \begin{bmatrix} (I_{yy_b} - I_{zz_b})qr + I_{yz_b}(q^2 - r^2) + I_{xz_b}pq - I_{xy_b}pr \\ (I_{zz_b} - I_{xx_b})pr + I_{xz_b}(r^2 - p^2) + I_{xy_b}qr - I_{yz_b}pq \\ (I_{xx_b} - I_{yy_b})pq + I_{xy_b}(p^2 - q^2) + I_{yz_b}pr - I_{xz_b}qr \end{bmatrix} - \begin{bmatrix} \dot{h}_{x_b} \\ \dot{h}_{y_b} \\ \dot{h}_{z_b} \end{bmatrix} \right\} \end{aligned} \tag{5.6.2}$$

$$\begin{bmatrix} \dot{x}_f \\ \dot{y}_f \\ \dot{z}_f \end{bmatrix} = \begin{bmatrix} e_0 \\ e_x \\ e_y \\ e_z \end{bmatrix} \otimes \left(\begin{bmatrix} 0 \\ u \\ v \\ w \end{bmatrix} \otimes \begin{bmatrix} e_0 \\ -e_x \\ -e_y \\ -e_z \end{bmatrix} \right) + \begin{bmatrix} V_{wx_f} \\ V_{wy_f} \\ V_{wz_f} \end{bmatrix} \tag{5.6.3}$$

$$\begin{bmatrix} \dot{e}_0 \\ \dot{e}_x \\ \dot{e}_y \\ \dot{e}_z \end{bmatrix} = \frac{1}{2} \begin{bmatrix} -e_x & -e_y & -e_z \\ e_0 & -e_z & e_y \\ e_z & e_0 & -e_x \\ -e_y & e_x & e_0 \end{bmatrix} \begin{bmatrix} p \\ q \\ r \end{bmatrix} \tag{5.6.4}$$

Equations (5.6.1)–(5.6.4) represent the equations of motion for a rigid-body vehicle including gravity relief from Earth's curvature. These equations form the foundation of the numerical examples shown in the remainder of the textbook. This set of equations can be used to model a wide variety of rigid-body vehicles. The dynamics of rigid-body vehicles will differ only in the pseudo aerodynamic forces and moments $\overline{\mathbf{F}}_b$ and $\overline{\mathbf{M}}_b$, the vehicle mass $\mathcal{M}$, and the vehicle gyroscopic effects $\mathbf{h}_b$. The reader is reminded that these relations can be used for vehicles with changing mass and inertia, as long as the pseudo aerodynamic force and moment vectors include the effects of mass entering or exiting the vehicle.

If the state of the vehicle at a point in time is known, along with required controls for the pseudo aerodynamic forces and moments, Eqs. (5.6.1)–(5.6.4) can be integrated forward to find the new state of the aircraft. Computing solutions to Eq. (5.6.2) requires computing the inverse of a 3×3 matrix. A closed-form solution for this inverse is included in Appendix D.

5.7 Numerical Integration

Given a current state of the vehicle, the quaternion formulation given in Eqs. (5.6.1)–(5.6.4) can be numerically integrated to predict the state of the vehicle at a future time. Various numerical integration techniques can be used for this purpose. The formulation and integration scheme should be carefully selected, since both can significantly impact solution accuracy. **To minimize error using modern digital computers, a quaternion formulation coupled with a fourth-order or higher integration scheme should always be used.**

Integration Error

All modern numerical integration techniques suffer from two types of numerical error, known as round-off error and truncation error. *Round-off error* arises due to limitations in machine precision from word size. The number of digits of accuracy retained throughout a computation has changed with time from single precision, which carried about 8 digits of accuracy, to double precision which carries about 16 digits of accuracy. Higher precision can be used on modern architectures to increase the number of digits even further, which reduces round-off error. *Truncation error* is present in any integration technique due to the selected order of Taylor-series expansion to the solution at a point. As higher-order integration schemes are used, the accuracy of the integration can increase, but also increases the computation time. Historically, first- and second-order integration techniques were common. However, today most common numerical integration schemes are fourth-order or higher, which generate significantly less truncation error in the process of numerical integration.

The presence of round-off error and truncation error in all modern numerical integration schemes causes errors in the simulation results. This error accumulates throughout the simulation, which incorrectly alters the state of the vehicle. The most prevalent non-physical change in the state of the vehicle is known as *drift error*. Drift error is the error in position, velocity, and orientation of the vehicle due to the numerical error in the integration scheme. For example, consider a fixed-wing aircraft trimmed to fly in steady-level flight. With no physical perturbations to the system, integration error would eventually build up from the numerical integration technique to show that the vehicle had drifted off course and had deviated from a steady-level orientation.

Although drift error is prevalent in all numerical simulations, it does not pose significant challenges to modern flight simulators because it has a physical counterpart. Drift error is similar to small infinitesimal disturbances to the aerodynamics of the vehicle. A human pilot or autopilot would naturally correct for these disturbances. Therefore, drift error due to numerical integration is naturally corrected by the pilot, and is not a significant concern.

A second form of error arises in the numerical integration of the quaternion formulation. During the process of numerically integrating the quaternion formulation, the length of the quaternion does not automatically remain unity. Deviation from the constraint given in Eq. (1.5.11) is called *orthogonality error* and is defined as

$$\varepsilon \equiv 1 - (e_0^2 + e_x^2 + e_y^2 + e_z^2) \tag{5.7.1}$$

In the early days of flight simulation, analog computers using first-order integration techniques were common. When first-order methods are used to integrate one form of the quaternion formulation, the magnitude of the quaternion grows rapidly, introducing significant orthogonality error. This phenomenon led to flight simulators requiring error-reduction methods for reducing the orthogonality error. One common method is known as *Corbett-Wright orthogonality control* [69–72]. This method was very useful for first-order integration techniques, and is still widely used today in flight simulation. However, with modern digital computers using higher-order integration schemes, use of the Corbett-Wright orthogonality control is not efficient, and can be a waste of computational power [2].

Modern fourth-order integration schemes produce significantly less orthogonality error in the quaternion formulation than first-order methods. However, even with these higher-order techniques, the orthogonality error must be periodically corrected. This can be done much more efficiently than Corbett-Wright orthogonality control by using the method of quaternion normalization discussed in Section 1.5. Orthogonality error can be corrected periodically throughout the simulation using Eq. (1.5.12) or (1.5.13). Typically, this correction is applied after each full time step. For an excellent discussion on orthogonality error, see Phillips et al. [2, 72].

Order and Efficiency of Integration Schemes

In general, truncation error can be reduced by increasing the order of the integration method or reducing the size of the time step. Increasing the order of the integration method typically increases the computation time required for a single time step. Additionally, reducing the size of the time step increases the number of steps that must be taken for a finite amount of time. Therefore, increasing the accuracy of a simulation typically requires increasing the computational time and power required for the simulation.

When digital computers were first used in flight simulation, methods for reducing computation time while maintaining accuracy were sought. Several methods were developed, some of which are still widely used today. For example, Pope [73] presented an algorithm called the *exponential method*, which gained popularity in the aerospace community. This method is often known as the *local linearization* (LL) method, and is sometimes incorrectly assumed to be an exact method. Although it is exact for the special case of constant rotation, in general it produces second-order integration accuracy [2].

A second integration method which has been widely used in the aerospace community is the Adams-Bashforth-Moulton [74] (ABM) method. This method uses information from the previous time step, along with only two evaluations of derivatives within the current time step, to produce fourth-order accurate results. The computation time of the method is similar to a second-order method, but provides fourth-order accuracy. Because information from the previous time step is needed to maintain fourth-order accuracy, a different fourth-order method must be used to start the simulation, after which the ABM method can be used.

Because truncation error can generally be reduced by reducing the step size or increasing the order of the numerical integration scheme, it is important to consider the trade-offs between decreasing step size and increasing the integration order. Considering the error only as a function of step size does not provide an accurate comparison, since for each time step, high-order integration schemes generally require more computational effort. **A more accurate comparison can be made by considering the error in the simulation as a function of computation cycles per simulated second.** A detailed study by Phillips et al. [2, 72] considered the total error as a function of computation cycles per simulated second using the second-order Runge-Kutta, second-order local linearization, fourth-order Runge-Kutta, and fourth-order ABM methods for the case

$$\begin{bmatrix} p \\ q \\ r \end{bmatrix} = \begin{bmatrix} \sin(t) \\ \sin(t) \\ \sin(t) \end{bmatrix} \tag{5.7.2}$$

Figure 5.7.1 shows digitized results from that study for the maximum total error in degrees as a function of computation cycles per simulated second. Results show that the fourth-order ABM method produces the smallest amount of error per computational effort. However, the fourth-order Runge-Kutta method is only slightly less efficient. The second-order Runge-Kutta and local linearization methods are significantly less efficient than the fourth-order methods. Because the Runge-Kutta method is nearly as efficient as the ABM method, and because it is widely used for numerical integration, this method will be used for the results presented in the remainder of this textbook.

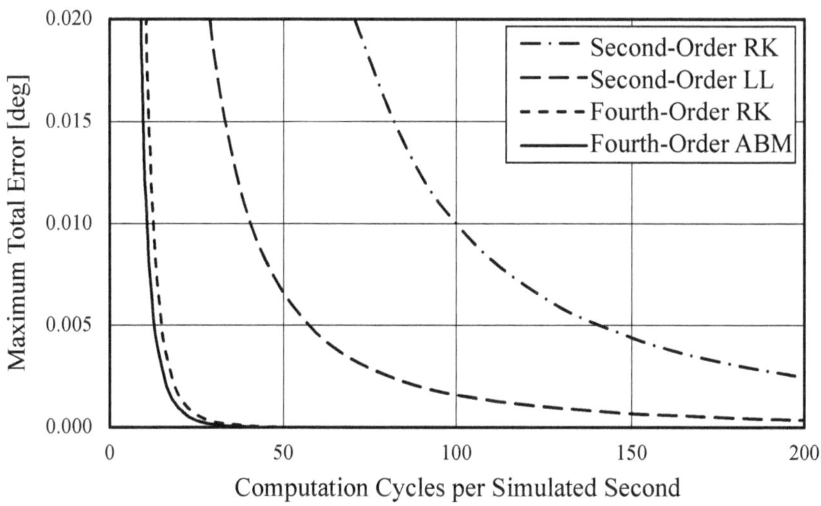

Figure 5.7.1. Simulation error for four numerical integration algorithms.

As a final note on integration accuracy, the step size for any simulation should be considered in the context of the simulation dynamics. From the results of their study, Phillips et al. [2, 72] suggest that the fourth-order Runge-Kutta integration method will give good results for flight simulation as long as the time step remains small enough to yield at least 10 integration steps per revolution of the body. A warning can easily be programmed into a code to signal that the time step is too large in the event that

$$\frac{|\omega_b|}{2\pi}\Delta t > 0.1 \tag{5.7.3}$$

where ω_b is the rotation rate of the vehicle and Δt is the time step used in the integration.

Fourth-Order Runge-Kutta Method

The Fourth-order Runge-Kutta (RK4) integration method is one of a family of methods developed by Carl Runge and Wilhelm Kutta, and is widely used for numerical integration. Details and theory of this method can be found in many numerical methods textbooks, and will not be repeated here. However, an overview how this method can be implemented will be provided here.

Let

$$\frac{dy}{dt} = f(t, y) \tag{5.7.4}$$

represent a differential equation where y is an unknown function of time t and $f(t, y)$ is a known function. Given an initial value for y_0 at time $t = t_0$, we wish to evaluate, to fourth-order accuracy, the value of y_1 at a future time $t_1 = t_0 + \Delta t$. The value for y_1 can be computed using the RK4 method as

$$y_1 = y_0 + \frac{\Delta t}{6}(k_1 + 2k_2 + 2k_3 + k_4) \tag{5.7.5}$$

where

$$k_1 = f(t_0, y_0) \tag{5.7.6}$$

$$k_2 = f\left(t_0 + \frac{\Delta t}{2}, y_0 + k_1\frac{\Delta t}{2}\right) \tag{5.7.7}$$

$$k_3 = f\left(t_0 + \frac{\Delta t}{2}, y_0 + k_2\frac{\Delta t}{2}\right) \tag{5.7.8}$$

$$k_4 = f(t_0 + \Delta t, y_0 + k_3\Delta t) \tag{5.7.9}$$

This process can be repeated for the next time step by setting the initial value of the next time step to the final value of the previous time step.

EXAMPLE 5.7.1 Use the fourth-order Runge-Kutta method to estimate the solution to the differential equation

$$\frac{dy}{dt} = 1 + \tan y$$

with initial conditions $t_0 = 0.0$ and $y_0 = 0.0$. Use a time step of $\Delta t = 0.025$ seconds, and compute the solution for the first 0.25 seconds.

Solution: For this problem, the differential equation function is $f(t, y) = 1 + \tan y$, which is only a function of y. Given the initial conditions $t_0 = 0.0$ and $y_0 = 0.0$, we can compute the following from Eqs. (5.7.6)–(5.7.9)

$$k_1 = 1 + \tan(0) = 1$$

$$k_2 = 1 + \tan\left[0 + (1)\frac{0.025}{2}\right] = 1.012500651082$$

$$k_3 = 1 + \tan\left[0 + (1.012500651082)\frac{0.025}{2}\right] = 1.012656933945$$

$$k_4 = 1 + \tan\left[0 + (1.012656933945)(0.025)\right] = 1.025321833347$$

Using these results in Eq. (5.7.5) yields

$$\begin{aligned} y_1 = y(0.025) &= 0 + \frac{0.025}{6}\left[1 + (2)1.01250065 + (2)1.01265693 + 1.02532183\right] \\ &= 0.025315154181 \end{aligned}$$

We can now repeat the process using the solution for $y(0.025)$ as the initial condition for the next step. Doing so produces the results shown in Table 5.7.1.

Table 5.7.1. Example numerical integration steps using the fourth-order Runge-Kutta method.

t	$y(t)$	k_1	k_2	k_3	k_4	$y(t+\Delta t)$
0.000	0.000000	1.000000	1.012501	1.012657	1.025322	0.025315
0.025	0.025315	1.025321	1.038150	1.038311	1.051318	0.051272
0.050	0.051272	1.051317	1.064502	1.064668	1.078046	0.077887
0.075	0.077887	1.078045	1.091618	1.091789	1.105571	0.105181
0.100	0.105181	1.105570	1.119565	1.119743	1.133967	0.133173
0.125	0.133173	1.133966	1.148423	1.148608	1.163317	0.161887
0.150	0.161887	1.163316	1.178282	1.178475	1.193719	0.191347
0.175	0.191347	1.193717	1.209245	1.209447	1.225283	0.221582
0.200	0.221582	1.225281	1.241432	1.241645	1.258138	0.252622
0.225	0.252622	1.258137	1.274981	1.275208	1.292435	0.284501
0.250	0.284501	1.292434	1.310056	1.310297	1.328350	0.317257

The fourth-order Runge-Kutta method can also be used to estimate the solution to a system of differential equations. For example, let

$$\frac{dy_1}{dt} = f_1(t, y_1, y_2, ..., y_N) \tag{5.7.10}$$

$$\frac{dy_2}{dt} = f_2(t, y_1, y_2, ..., y_N) \tag{5.7.11}$$

$$\vdots \tag{5.7.12}$$

$$\frac{dy_N}{dt} = f_N(t, y_1, y_2, ..., y_N) \tag{5.7.13}$$

represent a system of differential equations where $\mathbf{y} = (y_1, y_2, ..., y_N)$ is an array of unknown values as a function of time t, and the known differential-equation functions are $f_i(t, \mathbf{y})$ where $i = 1, 2, ..., N$. This system can be written in array form as

$$\frac{d\mathbf{y}}{dt} = \mathbf{f}(t, \mathbf{y}) \tag{5.7.14}$$

where $\mathbf{f}$ represents the array of differential functions $\mathbf{f} = (f_1, f_2, ..., f_N)$. Given an initial value for the array of values $\mathbf{y}_0$ and corresponding time $t = t_0$, the fourth-order Runge-Kutta method can be used to evaluate the value of the array $\mathbf{y}$ at a future time $t_1 = t_0 + \Delta t$ from the algorithm

$$\mathbf{k}_1 = \mathbf{f}(t_0, \mathbf{y}_0) \tag{5.7.15}$$

$$\mathbf{k}_2 = \mathbf{f}\left(t_0 + \frac{\Delta t}{2}, \mathbf{y}_0 + \mathbf{k}_1 \frac{\Delta t}{2}\right) \tag{5.7.16}$$

$$\mathbf{k}_3 = \mathbf{f}\left(t_0 + \frac{\Delta t}{2}, \mathbf{y}_0 + \mathbf{k}_2 \frac{\Delta t}{2}\right) \tag{5.7.17}$$

$$\mathbf{k}_4 = \mathbf{f}(t_0 + \Delta t, \mathbf{y}_0 + \mathbf{k}_3 \Delta t) \tag{5.7.18}$$

$$\mathbf{y}_1 = \mathbf{y}_0 + \frac{\Delta t}{6}(\mathbf{k}_1 + 2\mathbf{k}_2 + 2\mathbf{k}_3 + \mathbf{k}_4) \tag{5.7.19}$$

where $\mathbf{k}_1$, $\mathbf{k}_2$, $\mathbf{k}_3$, and $\mathbf{k}_4$ are arrays of length N.

EXAMPLE 5.7.2 Use the fourth-order Runge-Kutta method to estimate the solution to the system of differential equations

$$\frac{dx}{dt} = t + z^2 \sin x$$

$$\frac{dz}{dt} = t + x \cos z$$

with initial conditions $t_0 = x_0 = z_0 = 0.0$ and a time step size of 0.025.

Solution: We will define the vector $\mathbf{y} = (x, z)$ to contain the state variables x and z. Given the initial conditions $t_0 = x_0 = z_0 = 0.0$, we can compute the array values for $\mathbf{k}_1$ from Eq. (5.7.15), which is simply the value of the system of differential equations for $t = t_0$ and $y = y_0$. This yields

$$\mathbf{k}_1 = \begin{bmatrix} t_0 + z_0^2 \sin x_0 \\ t_0 + x_0 \cos z_0 \end{bmatrix} = \begin{bmatrix} (0) + (0)^2 \sin(0) \\ (0) + (0)\cos(0) \end{bmatrix} = \begin{bmatrix} 0 \\ 0 \end{bmatrix}$$

To compute the array of values $\mathbf{k}_2$ from Eq. (5.7.16), we must evaluate the differential equations at $t = t_0 + \Delta t/2 = 0.025/2 = 0.0125$ and

$$\mathbf{y} = \mathbf{y}_0 + \mathbf{k}_1 \frac{\Delta t}{2} = \begin{bmatrix} 0 \\ 0 \end{bmatrix} + \begin{bmatrix} (0)(0.025/2) \\ (0)(0.025/2) \end{bmatrix} = \begin{bmatrix} 0 \\ 0 \end{bmatrix}$$

Using these values for t and $\mathbf{y}$ in the differential equations yields

$$\mathbf{k}_2 = \begin{bmatrix} (0.0125) + (0)^2 \sin(0) \\ (0.0125) + (0)\cos(0) \end{bmatrix} = \begin{bmatrix} 0.0125 \\ 0.0125 \end{bmatrix}$$

To compute the array of values $\mathbf{k}_3$ from Eq. (5.7.17), we must evaluate the differential equations at $t = t_0 + \Delta t/2 = 0.025/2 = 0.0125$ and

$$\mathbf{y} = \mathbf{y}_0 + \mathbf{k}_2 \frac{\Delta t}{2} = \begin{bmatrix} 0 \\ 0 \end{bmatrix} + \begin{bmatrix} (0.0125)(0.025/2) \\ (0.0125)(0.025/2) \end{bmatrix} = \begin{bmatrix} 0.00015625 \\ 0.00015625 \end{bmatrix}$$

Using these values for t and $\mathbf{y}$ in the differential equations yields

$$\mathbf{k}_3 = \begin{bmatrix} (0.0125) + (0.00015625)^2 \sin(0.00015625) \\ (0.0125) + (0.00015625)\cos(0.00015625) \end{bmatrix} = \begin{bmatrix} 0.0125000000038 \\ 0.0126562499981 \end{bmatrix}$$

To compute the array of values $\mathbf{k}_4$ from Eq. (5.7.18), we must evaluate the differential equations at $t = t_0 + \Delta t = 0.025$ and

$$\mathbf{y} = \mathbf{y}_0 + \mathbf{k}_3 \Delta t = \begin{bmatrix} 0 \\ 0 \end{bmatrix} + \begin{bmatrix} (0.0125000000038)(0.025) \\ (0.0126562499981)(0.025) \end{bmatrix} = \begin{bmatrix} 0.000312500000095 \\ 0.000316406249952 \end{bmatrix}$$

Using these values for t and $\mathbf{y}$ in the differential equations yields

$$\mathbf{k}_4 = \begin{bmatrix} (0.025) + (0.0003164062)^2 \sin(0.0003125000) \\ (0.025) + (0.0003125000)\cos(0.0003164062) \end{bmatrix} = \begin{bmatrix} 0.02500000003129 \\ 0.02531249998445 \end{bmatrix}$$

Using these results in Eq. (5.7.19) yields

$$\begin{aligned} \mathbf{y}_1 &= \mathbf{y}(0.025) \\ &= \begin{bmatrix} 0 \\ 0 \end{bmatrix} + \frac{0.025}{6} \left(\begin{bmatrix} 0 \\ 0 \end{bmatrix} + 2 \begin{bmatrix} 0.0125 \\ 0.0125 \end{bmatrix} + 2 \begin{bmatrix} 0.01250000 \\ 0.01265625 \end{bmatrix} + \begin{bmatrix} 0.02500000003 \\ 0.02531249998 \end{bmatrix} \right) \\ &= \begin{bmatrix} 0.000312500000162 \\ 0.000315104166586 \end{bmatrix} \end{aligned}$$

Table 5.7.2. Example numerical solution using the fourth-order Runge-Kutta method.

t	x	z
0.000	0.00000000000000	0.00000000000000
0.025	0.00031250000016	0.00031510416659
0.050	0.00125000001488	0.00127083332598
0.075	0.00281250025077	0.00288281237691
0.100	0.00500000189605	0.00516666574203
0.125	0.00781250915667	0.00813801639653
0.150	0.01125003325601	0.01181248398877
0.175	0.01531259918476	0.01620568171840
0.200	0.02000025606020	0.02133321162142
0.225	0.02531309202710	0.02721065789732
0.250	0.03125125470781	0.03385357789978

We can now repeat the process using the solution for $\mathbf{y}(0.025)$ as the initial condition for the next step. Doing so produces the results shown in Table 5.7.2.

Integrating the Equations of Motion

The fourth-order Runge-Kutta algorithm given in Eqs. (5.7.15)–(5.7.19) can be used to integrate the quaternion formulation of the equations of motion given in Eqs. (5.6.1)–(5.6.4). To perform the integration, the initial condition of the vehicle must be given, and a step size for time must be provided. Additionally, the vehicle's mass, as well as the pseudo aerodynamic force and moment vectors must be provided as functions of the state variables. In the development of a simulation code, it is helpful to structure this integration process using five nested functions as outlined in Algorithm 5.7.2.

> The **`run_sim`** function initializes the time and sets the value of the state array to the initial values. It then runs a while loop. Each time through the loop, it calls the `rk4` function and passes as input the time at the beginning of the time step t_0, a corresponding array of state values $\mathbf{y}_0$, and a time step size Δt.
>
> > The **`rk4`** function accepts as input the time at the beginning of the time step, an array of corresponding state values, and a time step size. It performs the Runge-Kutta integration, which for fourth-order integration requires calling the `diff_eq` function four times. For each call to the `diff_eq` function, it passes as input a time and an array of state values.
> >
> > > The **`diff_eq`** function accepts as input a time and an array of state values. At the beginning of this function, it calls both the `pseudo_aero` and `mass_inertia` functions and passes as input to each, the current time and the array of state values.

Algorithm 5.7.2 Nested functions for integration of the equations of motion.

function `run_sim`
  Input
    t_0 (real) Start time of the simulation (default = 0.0)
    t_f (real) End (final) time of the simulation
    Δt (real) Time step size
    $\mathbf{y}_0$ (real array) Initial states of the vehicle
  while ($t_0 < t_f$) **do**
    $\mathbf{y}_1 =$ `rk4`$(t_0, \mathbf{y}_0, \Delta t)$
      $\dot{\mathbf{y}} =$ `diff_eq`$(t, \mathbf{y})$ (called within `rk4`)
        $\overline{\mathbf{F}}, \overline{\mathbf{M}} =$ `pseudo_aero`$(t, \mathbf{y})$ (called within `diff_eq`)
        $\mathcal{M}, [\mathbf{I}] =$ `mass_inertia`$(t, \mathbf{y})$ (called within `diff_eq`)
    $t_0 = t_0 + \Delta t$
    $\mathbf{y}_0 = \mathbf{y}_1$
  end while
end function

The **`pseudo_aero`** function accepts as input a time and an array of state values. It computes the corresponding pseudo aerodynamic force $\overline{\mathbf{F}}$ and moment $\overline{\mathbf{M}}$ vectors and returns these vectors.

The **`mass_inertia`** function accepts as input a time and an array of state values. It computes the mass $\mathcal{M}$, inertia $[\mathbf{I}]$, and inverse of the inertia tensor $[\mathbf{I}]^{-1}$ of the vehicle and returns these values. A closed-form solution for the inverse of a 3×3 matrix is included in Appendix D.

The **`diff_eq`** function uses the returned pseudo aerodynamic force and moment vectors, mass, and inertia tensor to compute the time derivative of the states $\dot{\mathbf{y}}$ as given in Eqs. (5.6.1)–(5.6.1). It returns this array of time derivatives to the `rk4` function.

The **`rk4`** function uses the time derivatives from four function calls to the `diff_eq` function to compute the states at the end of the time step $t = t_0 + \Delta t$. It returns these states to the `run_sim` function.

The **`run_sim`** function sets the returned states at the end of the previous time step to be the initial states of the next time step and increments the time. It then runs the loop again. This process is repeated until a final specified time is reached.

This modular structure allows the `pseudo_aero` and `mass_inertia` functions to be updated, altered, or replaced for different vehicles without requiring changes to the `rk4` and `diff_eq` functions. Methods for doing this are included in the following examples.

EXAMPLE 5.7.3 A ping-pong ball has a diameter of 0.13084 ft, a weight at sea level of 0.006 lbf, and a skin thickness of 0.00131 ft. The ball is hit such that it has an initial condition of 50 ft/s in a horizontal direction 300 ft above sea level. Using fourth-order Runge-Kutta integration and a time step size of 0.01, compute the trajectory of the ping-pong ball as a function of time for a total of 10 seconds. Use the aerodynamic drag model given in Section 3.6 to estimate the aerodynamic force on the ball throughout the simulation. Plot the resulting velocity components as a function of time, as well as the resulting trajectory of the ball.

Solution: Using the relationships in Chapter 2, the mass of the ball is

$$\mathcal{M} = 1.864857010294\text{E}{-04} \text{ slug}$$

and the inertia tensor for the ball is

$$[\mathbf{I}] = 5.215657263284\text{E}{-07} \begin{bmatrix} 1 & 0 & 0 \\ 0 & 1 & 0 \\ 0 & 0 & 1 \end{bmatrix} \text{ slug}\cdot\text{ft}^2$$

A code was written that uses Algorithm 5.7.2 to integrate the state of the ball forward in time and compute the state of the ball at each time step. As part of this code, a `pseudo_aero` function was written comprised of the following pseudo code:

```
1. Accept the state vector as an input
2. Compute total velocity using Eq. (3.4.3)
3. Compute the aerodynamic angles from the velocity
   components using Eqs. (3.4.4) and (3.4.5)
4. Obtain atmospheric properties at the altitude stored
   in the state array using the method outlined in
   Section 3.2
5. Compute the Reynolds number from the atmospheric
   properties, velocity, and sphere radius as given in
   Eq. (3.6.28)
6. Compute the drag coefficient from the Reynolds number
   using Eq. (3.6.29)
7. Compute the dimensional drag from Eq. (3.4.26)
8. Compute the total force in body-fixed coordinates
   based on relations given in Eq. (5.2.3)
9. Return the pseudo aerodynamic force and moment vector
   in body-fixed coordinates (all moments are zero)
```

Table 5.7.3 shows the resulting states of the ball at a few selected time steps. Detailed intermediate solution steps can be found on the book website.

Table 5.7.3. State values at sample time steps for the ping-pong ball.

State Variable	t = 0 [s]	t = 0.01 [s]	t = 1.0 [s]	t = 10.0 [s]
u [ft/s]	50	49.10311946	15.65456972	0.000899679
v [ft/s]	0	0	0	0
w [ft/s]	0	0.318844203	20.14489930	29.25779087
p [rad/s]	0	0	0	0
q [rad/s]	0	0	0	0
r [rad/s]	0	0	0	0
x_f [ft]	0	0.495489033	27.73961765	43.00922164
y_f [ft]	0	0	0	0
z_f [ft]	−300	−299.998401	−288.541371	−30.45432293
e_0	1	1	1	1
e_x	0	0	0	0
e_y	0	0	0	0
e_z	0	0	0	0

Figure 5.7.2 shows the trajectory of the ping-pong ball over the period of 10 seconds. Note that although the two axes have the same units, the axes are not equally scaled. The ball travels only a very short distance forward before entering a nearly vertical fall. Figure 5.7.3 shows the velocity components of the ping-pong ball as a function of time. The lateral velocity v is zero, and therefore not included in the plot. From the figure, it can be seen that the terminal velocity of the ping-pong ball is just below 30 ft/s.

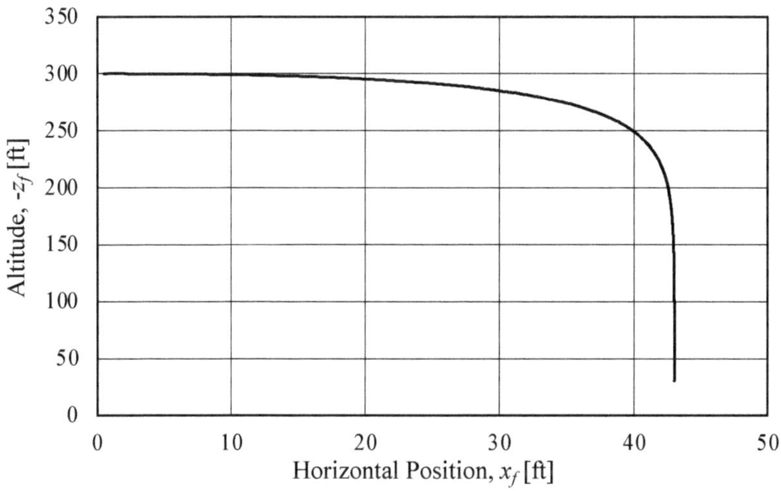

Figure 5.7.2. Position of the ping-pong ball as a function of time.

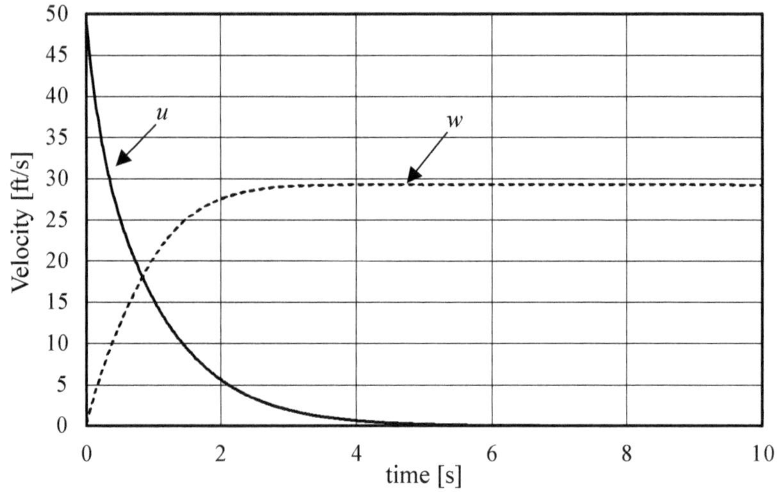

Figure 5.7.3. Velocity of the ping-pong ball as a function of time.

EXAMPLE 5.7.4 The mass and aerodynamics of an arrow with straight fletchings can be modeled as given in Appendix C.1. Use this model to predict the state of the arrow as a function of time after it is launched at 210 ft/s from an altitude of 5 feet with an initial elevation angle of 5 deg. Assume that all other initial states of the arrow are zero. Using fourth-order Runge-Kutta integration, compute the states of the arrow as a function of time until the arrow reaches sea-level altitude. Use a time step of 0.01 seconds for the integration.

Solution: The code from Example 5.7.3 was used with the mass and aerodynamic functions replaced with those for the arrow. The `pseudo_aero` function was written comprised of the following pseudo code:

```
1. Accept the state vector as an input
2. Compute total velocity using Eq. (3.4.3)
3. Compute the angle of attack using Eq. (3.4.4) and the
   flank angle using Eq. (3.4.13)
4. Obtain atmospheric properties at the altitude stored
   in the state array
5. Compute the aerodynamic force and moment coefficients
   using Eqs. (C.1.5)-(C.1.11)
6. Compute the dimensional forces and moments from Eq.
   (3.4.26)
7. Compute the total force in body-fixed coordinates
   based on relations given in Eq. (5.2.3)
8. Return the pseudo aerodynamic force and moment vector
   in body-fixed coordinates
```

Table 5.7.4. Longitudinal state values at sample time steps for the arrow with straight fletchings.

State Variable	$t = 0$ [s]	$t = 0.01$ [s]	$t = 0.1$ [s]	$t = 1.35$ [s]
u [ft/s]	210	209.703565	207.0341383	179.4804605
w [ft/s]	0	0.319717921	2.823165518	−0.259786977
q [rad/s]	0	−0.000534779	−0.047786676	−0.213684902
x_f [ft]	0	2.090671363	20.78630647	259.9796845
z_f [ft]	−5	−5.181302795	−6.659026232	−0.088303916
e_0	0.999048222	0.999048261	0.999083988	0.99761353
e_y	0.043619387	0.043618495	0.04279235	−0.069045244

Table 5.7.4 shows the resulting longitudinal states of the arrow at a few selected time steps. Note that all lateral components are zero. Detailed solution steps are available on the book website.

Figure 5.7.4 shows the trajectory of the arrow until it reaches sea level. Figure 5.7.5 shows the velocity and angle of attack of the arrow as a function of time. The oscillatory behavior of the angle of attack is representative of a common rigid-body dynamic mode of arrows. Repeating the simulation with the arrow at a bank angle of $\phi = 90$ deg produces similar results, but with differences in a few of the lateral and longitudinal components. The student should be able to deduce what differences ought to be present when the arrow is initialized with this bank angle, and test their code and intuition using this case.

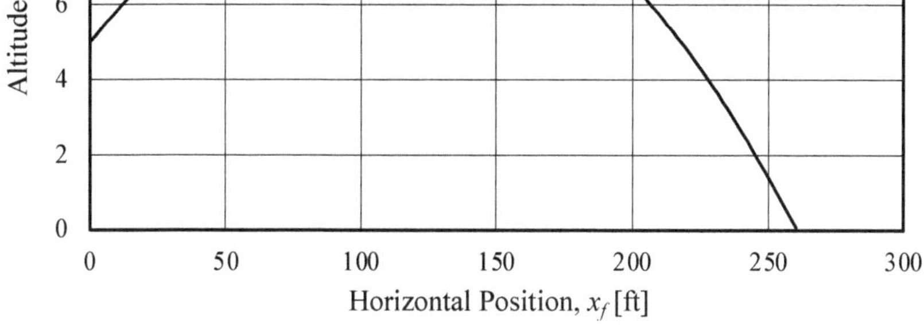

Figure 5.7.4. Position of the arrow with straight fletchings as a function of time.

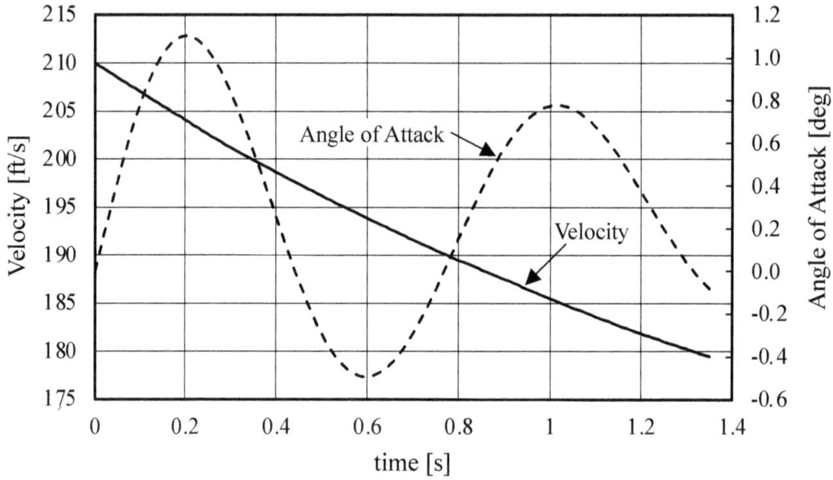

Figure 5.7.5. Velocity and angle of attack of the arrow with straight fletchings as a function of time.

EXAMPLE 5.7.5 Repeat Example 5.7.4 but for an arrow with angled fletchings. Assume there is no difference in the mass model of the two arrows, and the only difference in the aerodynamic models is that the arrow with angled fletchings produces a rolling moment at zero rolling rate. This is captured in the coefficient $C_{\ell_0} = 3.223$. Note that the side-force and yawing-moment coefficients given in Appendix C.1 are a function of the flank angle β_f rather than the traditional sideslip angle β. For this simulation, use a time step of 0.005 seconds.

Solution: The code from Example 5.6.4 was used with only two changes, which include a non-zero value for the aerodynamic coefficient C_{ℓ_0}, and a time step of 0.005 seconds. The time step was chosen based on the maximum rotation rate of the arrow. Recall from the discussion surrounding Eq. (5.7.3) that the fourth-order Runge-Kutta integration algorithm with periodic normalization of the quaternion will give good results as long as the time step is chosen to give about 10 steps per body revolution, which is equivalent to a maximum of 36 deg per time step. From trial simulations, it can be seen that the maximum rotation rate is around 6,000 deg per second, which would correlate to a maximum time step of 0.006 s to remain within the accuracy requirement. Using a time step of 0.005 is attractive to yield exactly 200 time steps per second. Table 5.7.5 shows the resulting states of the arrow at a few selected time steps. Detailed solution steps can be found on the book website.

Table 5.7.5. State values at sample time steps for the arrow with angled fletchings.

State Variable	t = 0.000 [s]	t = 0.005 [s]	t = 0.100 [s]	t = 1.350 [s]
u [ft/s]	210.00000000	209.85170442	207.03424373	179.25221074
v [ft/s]	0.00000000	0.00817751	2.65314885	1.25704075
w [ft/s]	0.00000000	0.15988289	−0.95371346	−2.12925482
p [rad/s]	0.00000000	19.77315801	106.19754827	93.66915982
q [rad/s]	0.00000000	−0.00013402	0.01292942	−0.06219867
r [rad/s]	0.00000000	0.00000679	0.04578577	0.18404015
x_f [ft]	0.00000000	1.04572311	20.78504950	259.78083772
y_f [ft]	0.00000000	0.00000037	−0.00003145	−0.18652519
z_f [ft]	−5.00000000	−5.09108681	−6.65912905	−0.08398141
e_0	0.99904822	0.99872223	−0.57612190	−0.75579281
e_x	0.00000000	0.02551979	−0.81624265	−0.65145052
e_y	0.04361939	0.04360504	−0.02464788	0.05451853
e_z	0.00000000	−0.00111422	0.03498530	−0.03764543

Figure 5.7.6 shows the angle of attack and flank angle of the arrow with angled fletchings as a function of time until it reaches sea-level altitude of $z_f = 0$. Note that the rotation rate of the arrow significantly alters the angle of attack relative to that from Example 5.7.4.

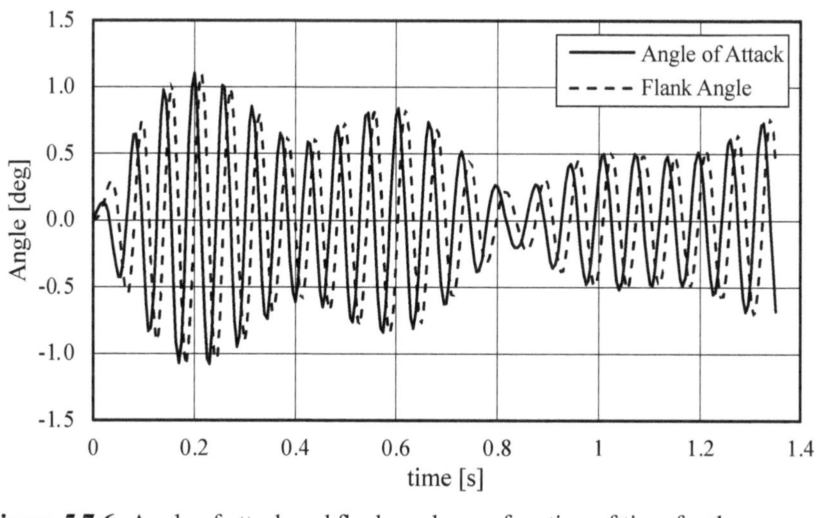

Figure 5.7.6. Angle of attack and flank angle as a function of time for the arrow with angled fletchings.

5.8 Fixed-Wing Examples

The following examples are based on the F-16 aircraft specified in Appendix C.3 and can be used to check computer code for accuracy.

EXAMPLE 5.8.1 Consider the F-16 aircraft specified in Appendix C.3 with the aerodynamic model F16-E given in Table C.3.15. Predict the state of the vehicle as a function of time from an initial velocity of 330 ft/s at an altitude of 1000 feet. Consider the case when $\theta = \alpha = 7.857407446414$ deg, $\delta_e = -4.940949432344$ deg, $\tau = 0.067931173301$, and all other initial states, including controls, are set to zero. Using fourth-order Runge-Kutta integration, compute the states of the vehicle as a function of time for 5 seconds using a time step of 0.01 seconds for the integration.

Solution: The code from Example 5.7.4 was used with the mass and pseudo aerodynamic functions replaced with those for the specified vehicle. The `pseudo_aero` function was written comprised of the following pseudo code:

```
1. Accept the state vector as an input
2. Compute total velocity using Eq. (3.4.3)
3. Compute the angle of attack using Eq. (3.4.4) and the
   sideslip angle using Eq. (3.4.5)
4. Obtain atmospheric properties at the altitude stored
   in the state array
5. Compute the aerodynamic force and moment coefficients
   using Eqs. (3.7.24)-(3.7.30)
6. Compute the dimensional forces and moments from Eq.
(3.4.26)
7. Compute the total force in body-fixed coordinates
   based on relations given in Eq. (5.2.3)
8. Return the pseudo aerodynamic force and moment vector
   in body-fixed coordinates
```

Table 5.8.1 shows the resulting non-zero states of the vehicle at a few selected time steps. Detailed solution steps can be found on the book website.

The state of the vehicle used for initialization in this example is called a *trim state* for steady level flight. Trim states will be discussed in more detail in Chapters 6 and 7. The important thing to note here is that for this particular trim state, the change in all components of the state vector are zero, except for the x_f positional state of the vehicle. Trim states are useful in the early stages of code development for debugging, since the `diff_eq` function should return zero for all values except for positional values. Because there is small machine error in the initialization and integration, the changes in the states for a trim condition will not remain perfectly zero. Rather, error will build up, and will eventually become large enough to be noticeable on a plot. For this example, only 5 seconds of simulation time was included with an integration time step of 0.01. After this number of steps, the largest errors in the state vector should be on the order of 1.0E−14 for this case.

Table 5.8.1. State values at sample time steps for Example 5.8.1.

State Variable	$t = 0$ [s]	$t = 0.01$ [s]	$t = 5.0$ [s]
u [ft/s]	326.90174983	326.90174983	326.90174983
w [ft/s]	45.11370036	45.11370036	45.11370036
x_f [ft]	0.00000000	3.30000000	1650.00000000
z_f [ft]	−1000.00000000	−1000.00000000	−1000.00000000
e_0	0.99765008	0.99765008	0.99765008
e_y	0.06851510	0.06851510	0.06851510

EXAMPLE 5.8.2 Consider the F-16 aircraft specified in Appendix C.3 with the aerodynamic model F16-E-1 given in Table C.3.15. Note that the reference location of the aerodynamic model is 1 ft aft of the vehicle CG. Predict the state of the vehicle as a function of time from an initial velocity of 330 ft/s and an altitude of 1000 feet. First consider the trim case when $\theta = \alpha = 8.442080225436$ deg, $\delta_e = -9.439213266902$ deg, $\tau = 0.074041048286$, and all other initial states, including controls, are set to zero. Using fourth-order Runge-Kutta integration, compute the states of the vehicle as a function of time for 60 seconds using a time step of 0.01 seconds for the integration. Next consider the case with all initial conditions set as specified here, but with $\delta_e = -9.0$ deg. Plot the angle of attack and sideslip angle for each case. Comment on the differences.

Solution: The code used for Example 5.8.1 was used for this simulation with one modification. Once the aerodynamic force and moment components were computed within the pseudo aerodynamic function, they were transformed to the center of gravity location using Eq. (5.2.7) with

$$\mathbf{P}_{\text{ref}} = \begin{bmatrix} -1.0 \\ 0.0 \\ 0.0 \end{bmatrix} \text{ ft}$$

This transformation was performed within the `pseudo_aero` function just prior to returning the forces and moments, such that the forces and moments are returned in the body-fixed coordinate system to the `diff_eq` function. Table 5.8.2 shows the resulting non-zero states of the vehicle at a few selected time steps for Case A, which is a trim case. Table 5.8.3 shows the resulting states of the vehicle at a few selected time steps for Case B, which is not a trim case. Detailed solution steps for each case can be found on the book website.

Table 5.8.2. State values at sample time steps for Case A in Example 5.8.2.

State Variable	$t = 0$ [s]	$t = 0.01$ [s]	$t = 60.0$ [s]
u [ft/s]	326.4243765	326.4243765	326.4243765
w [ft/s]	48.44715109	48.44715109	48.44715109
x_f [ft]	0.00000000	3.30000000	19800.00000
z_f [ft]	−1000.000000	−1000.000000	−1000.000000
e_0	0.99728752	0.99728752	0.99728752
e_y	0.07360443	0.07360443	0.07360443

Table 5.8.3. State values at sample time steps for Case B in Example 5.8.2.

State Variable	$t = 0$ [s]	$t = 0.01$ [s]	$t = 60.0$ [s]
u [ft/s]	326.4243765	326.4248079	336.8571850
v [ft/s]	0.00000000	0.00000001	0.00064362
w [ft/s]	48.44715109	48.44429682	44.31733839
p [rad/s]	0.00000000	0.00000000	−0.00001707
q [rad/s]	0.00000000	−0.00037170	−0.00659816
r [rad/s]	0.00000000	0.00000000	−0.00001405
x_f [ft]	0.00000000	3.30000003	21138.36529
y_f [ft]	0.00000000	0.00000000	−4.92830729
z_f [ft]	−1000.000000	−1000.000011	−1044.668698
e_0	0.99728752	0.99728758	0.99877829
e_x	0.00000000	0.00000000	−0.00005642
e_y	0.07360443	0.07360350	0.04941571
e_z	0.00000000	0.00000000	−0.00009230

Figure 5.8.1 shows the angle of attack (solid lines) and sideslip angle (dashed lines) as a function of time for both the trim case (Case A, black lines) and the non-trim case (Case B, gray lines). Note that the angle of attack and sideslip are constant for the trim case. However, there is noticeable oscillatory behavior in both the angle of attack and sideslip angle for the non-trim case. This oscillatory behavior is common when an aircraft that is statically stable experiences a perturbation in state.

Fighter aircraft such as the F-16 are typically not statically stable. However, assuming the aerodynamic reference location is 1 ft aft of the CG, as was done in this problem, results in a statically stable aircraft. Setting the elevator at a value slightly off of the trim value acts as a perturbation to the trim state in this simulation.

The high-frequency oscillation in the angle of attack, which damps out after about 4 seconds, is known as the short-period mode. The high-frequency oscillation in the sideslip angle, which damps out after about 10 seconds, is known as the Dutch

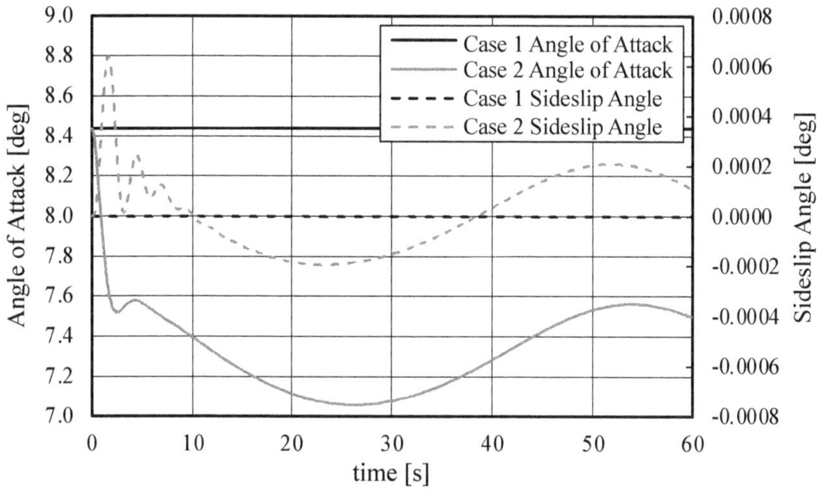

Figure 5.8.1. Angle of attack and sideslip angle as a function of time for the F-16 in a trim (Case A) and non-trim (Case B) case.

roll mode. The low-frequency oscillation that persists throughout the duration of the simulation as seen in both the angle of attack and sideslip angle is known as the long-period or phugoid mode. Each of these dynamic modes, along with two others, can be studied at various operating conditions using time-history simulations [75]. Very accurate estimates for these dynamic modes can be obtained by linearizing the equations of motion [75, 76].

EXAMPLE 5.8.3 Consider the F-16 aircraft specified in Appendix C.3 with the aerodynamic model F16-E-1 given in Table C.3.15. Predict the state of the vehicle as a function of time from an initial velocity of 330 ft/s, an altitude of 1000 feet, and the initial states

$$\begin{bmatrix}\phi\\ \theta\\ \psi\end{bmatrix} = \begin{bmatrix}45.0\\ 8.042590225317\\ 0.0\end{bmatrix} \text{ deg}, \qquad \begin{bmatrix}\alpha\\ \beta\end{bmatrix} = \begin{bmatrix}11.278107537468\\ 0.022727214169\end{bmatrix} \text{ deg}$$

$$\begin{bmatrix}p\\ q\\ r\end{bmatrix} = \begin{bmatrix}-0.766207504385\\ 3.834356702909\\ 3.834356702909\end{bmatrix} \text{ deg/s}, \qquad \begin{bmatrix}\delta_a\\ \delta_e\\ \delta_r\end{bmatrix} = \begin{bmatrix}0.123007387668\\ -11.389397552606\\ -0.871269845919\end{bmatrix} \text{ deg}$$

$$\tau = 0.129524457808$$

Using fourth-order Runge-Kutta integration, compute the states of the vehicle as a function of time for 60 seconds using a time step of 0.01 seconds for the integration.

Table 5.8.4. State values at sample time steps for a lateral trim state of the F-16.

State Variable	$t = 0$ [s]	$t = 0.01$ [s]	$t = 60.0$ [s]
u [ft/s]	323.6274954	323.6274954	323.6274954
v [ft/s]	0.13089935	0.13089935	0.13089935
w [ft/s]	64.53857060	64.53857060	64.53857060
p [rad/s]	−0.01337284	−0.01337284	−0.01337284
q [rad/s]	0.06692215	0.06692215	0.06692215
r [rad/s]	0.06692215	0.06692215	0.06692215
x_f [ft]	0.00000000	3.26863919	−1712.370814
y_f [ft]	0.00000000	−0.45386892	749.4962822
z_f [ft]	−1000.000000	−1000.000000	−1000.000000
e_0	0.92160499	0.92161771	−0.87993031
e_x	0.38174129	0.38171028	−0.38502680
e_y	0.06478912	0.06497155	0.04096830
e_z	−0.02683653	−0.02639608	0.27531546

Solution: The code used for Example 5.8.2 was used to predict the time history of the states of the vehicle beginning with the initial conditions listed here. Table 5.8.4 shows the resulting states of the vehicle at a few selected time steps for this lateral trim case. Detailed solution steps for each case can be found on the book website. This is an excellent case for debugging a simulation code, since the `diff_eq` function should always return something near machine zero for the first six components (u, v, w, p, q, r) in this lateral trim state.

EXAMPLE 5.8.4 Consider the F-16 aircraft specified in Appendix C.3 with the aerodynamic models F16-E-1 and F16-ES-1 given in Table C.3.15. Note that both models are identical, except that the F16-ES-1 model includes the stall model given in Example 3.8.1. Predict the state of the vehicle as a function of time using each model during a pull-up maneuver from an initial velocity of 350 ft/s, an altitude of 1000 feet, an elevator deflection of $\delta_e = -25.0$ deg, a throttle setting of $\tau = 0.4$, and all other initial states including aerodynamic angles, rates, Euler angles, and controls set to zero. Also set the gyroscopic vector to zero, i.e. $\mathbf{h} = 0$, to produce pure longitudinal motion. Using fourth-order Runge-Kutta integration, compute the states of the vehicle as a function of time for 30 seconds using a time step of 0.01 seconds for the integration. Compare the results of the simulation from each model and comment on the differences.

Solution: The code used for Example 5.8.2 was used to predict the time history of the states of the vehicle beginning with the initial conditions given here and with $\mathbf{h} = 0$. Table 5.8.5 shows the resulting longitudinal states of the vehicle at a few selected time steps for the F16-E-1 model, which does not include stall. Table 5.8.6 shows the

Table 5.8.5. Longitudinal state values at sample time steps using the F16-E-1 aerodynamic model (no stall) in a pull-up maneuver.

State Variable	$t = 0$ [s]	$t = 0.01$ [s]	$t = 30.0$ [s]
u [ft/s]	350.0000000	350.1152702	66.82042997
w [ft/s]	0.00000000	0.46260447	60.88159688
q [rad/s]	0.00000000	0.01791195	0.05140382
x_f [ft]	0.00000000	3.50057599	2347.610827
z_f [ft]	−1000.000000	−999.9978365	−1000.758110
e_0	1.00000000	1.00000000	−0.72735502
e_y	0.00000000	0.00004487	−0.68626138

Table 5.8.6. Longitudinal state values at sample time steps using the F16-ES-1 aerodynamic model (with stall) in a pull-up maneuver.

State Variable	$t = 0$ [s]	$t = 0.01$ [s]	$t = 30.0$ [s]
u [ft/s]	350.00000000	350.1158928	151.5439451
w [ft/s]	0.00000000	0.45851745	128.7227534
q [rad/s]	0.00000000	0.01606043	0.13662685
x_f [ft]	0.00000000	3.50057911	3549.073978
z_f [ft]	−1000.000000	−999.9978413	−384.7207231
e_0	1.00000000	1.00000000	0.93717504
e_y	0.00000000	0.00004022	0.34885948

resulting longitudinal states of the vehicle at a few selected time steps for the F16-ES-1 model, which includes stall. All states not included in these tables are exactly zero. Detailed solution steps for each case can be found on the book website.

Figure 5.8.2 shows the trajectory of the vehicle resulting from the two different aerodynamic models. Note that both simulations begin with the vehicle in horizontal flight, but performing a pull-up maneuver. The simulation without stall shows that the vehicle climbs into a loop, whereas the simulation including stall shows that the vehicle rapidly climbs in altitude, but does not have enough aerodynamic authority to complete a loop. The simulation without stall reaches a maximum altitude of about 2,050 ft at about 11.35 seconds, and the simulation with stall reaches a maximum altitude of about 1950 ft at about 13.6 seconds.

Figure 5.8.3 shows the time history of the elevation angle of each simulation. The simulation without stall performs a loop, reaching an elevation angle of 90 degrees at about 4 seconds, back to zero at about 12 seconds, −90 deg at about 16 seconds, and back to zero at about 21 seconds. In the simulation with stall, the vehicle maintains a high elevation angle through much of the climb, then rapidly pitches nose down after reaching the maximum altitude. It exhibits a minimum elevation angle of about −70 deg during the dive.

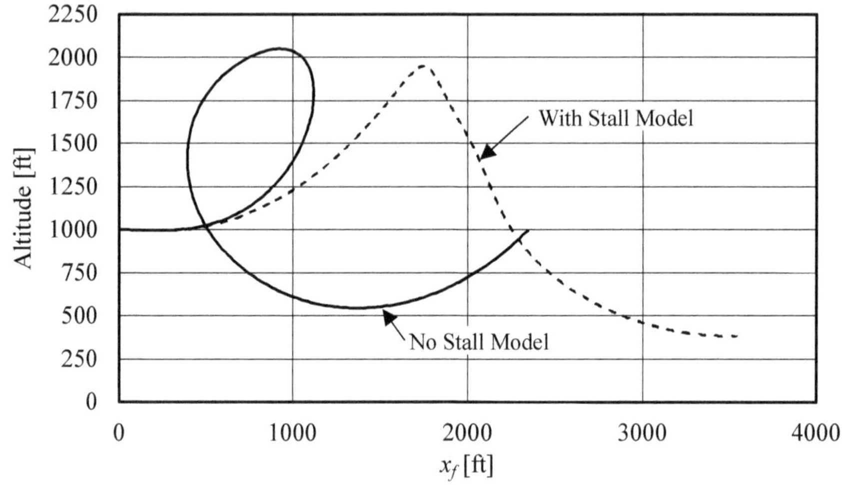

Figure 5.8.2. Earth-fixed coordinates showing the trajectory of the vehicle using two different aerodynamic models.

Figure 5.8.4 shows the angle-of-attack time history of each simulation. The simulation without stall holds a fairly consistent angle of attack through much of the maneuver, and dips significantly only near the top of the loop. The simulation with stall shows that the vehicle reaches a much higher angle of attack near the point of maximum altitude, and then drops to sub-stall angles through the dive.

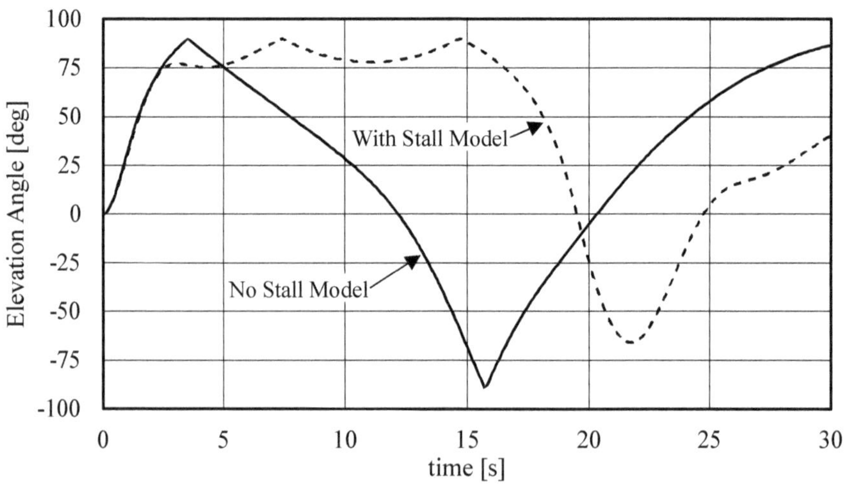

Figure 5.8.3. Time history of the Euler elevation angle of the vehicle using two different aerodynamic models.

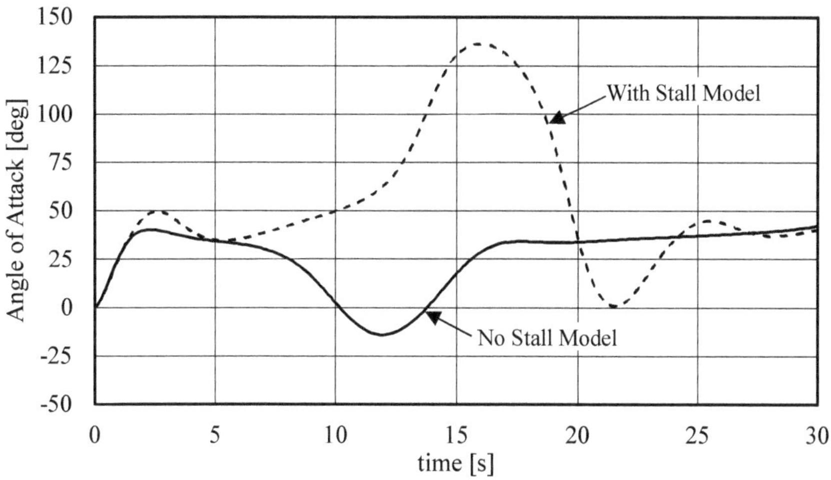

Figure 5.8.4. Time history of the angle of attack of the vehicle using two different aerodynamic models.

In summary, Figs. 5.8.2–5.8.4 show trajectories and time histories of the two simulations. The intuition that can be gained by comparison of these plots is assisted by recognizing that for the case of no wind, longitudinal flight requires $\theta = \alpha + \gamma$.

5.9 Real-Time Simulations

In the previous section, the step size for each integration time step was specified as a fixed constant. Such an approach is useful when complete control over the accuracy of the numerical integration is needed, and when the time in the simulation does not need to correspond to physical time. This allows the simulation to be run faster than real time, which is useful for running large batches of runs for Monte Carlo analyses, or slower than real time, which can be useful for studying a particular physical phenomenon in detail. However, in many situations, the change in time during the simulation must match physical time. Such simulations are said to run in *real time*.

Several methods can be used to ensure a simulation runs in real time. Here we will consider in detail only one method, which is flexible enough to maintain synchronization with physical time over multiple computer architectures. This method dynamically updates the time step size at the end of each loop in the main simulation loop.

Before beginning the main simulation loop, a time step size can be estimated by using the following steps:

1. Store the current system clock time.
2. Call the Runge-Kutta function to compute one step of the integration.
3. Store the new current system clock time.

The difference between the first call to the system clock time and the second call is then used as the initial time step size. Before the main integration loop is begun, the states of the vehicle are returned back to the initial states, and the simulation time is returned to the initial simulation time.

The main simulation loop then begins execution, beginning with a call to the Runge-Kutta function to compute one step of the integration. At the end of each loop, and following any additional commands within the loop such as printing information to the screen or performing any other calculations or communication, the system clock is called. This new system clock time is used to compute the amount of system time that has passed since the previous call to the system clock. This elapsed time is used as the time step size for the next call to the Runge-Kutta routine. Each time through the loop, the time step size is updated using the same method. After the simulation concludes, the simulated time duration can be compared to the actual system time elapsed to assess real-time performance accuracy. Algorithm 5.9.3 outlines a method that can be used for either a specified time step size, or a real-time simulation.

Algorithm 5.9.3 Method for running nearly real-time simulations.

Input
- t_0 — (real) Start (initial) time of the simulation (default = 0.0)
- t_f — (real) End (final) time of the simulation
- Δt — (real) Time step size (default = 0.0)
- $\mathbf{y}_0$ — (real array) Initial states of the vehicle
- t_R — (boolean) Real-time simulation (default = False)

if (t_R is True) **then**
- $\mathbf{y}_t \leftarrow \mathbf{y}_0$ — Store initial vehicle states in a temporary array
- $t_p \leftarrow$ Current system clock time — Store current system time
- $\mathbf{y}_1 =$ `rk4`$(t_0, \mathbf{y}_0, \Delta t)$ — Execute a single Runge-Kutta integration step
- $t_c \leftarrow$ Current system clock time — Get current system time
- $\Delta t = t_c - t_p$ — Compute elapsed system time
- $\mathbf{y}_0 \leftarrow \mathbf{y}_i$ — Reset state array back to the initial state of the vehicle
- $t_p \leftarrow t_c$ — Set previous system time to the current system time

end if

while ($t_0 < t_f$) **do**
- $\mathbf{y}_1 =$ `rk4`$(t_0, \mathbf{y}_0, \Delta t)$ — Execute a single Runge-Kutta integration step
- $t_0 = t_0 + \Delta t$ — Increment time based on time step size
- **if** (t_R is True) **then**
  - $t_c \leftarrow$ Current system clock time — Get current system time
  - $\Delta t = t_c - t_p$ — Compute elapsed system time since previous step
  - $t_p \leftarrow t_c$ — Set previous system time to the current system time
- **end if**

end while

5.10 Problems

Theory Problems

5.T1 Many aircraft are nearly symmetric when mirrored across the body-fixed x_b-z_b plane. Beginning with the Euler-angle formulation given in Eqs. (5.4.1)–(5.4.4), simplify the formulation for the special case of a vehicle that is perfectly symmetric about the x_b-z_b plane, has no gyroscopic effects, and is in pure longitudinal motion. Assume the vehicle is oriented with an azimuth angle of zero and that there is no wind.

5.T2 Repeat Problem 5.T1 using the quaternion formulation given in Eqs. (5.4.5)–(5.4.8).

5.T3 An arrow is nearly axisymmetric about the x_b axis. Beginning with the Euler-angle formulation given in Eqs. (5.4.1)–(5.4.4), simplify the formulation for the special case of a vehicle that is axisymmetric about the x_b axis, has no gyroscopic effects, and is in pure longitudinal motion. Assume the vehicle is oriented with an azimuth angle of zero and that there is no wind.

5.T4 Repeat Problem 5.T3 using the quaternion formulation given in Eqs. (5.4.5)–(5.4.8).

Coding Problems

5.C1 Write a 4th-order Runge-Kutta routine that can be used to estimate the solution to a differential equation of the form

$$\frac{dy}{dt} = f(t, y)$$

Your code should have three functions called `rk4`, `diff_eq`, and `test_main`. The `rk4` function should accept the parameters t_0 (time at the beginning of the step), y_0 (function value at the beginning of the step), Δt (step size), and return y (function value at the end of the step). The `diff_eq` function should accept the parameters t (time), and y (value of y at time t), and return the change in y with respect to t. The `test_main` function should contain a loop that calls the `rk4` function each time through the loop. Test your code by reproducing the case and results given in Example 5.7.1.

5.C2 Write a 4th-order Runge-Kutta routine that can be used to estimate the solution to a system of differential equations of the form

$$\frac{d\mathbf{y}}{dt} = \mathbf{f}(t, \mathbf{y})$$

This can be done by modifying your code for Problem 5.C1 such that the Runge-Kutta routine can accept and process the variable y as an array $\mathbf{y}$ rather than a scalar. Test your code by reproducing the case and results given in Example 5.7.2.

5.C3 Modify your code from Problem 5.C2 to the nested algorithm form outlined in Algorithm 5.7.2. For the `diff_eq` function, use the **quaternion formulation** of the equations of motion given in Eqs. (5.6.1)–(5.6.4). For now, you can create `pseudo_aero` and `mass_inertia` functions that are simply placeholders. Normalize the quaternion before beginning the integration routine, and renormalize the quaternion after each complete time step. Add a check in your code to produce a warning to the user if the rotation rate of the vehicle exceeds the limit given in Eq. (5.7.3).

5.C4 Use your code from Problem 5.C3 to solve for the state of a sphere as a function of time. Write the `pseudo_aero` and `mass_inertia` functions using the aerodynamic and mass models for spheres given in Chapters 2 and 3. Include the effects of altitude on density. Test your code by replicating Example 5.7.3.

5.C5 Repeat Problems 5.C3 and 5.C4 using the **Euler-angle formulation** of the equations of motion. Test your code by replicating Example 5.7.3. Because the quaternion formulation is more efficient and does not suffer from gimbal lock, the quaternion formulation should be used for the remainder of the problems.

5.C6 Select a sphere of interest to model, such as a basketball, tennis ball, etc. Modify the `pseudo_aero` and `mass_inertia` functions written for Problem 5.C4 to model the properties of your chosen sphere as closely as possible. You may find the aerodynamic and mass models given in Chapters 2 and 3 to be helpful. Include the effects of altitude on density. Use your code to predict the trajectory of the sphere based on initial conditions of your choice.

5.C7 Modify your code to solve for the state of an arrow with **straight fletchings** as a function of time. Use the aerodynamic and mass model for the arrow given in Appendix C.1. Include the effects of altitude on density. Assume the arrow is launched at zero angle of attack. Construct your code such that it can read the simulation time step and initial condition of the arrow from an input JSON file. Notes for reading a JSON file are included in the Notes on Code Development section at the beginning of this textbook. Test your code using the following input:

```
{
    "simulation" : {
        "time_step[s]" : 0.01
    },
    "initial" : {
        "airspeed[ft/s]" : 210.0,
        "altitude[ft]" : 5.0,
        "Euler_angles[deg]" : [0.0, 5.0, 0.0]
    }
}
```

Use your code to evaluate the states of the arrow as a function of time until the arrow hits the ground ($z_f = 0$). Compare your results to those shown in Example 5.7.4 including checking your answers out to the precision shown, and generating similar plots of the data. Detailed Runge-Kutta computational steps are included on the book website. Once your code is matching those solutions, test your code by running the same simulation with the arrow rotated at bank angles of 45 and 90 deg. How are the results similar and/or different from those shown in Example 5.7.4?

5.C8 Modify your computer code to model the trajectory of an arrow with **angled fletchings**. Use the aerodynamic and mass model for the arrow given in Appendix C.1. Include the effects of altitude on density. Assume the arrow is launched at zero angle of attack. Use your code to evaluate the longitudinal and lateral states of the arrow as a function of time until the arrow hits the ground ($z_f = 0$). Compare your results to those shown in Example 5.7.5 including checking your answers out to the precision shown, and generating similar plots of the data. Detailed Runge-Kutta computational steps are included on the book website. Once your code is matching those solutions, test your code by running the same simulation with the arrow rotated at bank angles of 45 and 90 deg. How are the results similar and/or different from those shown in Example 5.7.5?

5.C9 Modify your computer code such that it can read the simulation, vehicle, and initial conditions for the F-16 vehicle specified in Appendix C.3 from a JSON file formatted as follows:

```
{
    "simulation" : {
        "time_step[s]" : 0.01,
        "end_time[s]" : 5.0
    },
    "vehicle" : {Insert JSON information from Appendix C.3},
    "initial" : {
        "airspeed[ft/s]" : 330.0,
        "altitude[ft]" : 1000.0,
        "Euler_angles[deg]" : [0.0, 7.858650044026, 0.0],
        "state" : {
            "alpha[deg]" : 7.858650044026,
            "beta[deg]" : 0.0,
            "p[deg/s]" : 0.0,
            "q[deg/s]" : 0.0,
            "r[deg/s]" : 0.0,
            "aileron[deg]" : 0.0,
            "elevator[deg]" : -4.940749182476,
            "rudder[deg]" : 0.0,
            "throttle" : 0.06794815882517
        }
    }
}
```

Construct your code so that it creates an output file that stores the state of the aircraft after each time step with a precision of at least 12 significant figures.

Use your code to replicate Example 5.8.1 including checking your answers out to the precision shown.

5.C10 Modify your code so that it reads the reference location for the aerodynamic model within the `vehicle.aerodynamics.reference` key of the JSON file and uses it as shown in Example 5.8.2. Run your code with the following changes in the settings:

```
{
    "vehicle" : {
        "aerodynamics" : {
            "reference" : {
                "location[ft]" : [-1.0, 0.0, 0.0] (add this option)
            }
        }
    }
}
```

Use your code to replicate Example 5.8.2 including checking your answers out to the precision shown, and generating similar plots.

5.C11 Use the code developed for Problem 5.C10 to replicate Example 5.8.3 including checking your answers out to the precision shown, and generating similar plots.

5.C12 Add the option for the user to specify and use a blended stall model for the vehicle. The code should read the stall-model options from the JSON input file within the `vehicle.aerodynamics` key as follows:

```
{
    "vehicle" : {
        "aerodynamics" : {
            "stall" : {
                "include_stall" : true,
                "CL" : {
                    "alpha_b0[deg]" : 0.0,
                    "alpha_b[deg]" : 43.0,
                    "lambda_b" : 7.0
                },
                "CD" : {
                    "alpha_b0[deg]" : 5.0,
                    "alpha_b[deg]" : 45.0,
                    "lambda_b" : 7.0
                },
                "Cm" : {
                    "min" : -0.6184,
                    "alpha_b0[deg]" : 20.0,
                    "alpha_b[deg]" : 40.0,
                    "lambda_b" : 6.0
                }
            }
        }
    }
}
```

Use the code to replicate Example 5.8.4 including checking your answers out to the precision shown, and generating similar plots.

5.C13 Modify your code so that the location and direction of the thrust vector can be modified through the input file. The location should be in the body-fixed coordinate system relative to the vehicle CG, and the direction should be specified in Euler angles relative to the body-fixed coordinate system. The input to the code should be in the form:

```
{
    "vehicle" : {
        "thrust" : {
            "location[ft]" : [0.0, 0.0, 2.0] (add this option)
            "orientation[deg]" : [0.0, 5.0, 0.0] (add this option)
        }
    }
}
```

Run several test cases with your code to convince yourself that your implementation is correct. You may find it useful to begin with a trim case, such as one given in Example 5.8.2, and then to vary the location and orientation of the thrust vector in a set of test cases. Make sure the response of the vehicle matches your intuition of what should happen in each case.

5.C14 Modify your code so that the user has the option to run the simulation in real time, simply by specifying a time step size of 0.0 in the input file. Repeat Problem 5.C11 running the simulation in real-time, and compute the time error in the real-time simulation after 60 seconds.

Chapter 6

Static Trim

6.1 Introduction

When the vehicle translational and rotational velocities are not changing with time, the vehicle is in equilibrium. This equilibrium state is also known as *static trim* or just *trim*. Static trim occurs when the summation of forces and moments about the center of gravity is zero. In a trim state, the vehicle does not accelerate in translation or rotation, but maintains steady-state translational and rotational velocities. Two common examples of static trim maneuvers include steady level flight and steady coordinated turns. Understanding how to solve for the flight conditions and control inputs from the pilot that produce a trim state is an important step in understanding and solving more general vehicle control problems. This solution process also offers a valuable method for initializing a simulation, such that the vehicle is in a trim state at the beginning of the simulation.

In general, for a flight condition specified by the altitude, velocity, orientation, and trim maneuver, we wish to find the states of the vehicle that will produce a static trim condition. This includes solving for the aerodynamic angles, rotational rates, and pilot control inputs required to maintain trim. Solutions are obtained by considering the governing equations of motion for a rigid-body vehicle developed in Chapter 5.

The governing equations of motion for a rigid-body vehicle relative to the flat-Earth coordinate system including gravity relief can be expressed using the Euler-angle formulation in the form given in Eqs. (5.5.4), and (5.4.2)–(5.4.4). They are repeated here for convenience, and yield

$$\begin{bmatrix} \dot{u} \\ \dot{v} \\ \dot{w} \end{bmatrix} = \frac{1}{\mathcal{M}} \begin{bmatrix} \overline{F}_{x_b} \\ \overline{F}_{y_b} \\ \overline{F}_{z_b} \end{bmatrix} + (g - a_c) \begin{bmatrix} -S_\theta \\ S_\phi C_\theta \\ C_\phi C_\theta \end{bmatrix} + \begin{bmatrix} rv - qw \\ pw - ru \\ qu - pv \end{bmatrix} \tag{6.1.1}$$

$$\begin{aligned} \begin{bmatrix} \dot{p} \\ \dot{q} \\ \dot{r} \end{bmatrix} = \begin{bmatrix} I_{xx_b} & -I_{xy_b} & -I_{xz_b} \\ -I_{xy_b} & I_{yy_b} & -I_{yz_b} \\ -I_{xz_b} & -I_{yz_b} & I_{zz_b} \end{bmatrix}^{-1} & \left\{ \begin{bmatrix} \overline{M}_{x_b} \\ \overline{M}_{y_b} \\ \overline{M}_{z_b} \end{bmatrix} + \begin{bmatrix} 0 & -h_{z_b} & h_{y_b} \\ h_{z_b} & 0 & -h_{x_b} \\ -h_{y_b} & h_{x_b} & 0 \end{bmatrix} \begin{bmatrix} p \\ q \\ r \end{bmatrix} \right. \\ & \left. + \begin{bmatrix} (I_{yy_b} - I_{zz_b})qr + I_{yz_b}(q^2 - r^2) + I_{xz_b}pq - I_{xy_b}pr \\ (I_{zz_b} - I_{xx_b})pr + I_{xz_b}(r^2 - p^2) + I_{xy_b}qr - I_{yz_b}pq \\ (I_{xx_b} - I_{yy_b})pq + I_{xy_b}(p^2 - q^2) + I_{yz_b}pr - I_{xz_b}qr \end{bmatrix} - \begin{bmatrix} \dot{h}_{x_b} \\ \dot{h}_{y_b} \\ \dot{h}_{z_b} \end{bmatrix} \right\} \end{aligned} \tag{6.1.2}$$

$$\begin{bmatrix} \dot{x}_f \\ \dot{y}_f \\ \dot{z}_f \end{bmatrix} = \begin{bmatrix} C_\theta C_\psi & S_\phi S_\theta C_\psi - C_\phi S_\psi & C_\phi S_\theta C_\psi + S_\phi S_\psi \\ C_\theta S_\psi & S_\phi S_\theta S_\psi + C_\phi C_\psi & C_\phi S_\theta S_\psi - S_\phi C_\psi \\ -S_\theta & S_\phi C_\theta & C_\phi C_\theta \end{bmatrix} \begin{bmatrix} u \\ v \\ w \end{bmatrix} + \begin{bmatrix} V_{wx_f} \\ V_{wy_f} \\ V_{wz_f} \end{bmatrix} \tag{6.1.3}$$

$$\begin{bmatrix} \dot{\phi} \\ \dot{\theta} \\ \dot{\psi} \end{bmatrix} = \begin{bmatrix} 1 & S_\phi S_\theta / C_\theta & C_\phi S_\theta / C_\theta \\ 0 & C_\phi & -S_\phi \\ 0 & S_\phi / C_\theta & C_\phi / C_\theta \end{bmatrix} \begin{bmatrix} p \\ q \\ r \end{bmatrix} \tag{6.1.4}$$

For the purposes of considering a trim condition, we have chosen to apply the Euler-angle formulation. The quaternion formulation could also be applied. However, the Euler-angle formulation offers more intuition during the process of solving for the trim states of the aircraft.

In a trim state, the equations of motion must be satisfied such that **the vehicle translational and rotational velocities do not change with time**, i.e.

$$\dot{u} = \dot{v} = \dot{w} = \dot{p} = \dot{q} = \dot{r} = 0 \tag{6.1.5}$$

Hence, the left-hand side of Eqs. (6.1.1) and (6.1.2) must be identically zero. For this equilibrium development, we will also assume that **the gyroscopic effects do not change with time**, i.e.

$$\dot{\mathbf{h}} = 0 \tag{6.1.6}$$

Additionally, we will require that **the gravity vector does not change relative to the body-fixed coordinate system**. By inspection of Eq. (6.1.1), this requires that **the bank and elevation angles do not change with time**, i.e.

$$\dot{\phi} = \dot{\theta} = 0 \tag{6.1.7}$$

Therefore, the first two equations within the system of equations given in Eq. (6.1.4) must also be zero.

Within the approximations of the equations of motion given in Eqs. (6.1.1)–(6.1.4), the dynamics of the vehicle are completely independent of the x_f and y_f position of the vehicle. The dynamics of the vehicle depend on the z_f position of the vehicle through the change in gravity and atmospheric density with altitude. Furthermore, note that the change in z_f position of the vehicle is independent of the azimuth angle ψ. Therefore, **the dynamics of the vehicle are independent of the azimuth angle**, i.e.

$$\psi = \text{arbitrary} \tag{6.1.8}$$

In the context of initializing a flight simulator to a trim state, the azimuth angle ψ can be arbitrarily set by the user.

Applying the constraints given in Eqs. (6.1.5)–(6.1.7), Eqs. (6.1.1), (6.1.2), and (6.1.4) can be rearranged to yield

$$0 = \begin{bmatrix} \overline{F}_{x_b} \\ \overline{F}_{y_b} \\ \overline{F}_{z_b} \end{bmatrix} + \mathcal{M}(g - a_c) \begin{bmatrix} -S_\theta \\ S_\phi C_\theta \\ C_\phi C_\theta \end{bmatrix} + \mathcal{M} \begin{bmatrix} rv - qw \\ pw - ru \\ qu - pv \end{bmatrix} \tag{6.1.9}$$

$$0 = \begin{bmatrix} \overline{M}_{x_b} \\ \overline{M}_{y_b} \\ \overline{M}_{z_b} \end{bmatrix} + \begin{bmatrix} 0 & -h_{z_b} & h_{y_b} \\ h_{z_b} & 0 & -h_{x_b} \\ -h_{y_b} & h_{x_b} & 0 \end{bmatrix} \begin{bmatrix} p \\ q \\ r \end{bmatrix} + \begin{bmatrix} (I_{yy_b} - I_{zz_b})qr + I_{yz_b}(q^2 - r^2) + I_{xz_b}pq - I_{xy_b}pr \\ (I_{zz_b} - I_{xx_b})pr + I_{xz_b}(r^2 - p^2) + I_{xy_b}qr - I_{yz_b}pq \\ (I_{xx_b} - I_{yy_b})pq + I_{xy_b}(p^2 - q^2) + I_{yz_b}pr - I_{xz_b}qr \end{bmatrix} \tag{6.1.10}$$

$$p = -(qS_\phi + rC_\phi)S_\theta/C_\theta \tag{6.1.11}$$

$$q = rS_\phi/C_\phi \tag{6.1.12}$$

Equations (6.1.9)–(6.1.12) represent the core system of equations for trim. Each of these relations must be true for an aircraft to be in a trim state. Because there are eight equations, these can be used to solve for eight unknowns. Given mass, propulsion, gyroscopic, and aerodynamic information for a traditional fixed-wing aircraft with an aileron, elevator, rudder, and throttle, along with a specified vehicle velocity and altitude, the unknowns in these equations include the two aerodynamic angles (α, β), two remaining Euler angles (ϕ, θ), three rotation rates (p, q, r), three control-surface deflections $(\delta_a, \delta_e, \delta_r)$, and throttle setting τ. Since there are a total of 11 unknowns for a traditional fixed-wing aircraft, and only eight equations, additional information is required to close this system of equations.

Two of these additional relations usually come in the form of a specified orientation of the vehicle. With one exception, in this chapter we will consider cases when the Euler angles ϕ and θ are specified. Recall that the azimuth angle ψ is also user-specified, since the dynamics are independent of azimuth angle. Relationships between these Euler angles and other geographic or performance parameters are discussed in Chapter 7.

The final constraint required to close this system of equations comes in the form of a specified type of trim. In this chapter, we will consider four common trim conditions including **steady level flight**, a **steady coordinated turn**, **steady-heading sideslip**, and a **barrel roll**. As will be shown, each of these trim types is related to a single additional constraint that is applied to the rotation rates given in Eqs. (6.1.11) and (6.1.12).

Given a prescribed vehicle orientation and trim type, Eqs. (6.1.9)–(6.1.12) can be solved for the remaining six unknowns, which for a traditional fixed-wing aircraft include the aerodynamic angles α and β, and pilot controls δ_a, δ_e, δ_r, and τ. The relations given in Eqs. (6.1.11) and (6.1.12) are constraints that are related to the type of trim specified by the user. Therefore, this solution process includes solving Eqs. (6.1.9) and (6.1.10) for the six unknowns, subject to the constraints given in Eqs. (6.1.11) and (6.1.12).

Many solution techniques can be applied to solve the system of equations given in Eqs. (6.1.9) and (6.1.10), including non-linear solvers and optimization algorithms. In this text, we will demonstrate the use of Newton's method for solving this system of equations. Hence, before demonstrating this solution process, a brief review of Newton's method may be helpful to the reader.

6.2 Newton's Method

Newton's method is a powerful numerical method for finding the values of independent variables that provide a zero solution, or root, of dependent variables. Before considering a system of equations, let us first consider a scalar example.

Single Scalar Equation

Consider a scalar function $f(x)$. We wish to find a value of x that produces $f(x) = 0$. This is called a *root* of $f(x)$. If the function is simple and analytic, the root or roots can also be analytically obtained. For example, let $f(x) = x - 1$. The root for this function can be obtained by solving the equation $0 = x - 1$, which yields $x = 1$. However, for complex functions for which analytic solutions are difficult or not readily possible to obtain, the root can be found numerically using Newton's method.

Newton's method begins with an initial guess for the root, $x = x_0$. If the initial guess is sufficiently close to the root, a better approximation for the root can be obtained by applying a linear approximation of the function

$$f(x) \approx f(x_0) + (x - x_0)f'(x_0) \tag{6.2.1}$$

where

$$f'(x) = \frac{df}{dx} \tag{6.2.2}$$

The root of this approximate equation can be obtained by solving

$$f(x) \approx f(x_0) + (x_{\text{root}} - x_0)f'(x_0) = 0 \tag{6.2.3}$$

which yields

$$x_{\text{root}} = x_0 - \frac{f(x_0)}{f'(x_0)} \tag{6.2.4}$$

Because a linear approximation of the function was used in this method, the root x_{root} obtained in Eq. (6.2.4) is the exact root of the function only if the function is perfectly linear. However, for nonlinear functions, if the initial guess is sufficiently close to the root, the solution to Eq. (6.2.4) can be used as a better estimate for the root. Hence, this process can be repeated in an iterative form

$$x_{i+1} = x_i - \frac{f(x_i)}{f'(x_i)} \tag{6.2.5}$$

until $f(x_i)$ is sufficiently close to zero. This process is shown in Fig. 6.2.1.

Note that in order to apply this process, the derivative of the function at any value of x must be obtained. If the function is sufficiently simple, an analytic solution may be used. However, for more complex functions, the derivative must be computed numerically. Numerical finite-difference methods can be used, as developed in Appendix D. A simple second-order central-difference formula commonly used is

$$f'(x) \approx \frac{f(x + \delta) - f(x - \delta)}{2\delta} \tag{6.2.6}$$

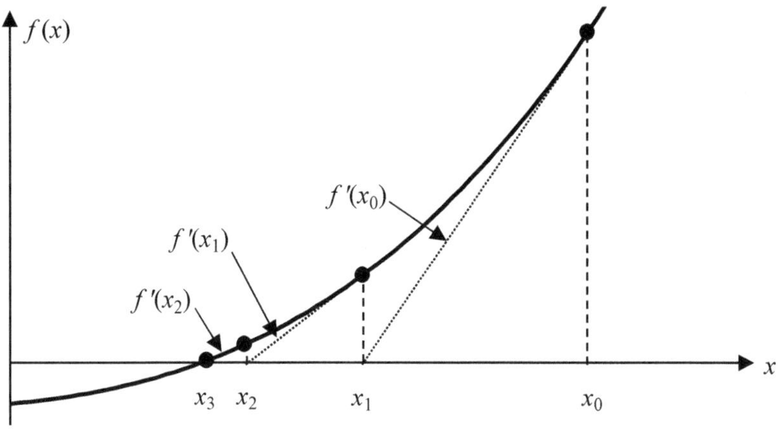

Figure 6.2.1. Visualization of Newton's method for a scalar function.

where δ is a finite-difference step size. Details on the development of this formula can be found in Appendix D.

EXAMPLE 6.2.1 Consider the scalar function

$$f(x) = x^3 - 8$$

The root of this function is $x = 2$. Solve for this root numerically by applying Newton's method with an initial guess of $x_0 = 4.0$. Use numerical finite-differencing with a step size of $\delta = 0.01$ to obtain the derivatives of the function. Continue iterating on the solution until the solution converges to within double precision of the correct value.

Solution: As instructed, we begin with an initial guess of $x_0 = 4.0$. The function value at this point is

$$f(4.0) = 4^3 - 8 = 56$$

Using central differencing as shown in Eq. (6.2.6) with a step size of $\delta = 0.01$, the first derivative of the function at $x = 4.0$ is

$$f'(4.0) \approx \frac{f(4.01) - f(3.99)}{2(0.01)} = \frac{56.481201 - 55.521199}{0.02} = 48.00009999999$$

Using these results in Eq. (6.2.5) gives

$$x_1 = 4.0 - \frac{56}{48.00009999999} = 2.8333357638838 \tag{6.2.7}$$

Repeating this process yields the results in Table 6.2.1. Note that seven iterations are required to obtain the root location to double precision accuracy.

Table 6.2.1. Newton's Method results for Example 6.2.1.

Iteration	x	$f(x)$	$f'(x)$
0	4.0000000000000	56.0000000000000	48.0000999999987
1	2.8333357638838	14.7454289061775	24.0834746527083
2	2.2210724124027	2.9569114925448	14.7995879834085
3	2.0212755332017	0.2580319186332	12.2567643433596
4	2.0002233279537	0.0026802347078	12.0027800850705
5	2.0000000267944	0.0000003215328	12.0001003215326
6	2.0000000000002	0.0000000000027	12.0001000000026
7	2.0000000000000	0.0000000000000	12.0000999999998

Before considering the case of a system of equations, it is helpful to recognize that Newton's method as given in Eq. (6.2.5) can be written in the form

$$x_{i+1} = x_i + \Delta x \tag{6.2.8}$$

where the step size of the independent variable x for each iteration is

$$\Delta x = -\frac{f(x_i)}{f'(x_i)} \tag{6.2.9}$$

For some highly non-linear equations, using the full step size as computed from Newton's method may cause the method to diverge rather than converging to the root. In such cases it can be helpful to use a relaxation factor to reduce the step size taken in each step. Here we will use the symbol Γ as a relaxation factor. Typically, the relaxation factor is limited to $0.0 < \Gamma \leq 1$ and applied such that

$$x_{i+1} = x_i + \Gamma \Delta x \tag{6.2.10}$$

Using a relaxation factor is often helpful in finding trim solutions for aircraft, as will be shown in later sections.

System of Equations

Newton's method can also be used to obtain the roots of a system of equations. To understand how Newton's method is typically formulated for a system of equations, notice that the function $f(x)$ used in the scalar example above can be thought of as a *residual*. In other words, the value of the function $f(x)$ is a value denoting the error in the solution. If the correct root of the function has been found, the value of the function will be zero, i.e. $f(x_{\text{root}}) = 0$. Furthermore, note that Eq. (6.2.9) can be rearranged in the form

$$f'(x_i)\Delta x = -f(x_i) \tag{6.2.11}$$

This form of Newton's method is useful for considering a system of equations. Here we will use $\mathbf{x}$ to represent a vector of independent variables, and define the residual vector of a system of equations as $\mathbf{R}$. We wish to find the values of $\mathbf{x}$ that produce the roots of the system of equations. In other words, we seek the solution for $\mathbf{x}$ such that

$$\mathbf{R}(\mathbf{x}) = 0 \tag{6.2.12}$$

The scalar equation given in Eq. (6.2.11) can be written in vector form as

$$[\mathbf{J}]\Delta\mathbf{x} = -\mathbf{R}(\mathbf{x}) \tag{6.2.13}$$

where $[\mathbf{J}]$ is called the *Jacobian* and is defined as

$$J_{ij} = \frac{\partial R_i}{\partial x_j} \tag{6.2.14}$$

The Jacobian matrix can be constructed from the partial derivatives

$$[\mathbf{J}] = \begin{bmatrix} \frac{\partial R_1}{\partial x_1} & \frac{\partial R_1}{\partial x_2} & \cdots & \frac{\partial R_1}{\partial x_{N-1}} & \frac{\partial R_1}{\partial x_N} \\ \frac{\partial R_2}{\partial x_1} & \frac{\partial R_2}{\partial x_2} & \cdots & \frac{\partial R_2}{\partial x_{N-1}} & \frac{\partial R_2}{\partial x_N} \\ \vdots & \vdots & \ddots & \vdots & \vdots \\ \frac{\partial R_{N-1}}{\partial x_1} & \frac{\partial R_{N-1}}{\partial x_2} & \cdots & \frac{\partial R_{N-1}}{\partial x_{N-1}} & \frac{\partial R_{N-1}}{\partial x_N} \\ \frac{\partial R_N}{\partial x_1} & \frac{\partial R_N}{\partial x_2} & \cdots & \frac{\partial R_N}{\partial x_{N-1}} & \frac{\partial R_N}{\partial x_N} \end{bmatrix} \tag{6.2.15}$$

where N is the number of equations in the system. If analytic partial derivatives are not readily available, the partial derivatives required in the Jacobian matrix can be estimated through finite differencing. For example, second-order central differencing can be performed to yield

$$J_{ij} \approx \frac{R_i(x_j + \delta) - R_i(x_j - \delta)}{2\delta} \tag{6.2.16}$$

where δ is the finite-difference step size.

Once the Jacobian matrix has been constructed, it is used in Eq. (6.2.13) to solve for the step in the independent variable vector

$$\Delta\mathbf{x} = -[\mathbf{J}]^{-1}\mathbf{R} \tag{6.2.17}$$

Lower-upper (LU) decomposition using partial pivoting is a fairly stable algorithm that can be used to solve Eq. (6.2.17). An excellent discussion on LU decomposition including a sample algorithm can be found in the well-known text *Numerical Recipes* by Press, Teukolsky, Vetterling, and Flannery [77]. Just as was shown in scalar form in Eq. (6.2.10), the array $\mathbf{x}$ is updated each iteration using a relaxation factor Γ as

$$\mathbf{x}_{i+1} = \mathbf{x}_i + \Gamma\Delta\mathbf{x} \tag{6.2.18}$$

This process is repeated until the solution converges. Convergence can be assessed by computing the absolute value of the largest component of $\mathbf{R}$, i.e. $\varepsilon = \max(\text{abs}(\mathbf{R}))$.

Algorithm 6.2.4 Newton's method for solving a system of equations.

```
Input
    x            (real array) initial guess for the vector of independent variables
    δ            (real) step size used for finite differencing
    Γ            (real) relaxation factor
    ε_t          (real) tolerance for convergence
R = calc_R(x)                          compute residual vector
ε = max(abs(R))                        find maximum error in the residual
while (ε > ε_t) do
    for j = 1 to N do                  construct Jacobian matrix
        x[j] = x[j] + δ
        R⁺ = calc_R(x)
        x[j] = x[j] − 2δ
        R⁻ = calc_R(x)
        for i = 1 to N do
            J[i][j] = (R⁺[i] − R⁻[i])/(2δ)
        end for
        x[j] = x[j] + δ
    end for
    Δx = −[J]⁻¹R      invert Jacobian and multiply by residual to compute new step
    x = x + ΓΔx                        apply relaxation factor for new step
    R = calc_R(x)                      compute residual vector
    ε = max(abs(R))                    find maximum error in the residual
end while
```

Once ε is below a user-defined tolerance, the solution is considered to be converged. Newton's method for solving a system of equations is outlined in Algorithm 6.2.4. Note that to avoid confusion in this algorithm, the evaluation of the system of equations or residual is denoted by the function `calc_R`($\mathbf{x}$).

EXAMPLE 6.2.2 Consider the system of equations

$$f_1 = y - x^2 + 5$$

$$f_2 = y - x^2 - (x - 1)$$

It can be quite simply shown that the values for the independent variables x and y that correspond to the root of this system are $x = -4$ and $y = 11$. Solve for the root numerically by applying Newton's method with an initial guess of $x_0 = y_0 = 0.0$. Use numerical finite-differencing with a step size of $\delta = 0.01$ to obtain the derivatives of the function. Continue iterating on the solution until the solution converges to within double precision of the correct value.

Solution: The residual vector for this system of equations can be written as

$$\mathbf{R}(x,y) = \begin{bmatrix} y - x^2 + 5 \\ y - x^2 - (x-1) \end{bmatrix}$$

We wish to find the solutions for x and y that produce $\mathbf{R} = 0$. Algorithm 6.2.4 can be used to solve for the root using an initial value for the vector $\mathbf{x} = [0,0]^T$, $\delta = 0.01$, and defining a `calc_R` function that accepts an $\mathbf{x}$ vector of two components and returns an array of length two containing the residual. The residual of the initial guess is

$$\mathbf{R}(0,0) = \begin{bmatrix} 5.0 \\ 1.0 \end{bmatrix}$$

with a maximum error value of $\varepsilon = 5.0$. For the first iteration, the resulting Jacobian matrix is

$$[\mathbf{J}] = \begin{bmatrix} 0.0 & 1.0 \\ -1.0 & 1.0 \end{bmatrix}$$

Inverting the Jacobian and multiplying it by the negative of the residual vector gives

$$\Delta\mathbf{x} = \begin{bmatrix} -4.0 \\ -5.0 \end{bmatrix}$$

Using a relaxation factor of $\Gamma = 1.0$ gives a new estimate for the vector of unknowns

$$\mathbf{x} = \begin{bmatrix} -4.0 \\ -5.0 \end{bmatrix}$$

The residual of this new estimate is

$$\mathbf{R}(-4,-5) = \begin{bmatrix} -16.0 \\ -16.0 \end{bmatrix}$$

with a maximum error value of $\varepsilon = 16.0$. Repeating this process yields the results in Table 6.2.2. Note that only three iterations are required to obtain the root location to double precision accuracy. If $\Gamma = 0.9$ is used, about 16 iterations are required to reach double-precision accuracy.

Table 6.2.2. Newton's Method results for Example 6.2.2.

Iteration	x	y	ε
0	0.000000000000	0.000000000000	5.00000000000
1	-4.000000000000	-5.000000000000	16.00000000000
2	-4.000000000000	11.000000000000	3.1823433E−12
3	-4.000000000000	11.000000000000	0.000000000000

Application of Newton's Method to the Equations of Motion

Newton's method can be used to solve the system of equations given in Eqs. (6.1.9) and (6.1.10). Recall that Eqs. (6.1.9) and (6.1.10) are a special case of Eqs. (6.1.1) and (6.1.2) for which $\dot{u} = \dot{v} = \dot{w} = \dot{p} = \dot{q} = \dot{r} = 0$. Therefore, the solution to Eqs. (6.1.9) and (6.1.10) represents the roots of Eqs. (6.1.1) and (6.1.2). Here we will define the residual vector as a six-element array containing the result of the six equations given in Eqs. (6.1.1) and (6.1.2)

$$\mathbf{R}(\mathbf{x}) = \begin{bmatrix} \dot{u} \\ \dot{v} \\ \dot{w} \\ \dot{p} \\ \dot{q} \\ \dot{r} \end{bmatrix} \tag{6.2.19}$$

We wish to find the set of independent variables $\mathbf{x}$ that results in $\mathbf{R}(\mathbf{x}) = 0$. The independent variables that will be used for most examples in this chapter are

$$\mathbf{x} = [\alpha, \beta, \delta_a, \delta_e, \delta_r, \tau] \tag{6.2.20}$$

For a given set of values $\mathbf{x}$, the error in the solution can be computed simply by reporting the values of $\dot{u}$, $\dot{v}$, $\dot{w}$, $\dot{p}$, $\dot{q}$, and $\dot{r}$ as computed from Eqs. (6.1.1) and (6.1.2). The `diff_eq` function written in Chapter 5 can be used for this purpose. Using the vector of unknowns given in Eq. (6.2.20), the Jacobian matrix is constructed from the following partial derivatives

$$[\mathbf{J}] = \begin{bmatrix} \frac{\partial \dot{u}}{\partial \alpha} & \frac{\partial \dot{u}}{\partial \beta} & \frac{\partial \dot{u}}{\partial \delta_a} & \frac{\partial \dot{u}}{\partial \delta_e} & \frac{\partial \dot{u}}{\partial \delta_r} & \frac{\partial \dot{u}}{\partial \tau} \\ \frac{\partial \dot{v}}{\partial \alpha} & \frac{\partial \dot{v}}{\partial \beta} & \frac{\partial \dot{v}}{\partial \delta_a} & \frac{\partial \dot{v}}{\partial \delta_e} & \frac{\partial \dot{v}}{\partial \delta_r} & \frac{\partial \dot{v}}{\partial \tau} \\ \frac{\partial \dot{w}}{\partial \alpha} & \frac{\partial \dot{w}}{\partial \beta} & \frac{\partial \dot{w}}{\partial \delta_a} & \frac{\partial \dot{w}}{\partial \delta_e} & \frac{\partial \dot{w}}{\partial \delta_r} & \frac{\partial \dot{w}}{\partial \tau} \\ \frac{\partial \dot{p}}{\partial \alpha} & \frac{\partial \dot{p}}{\partial \beta} & \frac{\partial \dot{p}}{\partial \delta_a} & \frac{\partial \dot{p}}{\partial \delta_e} & \frac{\partial \dot{p}}{\partial \delta_r} & \frac{\partial \dot{p}}{\partial \tau} \\ \frac{\partial \dot{q}}{\partial \alpha} & \frac{\partial \dot{q}}{\partial \beta} & \frac{\partial \dot{q}}{\partial \delta_a} & \frac{\partial \dot{q}}{\partial \delta_e} & \frac{\partial \dot{q}}{\partial \delta_r} & \frac{\partial \dot{q}}{\partial \tau} \\ \frac{\partial \dot{r}}{\partial \alpha} & \frac{\partial \dot{r}}{\partial \beta} & \frac{\partial \dot{r}}{\partial \delta_a} & \frac{\partial \dot{r}}{\partial \delta_e} & \frac{\partial \dot{r}}{\partial \delta_r} & \frac{\partial \dot{r}}{\partial \tau} \end{bmatrix} \tag{6.2.21}$$

The final component for obtaining a trim solution is the construction of the function denoted in Algorithm 6.2.4 as `calc_R`. As mentioned above, this function must accept the vector of independent variables $\mathbf{x}$ and return the residual vector $\mathbf{R}$, which is the result of the system of equations given in Eqs. (6.1.1) and (6.1.2). The `diff_eq` function can be used to compute $\mathbf{R}$. However, in order to call the `diff_eq` function, the vector of independent variables $\mathbf{x}$ must be used, along with user-defined states, to generate a full state vector for the vehicle. Defining all the components of the state vector typically requires the following steps:

Algorithm 6.2.5 Example residual function `calc_R` for the equations of motion.

Input
  $\mathbf{x} = [\alpha, \beta, \delta_a, \delta_e, \delta_r, \tau]$ — (real array) vector of independent variables
  V — (real) user-specified velocity magnitude
  x_f, y_f, z_f — (real) user-specified location
  ϕ, θ, ψ — (real) user-defined Euler angles
$\alpha = \mathbf{x}[1]$
$\beta = \mathbf{x}[2]$
$u = V\cos(\alpha)\cos(\beta)$
$v = V\sin(\beta)$
$w = V\sin(\alpha)\cos(\beta)$
$(p, q, r) =$ — set rotation rates depending on trim type
$(e_0, e_x, e_y, e_z) = f(\phi, \theta, \psi)$ — set quaternion from Euler angles using Eq. (1.6.2)
$\mathbf{y} = [u, v, w, p, q, r, x_f, y_f, z_f, e_0, e_x, e_y, e_z]$ — set state vector
controls $= \mathbf{x}[3:6]$ — set controls vector for use in `pseudo_aero` function
$\mathbf{R} =$ `diff_eq`$(\mathbf{y})$
return $\mathbf{R}[1:6]$ — return only the first 6 components

1. Compute and set the u, v, and w velocity components from the user-specified velocity magnitude V and the α and β values in the $\mathbf{x}$ vector
2. Compute and set the appropriate rotation rates (p, q, r), which can vary with trim type
3. Set the user-specified location and altitude of the vehicle
4. Set the quaternion orientation representation from the user-specified Euler angles
5. Set the control values to be used in the `pseudo_aero` function to those from the $\mathbf{x}$ vector

This process is outlined in Algorithm 6.2.5. With these definitions, let us now consider the solution process of finding the trim states for various trim conditions.

6.3 Steady Level Flight

Steady level flight is perhaps the most common trim condition of interest. It is a special case of more general cases to be considered in following sections. However, due to the importance of this case, and due to the fact that it serves as a valuable example for constructing an initial trim algorithm, we will consider it independently.

In steady level flight, the vehicle has zero rotational velocity ($p = q = r = 0$), a bank angle of $\phi = 0$ (upright flight) or $\phi = 180$ deg (inverted flight), and a user-specified azimuth angle. Steady level flight also requires that the vehicle is not ascending or descending, but maintaining a constant altitude. In order for this to be true, the elevation angle must be carefully prescribed. A method for selecting the particular required elevation angle that produces steady level flight is discussed in detail in Chapter 7. In this section, we will assume that the particular elevation angle that satisfies steady level flight is known and prescribed. These requirements can be used in the `calc_R` function along with the Newton's method outlined in the previous section to evaluate the required angle of attack, sideslip angle, aileron deflection, elevator deflection, rudder deflection, and throttle for steady level flight. This is demonstrated in the following example.

EXAMPLE 6.3.1 Consider the F-16 aircraft specified in Appendix C.3 with the aerodynamic model F16-E-1 given in Table C.3.15. Compute the aerodynamic angles α and β, control-surface deflections δ_a, δ_e, and δ_r, and throttle setting τ for the aircraft in **steady level flight** for the following cases with specified airspeed, altitude, bank angle, and elevation angle.

Case	Airspeed [ft/s]	Altitude [ft]	ϕ [deg]	θ [deg]
A	250.0	0.0	0.0	13.927546386874
B	330.0	5000.0	0.0	9.474974784737
C	250.0	0.0	180.0	13.305569851729
D	330.0	5000.0	180.0	9.130946229092

Solution: The trim solutions for these cases can be obtained using Newton's method as outlined in Algorithm 6.2.4, along with the `calc_R` function outlined in Algorithm 6.2.5. In order to ensure steady level flight as explained in this section, the user is allowed to specify a bank angle of $\phi = 0$ or $\phi = 180$ deg, and the rotation rates on line 11 of the `calc_R` function in Algorithm 6.2.5 are set to $p = q = r = 0$. Here we will use a finite-difference step of $\delta = 0.01$ for computing derivatives in the Jacobian, a relaxation factor of $\Gamma = 0.9$, and a convergence tolerance of $\varepsilon_t = 1.0\text{E}{-13}$. Making these modifications and running the code yields the iteration results for Case A shown in Table 6.3.1, where any independent variables not shown are simply zero. For this solution, an initial guess for all independent variables of $\alpha = \beta = \delta_a = \delta_e = \delta_r = \tau = 0$ was used. Intermediate solution steps for this case are included on the book website.

This solution process required 15 iterations to converge within the tolerance specified. Using a relaxation factor of $\Gamma = 1.0$ results in the method requiring only 4 iterations to converge to the same solution. In this example, all lateral independent variables remain zero ($\beta = \delta_a = \delta_r = 0$) throughout the iteration process due to the symmetry of the aircraft.

Results for the remaining three cases are shown in Table 6.3.2. Intermediate solution steps for each case are included on the book website, which can greatly assist in code development and solution checking. Each case represents a steady-level-flight condition of the vehicle. If the simulation is initialized to one of these cases, and then propagated forward in time, the vehicle will remain in a trim state at constant altitude and velocity.

Table 6.3.1. Iterative results using Newton's Method for Case A in Example 6.3.1.

Iteration	Residual	α [deg]	δ_e [deg]	τ
0	2.94E+01	0.00000000000	0.00000000000	0.0000000000000
1	3.01E+00	12.53026267088	-10.43040323440	0.1099329410473
2	2.90E−01	13.79258261616	-11.47198094741	0.1118947670039
3	2.89E−02	13.91409448101	-11.57689976629	0.1119653196320
4	2.89E−03	13.92620124344	-11.58739881081	0.1119712530842
5	2.89E−04	13.92741183318	-11.58844872924	0.1119718435074
6	2.89E−05	13.92753292713	-11.58855371545	0.1119719033456
7	2.89E−06	13.92754504046	-11.58856421343	0.1119719094199
8	2.89E−07	13.92754625219	-11.58856526317	0.1119719100365
9	2.90E−08	13.92754637340	-11.58856536813	0.1119719100991
10	2.90E−09	13.92754638553	-11.58856537863	0.1119719101054
11	2.90E−10	13.92754638674	-11.58856537968	0.1119719101061
12	2.90E−11	13.92754638686	-11.58856537978	0.1119719101062
13	2.89E−12	13.92754638687	-11.58856537980	0.1119719101062
14	2.95E−13	13.92754638687	-11.58856537980	0.1119719101062
15	2.49E−14	13.92754638687	-11.58856537980	0.1119719101062

Cases A and B represent upright flight, and Cases C and D represent an inverted orientation with the vehicle flying upside down. Note that for upright flight, the trim angle of attack has the same value as the elevation angle, i.e. $\alpha = \theta$, whereas for inverted flight, the trim angle of attack has the same magnitude but opposite sign of the elevation angle, i.e. $\alpha = -\theta$. These relationships between the angle of attack and elevation angle always hold for steady level flight. For the cases included here, specific elevation angles were specified for each flight condition to produce steady-level-flight conditions. A method for specifying the climb rate rather than the elevation angle in the solution process is discussed in Chapter 7.

Table 6.3.2. Converged results for Cases B, C, and D in Example 6.3.1.

Case	B	C	D
α [deg]	9.474974784737	-13.305569851729	-9.130946229092
β [deg]	0.0	0.0	0.0
δ_a [deg]	0.0	0.0	0.0
δ_e [deg]	-9.838952056990	-0.705529160256	-2.514937633656
δ_r [deg]	0.0	0.0	0.0
τ	0.088302015261	0.242205897172	0.245095066559

6.4 Steady Coordinated Turn

In a steady coordinated turn, the side force due to gravity and bank angle perfectly balance the side Coriolis force produced by angular velocity. From a passenger's perspective, this is a trim state that allows the vehicle to bank without the passenger experiencing any side force. For example, if a passenger has a drink on a tray, that drink will not spill because the *apparent gravity* is still in the x_b-z_b plane of the vehicle. From the second equation in Eq. (6.1.9), this requires that

$$(g - a_c)S_\phi C_\theta = ru - pw \tag{6.4.1}$$

Using this constraint in Eq. (6.1.9) and solving for the pseudo-aerodynamic side force yields

$$\overline{F}_{y_b} = 0 \tag{6.4.2}$$

In other words, the pseudo-aerodynamic side force on the vehicle is zero in a steady coordinated turn. **Because the side force must be zero, some have erroneously assumed that the sideslip velocity and sideslip angle must also be zero, i.e. $v = \beta = 0$. Although the sideslip angle is typically very small for the case of a steady coordinated turn, it is not exactly zero in general.**

Combining the relation given in Eq. (6.4.1) with Eqs. (6.1.11) and (6.1.12) gives three equations that can be solved for the rotation rates in the body-fixed coordinate system for a steady coordinated turn as a function of the freestream velocities and orientation. This gives

$$\begin{bmatrix} p \\ q \\ r \end{bmatrix} = \frac{(g - a_c)S_\phi C_\theta}{uC_\theta C_\phi + wS_\theta} \begin{bmatrix} -S_\theta \\ S_\phi C_\theta \\ C_\phi C_\theta \end{bmatrix} \tag{6.4.3}$$

Steady level flight is a special case of the steady coordinated turn for which the bank angle and climb angle are both zero. For this case, the resulting rotation rates are zero, as can be seen from Eq. (6.4.3) for the case when $\phi = 0$.

It may seem non-intuitive to the reader that a steady coordinated turn could have a non-zero value for the rolling rate as shown in Eq. (6.4.3). However, the reader is reminded that the rolling rate given in Eq. (6.4.3) is the rolling rate about the body-fixed x_b axis, not the wind axis. The roll, pitch, and yaw rates in the wind axis are related to the angle of attack and sideslip angle according to Eq. (3.4.10), which yields

$$\begin{bmatrix} p_w \\ q_w \\ r_w \end{bmatrix} = \begin{bmatrix} C_\alpha C_\beta & S_\beta & S_\alpha C_\beta \\ -C_\alpha S_\beta & C_\beta & -S_\alpha S_\beta \\ -S_\alpha & 0 & C_\alpha \end{bmatrix} \begin{bmatrix} p \\ q \\ r \end{bmatrix} \tag{6.4.4}$$

Substituting the required rotational rates for a steady coordinated turn given in Eq. (6.4.3) in Eq. (6.4.4) and retaining only the relation for the rolling rate in the wind coordinate system p_w yields

$$p_w = \frac{gS_\phi C_\theta}{uC_\theta C_\phi + wS_\theta}(-S_\theta C_\alpha C_\beta + S_\phi C_\theta S_\beta + C_\phi C_\theta S_\alpha C_\beta) \tag{6.4.5}$$

Applying the relations for the body-fixed velocity components (u, v, w) from Eq. (3.4.12) and simplifying gives

$$p_w = \left(\frac{gS_\phi C_\theta}{uC_\theta C_\phi + wS_\theta}\right)\left(\frac{1}{V}\right)(-uS_\theta + vS_\phi C_\theta + wC_\phi C_\theta) \tag{6.4.6}$$

Note that the terms in the parentheses on the far right-hand side of this expression are identical to the expression for $\dot{z}_f$ in the absence of a vertical components of wind from Eq. (6.1.3). Using the third equation in Eq. (6.1.3) in Eq. (6.4.6) yields

$$p_w = \left(\frac{gS_\phi C_\theta}{uC_\theta C_\phi + wS_\theta}\right)\left(\frac{1}{V}\right)(\dot{z}_f - V_{wz_f}) \tag{6.4.7}$$

From Eq. (6.4.7) it is clear that for the special case when there is no climb rate of the aircraft relative to the surrounding air, i.e. $\dot{z}_f - V_{wz_f} = 0$, the rolling rate of the vehicle about the wind axis in a steady coordinated turn is exactly zero. However, when the aircraft is climbing or descending, the rolling rate about the wind axis is, in general, non-zero.

EXAMPLE 6.4.1 Consider the F-16 aircraft specified in Appendix C.3 with the aerodynamic model F16-E-1 given in Table C.3.15 operating at an airspeed of 400 ft/s at an altitude of 3,000 ft. Compute the aerodynamic angles α and β, rotation rates p, q, and r, control-surface deflections δ_a, δ_e, and δ_r, and throttle setting τ for the aircraft in a steady coordinated turn with elevation angle and bank angles specified as follows:

Case	ϕ [deg]	θ [deg]
A	0.0	6.136079018241
B	20.0	6.084391247322
C	45.0	5.896183238190
D	60.0	5.713375671219

Solution: The trim solutions for these cases were obtained using Algorithms 6.2.4 and 6.2.5. Each time Algorithm 6.2.5 was called, the rotation rates on line 11 of that algorithm were set to those given in Eq. (6.4.3). Results are shown in the following tables. Intermediate solution steps for each case are included on the book website, which can greatly assist in code development and solution checking.

For each case, the elevation angle in the problem statement was carefully selected in order to produce a solution for a steady coordinated turn at constant altitude. Other elevation angles can be used, but they would result in solutions for which the vehicle is ascending or descending in altitude. If the vehicle is not trimmed to maintain constant altitude, it will not remain in a trim state due to the change in density and gravity with altitude. Note that Case A represents steady level flight with zero rotation rates.

Case	A	B
α [deg]	6.136079018241	6.469175474706
β [deg]	0.000000000000	0.006869386248
p [deg/s]	0.000000000000	-0.176566178251
q [deg/s]	0.000000000000	0.566536501132
r [deg/s]	0.000000000000	1.556546244156
δ_a [deg]	0.000000000000	-0.005953138247
δ_e [deg]	-8.551397626520	-8.782444258008
δ_r [deg]	0.000000000000	-0.292650717107
τ	0.071528613109	0.075802429806

Case	C	D
α [deg]	8.294194311108	11.278631391688
β [deg]	0.015240130888	0.021612044008
p [deg/s]	-0.468207559367	-0.778800198938
q [deg/s]	3.205813271196	6.741300942093
r [deg/s]	3.205813271196	3.892091913605
δ_a [deg]	0.022268256587	0.099479318648
δ_e [deg]	-9.963213712938	-11.763504697744
δ_r [deg]	-0.599694766085	-0.709452270024
τ	0.107726495132	0.191464868138

Repeating the process with negative bank angles rather than positive produces nearly identical results for the longitudinal parameters, and results with opposite sign for the lateral parameters. The difference in the magnitude of the results between a given positive and negative value for bank angle is due to the slight asymmetric dynamics of the aircraft caused by the engine angular momentum when the rotation rates are non-zero. Setting the angular momentum of the aircraft to zero such that the vehicle is perfectly symmetric, and running the solver for a set bank angle at both positive and negative values should produce results identical in magnitude. This is a worthwhile test for a simulation code and is left as an exercise for the student.

6.5 Steady-Heading Sideslip

Steady-heading sideslip is a trim condition in which the aircraft maintains a constant heading while sustaining a constant sideslip angle. This condition is very important for cross-wind landings, formation flight, and other precise navigational maneuvers. This case can be specified by either the bank angle ϕ or the sideslip angle β. However, once one of these angles is chosen, the other is constrained by the equilibrium condition and becomes a dependent variable.

For the case of steady-heading sideslip, the aircraft has no rotational velocity. Therefore,

$$\begin{bmatrix} p \\ q \\ r \end{bmatrix} = 0 \tag{6.5.1}$$

Note that steady level flight is a special case of steady-heading sideslip when the bank angle is $\phi = 0$. A knife-edge maneuver is a special case of a steady-heading sideslip with the bank angle set to $\phi \pm 90$ deg.

Newton's method can be used to solve for a steady-heading sideslip case when the bank angle ϕ is specified. This is demonstrated in the following example.

EXAMPLE 6.5.1 Consider the F-16 aircraft specified in Appendix C.3 with the aerodynamic model F16-E-1 given in Table C.3.15 operating at an airspeed of 520 ft/s at an altitude of 2,000 ft. Compute the aerodynamic angles α and β, rotation rates p, q, and r, control-surface deflections δ_a, δ_e, and δ_r, and throttle setting τ for the aircraft in a steady-heading sideslip with elevation angle and **bank angle** specified as follows:

Case	ϕ [deg]	θ [deg]
A	0.0	3.546530549415
B	10.0	4.021930248315
C	30.0	7.468988562820
D	90.0	60.0

Solution: The trim solutions for these cases were obtained using Algorithms 6.2.4 and 6.2.5. Each time Algorithm 6.2.5 was called, the rotation rates on line 11 of that algorithm were set to those given in Eq. (6.5.1). Results are shown in the following tables. Intermediate solution steps for each case are included on the book website, which can greatly assist in code development and solution checking. Note that, by definition, the values for the rotation rates are identically zero, i.e. $p = q = r = 0$, for steady-heading sideslip cases.

Cases A, B, and C represent cases where the vehicle is maintaining constant altitude. Case A also represents a steady-level-flight condition, since the bank angle is exactly zero. Case D represents a knife-edge maneuver, which results in a rudder deflection very near the maximum rudder deflection limit of 30 deg. For Case D, the vehicle is ascending and not maintaining constant altitude. Note that in order for the Newton's method solver to converge for Case D with an initial guess of zero for all unknowns, the relaxation factor must be dropped to a maximum of about $\Gamma < 0.67$, or the throttle must not be limited to a value of $0 \leq \tau \leq 1$ during the iteration process.

Case	A	B
α [deg]	3.546530549415	3.491998999388
β [deg]	0.000000000000	3.360704788348
δ_a [deg]	0.000000000000	-0.650133656273
δ_e [deg]	-7.557255188830	-7.535811623664
δ_r [deg]	0.000000000000	9.752904116735
τ	0.095317851844	0.099356995486

Case	C	D
α [deg]	3.060996275796	0.033639349744
β [deg]	9.611279123505	9.570768904630
δ_a [deg]	-1.856122468906	-1.822554090828
δ_e [deg]	-7.367103264587	-6.201782799256
δ_r [deg]	27.908032281271	27.916952178732
τ	0.129378123775	0.734903665720

The student is encouraged to test their code for each of these cases, but with a negative bank angle rather than positive. In each case, the solutions should match the magnitudes of those shown here, with longitudinal components matching in sign, and lateral components opposite in sign.

As noted previously, a steady-heading sideslip case can be specified by either a bank angle ϕ or sideslip angle β. Once either of those has been specified by the user, the other variable is no longer independent, but becomes a dependent variable. For the case when the sideslip angle is specified instead of the bank angle, Algorithms 6.2.4 and 6.2.5 can be used with one modification. Instead of using the array of independent variables $\mathbf{x}$ given in Eq. (6.2.20), β is specified and the set of independent variables for this case is

$$\mathbf{x} = [\alpha, \delta_a, \delta_e, \delta_r, \tau, \phi] \tag{6.5.2}$$

Some modifications to Algorithm 6.2.5 must be made to accommodate this change. The remainder of the solution process is identical. This is demonstrated in the following example.

EXAMPLE 6.5.2 Consider the F-16 aircraft and operating conditions specified in Example 6.5.1. Compute the bank angle ϕ, angle of attack α, rotation rates p, q, and r, control-surface deflections δ_a, δ_e, and δ_r, and throttle setting τ for the aircraft in a steady-heading sideslip with elevation angle and **sideslip angle** specified as follows:

Case	β [deg]	θ [deg]
A	0.0	3.546530549415
B	5.0	4.600266274412
C	10.0	7.796544790337
D	1.0	85.0

Solution: The trim solutions for these cases were obtained using Algorithms 6.2.4 and 6.2.5 with $\mathbf{x}$ given in Eq. (6.5.2). Because the sideslip angle was specified, the bank angle was an unknown value that was solved for in each case, along with the other unknowns. Results are shown in Tables 6.5.1 and 6.5.2. Intermediate solution steps for each case are included on the book website, which can greatly assist in code development and solution checking. Note that Cases C and D are near the limits of what the vehicle can sustain for a trim condition given the rudder and throttle limits. Note that the relaxation factor may need to be reduced in order for Newton's method to correctly converge for Case D.

Table 6.5.1. Solutions for Cases A and B in Example 6.5.2.

Case	A	B
ϕ [deg]	0.000000000000	14.985376829533
α [deg]	3.546530549415	3.424211628766
δ_a [deg]	0.000000000000	-0.967000824385
δ_e [deg]	-7.557255188830	-7.509188075475
δ_r [deg]	0.000000000000	14.511470212489
τ	0.095317851844	0.104300973042

Table 6.5.2. Solutions for Cases C and D in Example 6.5.2.

Case	C	D
ϕ [deg]	31.377170345835	36.691252073884
α [deg]	3.016229815185	0.278633532853
δ_a [deg]	-1.930838872548	-0.190742527547
δ_e [deg]	-7.349652293038	-6.296906953173
δ_r [deg]	29.038486865452	2.915357502462
τ	0.132308281709	0.851676005154

The student is encouraged to try negative sideslip angles and compare results to those shown here.

6.6 Barrel Roll

A barrel roll is a case in which the aircraft is rotating about the wind axis. The rotation rate is defined in wind coordinates by a roll rate of p_W, with the pitch and yaw rates in the wind axis exactly zero,

$$q_W = r_W = 0 \tag{6.6.1}$$

From Eq. (3.4.11), the rotation rates in the body-fixed coordinates can be related to the rotation rates in the wind coordinate system as

$$\begin{bmatrix} p \\ q \\ r \end{bmatrix} = \begin{bmatrix} p_W C_\alpha C_\beta - q_W C_\alpha S_\beta - r_W S_\alpha \\ p_W S_\beta + q_W C_\beta \\ p_W S_\alpha C_\beta - q_W S_\alpha S_\beta + r_W C_\alpha \end{bmatrix} \tag{6.6.2}$$

Applying Eq. (6.6.1) to Eq. (6.6.2) gives

$$\begin{bmatrix} p \\ q \\ r \end{bmatrix} = p_W \begin{bmatrix} C_\alpha C_\beta \\ S_\beta \\ S_\alpha C_\beta \end{bmatrix} \tag{6.6.3}$$

In light of Eq. (3.4.12), this can also be expressed in terms of the body-fixed velocity components

$$\begin{bmatrix} p \\ q \\ r \end{bmatrix} = \frac{p_W}{V} \begin{bmatrix} u \\ v \\ w \end{bmatrix} \tag{6.6.4}$$

In this chapter, we are concerned with trim cases for which the gravity vector does not change orientation relative to the body-fixed coordinate system. With this requirement, the only barrel-roll maneuver that is a trim condition is a vertical barrel roll in which the vehicle is ascending or descending vertically. Typical barrel rolls when the vehicle is not ascending or descending vertically could be considered as instantaneous conditions, but would not be trim conditions since the gravity vector changes throughout the maneuver. Note that a vertical climb is a special case of the vertical barrel roll for which $p_W = 0$. Trim conditions for which the vehicle has a specified climb angle, such as a vertical climb, are discussed in Section 7.2.

6.7 Problems

Theory Problems

6.T1 Beginning with Eqs. (6.1.1)–(6.1.4), show that for a trim case, these relations simplify to those given in Eqs. (6.1.9)–(6.1.12).

6.T2 Using Eqs. (6.4.1), (6.1.11), and (6.1.12), show that for a steady coordinated turn, the relationship in Eq. (6.4.3) must be true.

6.T3 Beginning with Eq. (6.4.3) for a steady coordinated turn, show that the rolling rate about the wind axis is zero if the vehicle is not climbing or descending relative to the surrounding air.

Coding Problems

6.C1 Write a computer code that uses Newton's method to solve for the root of a scalar function. Your code should be constructed such that it is easy to modify the function to be solved, the step size used for finite differencing, the relaxation factor, and the convergence tolerance. Test your code by reproducing Example 6.2.1.

6.C2 Write a computer code that uses Newton's method to solve for the root of a system of equations. Your code should be constructed such that it is easy to modify the system of equations to be solved, the step size used for finite differencing, the relaxation factor, and the convergence tolerance. Test your code by reproducing Example 6.2.2.

6.C3 Write a computer code that uses Newton's method to compute the required angle of attack, sideslip angle, control-surface deflections, and throttle setting for the case of **steady level flight** with the vehicle Euler angles specified. Construct your code such that it can read an input JSON file in the following format:

```
{
    "simulation" : {(simulation information contained here)},
    "vehicle" : {(vehicle mass and aerodynamic information contained here)},
    "initial" : {
        "airspeed[ft/s]" : 330.0,
        "altitude[ft]" : 1000.0,
        "Euler_angles[deg]" : [0.0, 0.0, 0.0], // [bank, elevation, azimuth]
        "type" : "trim", // this can be either "state" or "trim"
        "state" : {// if type=state, use this JSON section as shown in Ch.
 5 problems},
        "trim" : { // if type=trim, uses the following information
            "solver" : {
                "finite_difference_step_size" : 0.01,
                "relaxation_factor" : 0.9,
                "tolerance" : 1.0e-13,
                "max_iterations" : 100
            }
        }
    }
}
```

Test your code for all cases given in Example 6.3.1. Intermediate solution steps can be found on the book website that can greatly assist in code structure and debugging.

6.C4 Modify your code to compute the required angle of attack, sideslip angle, roll rate, pitch rate, yaw rate, control-surface deflections, and throttle setting for the case of a **steady coordinated turn** with the vehicle Euler angles specified. Construct your code such that it can read an input JSON file in the format given in Problem 6.C3, with one additional flag denoting the trim type:

```
{
    "initial" : {
        "trim" : {
```

```
                "type" : "sct", // "sct" = steady coordinated turn
            }
        }
}
```

Test your code for all cases given in Example 6.4.1. Intermediate solution steps can be found on the book website to help with code structure and debugging.

6.C5 Modify your code to compute the required angle of attack, sideslip angle, control-surface deflections, and throttle setting for the case of a **steady-heading sideslip** with the vehicle Euler angles specified.. Modify your code to read an input file with an option specifying steady-heading sideslip as follows:

```
{
    "initial" : {
        "trim" : {
            "type" : "shss" // "shss" = steady-heading sideslip
        }
    }
}
```

Test your code for the cases given in Example 6.5.1. Intermediate solution steps can be found on the book website.

6.C6 Modify your code so that it will compute the trim condition for the case of **steady-heading sideslip** given an elevation angle, azimuth angle, and **sideslip angle**. Modify your code to read an input file with an option to specify the sideslip angle as follows:

```
{
    "initial" : {
        "trim" : {
            "type" : "shss",
            "sideslip_angle[deg]" : 0.0 // add this option
        }
    }
}
```

If the `sideslip_angle[deg]` key is found in the dictionary, your code should automatically use that sideslip angle rather than the bank angle to specify the trim condition. You can test for the existence of a key in a JSON dictionary in most coding languages. Test your code for the cases given in Example 6.5.2. Intermediate solution steps can be found on the book website.

6.C7 Develop a trim algorithm based on the quaternion formulation rather than the Euler angle formulation for the equations of motion. Compare the results and numerical behavior to the Euler-based trim solutions in Problems 6.C3–6.C6.

Chapter 7

Geographic and Performance Relations

7.1 Introduction

The states of the vehicle used in the equations of motion discussed in Chapter 5 include the vehicle velocity relative to the surrounding atmosphere $\mathbf{V} = (u, v, w)$, the rotation rate of the vehicle about the body-fixed coordinate system $\omega_b = (p, q, r)$, the position of the vehicle relative to the Earth-fixed coordinate system (x_f, y_f, z_f), and the orientation of the vehicle relative to the Earth-fixed coordinate system specified in either Euler angles (ϕ, θ, ψ), or as components of the quaternion (e_0, e_x, e_y, e_z). These states were chosen for their convenience in formulating the equations of rigid-body motion from Newton's second law, as well as the kinematic transformation equations relating the orientation of the vehicle to the Earth-fixed coordinate system. Although these states are convenient for integrating the equations of motion, they are not particularly useful for describing the state of the vehicle in terms of intuitive geographic or performance parameters. For example, it is sometimes more intuitive to specify the velocity of the vehicle in terms of a ground speed, climb rate, and ground-track angle rather than the Earth-fixed components $(\dot{x}_f, \dot{y}_f, \dot{z}_f)$. Additionally, it is often important to know the location of the vehicle relative to geographic coordinates in addition to the local flat Earth-fixed coordinate system.

This chapter includes common geographic relations that can be helpful in a user interface to a simulation tool. These relations do not affect the physics directly, but can be used for user input or output to relate the states of the vehicle used in the equations of motion to geographic relations that are typically more intuitive.

7.2 Climb Rate

One important geographic characteristic of an aircraft, which is important for a pilot to know at any time, is the climb rate. The climb rate is defined as the change in vertical location with respect to time, i.e. $V_{\text{climb}} \equiv -\dot{z}_f$. Note that because the z axis of the Earth-fixed coordinate system points towards the center of the Earth, a positive climb rate is associated with a negative value for $\dot{z}_f$. It is often convenient to relate the climb rate to a climb angle. Two definitions exist for the climb angle associated with any climb rate.

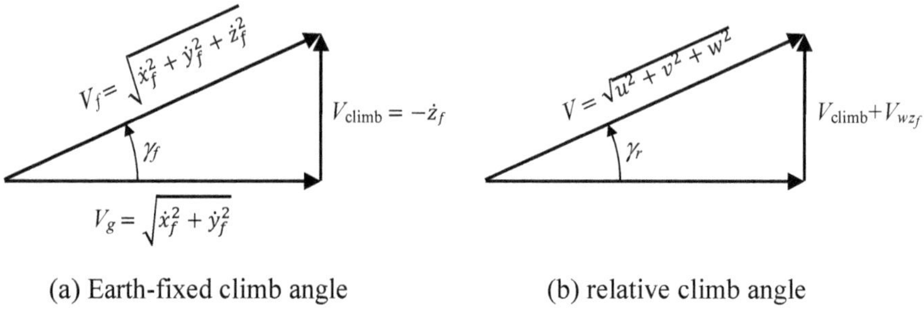

Figure 7.2.1. Vector diagram showing the relationship between climb rate and two ways to define the associated climb angle.

Earth-Fixed Climb Angle

The climb angle that is perhaps most intuitive is the angle between the total velocity vector of the vehicle relative to the Earth, and the horizontal, as shown in Fig. 7.2.1(a). In this text we will use the symbol γ_f to denote this climb angle, since it is the angle obtained from the Earth-fixed velocity vector of the vehicle. This yields a climb rate defined by

$$V_{\text{climb}} = -\dot{z}_f = V_f \sin \gamma_f \tag{7.2.1}$$

where V_f is the total velocity vector of the vehicle relative to the Earth, and is defined as

$$V_f \equiv \sqrt{\dot{x}_f^2 + \dot{y}_f^2 + \dot{z}_f^2} \tag{7.2.2}$$

Using Eq. (7.2.2) in Eq. (7.2.1) and solving for the climb angle yields

$$\gamma_f = \sin^{-1}\left(\frac{V_{\text{climb}}}{V_f}\right) = \sin^{-1}\left(\frac{-\dot{z}_f}{\sqrt{\dot{x}_f^2 + \dot{y}_f^2 + \dot{z}_f^2}}\right) \tag{7.2.3}$$

The velocity components $\dot{x}_f$, $\dot{y}_f$, and $\dot{z}_f$ are related to the body-fixed velocity components u, v, and w, as well as the constant wind vector $\mathbf{V}_{w_f}$ as shown in Eq. (5.4.3) for the Euler-angle formulation and Eq. (5.4.7) for the quaternion formulation. For example, the third equation within Eq. (5.4.3) yields the relation

$$\dot{z}_f = -S_\theta u + S_\phi C_\theta v + C_\phi C_\theta w + V_{wz_f} \tag{7.2.4}$$

Using Eq. (7.2.4) in (7.2.1) yields

$$V_{\text{climb}} = S_\theta u - S_\phi C_\theta v - C_\phi C_\theta w - V_{wz_f} \tag{7.2.5}$$

Since the state of the vehicle is known at each time step in a simulation, Eqs. (7.2.1) or (7.2.5) and (7.2.3) can be used at any point to report the climb rate and climb angle γ_f relative to the Earth-fixed coordinate system.

Relative Climb Angle

An alternate definition for the climb angle is the angle between the vehicle velocity vector relative to the surrounding air and the horizontal, as shown in Fig. 7.2.1(b). In this text we will use the symbol γ_r to denote this climb angle, since it is a relative angle to the surrounding air mass. Because the vertical component of wind V_{wz_f} may be nonzero, this yields

$$V_{\text{climb}} = V \sin\gamma_r - V_{wz_f} = -\dot{z}_f \tag{7.2.6}$$

where $V = \sqrt{u^2 + v^2 + w^2}$. Solving for the relative climb angle yields the relation

$$\gamma_r = \sin^{-1}\left(\frac{V_{\text{climb}} + V_{wz_f}}{V}\right) = \sin^{-1}\left(\frac{V_{wz_f} - \dot{z}_f}{V}\right) \tag{7.2.7}$$

Note that in the absence of wind, $\gamma_f = \gamma_r$.

The climb rate can be related to the aircraft orientation and velocity components using the Euler-angle formulation given in Eq. (5.4.3) or the quaternion formulation given in Eq. (5.4.7). For example, using the Euler-angle formulation given in Eq. (7.2.4) in Eqs. (7.2.6) and (7.2.7) yields

$$\gamma_r = \sin^{-1}\left(\frac{uS_\theta - (vS_\phi + wC_\phi)C_\theta}{\sqrt{u^2 + v^2 + w^2}}\right) \tag{7.2.8}$$

Alternate forms of Eqs. (7.2.4), (7.2.5), and (7.2.8) could be developed using the quaternion formulation rather than the Euler-angle formulation. This is left as an exercise for the student.

It is sometimes useful to compute elevation angle θ as a function of a specified relative climb angle γ_r. Using Eq. (7.2.4) in Eq. (7.2.6) gives a relationship between the relative climb angle, bank angle, elevation angle, and body-fixed velocity components

$$VS_{\gamma_r} = uS_\theta - (vS_\phi + wC_\phi)C_\theta \tag{7.2.9}$$

Rearranging and squaring gives

$$V^2S_{\gamma_r}^2 - 2uVS_\theta S_{\gamma_r} + u^2S_\theta^2 = (vS_\phi + wC_\phi)^2C_\theta^2 \tag{7.2.10}$$

Using the trigonometric identity $C_\theta^2 = 1 - S_\theta^2$ and rearranging gives a quadratic in S_θ

$$\left[u^2 + (vS_\phi + wC_\phi)^2\right]S_\theta^2 - 2uVS_{\gamma_r}S_\theta + V^2S_{\gamma_r}^2 - (vS_\phi + wC_\phi)^2 = 0 \tag{7.2.11}$$

This equation can be solved using the quadratic formula to yield

$$S_\theta = \frac{uVS_{\gamma_r} \pm (vS_\phi + wC_\phi)\sqrt{u^2 + (vS_\phi + wC_\phi)^2 - V^2S_{\gamma_r}^2}}{u^2 + (vS_\phi + wC_\phi)^2} \tag{7.2.12}$$

Equation (7.2.12) yields two solutions. The correct solution for θ is the root that satisfies Eq. (7.2.9). Given the aerodynamic velocity components and bank angle of the vehicle,

Eqs. (7.2.12) and (7.2.9) can be used to solve for the elevation angle for a specified relative climb angle. Again, an alternate form of Eq. (7.2.12) could be developed using the quaternion formulation rather than the Euler angle formulation.

EXAMPLE 7.2.1 An aircraft is flying at 330 ft/s with a relative climb angle of $\gamma_r = 10$ deg, an angle of attack of $\alpha = 5$ deg, a sideslip angle of $\beta = 5$ deg, and a bank angle of $\phi = 0$ deg. Compute the elevation angle of the vehicle.

Solution: From Eq. (3.4.12), the body-fixed velocity vector components (u, v, w) can be computed, which yields

$$\begin{bmatrix} u \\ v \\ w \end{bmatrix} = V \begin{bmatrix} C_\alpha C_\beta \\ S_\beta \\ S_\alpha C_\beta \end{bmatrix} = \begin{bmatrix} 327.493279247014 \\ 28.7613951067272 \\ 28.6519493150435 \end{bmatrix} \text{ ft/s}$$

Using these results in Eq. (7.2.12) yields two solutions

$$\theta^+ = 15.0385933182346 \text{ deg}$$
$$\theta^- = 5.03859331823459 \text{ deg}$$

Using the θ^+ solution in Eq. (7.2.9) gives an error of exactly zero. Using the θ^- solution in Eq. (7.2.9) gives an error of -57.0825. Therefore, the solution is

$$\theta = \theta^+ = 15.0385933182346 \text{ deg}$$

Use in a Trim Algorithm

The trim algorithm given in Chapter 6 requires the elevation angle to be specified, and uses an iterative method to solve for other states of the vehicle. This algorithm can be modified slightly to accommodate a user-specified climb angle γ_f or relative climb angle γ_r rather than a user-specified elevation angle. This can be done multiple ways.

One option is to solve for the elevation angle θ numerically just as is done with all other unknown variables given in the solution process outlined in Eqs. (6.2.19)–(6.2.21). This process includes increasing the length of the unknown $\mathbf{x}$ vector given in Eq. (6.2.20) to include the elevation angle θ to yield an unknown $\mathbf{x}$ vector of the form

$$\mathbf{x} = [\alpha, \beta, \delta_a, \delta_e, \delta_r, \tau, \theta] \tag{7.2.13}$$

Because this process includes one additional unknown, one additional equation or constraint must be added to the system of equations. The additional equation to be added to the system is the error in the actual climb angle for any set of $\mathbf{x}$ as compared to the user-specified climb angle. This is done by increasing the length of the residual vector $\mathbf{R}$ by one as computed within the `calc_R` function, and returning one additional error value. If γ_f was specified by the user, Eq. (7.2.3) can be used to compute the actual climb angle, and the difference reported as the error. Likewise, if γ_r was specified by the user, Eq. (7.2.7) or Eq. (7.2.8) can be used to compute the actual climb angle, and the difference reported as the error. Because the number of equations in the system and number of unknowns has increased by one, the size of the square Jacobian matrix must also increase by one. With this setup, Algorithm 6.2.4 can be used to solve for the values of the **x** vector, which includes the elevation angle θ as shown in Eq. (7.2.13).

Another option for obtaining the correct elevation angle for a specified relative climb angle is to use Eqs. (7.2.12) and (7.2.9) to compute the correct elevation angle within the `calc_R` function given in Algorithm 6.2.5, based on the current values of the translational velocity components (u, v, w) and bank angle. This can be accomplished by inserting code to solve for the appropriate elevation angle directly before line 12 in Algorithm 6.2.5. This method does not require the size of the system of equations to be increased, but simply adds an internal constraint on the elevation angle within each iteration.

The second option for obtaining the appropriate elevation angle for a specified climb angle may require slightly less computation to converge to the correct solution than the first option listed above. However, ease of implementation for either approach is likely the most important decision factor between the two methods listed.

EXAMPLE 7.2.2 Consider the F-16 aircraft specified in Appendix C.3 with the aerodynamic model F16-E-1 given in Table C.3.15. For the case with the vehicle at an altitude of 500 ft and a velocity of 220 ft/s, compute the elevation angle θ, aerodynamic angles α and β, rotation rates p, q, and r, control-surface deflections δ_a, δ_e, and δ_r, and throttle setting τ for the aircraft in a **steady coordinated turn** with the relative climb angle and bank angle specified as follows:

Case	γ_r [deg]	ϕ [deg]
A	0.0	0.0
B	0.0	60.0
C	10.0	60.0
D	-10.0	180.0

Solution: The trim solutions for these cases were obtained using Newton's method as outlined in Chapter 6, with the additional unknown value of the elevation angle included in the **x** array as shown in Eq. (7.2.13). This required adding one additional equation to the system of equations. This additional equation is the difference between the user-specified relative climb angle and that computed from Eq. (7.2.7) or (7.2.8). Results are shown in the following tables. Intermediate solution steps for each case are included on the book website, which can greatly assist in code development and solution checking.

Case	A	B
θ [deg]	17.765373007712	14.755393469546
α [deg]	17.765373007712	27.581823398298
β [deg]	0.000000000000	0.128243340506
p [deg/s]	0.000000000000	-3.270083803557
q [deg/s]	0.000000000000	10.752485898395
r [deg/s]	0.000000000000	6.207950627896
δ_a [deg]	0.000000000000	1.661204007392
δ_e [deg]	-13.148508618662	-21.198614208659
δ_r [deg]	0.000000000000	-1.028183227311
τ	0.148261787257	0.439141639686

Case	C	D
θ [deg]	23.825326227942	7.425811971054
α [deg]	24.751622022612	-17.425811971054
β [deg]	0.122081204522	0.000000000000
p [deg/s]	-4.587194658335	0.000000000000
q [deg/s]	8.996379149738	0.000000000000
r [deg/s]	5.194061923833	0.000000000000
δ_a [deg]	1.666889361137	0.000000000000
δ_e [deg]	-19.328417014984	1.193436735360
δ_r [deg]	-0.607789128152	0.000000000000
τ	0.471985445574	0.158767831368

EXAMPLE 7.2.3 Repeat Example 7.2.2 for the aircraft in a **steady-heading sideslip** with the relative climb angle and **bank angle** specified as follows:

Case	γ_r [deg]	ϕ [deg]
A	0.0	0.0
B	0.0	7.0
C	10.0	5.0

Solution: The trim solutions for these cases were obtained using Newton's method as outlined in Chapter 6 with the addition of an elevation-angle solver from a specified climb angle. Results are shown in the following table. Intermediate solution steps for each case are included on the book website. Note that for all cases, $p = q = r = 0$.

The student is encouraged to also test their code using negative values for the bank angle and comparing the results to those included here.

Case	A	B	C
θ [deg]	17.765373007712	18.553367730538	27.115658731404
α [deg]	17.765373007712	17.605477117481	16.605630635092
β [deg]	0.000000000000	9.184397827203	6.363110324246
δ_a [deg]	0.000000000000	-1.725908053107	-1.204461308730
δ_e [deg]	-13.148508618662	-13.057397816852	-12.659928977372
δ_r [deg]	0.000000000000	26.903345058472	18.596231066313
τ	0.148261787257	0.152704179577	0.258837440743

EXAMPLE 7.2.4 Repeat Example 7.2.2 for the aircraft in a **steady-heading sideslip** with the relative climb angle and **sideslip angle** specified as follows:

Case	γ_r [deg]	β [deg]
A	0.0	0.0
B	10.0	10.0
C	-10.0	-10.0

Solution: The trim solutions for these cases were obtained using Newton's method as outlined in Chapter 6 with the addition of an elevation angle solver from a specified climb angle. Results are shown in the following table. Intermediate solution steps for each case are included on the book website. Note that for all cases, $p = q = r = 0$. The student is encouraged to also test their code using opposite signs for the bank angle and comparing the results to those included here.

Case	A	B	C
ϕ [deg]	0.000000000000	7.879263950825	-7.486902376510
θ [deg]	17.765373007712	27.721124096129	9.294627260909
α [deg]	17.765373007712	16.455534543817	18.286429790561
δ_a [deg]	0.000000000000	-1.894832611856	1.868964579561
δ_e [deg]	-13.148508618662	-12.582850729505	-13.333712765057
δ_r [deg]	0.000000000000	29.215473521870	-29.342626312968
τ	0.148261787257	0.262429546227	0.036947458259

7.3 Load Factor

The term *load factor* is nearly universally used in the literature to mean the ratio of lift generated during a maneuver to the weight of the vehicle, i.e. L/W. Recall that the lift is defined as the component of the aerodynamic force vector that lies in the vehicle plane of symmetry and is perpendicular to the direction of flight, as shown in Fig. 3.4.1. Although the traditional definition of load factor is useful, in this section we will use a more formal definition and discussion of load factor.

Equation (5.5.4) gives an expression relating the three components of translational acceleration of the vehicle ($\dot{u}$, $\dot{v}$, $\dot{w}$) to the vehicle mass, pseudo aerodynamic force vector, Euler angles, and translational and rotational velocities. This equation can be written in a dimensionless form by dividing through by the acceleration of gravity, including gravity relief. Doing so and rearranging yields

$$\frac{1}{\mathcal{M}(g-a_c)}\begin{bmatrix}\overline{F}_{xb}\\ \overline{F}_{yb}\\ \overline{F}_{zb}\end{bmatrix} = \frac{1}{(g-a_c)}\begin{bmatrix}\dot{u}\\ \dot{v}\\ \dot{w}\end{bmatrix} - \begin{bmatrix}-S_\theta\\ S_\phi C_\theta\\ C_\phi C_\theta\end{bmatrix} - \frac{1}{(g-a_c)}\begin{bmatrix}rv-qw\\ pw-ru\\ qu-pv\end{bmatrix} \tag{7.3.1}$$

Equation (7.3.1) shows that the dimensionless pseudo aerodynamic force vector $\overline{\mathbf{F}}/[\mathcal{M}(g-a_c)]$ is equal to the sum of three components. The first component on the right-hand side is the dimensionless translational acceleration, the second component is the unit vector in the direction of gravity, and the third component is the dimensionless Coriolis acceleration. Here we will define the dimensionless vector on the left-hand side as the *load factor vector* and use the symbol $\bar{\mathbf{n}}$ to denote this vector

$$\text{load factor vector} \equiv \bar{\mathbf{n}} = \begin{bmatrix}\bar{n}_x\\ \bar{n}_y\\ \bar{n}_z\end{bmatrix} \equiv \frac{1}{\mathcal{M}(g-a_c)}\begin{bmatrix}\overline{F}_{xb}\\ \overline{F}_{yb}\\ \overline{F}_{zb}\end{bmatrix} \tag{7.3.2}$$

From this expression it can be seen that the load factor is a ratio of the pseudo aerodynamic force relative to the weight of the aircraft, $\bar{\mathbf{n}} = \overline{\mathbf{F}}/W$. Rather than having only a single component, as is typically implied from the traditional use of the term *load factor*, the load factor vector has three components as shown in Eq. (7.3.2). This vector is useful for evaluating the aerodynamic load that must be sustained by the vehicle structure in any direction during a maneuver. Although the load factor has three components, the component of most interest for fixed-wing aircraft is typically the component in the normal or $-z_b$ direction. This is called the *normal load factor* and is defined as

$$\text{normal load factor} \equiv \bar{n}_N \equiv -\bar{n}_z = -\frac{\overline{F}_{zb}}{\mathcal{M}(g-a_c)} \tag{7.3.3}$$

The normal load factor $\bar{n}_N$ differs from the traditional definition of the load factor in that the normal load factor is in the body-fixed coordinate system whereas the traditional load factor is typically defined in the wind coordinate system. Here we will use the symbol $\bar{n}_L$ to denote the traditional load factor. The definition for the traditional load factor arises from the universally accepted understanding that steady level flight is a *1-g maneuver*. In other words, in steady level flight, the load factor is exactly equal to unity, i.e. $\bar{n}_L = 1.0$. Considering the pseudo aerodynamic force normal to the direction of flight, which can be obtained from Table 3.4.4 and the definition of the axial and normal forces, we obtain the formal definition of the *load factor*

$$\text{load factor} \equiv \bar{n}_L \equiv \frac{\overline{F}_{x_b} \sin\alpha - \overline{F}_{z_b} \cos\alpha}{\mathcal{M}(g - a_c)} \tag{7.3.4}$$

In a simulation environment, the state of the vehicle is known at each time step, and the components of the load factor vector $\bar{\mathbf{n}}$ including the normal load factor $\bar{n}_N$ can easily be computed at any time step. Because the angle of attack is also known at any time step, the traditional load factor $\bar{n}_L$ can also be easily computed. It can be shown and should be noted that all the upright steady-level-flight trim cases included in Chapters 6 and 7 yield load factors of exactly 1.0.

It is sometimes useful to solve for the bank angle associated with a specified load factor for a steady coordinated turn. In the development of the steady-coordinated-turn trim condition shown in Section 6.4, the bank angle was used to specify the maneuver. Rather than specifying a bank angle for a steady coordinated turn, this trim condition can alternately be specified by a load factor. Using Eq. (7.3.1) in Eq. (7.3.4), recognizing that for a trim condition $\dot{u} = \dot{w} = 0$, and rearranging gives the relation

$$\phi = \cos^{-1}\left[\frac{C_\theta + C_\phi S_\theta w/u}{(\bar{n}_L - A)/C_\alpha + pv/(g - a_c)} - \frac{wS_\theta}{uC_\theta}\right] \tag{7.3.5}$$

where

$$A = \left[S_\theta + \frac{qw - rv}{g - a_c}\right] S_\alpha \tag{7.3.6}$$

Note that in this form, ϕ is a function of $\cos\phi$. Because Eq. (7.3.5) is an implicit equation of ϕ, and because the rotation rates in a steady coordinated turn are also functions of the bank angle as shown in Eq. (6.4.3), Eq. (7.3.5) must be solved iteratively.

Traditional Approximation

As has been shown here, the normal load factor $\bar{n}_N$ differs from the load factor $\bar{n}_L$. However, the reader is cautioned that these two terms are often used interchangeably in the aeronautics industry. These two terms are only equal in the special case when the body-fixed coordinate system is aligned with the direction of flight, i.e. $\alpha = 0$. Additionally, most definitions of the load factor use the simple expression $\bar{n}_L = L/W$ rather than the true definition given in Eq. (7.3.4). The simple expression replaces the pseudo aerodynamic force with the lift, which tacitly assumes that there is no thrust in the direction of lift, or that the thrust is aligned with the direction of flight. Although the angle of attack is typically small for fixed-wing aircraft in cruise conditions, and although the thrust is typically nearly aligned with the direction of flight, the angle of attack and thrust angle are not typically exactly zero. Therefore, the expression

$$\bar{n}_L \approx \frac{L}{W} \tag{7.3.7}$$

is only an approximation to the exact expression given in Eq. (7.3.4). The reader should understand that **this approximation is widely used throughout the aeronautics industry, and is even considered the definition of load factor in many textbooks.**

The approximate load factor given in Eq. (7.3.7) can be used to obtain an approximation for the bank angle associated with a given load factor for a vehicle in a steady coordinated turn. Beginning with the exact relation between bank angle and load factor given in Eq. (7.3.5), assuming the body-fixed coordinate system is aligned with the direction of flight, i.e. $v = w = 0$, and applying Eq. (7.3.7) yields the approximation

$$\phi \approx \cos^{-1}\left(\frac{\cos\theta}{\bar{n}_L}\right) \approx \cos^{-1}\left(\frac{\cos\theta}{L/W}\right) \tag{7.3.8}$$

Finally, for this special case, $u = V$ and Eq. (7.2.3) simplifies to $\theta = \gamma$. Therefore, the traditional approximation can be used to relate the bank angle directly to the climb angle

$$\phi \approx \cos^{-1}\left(\frac{\cos\gamma}{\bar{n}_L}\right) \approx \cos^{-1}\left(\frac{\cos\gamma}{L/W}\right) \tag{7.3.9}$$

Although Eqs. (7.3.7)–(7.3.9) are used in the industry, these are only all true for the special case when the body-fixed coordinate system is aligned with the direction of flight, there is no thrust in the body-fixed z_b direction, and the vehicle is in a steady coordinated turn. For most cases, the general solution given in Eqs. (7.3.4) should be used. If the vehicle is in a steady coordinated turn, the relation given in Eq. (7.3.5) also holds. Although Eqs. (7.3.8) and (7.3.9) are only approximations, these can be used as an initial guess for the bank angle in a trim solver where the load factor is specified by the user.

Use in a Steady-Coordinated-Turn Trim Algorithm

The trim algorithm given in Chapter 6 requires the bank angle to be specified for a coordinated turn, and uses an iterative method to solve for other states of the vehicle. This algorithm can be modified slightly to accommodate a user-specified load factor $\bar{n}_L$ rather than a user-specified bank angle. This can be done multiple ways.

One option is to solve for the bank angle ϕ numerically just as is done with all other unknown variables given in the solution process outlined in Eqs. (6.2.19)–(6.2.21). This process includes increasing the length of the unknown $\mathbf{x}$ vector given in Eq. (6.2.20) to include the bank angle ϕ to yield an unknown $\mathbf{x}$ vector of the form

$$\mathbf{x} = [\alpha, \beta, \delta_a, \delta_e, \delta_r, \tau, \phi] \tag{7.3.10}$$

Because this process includes one additional unknown, one additional equation or constraint must be added to the system of equations. The additional equation to be added is the error in the actual load factor for any set of $\mathbf{x}$ as compared to the user-specified load factor. This is done by increasing the length of the residual vector $\mathbf{R}$ by one as computed within the `calc_R` function, and returning one additional error value. Eq. (7.3.4) can be used to compute the actual load factor, and the difference reported as the error. Because the number of equations in the system and number of unknowns has increased by one, the size of the square Jacobian matrix must also increase by one. With this setup, Algorithm 6.2.4 can be used to solve for the values of the $\mathbf{x}$ vector, which includes the bank angle ϕ as shown in Eq. (7.3.10).

Another option for obtaining the correct bank angle for a specified load factor is to use Eqs. (7.3.5) and (7.3.6) to update the bank angle within the `calc_R` function given in Algorithm 6.2.5, based on the current estimates for the vehicle states, including the bank angle. This can be accomplished by inserting code to solve for the appropriate bank angle directly before line 12 in Algorithm 6.2.5. This method does not require the size of the system of equations to be increased, but simply adds an internal constraint on the elevation angle within each iteration.

The second option for obtaining the appropriate bank angle for a specified load factor may require slightly less computation to converge to the correct solution than the first option listed above. However, ease of implementation for either approach is likely the most important decision factor between the two methods listed.

EXAMPLE 7.3.1 Consider the F-16 aircraft specified in Appendix C.3 with the aerodynamic model F16-E-1 given in Table C.3.15 operating at an airspeed of 220 ft/s at sea level. Find the bank angle required for the vehicle to produce a load factor of $\bar{n}_L = 2.0$ in a steady coordinated turn with zero climb rate. Compare the solution to that from the traditional approximation given in Eq. (7.3.9).

Solution: If the vehicle does not change altitude, the climb angle must be $\gamma = 0$. Let us first consider the traditional approximation for the required bank angle. Using $\gamma = 0$ in Eq. (7.3.9) yields the traditional approximation for the bank angle

$$\phi = \cos^{-1}\left(\frac{\cos\gamma}{\bar{n}_L}\right) = \cos^{-1}\left(\frac{1}{2}\right) = 1.0471975511966 \text{ rad} = 60 \text{ deg}$$

The traditional approximation that a 60-deg bank angle yields a load factor of 2.0 is so widely accepted, that the student should commit this approximation to memory. Using a trim algorithm, it can be shown that the trim solution for the vehicle in a 60-degree bank steady coordinated turn with zero relative climb angle at sea level and a velocity of 220 ft/s, is

$$\begin{bmatrix}\phi\\ \theta\\ \psi\end{bmatrix} = \begin{bmatrix}60.0\\ 14.590465534558\\ \text{arbitrary}\end{bmatrix} \text{ deg}, \qquad \begin{bmatrix}\alpha\\ \beta\end{bmatrix} = \begin{bmatrix}27.306284728204\\ 0.127220723640\end{bmatrix} \text{ deg}$$

$$\begin{bmatrix}p\\ q\\ r\end{bmatrix} = \begin{bmatrix}-3.242673264733\\ 10.788348955512\\ 6.228656173577\end{bmatrix} \text{ deg/s}, \qquad \begin{bmatrix}\delta_a\\ \delta_e\\ \delta_r\end{bmatrix} = \begin{bmatrix}1.640139771416\\ -21.076527671178\\ -1.063768192225\end{bmatrix} \text{ deg}$$

$$\tau = 0.430006505340$$

The associated pseudo aerodynamic forces and moments are

$$\overline{F} = \begin{bmatrix} 17,237.9952674 \\ 0.0 \\ -33,389.0173706 \end{bmatrix} \text{lbf}, \quad \overline{M} = \begin{bmatrix} 159.604335916 \\ 338.732916458 \\ -503.611710945 \end{bmatrix} \text{ft-lbf}$$

Using these results in Eq. (7.3.3) and (7.3.4) yields load factors of

$$\text{normal load factor} = \bar{n}_N = 1.62884986082 \tag{7.3.11}$$

$$\text{load factor} = \bar{n}_L = 1.83312008030 \tag{7.3.12}$$

Note that the load factor $\bar{n}_L$ is about 8.3% below the specified load factor of 2.0, and the value of the normal load factor $\bar{n}_N$ is even less. Using an iterative trim solver capable of solving for the required bank angle for a user-specified load factor yields the trim solution

$$\begin{bmatrix} \phi \\ \theta \\ \psi \end{bmatrix} = \begin{bmatrix} 63.256219153032 \\ 14.126325295408 \\ \text{arbitrary} \end{bmatrix} \text{deg}, \quad \begin{bmatrix} \alpha \\ \beta \end{bmatrix} = \begin{bmatrix} 28.965072104211 \\ 0.145370650312 \end{bmatrix} \text{deg}$$

$$\begin{bmatrix} p \\ q \\ r \end{bmatrix} = \begin{bmatrix} -3.541870676699 \\ 12.567944310435 \\ 6.333054759149 \end{bmatrix} \text{deg/s}, \quad \begin{bmatrix} \delta_a \\ \delta_e \\ \delta_r \end{bmatrix} = \begin{bmatrix} 1.833231586461 \\ -22.479892020061 \\ -0.789191819674 \end{bmatrix} \text{deg}$$

$$\tau = 0.498910446604$$

The associated pseudo aerodynamic forces and moments are

$$\overline{F} = \begin{bmatrix} 19,853.9006629 \\ 0.0 \\ -35,868.9352142 \end{bmatrix} \text{lbf}, \quad \overline{M} = \begin{bmatrix} 189.968827329 \\ 375.707337749 \\ -639.347681780 \end{bmatrix} \text{ft-lbf}$$

Using these results in Eqs. (7.3.3) and (7.3.4) yields load factors of

$$\text{normal load factor} = \bar{n}_N = 1.74983017568$$

$$\text{load factor} = \bar{n}_L = 2.0$$

By comparing these solutions it can be seen that the actual trim condition that yields a load factor of 2.0 requires about 3.26 deg more bank angle than the traditional approximation. This difference between the traditional approximation and correct solution depends on the vehicle and flight condition.

EXAMPLE 7.3.2 Consider the F-16 aircraft specified in Appendix C.3 with the aerodynamic model F16-E-1 given in Table C.3.15. For the case with the vehicle at sea level and at a velocity of 550 ft/s and assuming there is no wind, compute the bank angle ϕ, elevation angle θ, aerodynamic angles α and β, rotation rates p, q, and r, control-surface deflections δ_a, δ_e, and δ_r, and throttle setting τ for the aircraft in a steady coordinated turn with relative climb angle and load factor specified as follows:

Case	γ_r [deg]	$\bar{n}_L$
A	0.0	5.0
B	45.0	2.0
C	10.0	1.0

Solution: The trim solutions for these cases were obtained using Newton's method as outlined in Chapter 6 with the addition of a bank-angle solver from a specified load factor. Results are shown in Table 7.3.1. Intermediate solution steps for each case are included on the book website, which can greatly assist in code development and solution checking.

Table 7.3.1. Trim solutions for Example 7.3.2.

Case	A	B	C
ϕ [deg]	78.772002653757	74.566943079126	10.104803046846
θ [deg]	2.669310248517	46.689158768600	12.916418426586
α [deg]	13.385522202954	5.376500419109	2.961524670994
β [deg]	0.016288863082	0.019994954091	0.001937949839
p [deg/s]	-0.764354907846	-6.451122228136	-0.132047295375
q [deg/s]	16.080937012798	5.862247281135	0.101022017752
r [deg/s]	3.192278023615	1.618371351194	0.566859130048
δ_a [deg]	0.090287106408	0.459875691234	-0.003293528506
δ_e [deg]	-13.482862420534	-9.039824441568	-7.345833364532
δ_r [deg]	-0.395861736823	-0.197962223337	-0.076503724317
τ	0.536416490602	0.619992555355	0.228586401614

Note that in the process of solving for the required bank angle that produces a specified load factor, we have considered only positive bank angles. A similar process can be used to solve for the negative bank angle that would result in a user-specified load factor. Developing such an algorithm is left as an exercise for the student.

7.4 Ground Speed and Ground Track

The vehicle velocity is typically defined in the body-fixed coordinate system as

$$\mathbf{V}_b \equiv \begin{bmatrix} u \\ v \\ w \end{bmatrix} \tag{7.4.1}$$

The wind velocity is typically defined in the Earth-fixed coordinate system as

$$\mathbf{V}_{w_f} \equiv \begin{bmatrix} V_{wx_f} \\ V_{wy_f} \\ V_{wz_f} \end{bmatrix} \tag{7.4.2}$$

It is often convenient to relate the vehicle body-fixed velocity $\mathbf{V}_b$ to a horizontal velocity of the vehicle relative to the Earth-fixed coordinate system. This is called the *ground velocity* and is defined as

$$\mathbf{V}_g \equiv \begin{bmatrix} \dot{x}_f \\ \dot{y}_f \end{bmatrix} \tag{7.4.3}$$

The ground velocity is constrained to the horizontal Earth-fixed x–y plane, and therefore has only two components. It is the sum of the Earth-fixed horizontal components of the vehicle and wind velocity vectors

$$\mathbf{V}_g = \mathbf{V}_h + \mathbf{V}_{wh} \tag{7.4.4}$$

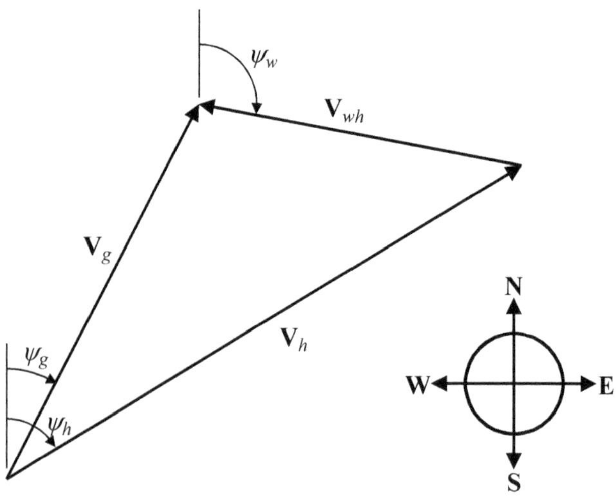

Figure 7.4.1. Vector diagram showing the relationship between ground velocity $\mathbf{V}_g$, horizontal wind velocity $\mathbf{V}_{wh}$, and horizontal vehicle velocity $\mathbf{V}_h$ in the Earth-fixed horizontal plane.

as shown in Figure 7.4.1. The horizontal velocity of the vehicle $\mathbf{V}_h$ is related to the vehicle velocity vector in body-fixed coordinates $\mathbf{V}_b$ and the vehicle orientation in Euler angles as demonstrated from the first two equations of Eq. (5.3.4), which yields

$$\mathbf{V}_h = \begin{bmatrix} C_\theta C_\psi & S_\phi S_\theta C_\psi - C_\phi S_\psi & C_\phi S_\theta C_\psi + S_\phi S_\psi \\ C_\theta S_\psi & S_\phi S_\theta S_\psi + C_\phi C_\psi & C_\phi S_\theta S_\psi - S_\phi C_\psi \end{bmatrix} \begin{bmatrix} u \\ v \\ w \end{bmatrix} \tag{7.4.5}$$

The horizontal velocity of the wind $\mathbf{V}_{wh}$ is

$$\mathbf{V}_{wh} = \begin{bmatrix} V_{wx_f} \\ V_{wy_f} \end{bmatrix} \tag{7.4.6}$$

Using Eqs. (7.4.5) and (7.4.6) in Eq. (7.4.4) yields

$$\mathbf{V}_g = \begin{bmatrix} C_\theta C_\psi & S_\phi S_\theta C_\psi - C_\phi S_\psi & C_\phi S_\theta C_\psi + S_\phi S_\psi \\ C_\theta S_\psi & S_\phi S_\theta S_\psi + C_\phi C_\psi & C_\phi S_\theta S_\psi - S_\phi C_\psi \end{bmatrix} \begin{bmatrix} u \\ v \\ w \end{bmatrix} + \begin{bmatrix} V_{wx_f} \\ V_{wy_f} \end{bmatrix} \tag{7.4.7}$$

The $\mathbf{V}_g$, $\mathbf{V}_h$, and $\mathbf{V}_{wh}$ velocities can also be expressed in terms of magnitudes and azimuth angles in the horizontal plane of the Earth-fixed coordinate system. The ground velocity magnitude $V_g = |\mathbf{V}_g|$ is called the *ground speed* and the associated angle ψ_g is called the *ground track*. The vehicle horizontal velocity magnitude $V_h = |\mathbf{V}_h|$ is called the *horizontal vehicle speed* and the associated angle ψ_h is called the *heading*. The horizontal wind velocity magnitude $V_{wh} = |\mathbf{V}_{wh}|$ is called the *horizontal wind speed* and the associated angle ψ_w is called the *wind direction*. Note that the wind direction is typically defined as the direction from which the wind is coming, as shown in Fig. 7.4.1. This is the customary meteorological definition, where wind direction indicates where the wind is coming *from*, which is opposite to the standard mathematical definition of vector direction.

Trigonometric relations of these three vectors can be useful in various situations. The x and y components of the vectors $\mathbf{V}_g$, $\mathbf{V}_h$, and $\mathbf{V}_{wh}$ are related to the associated angles ψ_g, ψ_h, and ψ_w through the relations

$$\psi_g = \tan^{-1}\left(\frac{V_{g_y}}{V_{g_x}}\right) \tag{7.4.8}$$

$$\psi_h = \tan^{-1}\left(\frac{V_{h_y}}{V_{h_x}}\right) \tag{7.4.9}$$

$$\psi_w = \pi + \tan^{-1}\left(\frac{V_{wh_y}}{V_{wh_x}}\right) \tag{7.4.10}$$

When computing these angles, it is necessary to use the `atan2` function to preserve the quadrant. From the law of sines, the magnitudes and angles of these vectors are related according to

$$\frac{V_h}{\sin(\psi_w - \psi_g)} = \frac{V_g}{\sin(\psi_w - \psi_h)} = \frac{V_{wh}}{\sin(\psi_h - \psi_g)} \tag{7.4.11}$$

From the law of cosines, the magnitudes and angles of these vectors are related according to

$$V_h = \sqrt{V_g^2 + V_{wh}^2 + 2V_g V_{wh}\cos(\psi_g - \psi_w)} \tag{7.4.12}$$

$$V_g = \sqrt{V_h^2 + V_{wh}^2 - 2V_h V_{wh}\cos(\psi_w - \psi_h)} \tag{7.4.13}$$

$$V_{wh} = \sqrt{V_h^2 + V_g^2 - 2V_h V_g\cos(\psi_h - \psi_g)} \tag{7.4.14}$$

These trigonometric relationships can be used to output the ground speed and ground track during simulation, or to find the necessary vehicle state to match an initial ground speed and ground track from user input.

During simulation, the state of the vehicle is known along with the wind velocity vector. Therefore, the ground velocity vector can be directly computed from Eq. (7.4.7). The magnitude of this result is the ground speed. The ground track can be computed from Eq. (7.4.8) where V_{g_x} and V_{g_y} are the x and y components of the ground velocity vector $\mathbf{V}_g$ respectively.

During simulation initialization, it is sometimes convenient for the user to be able to specify a ground speed and ground track for a given wind velocity vector as an initial condition. For example, this is commonly used for landing simulations. In this case, the relations shown here can be used in a trim solver to numerically solve for the appropriate vehicle azimuth angle to satisfy the specified conditions.

Heading and Euler Azimuth Angle

The heading ψ_h is the direction the vehicle is moving in the horizontal plane, and the azimuth angle ψ is the Euler angle, which, along with the elevation θ and bank ϕ Euler angles, specifies the orientation of the vehicle relative to the Earth-fixed coordinate system. In some texts, it is assumed that the heading ψ_h and the vehicle Euler azimuth angle ψ are identical. However, in general, this is not the case. Using Eq. (7.4.5) in Eq. (7.4.9) yields a **relation between the heading and Euler azimuth angle**

$$\psi_h = \tan^{-1}\left[\frac{uC_\theta S_\psi + v(S_\phi S_\theta S_\psi + C_\phi C_\psi) + w(C_\phi S_\theta S_\psi - S_\phi C_\psi)}{uC_\theta C_\psi + v(S_\phi S_\theta C_\psi - C_\phi S_\psi) + w(C_\phi S_\theta C_\psi + S_\phi S_\psi)}\right] \tag{7.4.15}$$

For the special case when both sideslip velocity and bank angle are zero, $v = \phi = 0$, Eq. (7.4.15) reduces to

$$\psi_h = \tan^{-1}\left(\frac{uC_\theta S_\psi + wC_\phi S_\theta S_\psi}{uC_\theta C_\psi + wC_\phi S_\theta C_\psi}\right) = \tan^{-1}\left[\left(\frac{S_\psi}{C_\psi}\right)\left(\frac{uC_\theta + wC_\phi S_\theta}{uC_\theta + wC_\phi S_\theta}\right)\right] = \psi \tag{7.4.16}$$

which results in the heading and azimuth angle being equivalent. Because many aircraft spend a majority of time with very small bank angles and sideslip angles, the terms heading and azimuth angle have become nearly synonymous. However, because in general the bank and sideslip angles may be nonzero, it is important to understand the difference when writing simulation software so that the correct values are being displayed for any flight condition.

EXAMPLE 7.4.1 An aircraft is landing from the south on a runway that is perfectly aligned with the Earth-fixed x_f axis. There is a cross-wind of 60 ft/s from the west. In order to maintain a descent angle of −10 deg at the landing speed of 220 ft/s, the vehicle is oriented with an elevation angle of $\theta = 8.180094921108$ deg and has an angle of attack of $\alpha = 18.18009492111$ deg. In order to maintain a zero bank angle for landing, this vehicle must also have zero sideslip. Find the azimuth angle orientation of the vehicle relative to the Earth-fixed coordinate system necessary for the vehicle to maintain a ground track aligned with the runway.

Solution: Because the runway is aligned with the Earth-fixed x_f axis, and because the vehicle is traveling from south to north, the ground-track angle required for the vehicle is $\psi_g = 0$, and $\tan\psi_g = 0$. The wind vector in the Earth-fixed coordinate system is

$$\begin{bmatrix} V_{wx_f} \\ V_{wy_f} \\ V_{wz_f} \end{bmatrix} = \begin{bmatrix} 0.0 \\ 60.0 \\ 0.0 \end{bmatrix} \text{ ft/s}$$

From Eq. (3.4.12), the body-fixed velocity vector components (u, v, w) can be computed, which yields

$$\begin{bmatrix} u \\ v \\ w \end{bmatrix} = V \begin{bmatrix} C_\alpha C_\beta \\ S_\beta \\ S_\alpha C_\beta \end{bmatrix} = \begin{bmatrix} 209.017710485761 \\ 0.0 \\ 68.6410715482391 \end{bmatrix} \text{ ft/s}$$

For the special case of zero bank angle and sideslip angle, Eq. (7.4.7) reduces to

$$\mathbf{V}_g = \begin{bmatrix} uC_\theta C_\psi + wS_\theta C_\psi + V_{wx_f} \\ uC_\theta S_\psi + wS_\theta S_\psi + V_{wy_f} \end{bmatrix}$$

We seek the solution that will result in the vehicle traveling directly to the north. This requires that the y component of the ground velocity vector be exactly zero

$$0 = uC_\theta S_\psi + wS_\theta S_\psi + V_{wy_f}$$

This expression can be solved for ψ to yield

$$\begin{aligned} \psi &= \sin^{-1}\left(\frac{-V_{wy_f}}{uC_\theta + wS_\theta}\right) \\ &= \sin^{-1}\left(\frac{-60}{209.01771\cos(8.1800949) + 68.641072\sin(8.1800949)}\right) \\ &= -16.0773327840103 \text{ deg} \end{aligned}$$

Because the Euler azimuth angle is typically defined in the range $0 \le \psi < 360$ deg, the final azimuth angle is $\psi = 343.92266721599$. Because this is a case of zero bank angle and zero sideslip, this angle is also equal to the heading ψ_h.

7.5 Geographic Coordinates

The Earth-fixed coordinate system (x_f, y_f, z_f) presented in Chapter 1 is appropriate only for very short flights that do not traverse a significant distance over the Earth's surface. Because the Earth is not flat, when longer flights are to be modeled in simulation, the position of the vehicle relative to a geographic coordinate system is needed. Geographic position on the surface of the Earth is typically defined in terms of latitude Φ, longitude Ψ, and altitude H.

Figure 7.5.1 shows the Earth visually oriented with the North Pole at the top and the South Pole at the bottom. Lines of constant latitude form circles parallel to the equator, as shown in Fig. 7.5.1. These lines run in an east-west direction and correspond to the angle relative to the equator, with a value of +90 degrees at the North Pole, and −90 degrees at the South Pole. Hence, latitude values have limits of

$$-90 \leq \Phi \leq 90 \text{ deg} \tag{7.5.1}$$

Lines of constant longitude are north-south lines that correspond to the angle relative to the Prime Meridian, which passes through Greenwich, England. Figure 7.5.2 shows the Earth visually oriented with the North Pole pointed out of the page. Lines of constant longitude are shown in Fig. 7.5.2 as solid lines. Typically, longitudinal position is considered positive in the east direction, and negative in the west direction, with limits of

$$-180 \leq \Psi \leq 180 \text{ deg} \tag{7.5.2}$$

Altitude is measured relative to a hypothetical smooth surface of the Earth, known as mean sea level (MSL). Any position on the surface of the Earth can be uniquely defined by the latitude, longitude, and altitude. For example, the position of the Los Angeles airport is $\Phi = 33.9424964$ deg, $\Psi = -118.4080486$ deg, $H = 127.8$ ft. The position of the Denver airport is $\Phi = 39.8616667$ deg, $\Psi = -104.6731667$ deg, $H = 5433.8$ ft.

Although the Earth is spherical in nature, it is not a perfect sphere. For example, the currently accepted polar radius, which is the distance from mean sea level at the poles to the center of the Earth, is

$$R_p = 6,356.7516 \text{ km} \tag{7.5.3}$$

The currently accepted equatorial radius, which is the distance from mean sea level at the equator to the center of the Earth, is

$$R_e \equiv 6,378.1363 \text{ km} \tag{7.5.4}$$

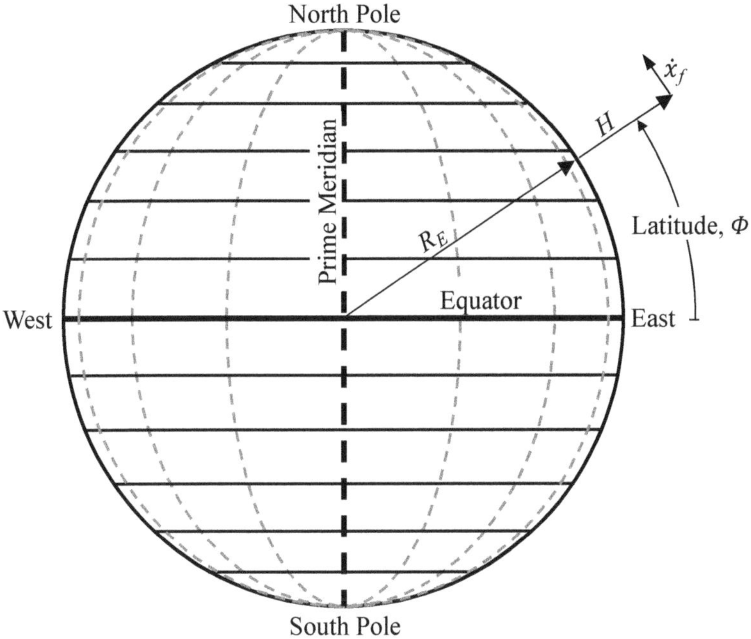

Figure 7.5.1. Geographic coordinates. Solid lines represent lines of constant latitude.

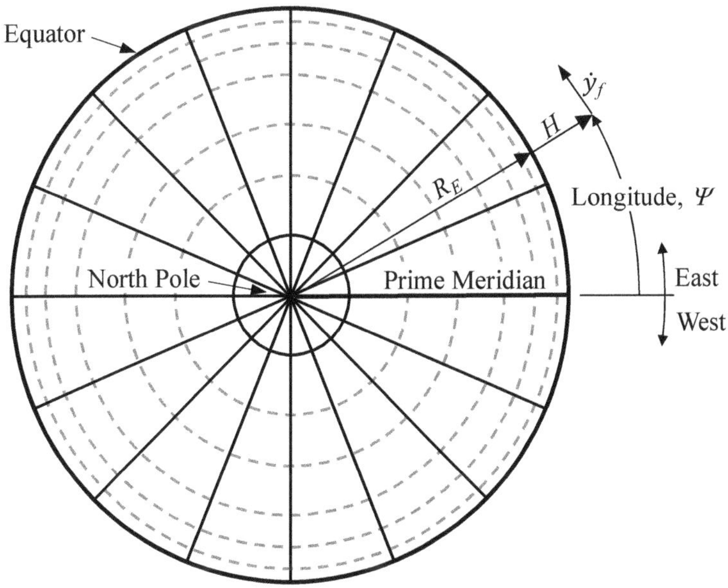

Figure 7.5.2. Geographic coordinates. Solid lines represent lines of constant longitude.

The difference between the polar radius and equatorial radius is about 21 km. Additionally, because mass is not distributed uniformly within the Earth, the gravity and mean-sea-level radius varies with position on the surface of the Earth. Multiple geographic models have been used in flight simulation to represent the shape of the Earth. Here we will discuss two commonly-used geographic models, which are the spherical-Earth approximation, and the ellipsoidal-Earth approximation. Higher-fidelity geographic models of the Earth are also sometimes used. However, given the accuracy of the aerodynamic models used in most flight simulators, higher fidelity models of the Earth are somewhat difficult to justify for simulation of typical aircraft.

Note that the equations of motion developed in Chapter 5 were developed relative to the flat Earth-fixed inertial coordinate system. These equations are appropriate when the distance traveled by the vehicle during a single time step is very small compared to the radius of the Earth. This is true for most typical aircraft. However, for vehicles at very high altitudes or that approach orbital speeds, such as launch vehicles or satellites, equations of motion relative to the Earth-centered inertial coordinate system that include Coriolis and centripetal accelerations due to the Earth's rotation, as well as gravitational effects from the sun, moon, or other planets, must be used.

The relations developed in this section can be used to relate changes in the Earth-fixed coordinate system $(\dot{x}_f, \dot{y}_f, \dot{z}_f)$ to changes in the geographic coordinate system including latitude, longitude, and altitude. These relations are independent of the physics model used in the simulation, and are strictly due to the geometric relation between an Earth-fixed coordinate system and the geographic coordinate system. In the Earth-fixed coordinate system, lines of zero acceleration are straight lines. However, in a geographic coordinate system, a path of zero acceleration is a curved line called a *geodesic*. The relations between the Earth-fixed Cartesian coordinates and the geographic coordinate system depend on the model used for the Earth.

Spherical-Earth Approximation

The difference between the currently accepted equatorial radius and polar radius is about 0.34%. Therefore, a close approximation for the Earth can be obtained by using a perfect sphere. Such a model significantly simplifies the transformation between the Earth-fixed coordinate system $(\dot{x}_f, \dot{y}_f, \dot{z}_f)$ and the geographic coordinates Φ, Ψ, and H.

The length of one meter was historically defined as $1/10,000,000$ times the length of the meridian arc from the North Pole to the equator. Therefore, the polar circumference of the Earth is about $40,000,000$ meters. One nautical mile (nm) was historically defined to be the length of the arc subtended by one minute of latitude. Because there are 60 minutes in one degree, this yields

$$1 \text{ nm} \equiv \frac{40,000,000 \text{ m}}{(60 \text{ min})(360 \text{ deg})} = 1,851.\overline{851} \text{ m} \tag{7.5.5}$$

The definition for one nautical mile is currently accepted to be exactly 1,852 meters. Using this as the definition for one nautical mile gives a circumference of the Earth of

$$(1852 \text{ m})(60 \text{ min})(360 \text{ deg}) = 40,003,200 \text{ m} \tag{7.5.6}$$

This corresponds to a radius of

$$R_E \approx \frac{40,003,200\text{m}}{2\pi} = 6,366,707.01949371 \text{ m} \tag{7.5.7}$$

Because this radius, based on the currently accepted definition of the nautical mile, is between the actual polar radius given in Eq. (7.5.3) and the equatorial radius given in Eq. (7.5.4), it is often used as the accepted radius of the Earth spherical model.

The arc length along the surface of a circle is equal to the radius of the circle multiplied by the subtended angle. Because the Earth-fixed coordinate system was defined with x_f oriented towards the north, a change in x_f at an altitude of H is related to a change in latitude according to

$$\Delta x_f = \Delta\Phi(R_E + H) \tag{7.5.8}$$

Differentiating with respect to time at constant altitude yields

$$\dot{x}_f = \dot{\Phi}(R_E + H) \tag{7.5.9}$$

Rearranging gives

$$\dot{\Phi} = \frac{\dot{x}_f}{(R_E + H)} \tag{7.5.10}$$

Because y_f is oriented towards the east, a change in y_f at an altitude of H and latitude of Φ is related to a change in longitude according to

$$\Delta y_f = \Delta\Psi(R_E + H)\cos\Phi \tag{7.5.11}$$

Differentiating with respect to time at constant altitude and latitude yields the relation

$$\dot{y}_f = \dot{\Psi}(R_E + H)\cos\Phi \tag{7.5.12}$$

Rearranging gives

$$\dot{\Psi} = \frac{\dot{y}_f}{(R_E + H)\cos\Phi} \tag{7.5.13}$$

Because the z_f coordinate is oriented towards the center of the Earth, the change in altitude with respect to time is simply

$$\dot{H} = -\dot{z}_f \tag{7.5.14}$$

Combining Eqs. (7.5.10), (7.5.13), and (7.5.14) yields the differential equations

$$\begin{bmatrix} \dot{\Phi} \\ \dot{\Psi} \\ \dot{H} \end{bmatrix} = \begin{bmatrix} \dot{x}_f/(R_E + H) \\ \dot{y}_f/[(R_E + H)\cos\Phi] \\ -\dot{z}_f \end{bmatrix} \tag{7.5.15}$$

The relationships given in Eq. (7.5.15) are exact for the spherical-Earth approximation.

In general, as a vehicle traverses the surface of the Earth, the geographic *bearing* or *ground-track angle* does not remain constant, even for a vehicle with zero acceleration.

Recall that the geodesic is a curve representing a path of zero acceleration on the surface of the Earth. As the vehicle traverses the surface of the Earth, the orientation of the geodesic relative to the local meridian changes. This is true except in the special case when the vehicle is moving directly east or west along the equator, or when the vehicle is moving directly north or south. The geodesic change in bearing will be denoted here as $\Delta\psi_g$. For the spherical-Earth approximation, the geodesic change in bearing between two points on the surface of the Earth can be approximated as [78]

$$\Delta\psi_g \approx (\Psi_2 - \Psi_1)\sin\left(\frac{\Phi_1 + \Phi_2}{2}\right) \tag{7.5.16}$$

Rather than incorporating the differential equations given in Eq. (7.5.15) into a numerical integration scheme, Phillips suggests applying the algorithm given in Algorithm 7.5.6 at the end of each time step, which he states is "mathematically exact" for a constant-curvature spherical-Earth model over a single time step. The step sizes in Earth-fixed coordinates Δx_f, Δy_f, and Δz_f over the time step, as well as the geographic coordinates Φ_1, Ψ_1, and H_1 at the beginning of the time step are the inputs to the algorithm. The algorithm yields the geographic coordinates at the end of the time step Φ_2, Ψ_2, and H_2, as well as the geodesic change in bearing $\Delta\psi_g$. The tolerance ε_t is a number on the order of the square root of machine precision, and is used to avoid an indeterminate calculation for vertical flight.

Recall that the Earth-fixed coordinate system is oriented with the x_f axis to the north, y_f axis to the east, and z_f axis towards the center of the Earth as shown in Fig. 1.1.1. Because of the curvature of the Earth, in general the orientation of the local Earth-fixed coordinate system relative to the vehicle will be different at the beginning and end of the integration time step. The change in the orientation of the vehicle relative to the local Earth-fixed coordinate system is captured in the geodesic change in bearing $\Delta\psi_g$. Therefore, at the end of each time step, the orientation of the vehicle relative to the local Earth-fixed coordinate system must be updated using the geodesic change in bearing $\Delta\psi_g$. This can be done by first computing the Euler angles from the current quaternion formulation, adding the geodesic change in bearing $\Delta\psi_g$ to the Euler azimuth angle ψ, and converting the result back to the quaternion formulation. However, a more efficient method is to simply rotate the quaternion by the azimuth angle $\Delta\psi_g$ as suggested by Phillips [78]. This yields

$$\begin{bmatrix} e_0 \\ e_x \\ e_y \\ e_z \end{bmatrix} = \cos(\Delta\psi_g/2)\begin{bmatrix} e_0 \\ e_x \\ e_y \\ e_z \end{bmatrix} + \sin(\Delta\psi_g/2)\begin{bmatrix} -e_z \\ -e_y \\ e_x \\ e_0 \end{bmatrix} \tag{7.5.17}$$

Equation (7.5.17) should be applied at the end of each time step. Additionally, the latitude and longitude values should be limited at the end of each time step to reflect the traditional limits given in Eqs. (7.5.1) and (7.5.2).

Most modern computers currently use double-precision for computation, which has about 14–15 digits of accuracy. For flat-Earth simulations, this double-precision accuracy

Algorithm 7.5.6 Spherical-Earth geographic solution for a single time step.

Input
- Δx_f (real) Step size in Earth-fixed x_f coordinate
- Δy_f (real) Step size in Earth-fixed y_f coordinate
- Δz_f (real) Step size in Earth-fixed z_f coordinate
- Φ_1 (real) Geographic latitude at beginning of step
- Ψ_1 (real) Geographic longitude at beginning of step
- H_1 (real) Geographic altitude at beginning of step
- ε_t (real) tolerance on the order of the square root of machine precision

$d = \sqrt{\Delta x_f^2 + \Delta y_f^2}$

if $(d < \varepsilon_t)$ **then**

    $\Phi_2 = \Phi_1$

    $\Psi_2 = \Psi_1$

    $\Delta\psi_g = 0$

else

    $\Theta = d/(R_E + H_1 - \Delta z_f/2)$

    $\psi_{g1} = \text{atan2}(\Delta y_f, \Delta x_f)$

    $\hat{x} = \cos\Phi_1 \cos\Theta - \sin\Phi_1 \sin\Theta \cos\psi_{g1}$

    $\hat{y} = \sin\Theta \sin\psi_{g1}$

    $\hat{z} = \sin\Phi_1 \cos\Theta + \cos\Phi_1 \sin\Theta \cos\psi_{g1}$

    $\hat{x}' = -\cos\Phi_1 \sin\Theta - \sin\Phi_1 \cos\Theta \cos\psi_{g1}$

    $\hat{y}' = \cos\Theta \sin\psi_{g1}$

    $\hat{z}' = -\sin\Phi_1 \sin\Theta + \cos\Phi_1 \cos\Theta \cos\psi_{g1}$

    $\hat{r} = \sqrt{\hat{x}^2 + \hat{y}^2}$

    $\Phi_2 = \text{atan2}(\hat{z}, \hat{r})$

    $\Psi_2 = \Psi_1 + \text{atan2}(\hat{y}, \hat{x})$

    $\hat{C} = \hat{x}^2 \hat{z}'$

    $\hat{S} = (\hat{x}\hat{y}' - \hat{y}\hat{x}') \cos^2\Phi_2 \cos^2(\Psi_2 - \Psi_1)$

    $\Delta\psi_g = \text{atan2}(\hat{S}, \hat{C}) - \psi_{g1}$

end if

$H_2 = H_1 - \Delta z_f$

is preserved throughout the simulation. However, computing the location of a vehicle in geographic coordinates requires computing small differences between large numbers in a single time step. The radius of the Earth is on the order of 10^8 ft. If double-precision computation is used, which retains about 14–15 digits of accuracy, this provides an accuracy on the surface of the Earth of about 10^{-6} ft. This is very nearly equivalent to single-precision accuracy on the surface of the Earth. Therefore, if geographic coordinates are needed, **a minimum of double-precision accuracy in the computations should be used in the simulation.**

As was discussed in Chapter 5, the accuracy of numerical integration schemes, such as those used in flight simulation, depends on the time step and how rapidly the state of the vehicle is changing. For double-precision computations using the Fourth-order Runge-Kutta integration routine when geographic coordinates are needed, Phillips [78] provides the rule of thumb

$$0.1 \text{ ft} < V\Delta t < 10.0 \text{ ft} \tag{7.5.18}$$

which he suggests is a good balance between the round-off error for smaller time steps using double-precision computations, and the truncation error from the fourth-order integration scheme resulting from larger time steps.

EXAMPLE 7.5.1 Consider a vehicle flying at an altitude of 60,000 ft at a speed of Mach 6.8. At this speed and altitude, it should take the vehicle a little less than 20,000 seconds (about 5.6 hours) to circumnavigate the globe. Use the spherical-Earth model to compute the geographic coordinates of the vehicle as a function of time using an appropriate time step for the integration. Begin the simulation with the vehicle located where the Prime Meridian crosses the Equator at the geographic coordinates of $\Phi = \Psi = 0$ and a heading of 45 deg. Run the simulation for a total 20,000 seconds and plot the geographic coordinates along with the geographic azimuth angle of the vehicle as a function of time.

Solution: Using standard-day atmospheric conditions, the speed of sound at an altitude of 60,000 ft is 968.0756480696 ft/s. Hence, the velocity at Mach 6.8 is

$$V = (968.0756480696)(6.8) = 6{,}582.91440687328 \text{ ft/s}$$

According to Eq. (7.5.18), the largest time step we can use and still expect reasonable accuracy in geographic coordinates is

$$\Delta t = \frac{10.0}{V} = \frac{10.0}{6{,}582.91440687328} = 0.001519084$$

Hence, for this example we will select a time step of $\Delta t = 0.0015$. Using these flight conditions along with Algorithm 7.5.6 gives the results in Table 7.5.1, which include only a few selected time steps for the first 100 seconds. Note that if an algorithm is used to trim the vehicle at this flight condition as was outlined in Chapters 6 and 7, the error in the trim solution must be driven as low as possible. For double-precision computations, an error tolerance of 1.0E−14 should be used.

Table 7.5.1. Solutions at selected time steps for Example 7.5.1.

Time [s]	Latitude [deg]	Longitude [deg]	Azimuth [deg]
0.0000	0.00000000000E+00	0.00000000000E+00	45.00000000000
0.0015	1.90972793883E−05	1.90972793883E−05	45.00000000000
0.0030	3.81945587766E−05	3.81945587766E−05	45.00000000001
0.1005	1.27951771891E−03	1.27951771923E−03	45.00000001429
1.0005	1.27378852471E−02	1.27378855618E−02	45.00000141593
10.0005	1.27321456894E−01	1.27321771257E−01	45.000141466
99.9990	1.27303444098E+00	1.27334878561E+00	45.014147205

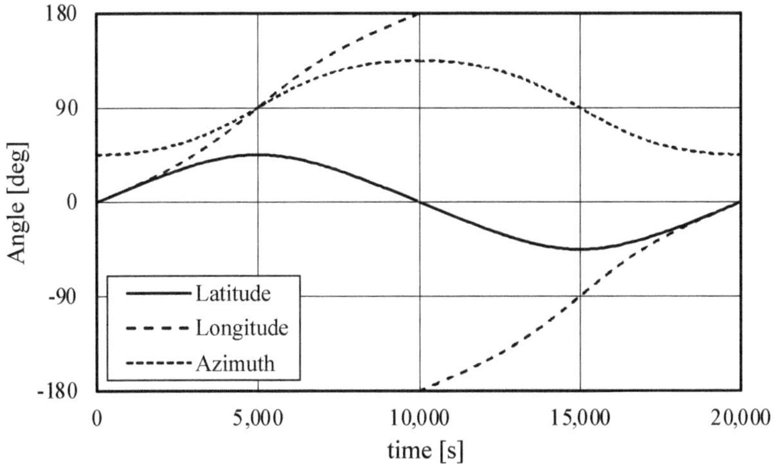

Figure 7.5.3. Geographic angles for the aircraft in Example 7.5.1.

Running this simulation for a total time of 20,000 seconds yields the results shown in Figs. 7.5.3 and 7.5.4. It is worth taking time to understand the information in these figures and to convince oneself that the relationships shown accurately depict the geographic angles of the vehicle as it traverses the Earth along a path of zero acceleration. Note that Fig. 7.5.4 shows the latitude and azimuth angles as a function of longitudinal angle, and that the simulation begins with the vehicle at the center of the x axis, at the point of $\Psi = 0$.

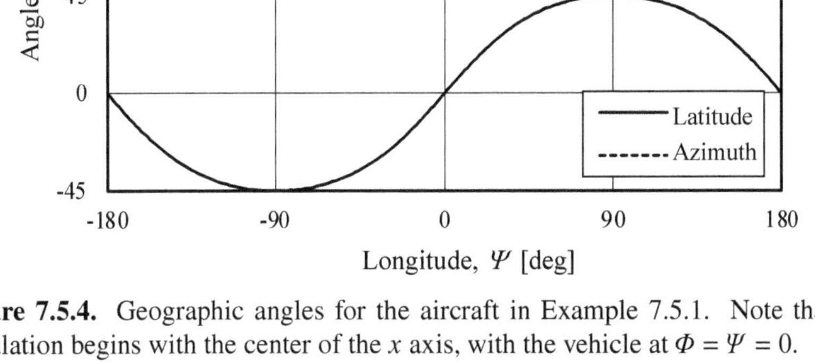

Figure 7.5.4. Geographic angles for the aircraft in Example 7.5.1. Note that the simulation begins with the center of the x axis, with the vehicle at $\Phi = \Psi = 0$.

EXAMPLE 7.5.2 We wish to consider the numerical precision of the spherical-Earth model given in Algorithm 7.5.6. To do so, we will consider a specific flight path that circumnavigates the Earth, and divide the circumference of the flight path into an integer number of steps. This setup allows us to evaluate the precision of the numerical algorithm after one complete revolution. Consider a vehicle flying at an altitude of 56,644.1214783759 ft. This altitude was specifically chosen such that the circumference of a flight path circumnavigating the spherical Earth is exactly 131,600,000 ft. According to the rule of thumb suggested in Eq. (7.5.18), the largest spatial step taken in each time step should be 10 ft. Here we will set the vehicle horizontal speed to 1,000 ft/s and use a time step of $\Delta t = 0.01$. With this setup, it should take exactly 13,160,000 integration steps to model the vehicle circumnavigating the Earth once. Use the spherical-Earth model to compute the geographic coordinates of the vehicle as a function of time. Begin the simulation with the vehicle located where the Prime Meridian crosses the Equator at the geographic coordinates of $\Phi = \Psi = 0$ and a heading of 45 deg. Run the simulation for a total of 131,600.00 seconds. Report the latitude, longitude, and azimuth angle of the vehicle at times $t = 0.0$, 32,900, 65,800, 98,700, and 131,600. The final time of $t = 131,600$ represents the time when the vehicle will have circumnavigated the Earth exactly once. Comment on the precision of the results.

Solution: A vehicle was trimmed to fly at the prescribed airspeed and altitude. The initial position of the vehicle, as well as the time step and total time of the simulation were also set as prescribed. Using these flight conditions and simulation settings along with Algorithm 7.5.6 gives the results shown in Table 7.5.2 out to about 7 significant figures at the listed time steps.

Results show that the vehicle returned to the original location after a single circumnavigation of the Earth to within about 7 digits of precision, which is near single precision. Repeating the process using quad precision yields results that match out to double precision after a single circumnavigation of the Earth.

Table 7.5.2. Solutions at selected time steps for Example 7.5.2.

Time [s]	Latitude [deg]	Longitude [deg]	Azimuth [deg]
0.0	0.0	0.0	45.0
32,900.0	45.0	90.0	90.0
65,800.0	0.0	180.0	135.0
98,700.0	-45.0	-90.0	90.0
131,600.0	0.0	0.0	45.0

Ellipsoidal-Earth Approximation

The Earth is not a perfect sphere, and can be better approximated as an ellipsoid. With the polar radius given in Eq. (7.5.3) and the equatorial radius given in Eq. (7.5.4), the eccentricity ε_E of the ellipsoid is

$$\varepsilon_E^2 \equiv 1 - \left(\frac{R_p}{R_e}\right)^2 = 0.0066943850 \tag{7.5.19}$$

Because the eccentricity is so small, a figure of the ellipsoidal model of the Earth would be virtually indistinguishable from a perfect sphere. To better understand this model, Fig. 7.5.5 shows a quarter of the ellipsoid with the eccentricity greatly exaggerated. For a perfect sphere, there is only one definition of latitude. However, for an ellipsoidal-Earth, there are two definitions for latitude. The angle between the equatorial plane and a line from the center of the Earth to the point of interest is called the *geocentric latitude*. This is shown in Fig. 7.5.5, and is the definition of latitude often used in astrodynamics. The angle between the equatorial plane and a line perpendicular to the surface of the ellipsoid at the point of interest is called the *geodetic latitude*. This is also shown in Fig. 7.5.5 and is the latitude typically displayed on maps and used in flight simulation and aircraft navigation. Here we will refer to the geodetic latitude simply as *latitude*.

At mean-sea level, the change in latitude, longitude, and altitude for an ellipsoidal Earth can be computed from [78]

$$\begin{bmatrix} \dot{\Phi} \\ \dot{\Psi} \\ \dot{H} \end{bmatrix} = \begin{bmatrix} \dot{x}_f(1 - \varepsilon_E^2 \sin^2\Phi)^{3/2}/[R_e(1 - \varepsilon_E^2)] \\ \dot{y}_f(1 - \varepsilon_E^2 \sin^2\Phi)^{1/2}/(R_e \cos\Phi) \\ -\dot{z}_f \end{bmatrix} \tag{7.5.20}$$

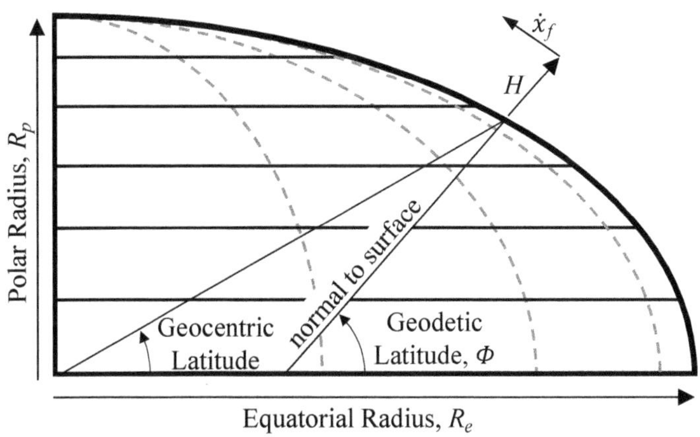

Figure 7.5.5. Geographic coordinates for the ellipsoid-Earth approximation. Solid lines represent lines of constant latitude.

Note that Eq. (7.5.20) is only accurate on the surface of the mean-sea-level ellipsoid. A surface offset from this ellipsoid by a constant altitude H is not a perfect ellipsoid, and differs in curvature from the ellipsoid at mean-sea level. At an altitude above mean-sea level of H, the change in latitude, longitude, and altitude can be computed from [78]

$$\begin{bmatrix} \dot{\Phi} \\ \dot{\Psi} \\ \dot{H} \end{bmatrix} = \begin{bmatrix} \dot{x}_f/(R_x + H) \\ \dot{y}_f/[(R_y + H)\cos\Phi] \\ -\dot{z}_f \end{bmatrix} \tag{7.5.21}$$

where R_x and R_y are the primary radii of curvature

$$R_x \equiv \frac{R_e(1-\varepsilon_E^2)}{(1-\varepsilon_E^2\sin^2\Phi)^{3/2}} \tag{7.5.22}$$

$$R_y \equiv \frac{R_e}{(1-\varepsilon_E^2\sin^2\Phi)^{1/2}} \tag{7.5.23}$$

The relationships given in Eq. (7.5.21) are exact for a vehicle at an altitude H relative to the ellipsoid-Earth approximation.

Algorithm 7.5.7 is a first-order approximation for the integration of Eq. (7.5.21) suggested by Phillips [78]. The steps sizes in Earth-fixed coordinates Δx_f, Δy_f, and Δz_f over the time step, as well as the geographic coordinates Φ_1, Ψ_1, and H_1 at the beginning of the time step are the inputs to the algorithm. The algorithm yields the geographic coordinates at the end of the time step Φ_2, Ψ_2, and H_2, as well as the geodesic change in bearing $\Delta\psi_g$. Again, the tolerance ε_t is a number on the order of the square root of machine precision, and is used to avoid an indeterminate calculation for vertical flight. This algorithm is most correctly applied at every step within a high-order integration scheme, such as a fourth-order Runge-Kutta integration method. However, if the distance traveled in a single complete time step is small compared to the local radii of curvature, this algorithm can be applied at the end of each complete time step with very good results. At the end of each time step, the geographic coordinates should be limited to reflect the traditional limits given in Eqs. (7.5.1) and (7.5.2).

In general, the geodesic change in bearing $\Delta\psi_g$ in a single time step is non-zero. Therefore, just as was the case for the spherical-Earth model, the orientation of the vehicle relative to the flat-Earth coordinate system must be updated at the end of each step. This can be done by rotating the quaternion by the azimuth angle $\Delta\psi_g$ using Eq. (7.5.17).

The equations and algorithm presented here for the ellipsoidal model can be used for any terrestrial body that resembles an ellipsoid by simply changing the primary polar and equatorial radii R_p and R_e to match that of the terrestrial body of interest. Note that for the special case when the polar and equatorial radii are equivalent, the eccentricity $\varepsilon_E = 0$, and Eq. (7.5.21) reduces to Eq. (7.5.15).

Algorithm 7.5.7 Ellipsoidal-Earth geographic first-order approximation for a single time step.

Input
- Δx_f — (real) Step size in Earth-fixed x_f coordinate
- Δy_f — (real) Step size in Earth-fixed y_f coordinate
- Δz_f — (real) Step size in Earth-fixed z_f coordinate
- Φ_1 — (real) Geographic latitude at beginning of step
- Ψ_1 — (real) Geographic longitude at beginning of step
- H_1 — (real) Geographic altitude at beginning of step
- ε_t — (real) tolerance on the order of the square root of machine precision

$d = \sqrt{\Delta x_f^2 + \Delta y_f^2}$

if $(d < \varepsilon_t)$ **then**

    $\Phi_2 = \Phi_1$

    $\Psi_2 = \Psi_1$

    $\Delta\psi_g = 0$

else

    $R_x = R_e(1 - \varepsilon_E^2)/[(1 - \varepsilon_E^2 \sin^2 \Phi_1)^{3/2}]$

    $R_y = R_e/[(1 - \varepsilon_E^2 \sin^2 \Phi_1)^{1/2}]$

    $\Theta_x = \Delta x_f/(R_x + H_1 - \Delta z_f/2)$

    $\Theta_y = \Delta y_f/(R_y + H_1 - \Delta z_f/2)$

    $\hat{x} = (1 - \varepsilon_E^2)[\cos(\Phi_1 + \Theta_x) - \cos \Phi_1] + (1 - \varepsilon_E^2 \sin^2 \Phi_1) \cos \Theta_y \cos \Phi_1$

    $\hat{y} = (1 - \varepsilon_E^2 \sin^2 \Phi_1) \sin \Theta_y$

    $\hat{z} = (1 - \varepsilon_E^2)[\sin(\Phi_1 + \Theta_x) - \sin \Phi_1] + (1 - \varepsilon_E^2 \sin^2 \Phi_1)(\cos \Theta_y - \varepsilon_E^2) \sin \Phi_1$

    $\hat{r} = \sqrt{\hat{x}^2 + \hat{y}^2}$

    $\Phi_2 = \text{atan2}[\hat{z}, (1 - \varepsilon_E^2)\hat{r}]$

    $\Psi_2 = \Psi_1 + \text{atan2}(\hat{y}, \hat{x})$

    $\Delta\psi_g = (\Psi_2 - \Psi_1) \sin[(\Phi_2 + \Phi_1)/2](1 - \varepsilon_E^2)/(1 - \varepsilon_E^2 \sin^2 \Phi_1)$

end if

$H_2 = H_1 - \Delta z_f$

EXAMPLE 7.5.3 Repeat Example 7.5.1 using the first-order approximation for the ellipsoidal-Earth given in Algorithm 7.5.7. Compare the results to those in Example 7.5.1.

Solution: Using the identical initial vehicle state, velocity, and integration step size along with Algorithm 7.5.7 to compute the geographic coordinates at the end of each time step yields results that are indistinguishable from those shown in Figs. 7.5.3 and 7.5.4 when compared on the same plot. However, the results are not identical out to precision. Table 7.5.3 shows example solutions at a few selected time steps for the first 100 seconds, which are slightly different than those reported in Example 7.5.1.

Table 7.5.3. Solutions at selected time steps for Example 7.5.3.

Time [s]	Latitude [deg]	Longitude [deg]	Azimuth [deg]
0.0000	0.00000000000E+00	0.00000000000E+00	45.00000000000
0.0015	1.91912623060E−05	1.90631559229E−05	45.00000000000
0.0030	3.83825246119E−05	3.81263118458E−05	45.00000000001
0.0045	5.75737869179E−05	5.71894677688E−05	45.00000000003
0.1005	1.28581457439E−03	1.27723144705E−03	45.00000001424
1.0005	1.28005718509E−02	1.27151252100E−02	45.00000141085
10.0005	1.27948038739E−01	1.27094269690E−01	45.00014095816
99.9990	1.27929764005E+00	1.27107348777E+00	45.01409642870

7.6 Navigation

If a vehicle follows a path of zero acceleration at an altitude H above the surface of the Earth, the vehicle will follow a path called a *geodesic*. As discussed in the previous section, along the path of a geodesic, the ground-track angle or bearing constantly changes, except for the special case when the vehicle is traveling due north or south, or when the vehicle is traveling due east or west along the equator. A path of constant ground-track angle or bearing is called a *loxodrome*. Except for the north-south and equatorial east-west case mentioned, a loxodrome is a spiral of theoretically infinite length that asymptotically approaches the poles. As it approaches the pole, the spiral tightens.

The shortest distance between two points on the surface of the Earth is a line passing through the surface of the Earth. However, if a vehicle is constrained to remain on the external surface of the Earth, the shortest distance between two points on the surface is a geodesic. For the spherical-Earth model, the geodesic is a segment of what is called a *great circle*. A great circle is the intersection of the surface of the spherical Earth with a

Algorithm 7.6.8 Spherical-Earth great-circle navigation.

Input
- Φ (real) Geographic latitude at beginning of navigation
- Ψ (real) Geographic longitude at beginning of navigation
- Φ_d (real) Geographic latitude at destination
- Ψ_d (real) Geographic longitude at destination
- H (real) Constant geographic altitude during navigation

$\Theta = 2\sin^{-1}\left[\sqrt{\sin^2\left(\frac{\Phi_d-\Phi}{2}\right) + \cos\Phi\cos\Phi_d\sin^2\left(\frac{\Psi_d-\Psi}{2}\right)}\right]$

$\psi_g = \cos^{-1}[(\sin\Phi_d - \sin\Phi\cos\Theta)/(\sin\Theta\cos\Phi)]$

if $[\sin(\Psi_d - \Psi) \leq 0.0]$ $\psi_g = 2\pi - \psi_g$

$d = (R_E + H)\Theta$

plane passing through the center of the Earth. In order to define a plane, three points are needed. To define the plane for the great circle, the three points are the beginning location, the ending location, and the center of the Earth. These three points uniquely define the great circle, and therefore, the geodesic, that represents the shortest distance along the surface of the Earth between two points.

For the spherical-Earth approximation, Algorithm 7.6.8 [78] can be used to compute the initial ground-track angle ψ_g or bearing from an initial position (Φ, Ψ) on the surface of the Earth to the destination (Φ_d, Ψ_d). The algorithm also yields the distance d between the two points if a path of constant altitude H is followed.

For very short distances along the surface of the Earth, *rhumb-line navigation* is sometimes used, which is a path of constant bearing along a loxodrome. For the spherical-Earth approximation, Algorithm 7.6.9 [78] can be used to compute the initial ground-track angle ψ_g or bearing from an initial position (Φ, Ψ) on the surface of the Earth to the destination (Φ_d, Ψ_d). The algorithm also yields the distance d between the two points if a path of constant altitude H is followed.

Algorithm 7.6.9 Spherical-Earth rhumb-line navigation.

Input
- Φ — (real) Geographic latitude at beginning of navigation
- Ψ — (real) Geographic longitude at beginning of navigation
- Φ_d — (real) Geographic latitude at destination
- Ψ_d — (real) Geographic longitude at destination
- H — (real) Constant geographic altitude during navigation

$\Delta\Psi_E = \mathrm{mod}(\Psi_d - \Psi, 2\pi)$ — mod is the modulus operator
if $(\Delta\Psi_E < 0.0)$ $\Delta\Psi_E = \Delta\Psi_E + 2\pi$
$\Delta\Psi_W = \mathrm{mod}(\Psi - \Psi_d, 2\pi)$ — mod is the modulus operator
if $(\Delta\Psi_W < 0.0)$ $\Delta\Psi_W = \Delta\Psi_W + 2\pi$
$\Delta\Phi = \ln\left[\tan(\Phi_d/2 + \pi/4)/\tan(\Phi/2 + \pi/4)\right]$
if $(|\Phi_d - \Phi| < \varepsilon_t)$ **then**
    $r = \cos\Phi$
else
    $r = (\Phi_d - \Phi)/\Delta\Phi$
end if
if $(\Delta\Psi_W < \Delta\Psi_E)$ **then**
    $\psi_g = \mathrm{atan2}(-\Delta\Psi_W, \Delta\Phi)$
    $d = (R_E + H)\sqrt{(r\Delta\Psi_W)^2 + (\Phi_d - \Phi)^2}$
else
    $\psi_g = \mathrm{atan2}(\Delta\Psi_E, \Delta\Phi)$
    $d = (R_E + H)\sqrt{(r\Delta\Psi_E)^2 + (\Phi_d - \Phi)^2}$
end if
if $(\psi_g < 0.0)$ $\psi_g = \psi_g + 2\pi$

EXAMPLE 7.6.1 The John F. Kennedy (JFK) Airport in New York is located at $(\Phi, \Psi) = (40.643024, -73.77434)$ deg, and the London-Heathrow Airport in the United Kingdom is located at $(\Phi, \Psi) = (51.47002, -0.454295)$ deg. Compute the initial bearing angle for both great-circle and rhumb-line navigation from the JFK Airport to the London-Heathrow Airport. If an aircraft is to travel between the two airports at an altitude of 30,000 ft, compute the total distance the aircraft must fly based on both great-circle and rhumb-line navigation.

Solution: Using Algorithm 7.6.8 to compute the initial heading based on great-circle navigation yields

$$\Theta = 0.8695092559632 = 49.8192106143 \text{ deg}$$

$$\psi_g = 0.896306439429 = 51.3545761297 \text{ deg}$$

This yields a total distance along great-circle navigation of

$$d = 18,188,521.90319 \text{ ft} = 3,444.795814998 \text{ miles}$$

Using Algorithm 7.6.9 to compute the initial heading based on rhumb-line navigation yields

$$\Delta\Psi_E = 1.27967619296 = 73.320045 \text{ deg}$$

$$\Delta\Psi_W = 5.00350911422 = 286.679955 \text{ deg}$$

$$\Delta\Phi = 0.273595602002 = 15.6758732881 \text{ deg}$$

$$r = 0.69067896895$$

$$\psi_g = 1.36016680829 = 77.9318175491 \text{ deg}$$

This yields a total distance along rhumb-line navigation of

$$d = 18,906,245.80926 \text{ ft} = 3,580.72837297 \text{ miles}$$

These results show that the initial heading specified by the two types of navigation differ by about 26.6 deg. Using great-circle navigation results in a total distance that is about 136 miles less than that from rhumb-line navigation.

7.7 Problems

Theory Problems

7.T1 Derive the quaternion form of Eq. (7.2.5).

7.T2 Derive the quaternion form of Eq. (7.2.8).

7.T3 Beginning with Eq. (7.2.8), show the derivation of Eq. (7.2.12).

7.T4 Starting from the definition of climb rate given in Eq. (7.2.1), show that for the case of zero wind, zero sideslip angle, and zero bank angle

$$\theta = \gamma + \alpha$$

7.T5 Beginning with the definition of the load-factor vector given in Eq. (7.3.2), use the coordinate-transformations between the body-fixed and wind coordinate systems outlined in Chapter 3 to show that the load factor in the direction of lift is given in Eq. (7.3.4).

7.T6 Use Eqs. (7.3.1) and (7.3.4) to show that in a coordinated turn, the bank angle is related to the load factor according to Eq. (7.3.5).

7.T7 Beginning with the definition of the load factor given in Eq. (7.3.4), develop the traditional relationship for the load factor given in Eq. (7.3.8). State what assumptions must be made in order for Eq. (7.3.8) to be true.

7.T8 Beginning with Eqs. (6.1.3) and (7.4.4), develop an algorithm that can be used to compute the initial azimuth angle given a user-specified ground-track angle and bank angle along with a specified wind velocity vector.

7.T9 Beginning with Eqs. (6.1.3) and (7.4.4), develop an algorithm that can be used to compute the initial bank angle given a user-specified ground-track angle and azimuth angle along with a specified wind velocity vector.

Coding Problems

7.C1 Modify your trim code developed for the problems in Chapter 6 so that the user can specify a relative climb angle rather than an elevation angle in the JSON file as follows:

```
{
    "initial" : {
        "trim" : {
            "relative_climb_angle[deg]" : 10.0 // add this option
        }
    }
}
```

Set up your code so that by default it uses the specified relative climb angle if available. You can test for the existence of a key in a JSON dictionary in most coding languages.

7.C2 Test your code developed for Problem 7.C1 using all the cases given in Examples 7.2.2, 7.2.3, and 7.2.4. This will test your code for both steady coordinated turn and steady-heading sideslip cases. Intermediate solution steps can be found on the book website.

7.C3 Consider the F-16 aircraft specified in Appendix C.3 with the aerodynamic model F16-E-1 given in Table C.3.15. For the case when the vehicle is at an altitude of 2,000 ft and a velocity of 520 ft/s, compute the aerodynamic angles, vehicle rotation rates, control-surface deflections, and throttle setting for the aircraft in a steady coordinated turn with relative climb angles and bank angles specified in the following two cases:

Case	γ_r [deg]	ϕ [deg]
A	0.0	80.0
B	20.0	80.0

Assuming the aircraft begins flight in these trimmed positions at a location of $x = y = 0$, plot the Earth-fixed x and y coordinates of the aircraft for 20 seconds of flight using a fixed time step of $\Delta t = 0.01$ for both cases on a single plot. Comment on the differences and explain the physics that drives the differences between the two solutions.

7.C4 Modify your trim code so that the user can specify a **fixed climb angle** rather than an elevation angle in the JSON file as follows:

```
{
    "initial" : {
        "trim" : {
            "fixed_climb_angle[deg]" : 10.0 // add this option
        }
    }
}
```

Set up your code so that by default it uses the specified fixed climb angle if available. Repeat Problems 7.C1–7.C3 using the fixed climb angle rather than the relative climb angle.

7.C5 Consider the F-16 aircraft specified in Appendix C.3 with the aerodynamic model F16-E-1 given in Table C.3.15. Use your code to plot the required angle of attack and elevator deflection to trim the aircraft in steady-level flight as a function of speed at an altitude of 1,000 ft. Plot your results from the stall speed up to a speed of 2 times the stall speed. Assume the aircraft has a maximum lift coefficient of 1.4.

7.C6 The F-16 has the ability to deflect the aileron and rudder within the ranges $\text{abs}(\delta_a) \leq \pm 21.5$ deg and $\text{abs}(\delta_r) \leq \pm 30$ deg. Use your trim code to produce a minimum-control airspeed plot for the F-16 in a constant-altitude maneuver at an altitude of 500 ft.

7.C7 Modify your code so that it can correctly compute a trim solution for the case when the aircraft is ascending or descending vertically with all body-fixed angular rates set to zero.

7.C8 Use your code from Problem 7.C7 to estimate the trim state of an F-16 as it does a vertical climb just after takeoff at Hill Air Force Base. Assume it begins the vertical climb at an altitude of 4,774 feet and a velocity of 280 ft/s. Report the trim parameters. If the aircraft is trimmed in this state and the pilot does not touch the controls, compute the maximum altitude the aircraft will reach.

7.C9 Modify your code so that it can correctly compute a trim solution for the case of a vertical barrel roll.

7.C10 Modify your trim code so that the user can specify a load factor rather than a bank angle to initialize the vehicle in an steady coordinated turn. This option should be added in the JSON file as follows:

```
{
    "initial" : {
        "trim" : {
            "load_factor" : 2.0 // add this option
        }
    }
}
```

Set up your code so that by default it uses the specified load factor if available.

7.C11 Test your code developed for Problem 7.C10 using all the cases given in Examples 7.3.1 and 7.3.2. Intermediate solution steps can be found on the book website.

7.C12 Modify your trim code such that the user can specify that the negative bank angle solution be found that satisfies a user-specified load factor rather than the positive bank angle solution.

7.C13 Modify your trim code so that a ground-track angle can be specified rather than an azimuth angle for the initial condition of a vehicle in the JSON file as follows:

```
{
    "initial" : {
        "trim" : {
            "ground_track_angle[deg]" : 10.0 // add this option
        }
    }
}
```

Set up your code so that by default it uses the specified ground-track angle if available. Show that it works over a range of specified conditions.

7.C14 The F-16 is landing on runway 32 at Hill Air Force Base in a crosswind of 40 kts. The aircraft is traveling at a speed 10% above the stall speed and descending at a rate of 100 fpm. Assume the maximum lift coefficient of the F-16 in landing configuration is 1.4. Compute the heading angle and trim settings of the aircraft for the case of zero bank angle. Repeat these computations for the case of a 5° bank angle.

7.C15 Modify your code to accept geographic latitude and longitude coordinates for the initial location of the vehicle to be specified by the user in the JSON file as follows:

```
{
    "initial" : {
        "latitude[deg]" : 0.0,
        "longitude[deg]" : 0.0,
    }
}
```

Use Algorithm 7.5.6 to compute the geographic coordinates of the vehicle at the end of each time step using the spherical-Earth approximation. Print this information at each time step to an output file. Check your code by comparing the solutions to those given in Examples 7.5.1 and 7.5.2.

7.C16 Consider the F-16 aircraft specified in Appendix C.3 with the aerodynamic model F16-E-1 given in Table C.3.15. Consider the case when the vehicle is at an altitude of 30,000 ft, an airspeed of 550 ft/s, and located at latitude $\Phi = 80$ deg and longitude $\Psi = 90$ deg and an initial Euler azimuth angle of $\psi = 270$ deg. Find the trim state of the vehicle in a steady coordinated turn with a load factor of $\bar{n}_L = 2.2$ at constant altitude and positive bank angle. Using that trim state as an initial condition, propagate the vehicle forward in time using time steps of $\Delta t = 0.01$ for a total time of 50 seconds. Using the spherical-Earth model, report the latitude, longitude, and azimuth angle of the vehicle out to 8 significant digits at the times listed in the following table:

time [s]	Latitude Φ [deg]	Longitude Ψ [deg]	Azimuth ψ [deg]
0.0			
15.0			
30.0			
45.0			
60.0			

7.C17 Consider the F-16 aircraft specified in Appendix C.3 with the aerodynamic model F16-E-1 given in Table C.3.15. Consider the case when the vehicle is flying due North at an altitude of 30,000 ft, an airspeed of 550 ft/s, and located at latitude $\Phi = 89.99$ deg and longitude $\Psi = 45.0$ deg. Find the trim state of the vehicle in steady level flight. Using that trim state as an initial condition, propagate the vehicle forward in time using time steps of $\Delta t = 0.01$ for a total time of 10 seconds. Using the spherical-Earth model, report the latitude, longitude, and azimuth angle of the vehicle out to 8 significant digits at the times listed in the following table:

time [s]	Latitude Φ [deg]	Longitude Ψ [deg]	Azimuth ψ [deg]
0.0			
6.63			
6.64			
10.0			

7.C18 Repeat Problem 7.C15 using Algorithm 7.5.7 to compute the geographic coordinates using the ellipsoidal-Earth model. Check your code by comparing the solutions to those given in Example 7.5.3.

7.C19 Modify your code so that the user can specify whether the spherical-Earth model or the ellipsoidal-Earth model is used for the geographic coordinates by adding a `geographic_model` option in the `simulation` dictionary in the JSON file. Allow the user to specify "sphere", "ellipse", or "none" for the geographic model in the form

```
{
        "simulation": {
            "geographic_model" : "ellipse"
        }
{
```

Note that for the case of "none", the simulation should simply use a flat Earth model.

7.C20 Modify your code so that in addition to the initial geographic coordinates of the vehicle, the user can also specify a desired destination in geographic coordinates. Your code should compute the initial bearing angle for great-circle navigation to navigate from the initial location to the final destination.

7.C21 Consider the F-16 aircraft specified in Appendix C.3 with the aerodynamic model F16-AD-1 given in Table C.3.15. Repeat the cases in Examples 7.2.2–7.2.4 with the vehicle at an altitude of 3,000 ft and an airspeed of 350 ft/s. Use leading-edge flap and speed-brake deflections of 0.

7.C22 Consider the F-16 aircraft specified in Appendix C.3 with the aerodynamic model F16-AD-1 given in Table C.3.15. Repeat the cases in Example 7.3.2 with the vehicle at an altitude of 3,000 ft and an airspeed of 350 ft/s. Use leading-edge flap and speed-brake deflections of 0. For Case A, use a load factor of 3.0 instead of 5.0 to ensure the solution remains within the data table limits.

Chapter 8

Atmospheric Disturbances

8.1 Introduction

In the development of the equations of motion, we considered the wind as a vector with a coordinate system aligned with the Earth-fixed coordinate system, and with a magnitude that is independent of time

$$\mathbf{V}_{w_f} = \begin{bmatrix} V_{wx_f} \\ V_{wy_f} \\ V_{wz_f} \end{bmatrix} \tag{8.1.1}$$

This vector represents the velocity of the atmosphere relative to the Earth-fixed coordinate system. This treatment allowed us to use the atmospheric coordinate system as an inertial coordinate system, since it moves at a constant velocity relative to the Earth-fixed coordinate system. In reality, the wind is not constant in direction or magnitude relative to the Earth-fixed coordinate system. Aircraft are constantly exposed to disturbances in the atmosphere such as changing wind magnitude and direction, intermittent gusts, and thermals. Hence, the wind velocity, whether measured relative to a fixed point in Earth-fixed coordinates or relative to a vehicle moving through the atmosphere, is always changing with time. Including these types of disturbances in flight simulation can be important for assessing vehicle performance, control algorithms, aero-structural response, and pilot opinion of vehicle handling qualities.

Capturing the physics of the turbulent fluctuations in the atmosphere and their effect on the vehicle dynamics to a high degree of fidelity is extremely challenging. For example, many atmospheric disturbances have stochastic characteristics and are difficult to replicate in controlled testing environments or simulation. Furthermore, because the atmosphere is a three-dimensional fluid through which the vehicle is traversing, the variations in the atmosphere are not constant over the external surface of the vehicle.

High-fidelity simulations of atmospheric disturbances are rarely necessary for most flight-simulation scenarios. Instead, flight simulations typically employ approximate models that induce vehicle responses similar to those observed in flight. For example, rather than modeling spatially distributed disturbances over the vehicle's surface, these disturbances are often simply modeled as equivalent perturbations in translational and rotational flow quantities evaluated at the vehicle center of gravity. Furthermore, the variations are typically applied as disturbances to the aerodynamic properties of the vehicle, such that the mean wind vector is constant and can be used as an inertial coordinate system. Although these typical approaches are not exactly physically correct, they are sufficient for the purposes of most flight-simulation scenarios such as those built to test controller designs and/or vehicle handling qualities.

A significant amount of data has been collected to assess the atmospheric properties over a range of altitudes and various atmospheric disturbance models have been developed based on these data [79]. Disturbances in the atmosphere are often modeled as random processes with correlated spectral content. Hence, the fluctuations can be analyzed, categorized, and modeled using mathematical tools for stochastic processes and signal processing. Hence, in order to model atmospheric properties, one must first understand some basic principles of stochastic processes and signal processing.

Fundamentals of Stochastic Processes and Signal Processing

Consider a continuous signal as a function of time as shown in Fig. 8.1.1. This signal could represent many analog signals. However, for the purposes of this chapter, let this signal represent the wind speed at a specific geographic location as a function of time. We wish to analyze the properties of this signal using numerical methods. This can be done by first sampling the signal at specific instances in time over a finite time interval or duration. To simplify the following analysis, we will constrain our discussion to cases in which the sampling is uniformly spaced in time with a constant Δt between each sample, with a total number of samples N. The time between each sample is more commonly expressed as a sampling frequency

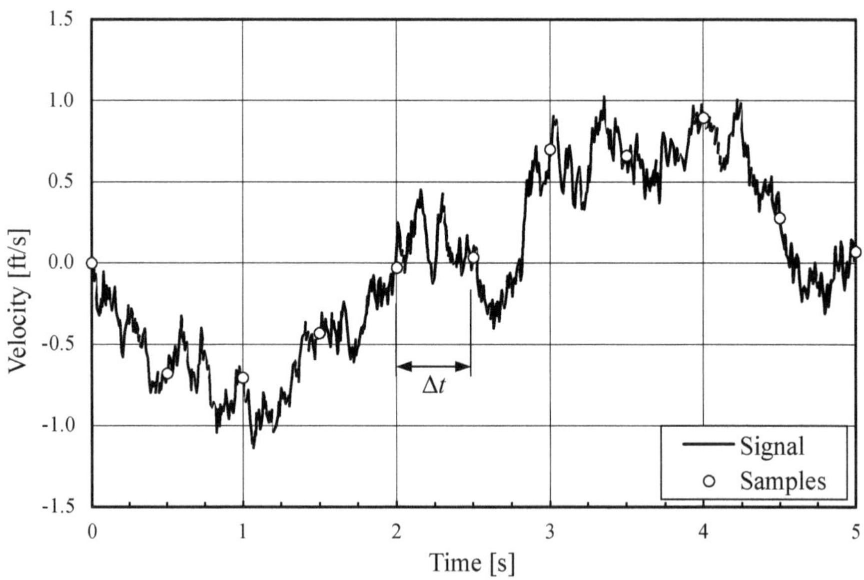

Figure 8.1.1. Example velocity fluctuations as a function of time at a single point in space.

$$\text{sampling frequency} \equiv f_s = \frac{1}{\Delta t} \tag{8.1.2}$$

which has the units of samples per second (Hz).

At this point, it may be natural to inquire at what rate the signal must be sampled, and how many points N are needed to ensure that our analysis of the sampled data points accurately represents the properties of the original signal. To select an appropriate sampling rate, it is helpful to understand that any function or signal can be represented as a superposition of sinusoidal components in the form

$$x(t) = \bar{x} + \sum_{i=1}^{\infty} A_i \sin(2\pi f_i t + \phi_i) \tag{8.1.3}$$

where $\bar{x}$ is the mean of the signal and A_i is the amplitude, f_i is the frequency, and ϕ_i is the phase of the ith sine wave[1]. The frequency f is in units of Hz, which is typical in the field of signal processing. Hence, f is multiplied by 2π radians per cycle to yield the proper units within the sine term. The highest frequency that can be uniquely represented in the sampled signal is a frequency at half the sampling rate or $f_s/2$. This frequency is called the *Nyquist frequency*. Therefore, the signal must be sampled at a rate at least twice the highest frequency to be captured in the analysis.

Since the time intervals between samples is constant, the number of samples N is directly related to the total time over which the signal is sampled. The duration over which the samples should be taken depends on the properties of the signal and types of information needed. A signal for which the statistical properties do not change with time is referred to as a *stationary* signal. True properties of stationary signals can be obtained by increasing the duration over which the signal is sampled until the properties converge. Properties of non-stationary signals must be obtained by ensemble averaging many instances of the signal. In the following analysis, we will assume the signals representing atmospheric turbulence are stationary. Hence, for ergodic stationary processes, ensemble averaging and time averaging will converge to the same result as N becomes large.

Sampled signals contain only a finite subset of the information contained in the underlying continuous signal. Statistical properties obtained from the sampled signal are typically referred to as *estimated* properties, since they are only estimates of the *true* properties of the continuous signal. The following numerical analysis of signal samples does not yield the true statistical properties of the actual signal. Instead, the numerical methods here can be used to estimate the statistical properties of the true signal. To denote this, the approximation symbol ($\approx$) will be used in some expressions instead of the equal sign ($=$). As the size or number of the samples becomes large, the estimated values approach the true values.

---

[1] Note that this summation is an approximation of the continuous integral $x(t) = \int_{-\infty}^{\infty} A(f) \sin(2\pi f t + \phi_f) df$, where $A(f)$ is related to the Fourier Transform of $x(t)$.

Mean and Variance

The sample *mean* of the signal is estimated as the average value of the sampled data

$$\text{mean} \equiv \bar{x} \equiv \frac{1}{N}\sum_{i=1}^{N} x_i \tag{8.1.4}$$

Subtracting the average value $\bar{x}$ from the measured values results in the *deviation* of any single value from the mean

$$\text{deviation} \equiv \tilde{x}_i \equiv x_i - \bar{x} \tag{8.1.5}$$

In the present analysis, we will use the tilde symbol (˜) to represent a fluctuation or a deviation from the mean value. The sample *standard deviation* is the square root of the unbiased sample variance and can be estimated as

$$\text{standard deviation} \equiv \sigma \approx \sqrt{\frac{1}{N-1}\sum_{i=1}^{N}(x_i - \bar{x})^2} \tag{8.1.6}$$

Note that $N - 1$ is used in the denominator rather than N. This is known as *Bessel's correction* and is typically more accurate for estimating the true standard deviation when N is small. As N becomes large, the results from using N or $N - 1$ converge. The *variance* is defined as the square of the standard deviation, i.e.

$$\text{variance} \equiv \sigma^2 \tag{8.1.7}$$

EXAMPLE 8.1.1 Use a random-number generator to produce a set of random numbers with a mean of zero and a standard deviation of 1.0. Compute the mean and standard deviation of the set. Repeat the process with different sizes of sets to consider the effects of N on the mean and standard deviation.

Solution: In reality, no data produced by a deterministic pseudorandom-number generator is truly random. However, most programming languages have methods built in for producing a set of numbers that have properties similar to a set of random numbers. In C++ this can be done with the following code:

```
#include <random>
#include <cstdio>
int main () {
    int N = 10000;
    double data[N];
    double sum = 0.0;
    double average, sum2, stdev;
    FILE* datafile = fopen("8.1.1_data.csv", "w");
    fprintf(datafile,"sample_number,data,sum,average,stdev,variance\n");
```

```
    // generate the data
    unsigned seed = 0;
    std::default_random_engine generator(seed);
    std::normal_distribution<double> distribution{0.0, 1.0};
    for(int i = 0; i < N; i++) {
        data[i] = distribution(generator);
        sum += data[i];
        average = sum/double(i+1);
        sum2 = 0.0;
        // the std dev needs a difference between values and current average
        for(int j = 0; j <= i; j++) sum2 += (data[j]-average)*(data[j]-average);
        stdev = sqrt(sum2/double(i));
        fprintf(datafile,"%d,%20.12e,%20.12e,%20.12e,%20.12e,%20.12e\n",\
                i+1,data[i],sum,average,stdev,stdev*stdev);
    }fclose(datafile);
}
```

Note that the `std::default_random_engine` requires an unsigned integer seed when initializing. This seed is used to initialize the set of random numbers that will be produced by the generator. Using the same value for the seed each time the code runs will produce repeatable results. To produce different results each time the code runs, the seed can be set to a value associated with the computer clock at runtime. For example, the following code can be used:

```
#include <chrono>
seed = std::chrono::system_clock::now().time_since_epoch().count();
```

Also note that the initialization of the `normal_distribution` requires two parameters. These are the mean and standard deviation. In the limit as the sample size approaches infinity, the sample mean and standard deviation will converge to the values used to initialize the distribution.

Running this code produces a CSV file with 10,000 data points. Also stored in the CSV file is the sum, mean, standard deviation, and variance of the data if only the data on that line and above were included in the analysis. Hence, results from this example can be used to evaluate how the estimated mean, standard deviation, and variance depend on the number of samples N. Fig. 8.1.2 shows the estimated mean and standard deviation as a function of N. Note that the plot uses logarithmic scales for the horizontal axis to emphasize the variation in the data for smaller sample sizes. As shown above, the *population* standard deviation was specified as 1.0 in the code. In this particular dataset, it requires 7,676 samples to yield a consistent *estimated* standard deviation within 1% of the population standard deviation. Compiling and running the code listed here on different computers should produce similar behavior, but not necessarily identical results since the underlying architecture for the random-number generator may differ by computer.

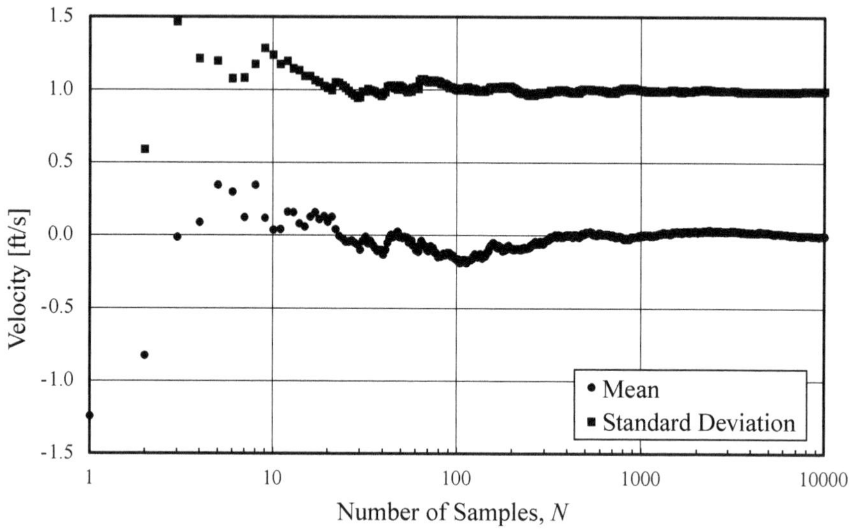

Figure 8.1.2. Estimated mean and standard deviation as a function of sample size for a set of random numbers generated to have a population mean of zero and a population standard deviation of 1.0.

The previous example showed how C++ can be used to generate random numbers with a normal distribution. In some cases, a random number generator that produces a normal distribution may not be readily available. For example, the `random_number()` intrinsic function in Fortran returns a uniform distribution of random numbers with values between 0 and 1. A normal distribution of random numbers can be generated from a uniform distribution using the Box-Muller transform. The following Fortran code shows an example for computing random numbers with a normal distribution, with a mean of zero and a standard deviation of 1, from the intrinsic random number generator.

```
function rand_normal() result(z)
    ! Returns a single normally distributed random number (mean=0, std=1)
    real :: z
    real, save :: z2 = 0.0
    logical, save :: has_saved = .false.
    real :: u1, u2, r, theta

    if (has_saved) then
        ! Return the saved value from last time
        z = z2
        has_saved = .false.
    else
        ! Generate two uniforms
        call random_number(u1)
        call random_number(u2)

        ! Guard against u1 = 0
        if (u1 <= 0.0) u1 = 1.0e-12
        if (u1 >  1.0) u1 = 1.0
```

```
        ! Box-Muller transform
        r = sqrt(-2.0*log(u1))
        theta = 2.0*PI*u2
        z  = r * cos(theta)
        z2 = r * sin(theta)

        has_saved = .true.
    end if
end function rand_normal
```

Note that this generator must first be seeded using the intrinsic function `random_seed()` prior to calling the function listed here. The following code shows an example of how to seed the random number generator.

```
    integer :: i, n, seed
    integer, allocatable :: seed_array(:)

    ! set up random number generator
    if(turb_repeatable) then
        seed = 12345
    else
        call system_clock(count=seed)
    end if
    call random_seed(size=n)
    allocate(seed_array(n))
    seed_array = seed + 7 * [(i-1, i=1,n)] ! any small odd number will do
    call random_seed(put=seed_array)
    deallocate(seed_array)
```

Power Spectral Density

The *power-spectral-density* (PSD) of a stationary signal describes how the signal variance (or power) is distributed across frequency. The PSD Φ of a stationary process is formally defined as the Fourier transform of the autocorrelation function. For finite-sampled data, it is commonly estimated using a periodogram. Because we are limiting our analysis to data points sampled at equal intervals, the periodogram can be rapidly computed using a discrete Fourier transform. The value of the periodogram P_{xx} at any given frequency f can be computed from

$$\Phi(f) \approx P_{xx}(f) = \frac{\Delta t}{N}\left|\sum_{i=1}^{N} x_i e^{-j2\pi(i-1)f\Delta t}\right|^2 \tag{8.1.8}$$

where Δt is the time between samples and $j \equiv \sqrt{-1}$. The range of frequencies over which the periodogram can be computed is bounded by the Nyquist frequency and yields $-f_s/2 \le f \le f_s/2$. Because the periodogram is symmetric about $f = 0$, it is customary to consider the periodogram only in the range $0 \le f \le f_s/2$. The one-sided periodogram can be obtained by multiplying the results from Eq. (8.1.8) in the range $0 < f < f_s/2$ by 2, except the cases $f = 0$ and $f = f_s/2$. This conserves the total power in the signal.

The frequency resolution, or bin size, of a one-sided PSD is typically determined from the number of samples, N, and is f_s/N. Therefore, the set of frequencies for which the periodogram is typically computed are

$$f = \left[0, \frac{f_s}{N}, 2\frac{f_s}{N}, \cdots, \frac{f_s}{2}\right] \tag{8.1.9}$$

This can also be expressed as

$$f_k = k\frac{f_s}{N} \tag{8.1.10}$$

where k is an integer in the range $0 \le k \le N/2$. Using Eqs. (8.1.2), (8.1.9), and (8.1.10) in Eq. (8.1.8) gives an alternate form for computing the one-sided periodogram

$$P_{xx}\left(\frac{kf_s}{N}\right) = \frac{1}{f_s N}\left|\sum_{i=1}^{N} x_i e^{-j2\pi(i-1)k/N}\right|^2 \begin{cases} 1, k = 0 \\ 2, 1 \le k \le N/2 - 1 \\ 1, k = N/2 \end{cases} \tag{8.1.11}$$

where k is indexed from zero. For a stationary zero-mean process, the integral of the PSD is equal to the variance. Hence, applying the trapezoidal rule, the variance can be estimated from a one-sided periodogram

$$\sigma^2 \approx \frac{f_s}{N}\left(\frac{P_{xx}(0) + P_{xx}(f_s/2)}{2} + \sum_{k=1}^{N/2-1} P_{xx}(f_k)\right) \tag{8.1.12}$$

where $P_{xx}(f_k)$ is computed from Eq. (8.1.11) and f_s/N is the frequency bin size. The *normalized PSD* is typically defined as the ratio of the PSD to the variance

$$||\Phi|| \equiv \frac{\Phi}{\sigma^2} \approx \frac{P_{xx}}{\sigma^2} \tag{8.1.13}$$

It is important to note from Eq. (8.1.2) that the frequency f used in the periodogram has units of Hz. From Eq. (8.1.11) it can be seen that the resulting periodogram P_{xx} has units of the signal squared per Hz. Because the variance σ^2 has units of the signal squared, the normalized PSD $||\Phi||$ has units of 1/Hz.

EXAMPLE 8.1.2 Generate a signal for velocity in ft/s consisting of two sine waves with frequencies of 0.5 and 2.0 Hz, each with an amplitude of unity. Sample the signal at 100 Hz for a total of 10,000 samples. Use the sampled data to compute the estimated PSD of the signal, as well as the estimated variance from both Eqs. (8.1.6) and (8.1.12). Repeat the process adding in white noise with a mean of zero and standard deviation of 1.0.

Solution: The signal equation is

$$V(t) = \sin[(0.5)2\pi t] + \sin[(2.0)2\pi t]$$

The following C++ code can be used to generate the clean signal without any noise, sample the signal at 100 Hz over 10,000 samples, compute the periodogram of the sampled data, and compute the estimated variance using Eqs. (8.1.6) and (8.1.12).

```
#include <cstdio>
#include <cmath>
#include <complex>
// #include <random> // add noise

int main () {
    int N = 10000;
    double samples[N];
    double fs = 100.0; // samples per second (Hz)
    double pi = 2.0*asin(1.0);
    double time;
    FILE* datafile = fopen("8.1.2_data.csv", "w");
    fprintf(datafile,"sample_number,data,sum,mean,stdev,variance\n");
    // unsigned seed = 0; // add noise
    // std::default_random_engine generator(seed); // add noise
    // std::normal_distribution<double> distribution(0.0, 1.0); // add noise

    // sample the signal
    for(int i = 0; i < N; i++) {
        time = double(i)/fs;
        samples[i] = sin(2.0*pi*0.5*time) + sin(2.0*pi*2.0*time);
        // samples[i] += distribution(generator); // add noise
        fprintf(datafile,"%d,%20.12e,%20.12e\n",i+1,time,samples[i]);
    }
    fclose(datafile);

    // compute the periodogram
    std::complex<double> eye(0.0,1.0);
    double Pxx[N/2+1];
    FILE* psdfile = fopen("8.1.2_psd.csv", "w");
    fprintf(psdfile,"fs[Hz],PSD[units^2/Hz]\n");
    for(int k = 0; k < N/2+1; k++) {
        double f = double(k)*fs/double(N);
        std::complex<double> temp(0.0,0.0);
        for(int i = 0; i < N; i++) {
            temp += (samples[i]-average)*exp(-eye*2.0*pi*double(i*k)/double(N));
        }
        Pxx[k] = 1.0/(fs*double(N))*abs(temp)*abs(temp);
        if((k!=0) && (k!=N/2)) Pxx[k] *= 2.0;
        fprintf(psdfile,"%20.12e,%20.12e\n",f,Pxx[k]);
    }
    fclose(psdfile);

    // compute estimated variance from data
    double sum = 0.0;
    for(int i = 0; i < N; i++) sum += samples[i];
    double average = sum/double(N);
    double sum2 = 0.0;
    for(int i = 0; i < N; i++) sum2 +=
     (samples[i]-average)*(samples[i]-average);
    double stdev = sqrt(sum2/double(N-1));
    printf("Estimated variance from data = %20.12e\n",stdev*stdev);

    // compute estimated variance from periodogram
    double est_variance = 0.5*(Pxx[0] + Pxx[N/2]);
    for(int k = 1; k < N/2; k++) est_variance += Pxx[k];
    est_variance *= fs/double(N);
    printf("Estimated Variance from periodogram = %20.12e\n",est_variance);
}
```

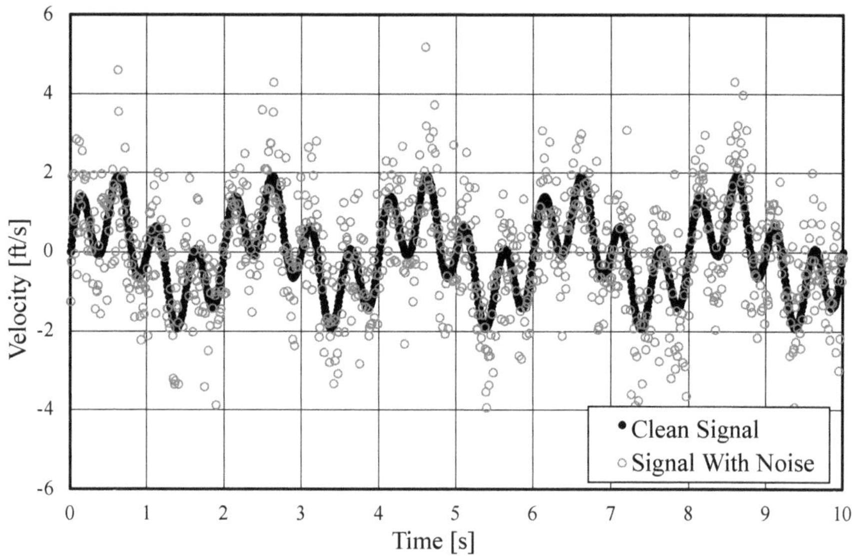

Figure 8.1.3. Sampled signals generated in Example 8.1.2.

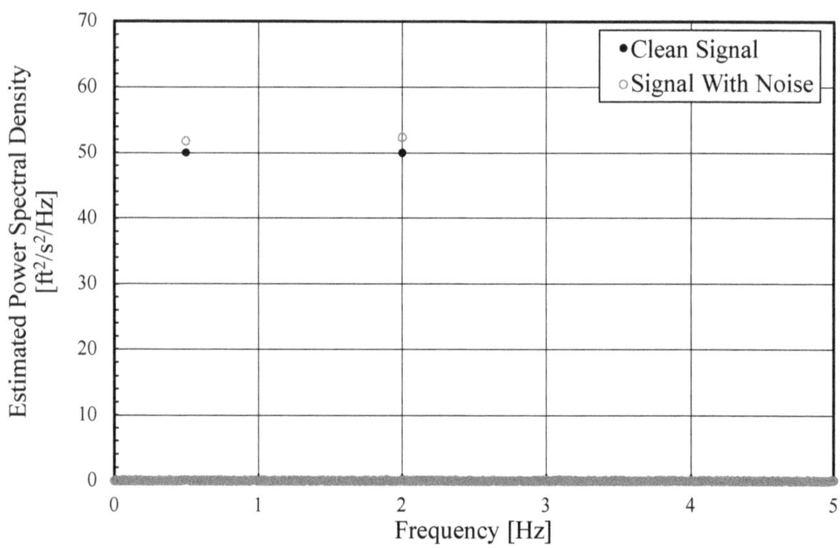

Figure 8.1.4. Periodogram of sampled signals in Example 8.1.2.

Figure 8.1.3 shows the sampled clean signal, and Figure 8.1.4 shows the resulting periodogram of the sampled clean signal, each denoted by black filled dots. The values for the PSD are all on the order of machine precision for all frequencies except for 0.5 and 2.0 Hz, which are each 50.0 out to machine precision.

White noise with a mean of zero and a standard deviation of 1.0 can be added to the signal by uncommenting all lines in the code above that end in `// add noise`. Doing so and running the code produces the results shown in the gray open circles in Figures 8.1.3 and 8.1.4. The values for the PSD of the sampled signal with noise are on the order of about 10^{-2} for all frequencies except 0.5 and 2.0 Hz, which are each a little over the true value of 50.0.

The variance of the sampled signal can be estimated by squaring the results of Eq. (8.1.6) or from the periodogram using Eq. (8.1.12). For the clean signal, the computed variance using Eq. (8.1.6) is

$$\sigma^2 = 1.000100010001 \ \text{ft}^2/\text{s}^2$$

and the variance using the results from the periodogram is

$$\sigma^2 = 1.000000000000 \ \text{ft}^2/\text{s}^2$$

These agree to at least 4 significant digits. For the signal with noise, the computed variance using Eq. (8.1.6) is

$$\sigma^2 = 2.029292999299 \ \text{ft}^2/\text{s}^2$$

and the variance using the results from the periodogram is

$$\sigma^2 = 2.029035326198 \ \text{ft}^2/\text{s}^2$$

These also agree to at least 4 significant digits.

The previous example demonstrated how C++ can be used to compute a periodogram. The following is an example Fortran function that computes a normalized periodogram.

```
subroutine psd(samples, dt, psd_norm)
    implicit none
    real, intent(in) :: samples(:), dt
    real, intent(out) :: psd_norm(:,:) ! must be size (n/2+1,2)
    integer :: n, k, i
    real :: fs, f, mean, stdev
    real, allocatable :: Pxx(:)
    complex :: eye, temp

    n = size(samples)
    if (mod(n,2) /= 0) error stop "periodogram: N must be even for N/2 bin."
    allocate(Pxx(n/2+1))
    eye = cmplx(0.0, 1.0)

    ! compute mean and stdev
    mean = 0.0
    stdev = 0.0
```

```
        do i=1,n
            mean = mean + samples(i)/n
        end do
        do i=1,n
            stdev = stdev + (samples(i)-mean)**2
        end do
        stdev = sqrt(stdev/(n-1))

        Pxx(:) = 0.0
        fs = 1.0/dt
        do k=0,n/2
            f = real(k)*fs/real(n)
            temp = cmplx(0.0, 0.0)
            do i=1,n
                temp = temp +
     (samples(i)-mean)*exp(-eye*2.0*PI*real(i-1)*real(k)/real(n))
            end do
            Pxx(k+1) = 1/fs/real(n)*abs(temp)**2
            if (k /= 0 .and. k /= n/2) Pxx(k+1) = 2.0*Pxx(k+1)
            psd_norm(k+1,1) = f
            psd_norm(k+1,2) = Pxx(k+1)/stdev**2
        end do
    end subroutine psd
```

8.2 Characteristics of Atmospheric Turbulence Models

Atmospheric disturbances have been measured for many years, and the stochastic properties of these disturbances have been analyzed extensively. Analyses show that the probability of encountering turbulence with a given standard deviation σ depends on altitude. Note that because turbulent fluctuations are defined relative to the mean wind, the standard deviation of the turbulent fluctuations is often referred to as the *root-mean-square (RMS) turbulence amplitude* in the literature. Analyses also show that the normalized PSD Φ/σ^2 of the turbulent fluctuations are characterized by one or more length scales, which vary with altitude. In order to generate a realistic signal of atmospheric turbulence, the standard deviation and length scales of the turbulence must be known. Methods for predicting the appropriate standard deviation and length scale as a function of altitude form the foundation of modern atmospheric turbulence models.

Correlations of atmospheric turbulence properties are typically divided into two categories known as *high-altitude* for altitudes greater than 2,000 ft above ground, or *low-altitude* for altitudes less than 1,000 ft above ground. For altitudes between 1,000 and 2,000 ft above ground, the standard deviation and length scales can be linearly interpolated between those presented here. Two main differences exist in the high-altitude turbulence characteristics relative to the low-altitude characteristics. First, high-altitude turbulence is assumed to be isotropic, where the low-altitude turbulence is not. Second, the high-altitude turbulence characteristics are typically expressed in terms of body-fixed coordinates, where the low-altitude turbulence characteristics are expressed in Earth-fixed coordinates.

High-Altitude Characteristics

Atmospheric turbulence at altitudes above 2,000 ft above ground falls in a category termed *high-altitude*. At these altitudes, the turbulence can be considered isotropic. Among other things, this means that the standard deviation in turbulent fluctuations σ experienced by the vehicle is independent of vehicle direction and orientation. Therefore, it is convenient to generate the turbulent fluctuations directly in the body-fixed coordinate system $\tilde{\mathbf{V}}_b = (\tilde{u}, \tilde{v}, \tilde{w})$. The translational turbulent fluctuations are typically treated as mutually uncorrelated, and the standard deviation in the turbulent fluctuations is the same for each component, i.e. $\sigma = \sigma_{u_b} = \sigma_{v_b} = \sigma_{w_b}$.

Atmospheric measurements show that the standard deviation σ varies with altitude. The probability of exceeding a particular standard deviation is given in Figure 8.2.1 as currently accepted and published in military specification documents [79, 80]. The dashed lines are labeled with the corresponding probability of exceeding a particular standard deviation as a function of altitude.

Three levels of turbulence intensity have been defined and are widely used in the simulation of atmospheric turbulence. These are known as *light*, *moderate*, and *severe* turbulence intensity levels. The standard deviation for each level of intensity is defined in Tables 8.2.1–8.2.3 for altitudes greater than 2,000 feet above ground level. These definitions are also included as solid lines in Fig. 8.2.1.

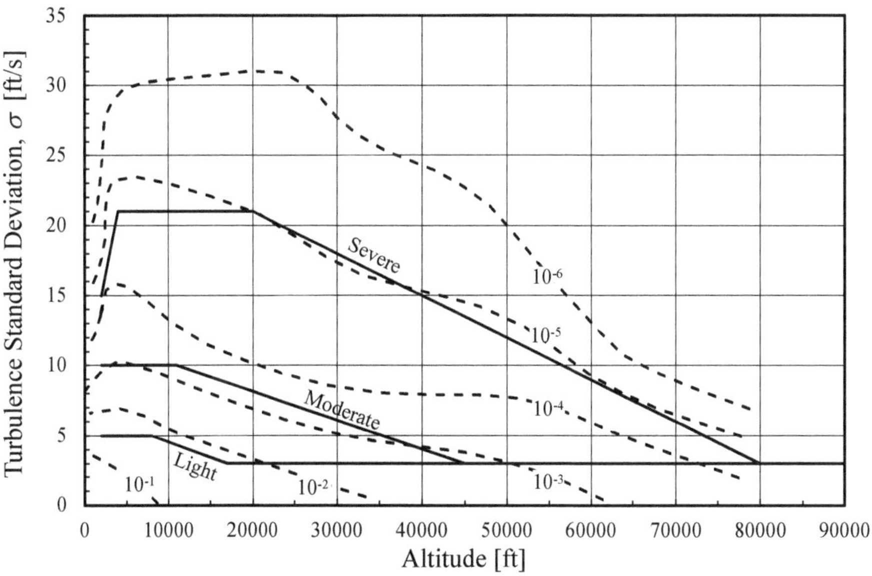

Figure 8.2.1. Probability of exceeding a given turbulence standard deviation as a function of altitude. Standard deviation in ft/s true airspeed.

Table 8.2.1. Light turbulence intensity definitions at high altitudes.

Altitude [ft]	Turbulence Standard Deviation, σ [ft/s]
2,000	5.0
8,000	5.0
≥ 17,000	3.0

Table 8.2.2. Moderate turbulence intensity definitions at high altitudes.

Altitude [ft]	Turbulence Standard Deviation, σ [ft/s]
2,000	10.0
11,000	10.0
≥ 45,000	3.0

Table 8.2.3. Severe turbulence intensity definitions at high altitudes.

Altitude [ft]	Turbulence Standard Deviation, σ [ft/s]
2,000	15.0
4,000	21.0
20,000	21.0
≥ 80,000	3.0

The length scales associated with the PSD used for modern atmospheric turbulence models vary with turbulence model [79, 80]. However, the two most-widely accepted models employ length scales related by

$$L_{u_b} = 2L_{v_b} = 2L_{w_b} \tag{8.2.1}$$

Note that **these length scales are associated with the turbulent properties encountered in the body-fixed coordinate system**, which has been denoted with the subscript b. Specific values for the length scales for each model will be addressed in the following sections.

Low-Altitude Characteristics

Atmospheric characteristics at altitudes below 1,000 ft above ground fall in a category termed *low-altitude*. At these altitudes, military specifications suggest that the turbulence standard deviation be correlated to the mean wind speed at 20 ft above ground [79, 80]. Here we will denote the mean wind speed at 20 feet above the ground as V_{20}. The suggested value for V_{20} as a function of probability of exceedance can be approximated by the relation

$$V_{20} = \begin{cases} -5.53\ln(P_E) + 13.3, & P_E \le 0.1 \\ 49.18P_E^4 - 136.1P_E^3 + 133.9P_E^2 - 79.71P_E + 32.77, & P_E > 0.1 \end{cases} \tag{8.2.2}$$

where P_E is the probability of exceedance and V_{20} is in units of ft/s. Definitions for mean wind speeds at 20 ft above ground have been given for the three accepted levels of turbulence intensity [80]. These definitions are specified in knots and are given in Table 8.2.4. A column including these definitions in units of ft/s rounded to four significant figures is included in the table.

The turbulence intensity at high altitudes is considered isotropic, and therefore typically expressed directly in the body-fixed coordinate system. However, at low altitudes, the presence of the ground influences the directionality of the turbulence. Hence, at low altitudes, the turbulence standard deviation is specified in a coordinate system aligned with the mean wind direction. The subscripts u_a, v_a, and w_a will be used to denote standard-deviation and length-scale components in this coordinate system. The symbol u_a denotes horizontal directions in the Earth-fixed coordinate system aligned with the horizontal mean wind direction. The subscript w_a denotes the vertical direction, and v_a completes the right-hand coordinate system. For a given wind speed at 20 ft above ground, the accepted turbulence standard deviation in the vertical direction is

$$\sigma_{w_a} = 0.1V_{20} \tag{8.2.3}$$

The standard deviations in the horizontal directions are

$$\sigma_{u_a} = \sigma_{v_a} = \begin{cases} \frac{\sigma_{w_a}}{(0.177+0.000823h)^{0.4}}, & h < 1,000 \text{ ft} \\ \sigma_{w_a}, & h \geq 1,000 \text{ ft} \end{cases} \tag{8.2.4}$$

where h is the height above ground.

The length scales at altitudes below 2,000 ft above ground for the two most widely accepted turbulence models are given by

$$L_{u_a} = \begin{cases} 75.64, & 0 \leq h < 10 \text{ ft} \\ h/(0.177 + 0.000823h)^{1.2}, & 10 \leq h < 1,000 \text{ ft} \\ 1000, & 1,000 \leq h < 2,000 \text{ ft} \end{cases} \tag{8.2.5}$$

$$L_{v_a} = L_{u_a}/2 \tag{8.2.6}$$

$$L_{w_a} = \begin{cases} 5, & 0 \leq h < 10 \text{ ft} \\ h/2, & 10 \leq h < 1,000 \text{ ft} \end{cases} \tag{8.2.7}$$

Table 8.2.4. Mean wind speed definitions at 20 ft above ground for defined levels of turbulence intensity at low altitudes.

Turbulence Intensity	V_{20} [knots]	V_{20} [ft/s]
Light	15.0	≈ 25.32
Moderate	30.0	≈ 50.63
Severe	45.0	≈ 75.95

Characteristics at Altitudes Between 1,000 and 2,000 ft Above Ground

For altitudes between 1,000 and 2,000 ft above the ground, it is customary to linearly interpolate the atmospheric characteristics between the low-altitude characteristics at 1,000 ft and the high-altitude characteristics at 2,000 ft. Hence, at these altitudes, characteristics at 1,000 ft and 2,000 ft must be independently computed before linearly interpolating. The reader is reminded that the turbulence standard deviations at low altitude are in the Earth-fixed coordinate system, and the length scales at low altitude are associated with the constant-wind coordinate system. Therefore, the low-altitude characteristics must be rotated to the vehicle body-fixed coordinate system before being used in the interpolation process with the high-altitude turbulence standard deviation and length scales.

Frozen-Field Turbulence Models

The atmospheric turbulence models discussed in this chapter fall under a category of models often called *frozen-field* models, in which the turbulent fluctuations are assumed to vary with vehicle location, but not time. These models are one-dimensional location models, meaning that the turbulence depends only on a translational location of the vehicle within the frozen field. The symbol x_{ff} will be used to denote the one-dimensional translational location in the frozen field. The $\tilde{u}$, $\tilde{v}$, and $\tilde{w}$ translational velocity fluctuations are modeled independently, but each depends only on x_{ff}. If x_{ff} is not known directly, it can be computed by integrating a step size related to the vehicle velocity and size of the time step

$$\Delta x_{ff} = \Delta t V = \Delta t \sqrt{V_{x_b}^2 + V_{y_b}^2 + V_{z_b}^2} = \Delta t \sqrt{u^2 + v^2 + w^2} \tag{8.2.8}$$

where Δt is the time step between samples.

We will now consider the two most widely used frozen-field turbulence models, which include the Dryden model and the von Kármán model. The atmospheric turbulent properties presented in the previous section can be used in these models to generate a spatial realization of turbulent fluctuations at a given altitude and turbulence intensity or probability of exceedance. Each model requires the length scales L_u, L_v, and L_w, and the standard deviations σ_u, σ_v, and σ_w. **At altitudes greater than 2,000 ft above ground,**

$$\begin{aligned} L_u &= L_{u_b}, \quad L_v = L_{v_b}, \quad L_w = L_{w_b}, \\ \sigma_u &= \sigma_{u_b}, \quad \sigma_v = \sigma_{v_b}, \quad \sigma_w = \sigma_{w_b} \end{aligned} \tag{8.2.9}$$

At altitudes below 1,000 ft above ground, the turbulence properties are first defined in the wind-aligned coordinate system and then transformed to body-fixed coordinates before being used in the turbulence model. **At altitudes between 1,000 and 2,000 ft above ground**, the length scales and standard deviations must be interpolated as discussed previously.

8.3 Dryden Atmospheric Turbulence Model

The Dryden atmospheric turbulence model is perhaps the most widely used model for aerospace simulations due to its simplicity of implementation. Military standards state that this model is sufficient for many simulation scenarios, as long as the effects of structural modes are not of importance [79, 80]. The Dryden atmospheric turbulence model is defined by the power-spectral density functions

$$\Phi_u(\omega) = \sigma_u^2 \frac{2L_u}{\pi} \frac{1}{1 + (L_u\omega)^2} \tag{8.3.1}$$

$$\Phi_v(\omega) = \sigma_v^2 \frac{2L_v}{\pi} \frac{1 + 12(L_v\omega)^2}{[1 + 4(L_v\omega)^2]^2} \tag{8.3.2}$$

$$\Phi_w(\omega) = \sigma_w^2 \frac{2L_w}{\pi} \frac{1 + 12(L_w\omega)^2}{[1 + 4(L_w\omega)^2]^2} \tag{8.3.3}$$

with units $\mathrm{ft}^3/\mathrm{s}^2$ where ω is the spatial frequency (wave number) in rad/ft. Note that the units for ω denote this is a frozen-field turbulence model, which has fluctuations that vary spatially but not temporally. The standard deviations σ_u, σ_v, and σ_w have units of ft/s and depend on the turbulence intensity or probability of exceedance and height above ground, as given in the previous section. The length scales L_u, L_v, and L_w have units of ft and depend on height above ground. The suggested length scales at altitudes greater than 2,000 ft above ground for this model are

$$L_{u_b} = 1750 \text{ ft} \tag{8.3.4}$$

$$L_{v_b} = L_{w_b} = 875 \text{ ft} \tag{8.3.5}$$

The suggested length scales for altitudes less than 1,000 ft above ground is given in the previous section, along with a discussion for computing the length scales between 1,000 ft and 2,000 ft above ground. Figure 8.3.1 shows plots of the normalized power-spectral density Φ/σ^2 of the Dryden turbulence model using the high-altitude length scales.

Whereas Eqs. (8.3.1)–(8.3.3) express the turbulent fluctuation properties in the spatial frequency domain, for flight simulation we must generate a frozen-field turbulence realization in a spatial domain. This can be accomplished by passing white noise through a filter designed to produce an output signal that has the frequency content specified by Eqs. (8.3.1)–(8.3.3). To obtain such a filter for the Dryden turbulence model, it is helpful to express Eqs. (8.3.1)–(8.3.3) in the alternate form

$$\Phi_u(\omega) = \sigma_u^2 \frac{2L_u}{\pi} \left(\frac{1}{1 + L_u j\omega}\right)\left(\frac{1}{1 - L_u j\omega}\right) \tag{8.3.6}$$

$$\Phi_v(\omega) = \sigma_v^2 \frac{2L_v}{\pi} \left[\frac{1 + 2\sqrt{3}L_v j\omega}{(1 + 2L_v j\omega)^2}\right]\left[\frac{1 - 2\sqrt{3}L_v j\omega}{(1 - 2L_v j\omega)^2}\right] \tag{8.3.7}$$

$$\Phi_w(\omega) = \sigma_w^2 \frac{2L_w}{\pi} \left[\frac{1 + 2\sqrt{3}L_w j\omega}{(1 + 2L_w j\omega)^2}\right]\left[\frac{1 - 2\sqrt{3}L_w j\omega}{(1 - 2L_w j\omega)^2}\right] \tag{8.3.8}$$

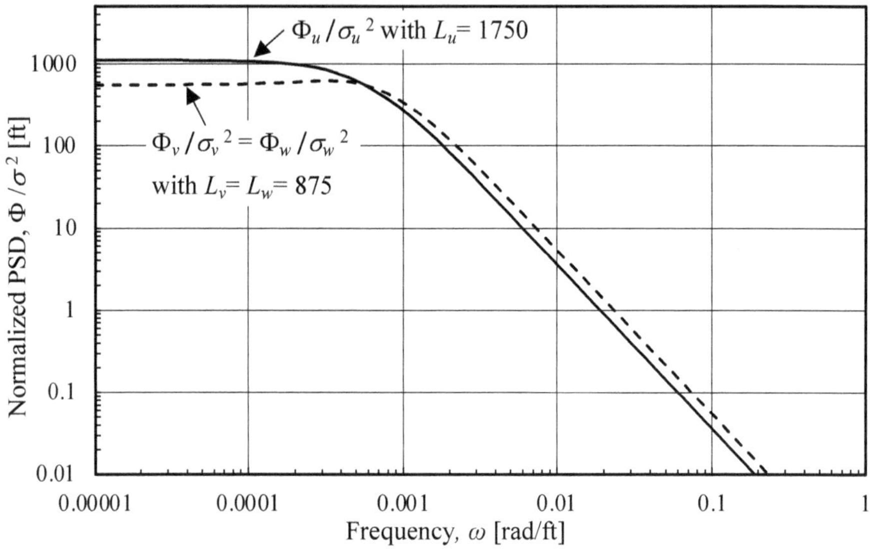

Figure 8.3.1. Normalized power spectral densities for the Dryden atmospheric turbulence model using high-altitude length scales.

The PSD of a white-noise signal Φ_n is a constant value, independent of frequency. When white noise is passed through a filter defined by a Laplace-domain transfer function $F(s)$, a response function is produced that has the power spectral density

$$\Phi(\omega) = |F(j\omega)|^2\Phi_n \tag{8.3.9}$$

Note that Eq. (8.3.9) can alternately be written as

$$\Phi(\omega) = F(j\omega)F(-j\omega)\Phi_n \tag{8.3.10}$$

In order to produce the desired power spectral densities defined in Eqs. (8.3.1)–(8.3.3), we can equate Eq. (8.3.10) to each PSD in Eqs. (8.3.6)–(8.3.8) to yield

$$F_u(j\omega) = \frac{1}{1 + L_u j\omega} \tag{8.3.11}$$

$$F_v(j\omega) = \frac{1 + 2\sqrt{3}L_v j\omega}{(1 + 2L_v j\omega)^2} \tag{8.3.12}$$

$$F_w(j\omega) = \frac{1 + 2\sqrt{3}L_w j\omega}{(1 + 2L_w j\omega)^2} \tag{8.3.13}$$

where the required white-noise PSD for each component must be

$$\Phi_{n_u} = \sigma_u^2\frac{2L_u}{\pi}, \quad \Phi_{n_v} = \sigma_v^2\frac{2L_v}{\pi}, \quad \Phi_{n_w} = \sigma_w^2\frac{2L_w}{\pi} \tag{8.3.14}$$

This yields the well-known transfer functions for the Dryden model

$$F_u(s) = \frac{1}{1 + L_u s} \tag{8.3.15}$$

$$F_v(s) = \frac{1 + 2\sqrt{3}L_v s}{(1 + 2L_v s)^2} \tag{8.3.16}$$

$$F_w(s) = \frac{1 + 2\sqrt{3}L_w s}{(1 + 2L_w s)^2} \tag{8.3.17}$$

It can be shown that the resulting filter differential equations are

$$\tilde{u}' = \frac{\eta_u - \tilde{u}}{L_u} \tag{8.3.18}$$

$$\tilde{v}' = \sqrt{3}f' + \frac{f - \tilde{v}}{2L_v} \tag{8.3.19}$$

$$\tilde{w}' = \sqrt{3}g' + \frac{g - \tilde{w}}{2L_w} \tag{8.3.20}$$

where

$$f' = \frac{\eta_v - f}{2L_v} \tag{8.3.21}$$

$$g' = \frac{\eta_w - g}{2L_w} \tag{8.3.22}$$

and the prime denotes differentiation in space along the frozen field x_{ff}. The symbols η_u, η_v, and η_w represent samples of white noise with PSD properties specified in Eq. (8.3.14), and are functions of location along the frozen field x_{ff}.

Equations (8.3.18)–(8.3.22) are differential equations that can be integrated forward in space. For purposes of numerical integration, the step size in the frozen field is related to time and vehicle velocity through Eq. (8.2.8), $\Delta x_{ff} = V\Delta t$ where V is the velocity of the vehicle. A number of methods have been suggested and implemented by various authors for implementing the Dryden turbulence model for flight simulation or other analyses [79–82]. Here we will consider two well-known implementations.

Beal Implementation

One common method for implementing the Dryden turbulence model is that given by Beal [81]. Recall that the spatial frequency ω used in Eqs. (8.3.1)–(8.3.3), (8.3.6)–(8.3.8), and (8.3.11)–(8.3.13) has units of radians per foot. Hence it is a frequency in space rather than time. It is customary to use a random-number generator to produce a white-noise signal at each new location in the frozen field at a distance of Δx_{ff} from the previous location. Beal suggests considering the white-noise function as an analog step function that takes on a constant value between the x_{ff} locations at which the turbulent properties are needed. For this case, Beal has shown that when the quantity

$\omega \Delta x_{ff}$ is sufficiently small, the PSD of the white-noise function so defined can be closely approximated as

$$\Phi_n = \sigma_n^2 \frac{\Delta x_{ff}}{\pi} \tag{8.3.23}$$

Hence, equating Eq. (8.3.23) to each component in Eq. (8.3.14) gives the required white-noise standard deviations for each component

$$\sigma_{n_u} = \sigma_u \sqrt{\frac{2L_u}{\Delta x_{ff}}}, \quad \sigma_{n_v} = \sigma_v \sqrt{\frac{2L_v}{\Delta x_{ff}}}, \quad \sigma_{n_w} = \sigma_w \sqrt{\frac{2L_w}{\Delta x_{ff}}} \tag{8.3.24}$$

In addition to the white-noise step function, Beal [81] suggests applying Eqs. (8.3.18)–(8.3.22) at the midpoint between the points $x_{ff_{i-1}}$ and x_{ff_i}. Second-order central differencing at the midpoint gives the approximations

$$\begin{aligned} \tilde{u} &\approx \frac{\tilde{u}_i + \tilde{u}_{i-1}}{2}, \quad \tilde{v} \approx \frac{\tilde{v}_i + \tilde{v}_{i-1}}{2}, \quad \tilde{w} \approx \frac{\tilde{w}_i + \tilde{w}_{i-1}}{2}, \\ \tilde{u}' &\approx \frac{\tilde{u}_i - \tilde{u}_{i-1}}{\Delta x_{ff}}, \quad \tilde{v}' \approx \frac{\tilde{v}_i - \tilde{v}_{i-1}}{\Delta x_{ff}}, \quad \tilde{w}' \approx \frac{\tilde{w}_i - \tilde{w}_{i-1}}{\Delta x_{ff}} \end{aligned} \tag{8.3.25}$$

Using Eq. (8.3.25) in Eqs. (8.3.18)–(8.3.22) gives the final algorithm suggested by Beal

$$\tilde{u}_i = \frac{(1 - A_u)\tilde{u}_{i-1} + 2A_u \eta_u}{1 + A_u} \tag{8.3.26}$$

$$\tilde{v}_i = \frac{(1 - A_v)\tilde{v}_{i-1} + A_v(f_i + f_{i-1}) + \sqrt{3}(f_i - f_{i-1})}{1 + A_v} \tag{8.3.27}$$

$$\tilde{w}_i = \frac{(1 - A_w)\tilde{w}_{i-1} + A_w(g_i + g_{i-1}) + \sqrt{3}(g_i - g_{i-1})}{1 + A_w} \tag{8.3.28}$$

where

$$A_u = \frac{\Delta x_{ff}}{2L_u}, \quad A_v = \frac{\Delta x_{ff}}{4L_v}, \quad A_w = \frac{\Delta x_{ff}}{4L_w} \tag{8.3.29}$$

and

$$f_i = \frac{(1 - A_v)f_{i-1} + 2A_v \eta_v}{1 + A_v}, \tag{8.3.30}$$

$$g_i = \frac{(1 - A_w)g_{i-1} + 2A_w \eta_w}{1 + A_w} \tag{8.3.31}$$

The values η_u, η_v, and η_w are random numbers generated from a normal distribution with standard deviations given by Eq. (8.3.24). Note that the previous value for the gust as well as f and g must be known, and therefore stored at the end of each step for use in the next step.

The validity of any atmospheric-turbulence-model implementation can be tested by comparing results from the model to the PSD relations defined for the model. The reader is reminded that the frequency f used in a periodogram has units of Hz, the resulting periodogram magnitude P_{xx} has units of the signal squared per Hz, and a normalized PSD $\|\Phi\|$ has units of 1/Hz. However, frozen-field models typically use radians per foot as units of spatial frequency, and a normalized PSD with units of ft. Hence, **the frequency output of a periodogram must be multiplied by 2π, and the magnitude of the periodogram must be divided by 2π in order to yield results in the units consistent with Fig. 8.3.1 and Eqs. (8.3.1)–(8.3.3) for comparison.**

EXAMPLE 8.3.1 Apply the implementation of the Dryden atmospheric turbulence model suggested by Beal [81]. Produce $\tilde{u}$, $\tilde{v}$, and $\tilde{w}$ turbulent fluctuation components for light turbulence at an altitude of 2,000 ft above ground. Produce and plot a spatial distribution of turbulent fluctuations using 1,000 points spaced 20.0 ft apart in the frozen field. Test your implementation by generating a spatial distribution with 20,000 points spaced 20.0 ft apart and computing the periodogram of the distribution. Repeat the process up to 1,000 times using different seeds for the random number generator. Average the periodogram results and plot the ensemble-averaged results in comparison to the model PSD given in Eqs. (8.3.1)–(8.3.3).

Solution: At an altitude of 2,000 ft above the ground, the length scales are given from Eqs. (8.3.4) and (8.3.5)

$$L_u = 1750 \text{ ft}, \qquad L_v = L_w = 875 \text{ ft}$$

The RMS turbulent amplitude, or standard deviation σ, of the fluctuations at 2,000 ft for light turbulence can be obtained from Fig. 8.2.1 or Table 8.2.1, which yields

$$\sigma_u = \sigma_v = \sigma_w = 5.0 \text{ ft/s}$$

Using these values along with $\Delta x_{ff} = 20$ ft in Eqs. (8.3.24) and (8.3.29) gives

$$\sigma_{\eta_u} = \sigma_u\sqrt{2L_u/\Delta x_{ff}} = 5\sqrt{2(1750)/20} = 66.1437827766148 \text{ ft/s}$$

$$\sigma_{\eta_v} = \sigma_{\eta_w} = \sigma_v\sqrt{2L_v/\Delta x_{ff}} = 5\sqrt{2(875)/20} = 46.7707173346743 \text{ ft/s}$$

and

$$A_u = A_v = A_w = \frac{\Delta x_{ff}}{2L_u} = \frac{\Delta x_{ff}}{4L_v} = \frac{20}{2(1750)} = 0.00571428571428571$$

Applying the method suggested by Beal to generate 1,000 points spaced 20 ft apart yields a sample spatial distribution shown in Fig. 8.3.2. Note that results produced by an independent code will not produce the exact same solutions. However, it is helpful to visualize the spatial frequency and magnitude of the fluctuations that should be generated by any code implementing the Dryden atmospheric turbulence model.

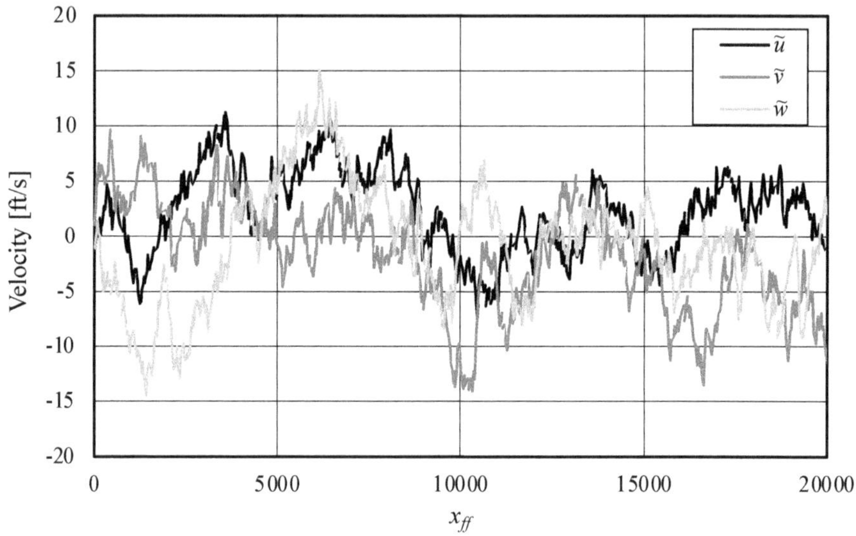

Figure 8.3.2. Sample results for $\tilde{u}$, $\tilde{v}$, and $\tilde{w}$ for the Dryden turbulence model with data points spaced 20 ft apart.

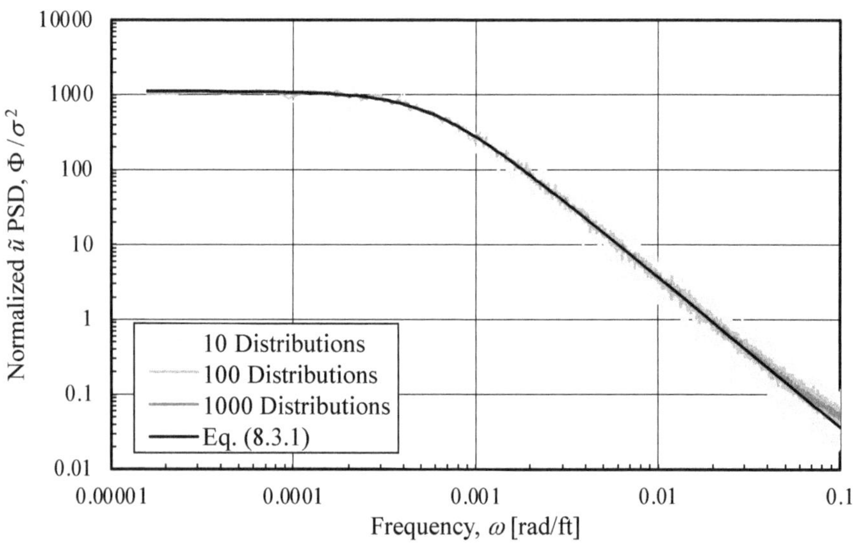

Figure 8.3.3. Ensemble-averaged normalized power spectral densities for $\tilde{u}$ from the Dryden turbulence model using the implementation suggested by Beal.

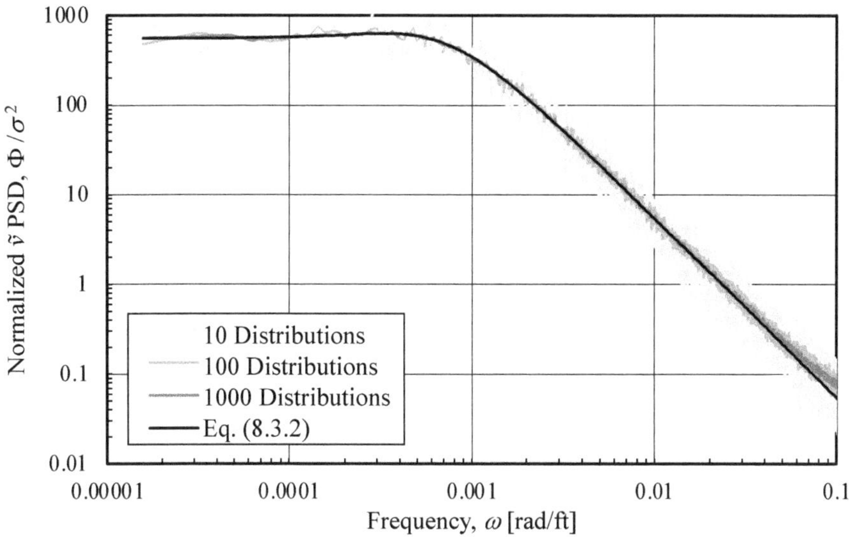

Figure 8.3.4. Ensemble-averaged normalized power spectral densities for $\tilde{v}$ from the Dryden turbulence model using the implementation suggested by Beal.

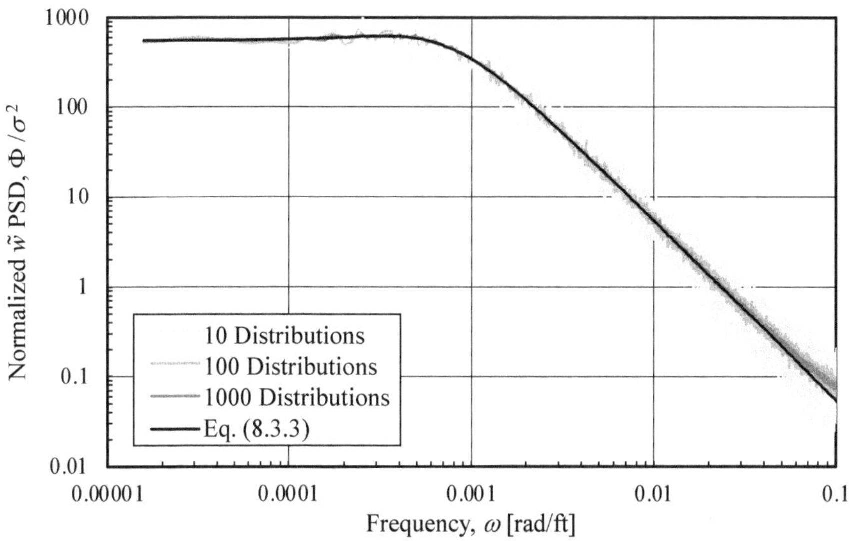

Figure 8.3.5. Ensemble-averaged normalized power spectral densities for $\tilde{w}$ from the Dryden turbulence model using the implementation suggested by Beal.

Repeating the process using 20,000 points in each independently generated spatial distribution and computing the ensemble-averaged periodogram yields the results shown in Figs. 8.3.3–8.3.5. As can be seen in the figures, as more spatial distributions are considered, the ensemble-averaged normalized periodogram approaches the PSD functions given in Eqs. (8.3.1)–(8.3.3). Note that at high spatial frequencies, the periodogram results are higher than the prescribed Dryden PSD functions. This is likely due to the finite spatial step size and the step-function approximation used for the white-noise input. Decreasing the step size Δx_{ff} yields more accurate results at higher spatial frequencies.

Again the reader is reminded that an independent code will not produce results identical to those shown in these figures, since the random number generator for each code may be different. However, any correct implementation of this model should demonstrate that the ensemble-averaged normalized periodogram results approach the PSD functions given in Eqs. (8.3.1)–(8.3.3) as more independently generated spatial distributions are considered.

Military Specification Implementation

The military standard specification MIL-STD-1797A [80] suggests a digital filter implementation for the Dryden atmospheric turbulence model. Because this military standard specification document is so widely referenced, this method is included here for completeness. However, the implementation given here treats all three components as first-order, single-pole filters. This reproduces the correct $\tilde{u}$ Dryden spectrum, but does not accurately reproduce the $\tilde{v}$ and $\tilde{w}$ Dryden spectra specified in Eqs. (8.3.2) and (8.3.3). According to MIL-STD-1797A, given the gust components $\tilde{u}_{i-1}$, $\tilde{v}_{i-1}$, and $\tilde{w}_{i-1}$ at a previous location in the frozen field, the gust components $\tilde{u}_i$, $\tilde{v}_i$, and $\tilde{w}_i$ at a neighboring location of distance Δx_{ff} from the previous location in the one-dimensional frozen field can be computed from

$$\tilde{u}_i = (1 - \Delta x_{ff}/L_u)\tilde{u}_{i-1} + \sigma_u \eta_u \sqrt{2\Delta x_{ff}/L_u} \tag{8.3.32}$$

$$\tilde{v}_i = (1 - \Delta x_{ff}/L_v)\tilde{v}_{i-1} + \sigma_v \eta_v \sqrt{2\Delta x_{ff}/L_v} \tag{8.3.33}$$

$$\tilde{w}_i = (1 - \Delta x_{ff}/L_w)\tilde{w}_{i-1} + \sigma_w \eta_w \sqrt{2\Delta x_{ff}/L_w} \tag{8.3.34}$$

where η_u, η_v, and η_w are independent random numbers generated from a normal distribution with a standard deviation of 1.0.

EXAMPLE 8.3.2 Apply the implementation of the Dryden atmospheric turbulence model suggested in MIL-STD-1797A [80]. Produce $\tilde{u}$, $\tilde{v}$, and $\tilde{w}$ turbulent fluctuation components for light turbulence at an altitude of 2,000 ft above ground. Test your implementation by generating a spatial distribution with 20,000 points spaced 20.0 ft apart and computing the periodogram. Repeat the process up to 1,000 times using different seeds for the random number generator. Average the periodogram results and plot the ensemble-averaged results in comparison to the model PSD given in Eqs. (8.3.1)–(8.3.3).

Solution: At an altitude of 2,000 ft above the ground, the length scales are given from Eqs. (8.3.4) and (8.3.5)

$$L_u = 1750 \text{ ft}, \qquad L_v = L_w = 875 \text{ ft}$$

The RMS turbulent amplitude, or standard deviation σ, of the fluctuations at 2,000 ft for light turbulence can be obtained from Fig. 8.2.1 or Table 8.2.1, which yields

$$\sigma_u = \sigma_v = \sigma_w = 5.0 \text{ ft/s}$$

Using these values along with $\Delta x_{ff} = 20$ ft in the method given in Eqs. (8.3.32)–(8.3.34) yields the results shown in Figs. 8.3.6–8.3.8. As can be seen in Fig. 8.3.6, as more spatial distributions are considered, the ensemble-averaged normalized periodogram approaches the PSD function given in Eq. (8.3.1) for $\tilde{u}$. However, the same is not true for the $\tilde{v}$ and $\tilde{w}$ results.

Although the results in Figs. 8.3.7 and 8.3.8 are close to approaching the PSD functions given in Eqs. (8.3.2) and (8.3.3), there is a range of spatial frequencies $0.0001 < \omega < 0.001$ rad/ft for which the PSD results generated using the method suggested in MIL-STD-1797A are very inaccurate. This can be more clearly seen by not using logarithmic scales when plotting. Figure 8.3.9 shows the same data plotted in Fig. 8.3.7 using 1,000 spatial distributions, but without using logarithmic scaling and truncated to show only data in the range $\omega < 0.005$ rad/ft. Note the increase in the PSD value from Eq. (8.3.2) in the range $0.0 < \omega < 0.0005$ rad/ft, which is not captured using the method suggested in MIL-STD-1797A. Additionally, note the vertical deviation between the results generated from the MIL-STD-1797A method and Eq. (8.3.2) at higher spatial frequencies. Results from the method suggested by Beal [81] are included for comparison. Note that Beal's method accurately reproduces the correct PSD distribution over this range of spatial frequencies. Similar trends are exhibited by both models for the solutions of $\tilde{w}$, and are not shown in detail here.

In summary, the MIL-STD-1797A implementation of the Dryden atmospheric turbulence model does not accurately reproduce the model PSD distributions given in Eqs. (8.3.2) and (8.3.3), and should not be used when accurate reproduction of the $\tilde{v}$ and $\tilde{w}$ Dryden spectra is required. However, the MIL-STD-1797A implementation does accurately produce $\tilde{u}$ frozen-field distributions that match the Dryden atmospheric turbulence model.

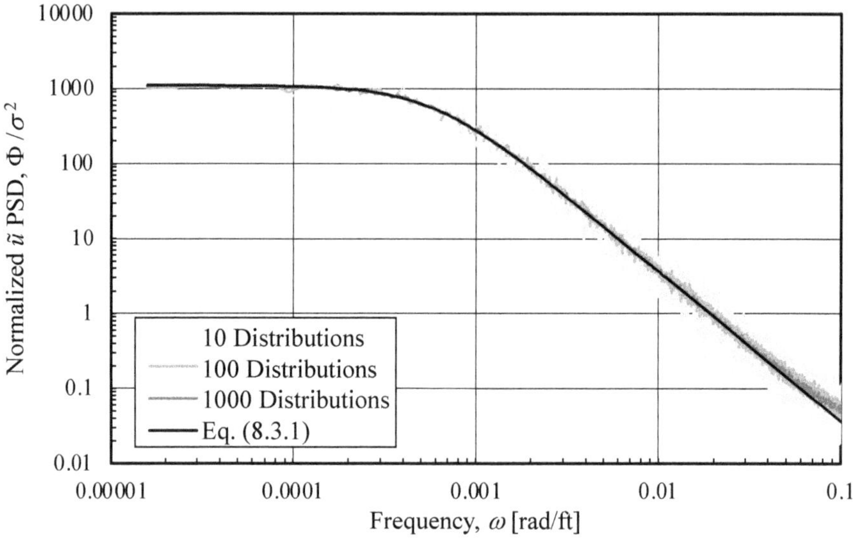

Figure 8.3.6. Ensemble-averaged normalized power spectral densities for $\tilde{u}$ from the Dryden turbulence model using the implementation suggested in MIL-STD-1797A.

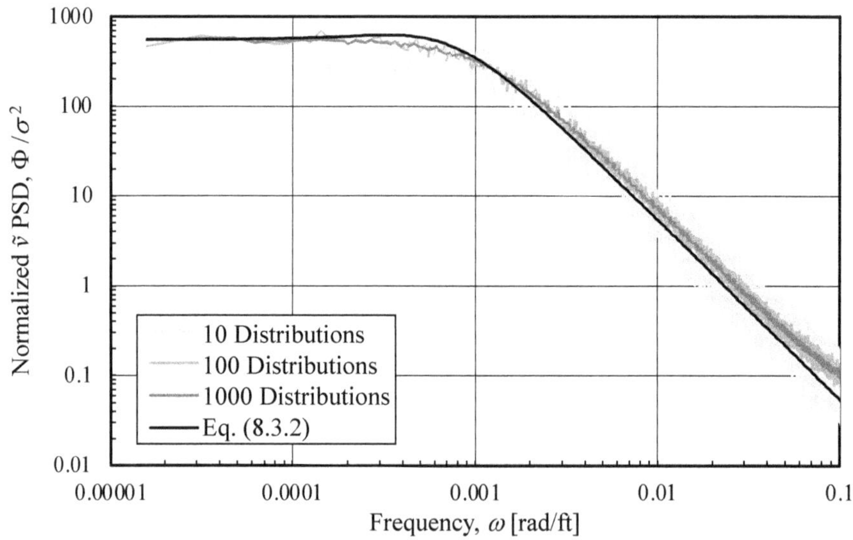

Figure 8.3.7. Ensemble-averaged normalized power spectral densities for $\tilde{v}$ from the Dryden turbulence model using the implementation suggested in MIL-STD-1797A.

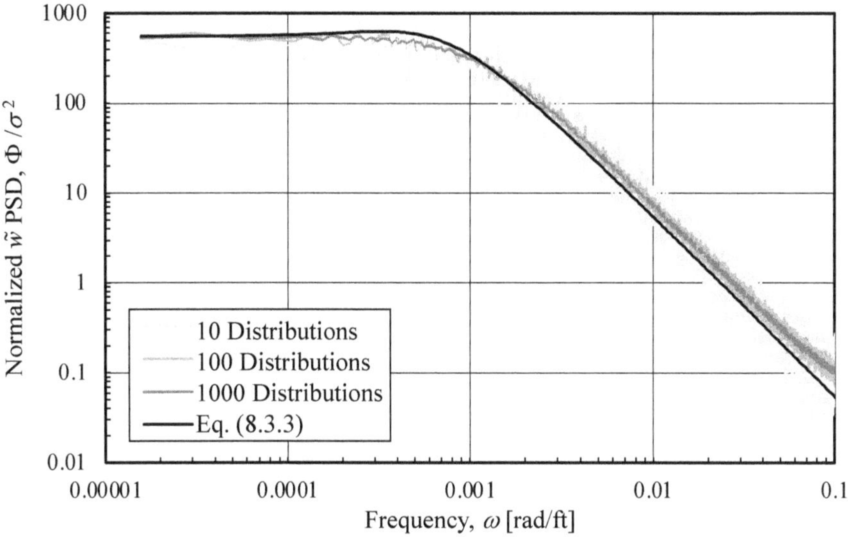

Figure 8.3.8. Ensemble-averaged normalized power spectral densities for $\tilde{w}$ from the Dryden turbulence model using the implementation suggested in MIL-STD-1797A.

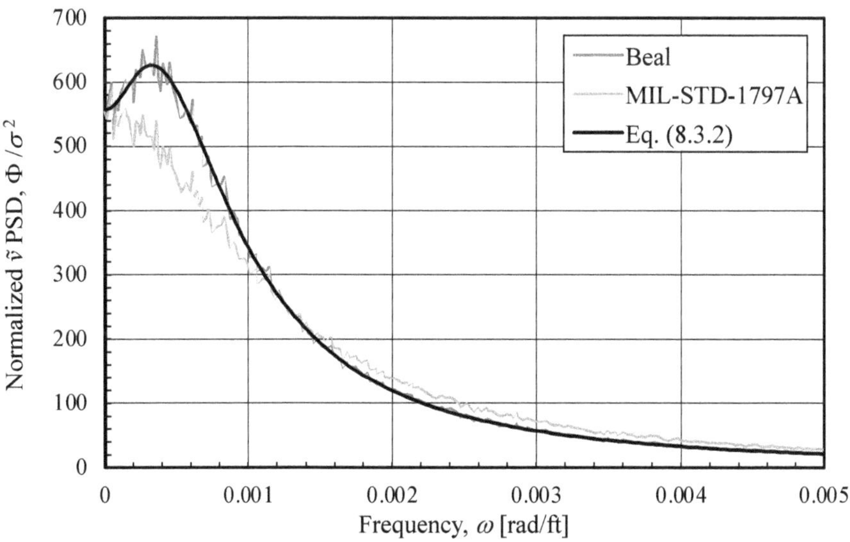

Figure 8.3.9. Ensemble-averaged normalized power-spectral-density results using 1000 distributions from Figs. 8.3.4 and 8.3.7 plotted without logarithmic scaling.

8.4 von Kármán Atmospheric Turbulence Model

Another well-known atmospheric turbulence model is the von Kármán model. This model generally provides a more realistic representation of measured atmospheric turbulence spectra than the Dryden model, particularly at high spatial frequencies. However, it is not nearly as simple to implement as the Dryden model, and is therefore often reserved for higher-fidelity simulations wheere accurate gust spectra are important, such as aero-structural coupling.

The power spectral density functions for the von Kármán model can be written as

$$\Phi_u(\omega) = \sigma_u^2 \frac{2L_u}{\pi} \frac{1}{[1+(1.339L_u\omega)^2]^{5/6}} \tag{8.4.1}$$

$$\Phi_v(\omega) = \sigma_v^2 \frac{2L_v}{\pi} \frac{1+(8/3)(2.678L_v\omega)^2}{[1+(2.678L_v\omega)^2]^{11/6}} \tag{8.4.2}$$

$$\Phi_w(\omega) = \sigma_w^2 \frac{2L_w}{\pi} \frac{1+(8/3)(2.678L_w\omega)^2}{[1+(2.678L_w\omega)^2]^{11/6}} \tag{8.4.3}$$

with units $\mathrm{ft}^3/\mathrm{s}^2$ where ω is the spatial frequency (wave number) in rad/ft. The standard deviations σ_u, σ_v, and σ_w have units of ft/s and depend on the turbulence intensity or probability of exceedance and height above ground, as given in Section 8.2. The length scales L_u, L_v, and L_w have units of ft and depend on height above ground. The suggested length scales at altitudes greater than 2,000 ft above ground for this model are

$$L_{u_b} = 2500 \text{ ft} \tag{8.4.4}$$

$$L_{v_b} = L_{w_b} = 1250 \text{ ft} \tag{8.4.5}$$

The suggested length scales for altitudes less than 1,000 ft above ground is given in Section 8.2, along with a discussion for computing the length scales between 1,000 ft and 2,000 ft above ground. Figure 8.4.1 shows plots of the normalized power-spectral densities of the von Kármán turbulence model using the high-altitude length scales. The normalized power-spectral densities of the Dryden turbulence model using the high-altitude length scales for the Dryden model are also included in the plot for comparison.

Unlike the Dryden model, there is no exact low-order filter through which white noise can be passed to generate a signal with the PSD distributions given in Eqs. (8.4.1)–(8.4.3). However, frozen-field turbulence matching the von Kármán model can be generated as suggested by [81] by using a summation of sine waves of the form

$$\tilde{u}(x_{ff}) = \sum_{i=1}^{N} A_{u_i} \sin(\omega_{u_i} x_{ff} + \phi_{u_i}) \tag{8.4.6}$$

$$\tilde{v}(x_{ff}) = \sum_{i=1}^{N} A_{v_i} \sin(\omega_{v_i} x_{ff} + \phi_{v_i}) \tag{8.4.7}$$

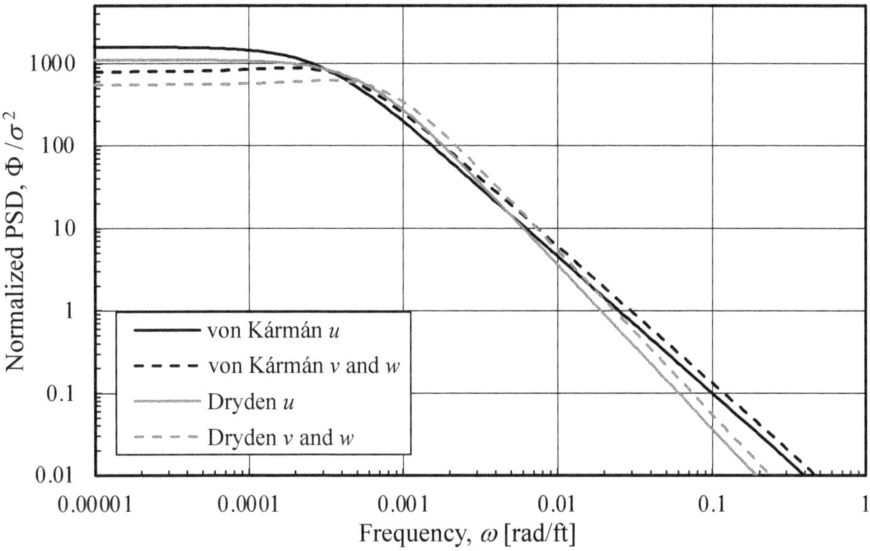

Figure 8.4.1. Normalized power spectral densities for the von Kármán and Dryden atmospheric turbulence models using high-altitude length scales.

$$\tilde{w}(x_{ff}) = \sum_{i=1}^{N} A_{w_i} \sin(\omega_{w_i} x_{ff} + \phi_{w_i}) \tag{8.4.8}$$

where N is the number of sine waves used to generate the frozen-field turbulence distribution, A_{u_i}, A_{v_i}, and A_{w_i} are the amplitudes of individual sine waves, ω_{u_i}, ω_{v_i}, ω_{w_i} are the spatial frequencies of individual sine waves, and ϕ_{u_i}, ϕ_{v_i}, and ϕ_{w_i} are the phases of individual sine waves. The phase of each sine wave is randomly generated at the beginning of the simulation using a uniform distribution between $-\pi$ and π. The amplitudes and spatial frequencies of the N sine waves are carefully selected in order to approximate the PSD distributions given in Eqs. (8.4.1)–(8.4.3). A method for doing this is discussed in detail by Beal [81].

Although the summation of sines method for implementing the von Kármán model is simple, it has two potential drawbacks. First, because the summation of sines must be computed at each time step, it is typically more computationally expensive than any model that can be implemented using filtered white noise, such as the Dryden model. Beal [81] offers some suggestions for improving the computational efficiency. Second, it is important to note that using a summation of sines results in a frozen field that is identically periodic after some distance. This may not be an issue for many simulation scenarios, but should be recognized by the user.

8.5 Rotational Disturbances

The atmospheric disturbance models given in Sections 8.3 and 8.4 considered only the translational disturbances in the flow relative to the vehicle. Application of only the translational disturbances is equivalent to assuming that the entire vehicle surface is exposed to a uniform translational flow disturbance. In other words, application of only the translation components neglects any variation in the flow disturbances along the surface of the vehicle. In reality, the disturbances vary along the surface of the vehicle, which introduces disturbances in the effective angular rates of the vehicle in addition to translational disturbances. As with all atmospheric disturbances, exact modeling of these effects is extremely difficult. However, for simulation purposes, when the effect on structural modes is not significant, it is typically adequate to apply rotational disturbances only at the vehicle center of gravity.

In order to understand how angular disturbances might be modeled in simulation, it is important to understand how these disturbances are applied in a simulation framework. According to guidance from the MIL-STD-1797A [80], the turbulent fluctuations $\tilde{u}$, $\tilde{v}$, $\tilde{w}$, $\tilde{p}$, $\tilde{q}$, and $\tilde{r}$ should be applied to the aerodynamics of the vehicle only. This is done by adding these disturbances to the u, v, w, p, q, and r states of the vehicle when computing the aerodynamic properties of the vehicle. With this understanding, a positive $\tilde{w}$ disturbance will emulate a positive disturbance in the vehicle angle of attack, and a positive $\tilde{p}$ will emulate a positive disturbance in vehicle rolling rate. These disturbances in vehicle aerodynamics indirectly affect the vehicle dynamics. This is discussed in detail in the following section, but must be understood at a cursory level prior to a discussion on modeling angular disturbances.

With this framework in mind for applying the translational and angular disturbances in flight simulation, an approximation for the angular disturbances experienced by the vehicle can be obtained by differentiating the translational disturbances

$$\tilde{p} \approx \frac{\partial \tilde{w}}{\partial y}, \quad \tilde{q} \approx -\frac{\partial \tilde{w}}{\partial x}, \quad \tilde{r} \approx \frac{\partial \tilde{v}}{\partial x} \tag{8.5.1}$$

These approximations are only accurate at very low spatial frequencies or if the vehicle length scale is small compared to the length scales associated with the turbulent fluctuations. The signs used in Eq. (8.5.1) can be determined by considering the relationship between translational disturbances and rotational disturbances. For example, if $\tilde{w}$ increases across the span of the wing in the positive body-fixed y direction, $\partial \tilde{w}/\partial y$ is positive. This positive gradient in $\tilde{w}$ corresponds to the aerodynamic equivalent of a positive disturbance in rolling rate or $\tilde{p}$. Hence, $\tilde{p} \approx \partial \tilde{w}/\partial y$. Similar logic can be used to obtain the correct signs for the other two relationships in Eq. (8.5.1). Note that there is little agreement in the literature as to the signs that should be used in Eq. (8.5.1) [79–81]. The logic presented here for obtaining the signs used in Eq. (8.5.1) appears to most closely follow the correct physics for implementing rotational disturbances.

MIL-F-8785C [79] suggests that application of these partial derivatives yields the following PSD for disturbances in roll, pitch, and yaw

$$\Phi_p(\omega) = \frac{\sigma_w^2}{L_w} \frac{0.8(L_w/L_b)^{1/3}}{1 + (L_b\omega)^2} \tag{8.5.2}$$

$$\Phi_q(\omega) = \frac{\omega^2}{1 + (L_b\omega)^2} \Phi_w(\omega) \tag{8.5.3}$$

$$\Phi_r(\omega) = \frac{\omega^2}{1 + (3L_b\omega/4)^2} \Phi_v(\omega) \tag{8.5.4}$$

where

$$L_b = \frac{4b}{\pi} \tag{8.5.5}$$

and b is the wingspan of the vehicle. Note that direct application of the roll disturbance given in Eq. (8.5.1) would require knowledge of the vertical disturbance $\tilde{w}$ as a function of y in addition to x, whereas the previous section only addressed solutions for obtaining translational disturbances as a function of x. Hence, the roll disturbance must be produced by an independently generated function to $\tilde{u}$, $\tilde{v}$, and $\tilde{w}$. In contrast, the pitch and yaw disturbances should be directly correlated to the translational disturbances $\tilde{w}$ and $\tilde{v}$ respectively.

MIL-STD-1979A [80] implies that best practice for modeling rotational disturbances is to use Eq. (8.5.2) for roll perturbations, but estimate the disturbances in pitch and yaw by considering the influence of the horizontal and vertical stabilizers along with the time-history of the translational disturbances. This is the approach taken here.

Disturbances in Roll Rate

By integrating Eq. (8.5.2), it can be shown that the roll variance is

$$\sigma_p^2 = 0.4\pi\sigma_w^2 \frac{(L_w/L_b)^{1/3}}{L_w L_b} \tag{8.5.6}$$

With this result, Eq. (8.5.2) can alternately be written as

$$\Phi_p(\omega) = \sigma_p^2 \frac{2L_b}{\pi} \frac{1}{1 + (L_b\omega)^2} \tag{8.5.7}$$

Following the same process as described in Section 8.3, it can be shown that the appropriate transfer function for $\tilde{p}$ is

$$F_p(s) = \frac{1}{1 + L_b s} \tag{8.5.8}$$

with a white-noise source having the PSD

$$\Phi_{n_p} = \sigma_w^2 \frac{0.8(L_w/L_b)^{1/3}}{L_w} \tag{8.5.9}$$

The filter differential equations that must be solved is then

$$\tilde{p}' = \frac{\eta_p(x_{ff}) - \tilde{p}}{L_b} \tag{8.5.10}$$

Again, this equation could be directly integrated forward in space, and authors have suggested various techniques. Here we will consider the most common methods for solving this.

Beal's Implementation

Following the implementation of Beal [81], the white noise can be approximated as a step function that holds a constant value between sample locations in the frozen field. This yields the approximation for the PSD of the white-noise signal given in Eq. (8.3.23). Setting Eq. (8.3.23) equal to Eq. (8.5.9) and rearranging yields the required standard deviation to be used in the white-noise generator

$$\sigma_{n_p} = \sigma_w \sqrt{\frac{0.8\pi(L_w/L_b)^{1/3}}{L_w \Delta x_{ff}}} \tag{8.5.11}$$

Second-order central difference approximations at the center of the step yield

$$\tilde{p} \approx \frac{\tilde{p}_i + \tilde{p}_{i-1}}{2}, \quad \tilde{p}' \approx \frac{\tilde{p}_i - \tilde{p}_{i-1}}{\Delta x_{ff}} \tag{8.5.12}$$

Using Eq. (8.5.12) in Eq. (8.5.10) and rearranging yields the algorithm

$$\tilde{p}_i = \frac{(1 - A_p)\tilde{p}_{i-1} + 2A_p\eta_p}{1 + A_p} \tag{8.5.13}$$

where

$$A_p = \frac{\Delta x_{ff}}{2L_b} \tag{8.5.14}$$

and η_p is a random number generated from a normal distribution with standard deviation given in Eq. (8.5.11).

EXAMPLE 8.5.1 Apply the atmospheric turbulence model for roll disturbances given in Eq. (8.5.7) using the implementation suggested by Beal [81]. Use the L_w length scale associated with the Dryden atmospheric turbulence model. Use your code to produce $\tilde{p}$ turbulent fluctuation components for light turbulence at an altitude of 2,000 ft above ground for an aircraft with a wingspan of 30 ft. Produce a dataset with 20,000 points spaced 20.0 ft apart in the frozen field. Test your implementation by repeating the process up to 1,000 times using different seeds for the random number generator. Average the periodogram results and plot the ensemble-averaged results against the model PSD given in Eq. (8.5.7).

Solution: At an altitude of 2,000 ft above the ground, the length scales are given from Eqs. (8.3.5) and (8.5.5)

$$L_w = 875 \text{ ft}$$

$$L_b = \frac{4b}{\pi} = \frac{4(30)}{\pi} = 38.197186342055 \text{ ft}$$

The RMS turbulent amplitude, or standard deviation σ, of the fluctuations at 2,000 ft for light turbulence can be obtained from Fig. 8.2.1 or Table 8.2.1, which yield

$$\sigma_w = 5.0 \text{ ft/s}$$

Using these values along with $\Delta x_{ff} = 20$ ft in Eqs. (8.5.11) and (8.5.14) gives

$$\begin{aligned}
\sigma_{n_p} &= \sigma_w \sqrt{\frac{0.8\pi(L_w/L_b)^{1/3}}{L_w \Delta x_{ff}}} \\
&= 5\sqrt{\frac{0.8\pi(875/38.19718634)^{1/3}}{(875)(20)}} = 0.100979610171 \text{ rad/s}
\end{aligned}$$

and

$$A_p = \frac{\Delta x_{ff}}{2L_b} = \frac{20}{2(38.19718634)} = 0.2617993878$$

Applying the method suggested by Beal in Eq. (8.5.13) yields results shown in Fig. 8.5.1. As can be seen in the figures, as more spatial distributions are considered, the averaged normalized periodogram approaches the PSD function given in Eq. (8.5.7). As was the case with the translational results included in Example 8.3.1, the periodogram results at high spatial frequencies are higher than the prescribed PSD function. Decreasing the step size Δx_{ff} yields more accurate results at higher spatial frequencies.

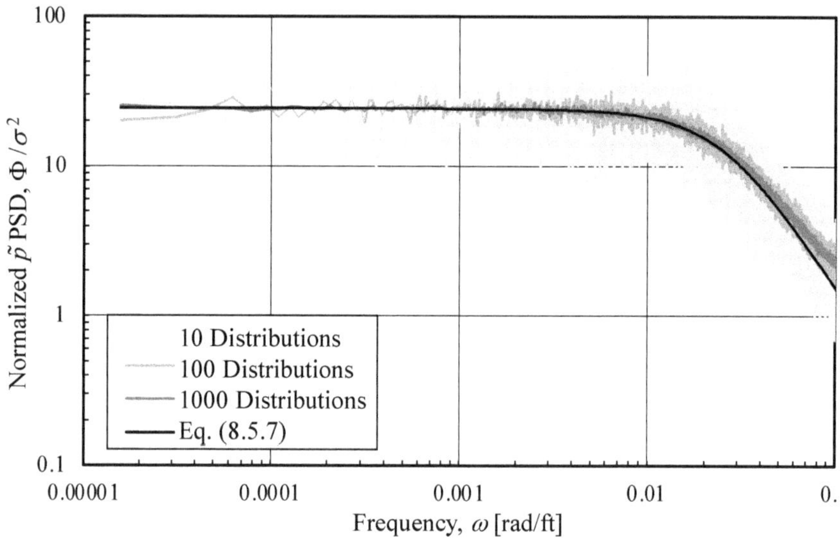

Figure 8.5.1. Normalized $\tilde{p}$ power spectral densities for the Dryden turbulence model using various numbers of distributions.

Military Specification Implementation

The military standard specification MIL-STD-1797A [80] suggests a digital filter implementation for rolling-rate disturbances. According to MIL-STD-1797A, given the gust component $\tilde{p}_{i-1}$ at a previous location in the frozen field, the gust component $\tilde{p}_i$ at the next location in the one-dimensional frozen field can be computed from

$$\tilde{p}_i = (1 - A_p \Delta t)\tilde{p}_{i-1} + \sigma_p \eta_p \sqrt{2A_p \Delta t} \tag{8.5.15}$$

where

$$\sigma_p = \frac{1.9}{\sqrt{2L_w b}}\sigma_w \tag{8.5.16}$$

$$A_p = \frac{2.6}{\sqrt{2L_w b}}$$

and η_p is a random number generated from a normal distribution with a standard deviation of 1.0. Notice that this implementation requires a step in time Δt rather than space Δx_{ff}.

Disturbances in Pitch and Yaw Rates

Whereas the roll disturbance is not directly correlated to the translational disturbances $\tilde{u}$, $\tilde{v}$, and $\tilde{w}$ as a function of x_{ff}, the pitch and yaw disturbances are. Consider an aircraft

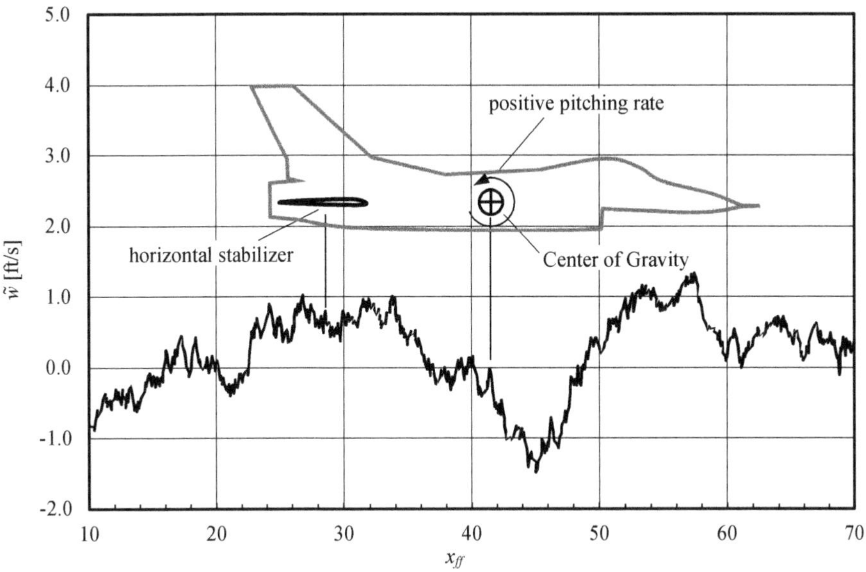

Figure 8.5.2. Effect of vertical velocity disturbances on pitch rate.

traversing a frozen-field of perturbations in vertical velocity $\tilde{w}$ as shown in Fig. 8.5.2. At any point in time or space, the disturbance in pitch rate $\tilde{q}$ depends on the distribution of vertical disturbances $\tilde{w}$ along the surface of the vehicle. Because the horizontal tail is typically the primary aft lifting surface with a significant moment arm relative to the center of gravity, the main contributor to a disturbance in pitch rate typically depends on the difference between the vertical disturbance at the center of gravity $\tilde{w}_{CG}$ and that at the horizontal tail $\tilde{w}_h$. The pitch-rate disturbance can be approximated by the difference in disturbances at the center of gravity and horizontal tail, divided by the horizontal distance between them, i.e.

$$\tilde{q}_i = -\frac{\tilde{w}_{CG} - \tilde{w}_h}{l_h} \tag{8.5.17}$$

where l_h is the axial distance between the aerodynamic center of the horizontal tail and the center of gravity and is positive when the horizontal tail is aft of the center of gravity. If $\tilde{w}$ at the horizontal tail is greater than that at the center of gravity, as shown in Fig. 8.5.2, the resulting aerodynamics are similar to a positive disturbance in the vehicle pitching rate. This disturbance in pitch rate is applied to the aerodynamics, which, for this fixed-wing example with an aft tail, will produce an aerodynamic nose-down or negative pitching moment. Hence, for a fixed-wing aircraft with an aft horizontal tail, a positive disturbance in vertical velocity $\tilde{w}$ at one moment in time will contribute to a nose-down pitching moment a short time later. Because the pitch disturbance depends on the history of the vertical disturbance, $\tilde{w}$ and $\tilde{q}$ are correlated.

Similar logic can be used to develop an approximation for the yaw-rate disturbance $\tilde{r}$ as a function of $\tilde{v}$ and the distance between the center of gravity and the vertical tail. This yields

$$\tilde{r}_i = \frac{\tilde{v}_{CG} - \tilde{v}_v}{l_v} \tag{8.5.18}$$

where $\tilde{v}_v$ is the $\tilde{v}$ disturbance at the location of the vertical tail and l_v is the horizontal distance between the aerodynamic center of the vertical tail and the center of gravity. Note that l_v is positive when the vertical tail is aft of the center of gravity. The sign of the relationship in Eq. (8.5.18) results from the fact that if $\tilde{v}$ at the vertical tail is greater than that at the center of gravity, this disturbance can be modeled as a negative yawing rate in the aerodynamics. For a vehicle with an aft vertical tail, this will produce a positive yawing moment.

The translational disturbances $\tilde{v}_{CG}$ and $\tilde{w}_{CG}$ at the center of gravity at any point in time or space can be computed as discussed in the previous sections. However, in order to compute the translational disturbances $\tilde{v}_v$ and $\tilde{w}_h$, a history of the frozen-field vertical disturbances $\tilde{v}$ and $\tilde{w}$ must be stored as a function of position x_{ff} along the frozen field. This history can be stored as a two-dimensional array with three columns. The first column stores values for the distance behind the CG for which turbulent properties have been computed in the frozen field. The second and third columns store values for the lateral and vertical disturbances, $\tilde{v}$ and $\tilde{w}$ respectively, at the corresponding location in the frozen-field. Linear interpolation can then be used within this array to estimate the $\tilde{v}$ and $\tilde{w}$ disturbances at locations l_h and l_v behind the center of gravity.

EXAMPLE 8.5.2 Apply Eq. (8.5.17) to model atmospheric disturbances in pitch. Use your code to produce $\tilde{q}$ turbulent fluctuation components for light turbulence at an altitude of 2,000 ft above ground for an aircraft with a horizontal stabilizer located 15 ft aft of the center of gravity. Produce a spatial distribution of $\tilde{q}$ with 80,000 points spaced 5.0 ft apart in the frozen field. Test your implementation by repeating the process up to 100 times using different seeds for the random number generator. Average the periodogram results and plot the normalized ensemble-averaged periodogram results. Also report the ensemble average variance of the $\tilde{q}$ distributions.

Solution: Results for $\tilde{w}$ turbulent fluctuations were produced using Beal's implementation as outlined in Section 8.3 and using the same methods demonstrated in Example 8.3.1. Positional history of the turbulent fluctuations was stored in an array containing enough history to include that at the location of the horizontal stabilizer. Linear interpolation was used to compute the turbulent fluctuations at the locations of the horizontal stabilizer. Figure 8.5.3 shows example translational $\tilde{w}$ and rotational $\tilde{q}$ disturbances generated over a distance of 1,000 ft in the frozen field. Note that the translational fluctuations contain some energy at relatively low spatial frequencies whereas the rotational fluctuations mainly contain energy at relatively high spatial frequencies.

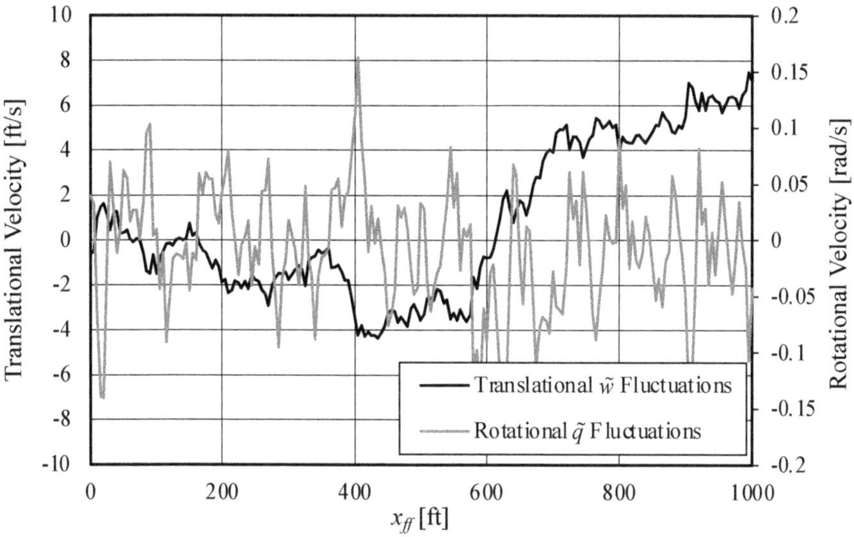

Figure 8.5.3. Example translational $\tilde{w}$ and rotational $\tilde{q}$ distributions as a function of position in the frozen field.

Figure 8.5.4 shows results for the averaged normalized periodograms using 2, 10, and 100 distributions. The ensemble average variance over 100 distributions was computed to be

$$\sigma_q^2 \approx 0.002841 \ \text{(rad/s)}^2$$

The square root of this ensemble average yields a representative RMS or standard deviation of

$$\sigma_q \approx 0.0533 \ \text{rad/s}$$

Note that these values could vary with implementation and sampling rate.

By comparing the data in Fig. 8.5.4 to Figs. 8.3.3–8.3.5 and 8.5.1, it can be seen that the pitching-rate disturbances include little low-frequency energy compared to the translational and rolling-rate disturbances. Additionally, it can be seen that there is a notch in the signal at about ω = 0.42 rad/ft. This corresponds to a spatial frequency of $1/15$ cycles/ft. This results from the fact that any $\tilde{w}$ disturbance with a period of exactly 15 ft, which is the distance between the horizontal stabilizer and the center of gravity, will generate no pitching-moment disturbance since the disturbance at the horizontal stabilizer would be the same as that at the center of gravity. This notch in frequency content would appear for any frequency that is an integer multiple of $1/15$ Hz.

Results in this example have focused on using a $\tilde{w}$ history to predict the $\tilde{q}$ fluctuations. An identical process can be followed using a $\tilde{v}$ history to predict the $\tilde{r}$ fluctuations. The student should repeat this example to verify their implementation of $\tilde{r}$ fluctuations is correct.

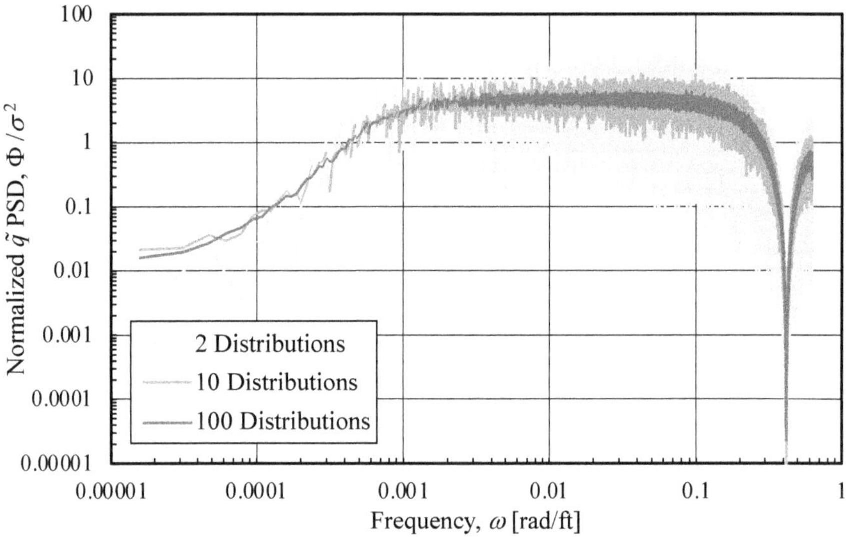

Figure 8.5.4. Average normalized $\tilde{q}$ power spectral densities using various numbers of distributions.

8.6 Application of Turbulence Models to Flight Simulation

The equations of motion given in Chapter 5 were developed under the assumption that the atmospheric coordinate system is not accelerating relative to the Earth-fixed coordinate system. This assumption allowed the atmospheric coordinate system to be used as an inertial coordinate system. Such a development naturally accounts for the effects of constant wind velocity relative to the Earth-fixed coordinate system, and the appropriate constant-wind terms were included in that formulation.

Although atmospheric disturbances can result from a number of sources such as thermals, turbulence, and gusts, each of these disturbances can be modeled in a similar manner. As discussed in the introduction, accurately capturing the physics of atmospheric disturbances is challenging. However, the effects of disturbances can be approximated quite simply by including the disturbances only in the aerodynamic model of the vehicle [79–81]. Such a formulation assumes that the disturbances are zero-mean relative to the average wind. This allows the average wind atmospheric coordinate system to be used for the inertial coordinate system. As stated in section 4.9.3 of MIL-STD-1797, "The gust and turbulence velocities shall be applied to the aircraft equations of motion through the aerodynamic terms only, and the direct effect on the aerodynamic sensors shall be included when such sensors are part of the aircraft augmentation system."

In the aerodynamic models presented in Chapter 3, the pseudo aerodynamic force and moment vectors are functions of the body-fixed velocities u, v, w as well as the body-fixed rotation rates p, q, r

$$\overline{F} = \begin{bmatrix} \overline{F}_{x_b} \\ \overline{F}_{y_b} \\ \overline{F}_{z_b} \end{bmatrix} = f(u, v, w, p, q, r) \tag{8.6.1}$$

$$\overline{M} = \begin{bmatrix} \overline{M}_{x_b} \\ \overline{M}_{y_b} \\ \overline{M}_{z_b} \end{bmatrix} = f(u, v, w, p, q, r) \tag{8.6.2}$$

The suggested approach is to replace the velocity and rate components u, v, w, p, q, r within these functions with the components plus a fluctuating component due to the gust $\tilde{u}, \tilde{v}, \tilde{w}, \tilde{p}, \tilde{q}, \tilde{r}$, to yield

$$\overline{F} = \begin{bmatrix} \overline{F}_{x_b} \\ \overline{F}_{y_b} \\ \overline{F}_{z_b} \end{bmatrix} = f(u + \tilde{u}, v + \tilde{v}, w + \tilde{w}, p + \tilde{p}, q + \tilde{q}, r + \tilde{r}) \tag{8.6.3}$$

$$\overline{M} = \begin{bmatrix} \overline{M}_{x_b} \\ \overline{M}_{y_b} \\ \overline{M}_{z_b} \end{bmatrix} = f(u + \tilde{u}, v + \tilde{v}, w + \tilde{w}, p + \tilde{p}, q + \tilde{q}, r + \tilde{r}) \tag{8.6.4}$$

Hence, the gusts are modeled as an aerodynamic disturbance to the vehicle, not a disturbance to the inertial coordinate system.

Practical implementation of Eqs. (8.6.3) and (8.6.4) can be achieved by calling a turbulence function at the beginning of the `pseudo_aero` function discussed in Chapter 5. The turbulence function accepts the current state of the vehicle and returns the six components of turbulent fluctuations including $\tilde{u}$, $\tilde{v}$, $\tilde{w}$, $\tilde{p}$, $\tilde{q}$, and $\tilde{r}$. Within the turbulence function, the appropriate length scales L_u, L_v, and L_w as well as the appropriate RMS turbulent amplitudes σ_u, σ_v, and σ_w are computed from user settings and the current vehicle height above ground. The step size in the frozen field Δx_{ff} can be computed from

$$\Delta x_{ff} = \sqrt{(x_{f_i} - x_{f_{i-1}})^2 + (y_{f_i} - y_{f_{i-1}})^2 + (z_{f_i} - z_{f_{i-1}})^2} \tag{8.6.5}$$

where the subscript i represents the current location of the vehicle and the subscript $i-1$ represents the location of the vehicle the last time the function was called. Note that upon simulation initialization, it is helpful to set the previous location of the vehicle in this function to the initial state of the vehicle. As will be discussed in Section 9.4, the translational disturbances $\tilde{u}$, $\tilde{v}$, and $\tilde{w}$ should be included when modeling outputs from the air data system and aerodynamic-angle sensors.

EXAMPLE 8.6.1 Consider the F-16 aircraft specified in Appendix C.3 with the aerodynamic model F16-E-1 given in Table C.3.15. Compute the trim state of the vehicle in steady-level flight at an altitude of 3,000 ft and an airspeed of 350 ft/s. Apply the Dryden turbulence model and compute the state of the vehicle as a function of time using a time step of $\Delta t = 0.01$ seconds with all controls held constant. Repeat this 100,000 times and compute the mean and standard deviation in the final u, v, w, p, q, r, x_f, y_f, z_f states of the vehicle after 10 seconds of flight. Also plot the resulting probability density of the x_f, y_f, and z_f coordinates using 70 bins spanning ± 4 standard deviations about the mean in comparison to a normal distribution with the same standard deviation. Repeat this study using light, moderate, and severe turbulence and compare the results. Assume a flat-Earth model for all simulations.

Solution: The trim solution for this vehicle at this flight condition is $\theta = \alpha = 7.973921791130$ deg, $\delta_e = -9.258535573716$ deg, $\tau = 0.075538674855$, and $\beta = p = q = r = \delta_a = \delta_r = 0$. One example of running 100,000 10-second simulations as prescribed in the problem statement yields the mean values shown in Table 8.6.1 and the standard deviation values shown in Table 8.6.2. Note the significant increase in standard deviation of the states at higher turbulence levels. Repeating the process yields results that are similar but not identical to those shown in these tables. The student is encouraged to make sure their code matches the values reported here reasonably well, but not out to the precision shown.

Probability density plots for the final x_f, y_f, and z_f coordinates after 10 seconds are shown in Figs. 8.6.1–8.6.3. These plots were generated using 70 bins over a range of ± 4 standard deviations about the mean. Data is shown as data points, and the lines represent a normal probability density function with the same standard deviation. The plots show that the results are very nearly normally distributed. Note that the mean of the z_f coordinate shifts to the right, representing an average lower final altitude at higher turbulence levels.

Table 8.6.1. Approximate mean state values after 10 seconds over 100,000 simulations with varying turbulence levels.

State	Light	Moderate	Severe
u [ft/s]	346.7	346.8	347.1
v [ft/s]	0.007	0.018	-0.046
w [ft/s]	48.47	48.32	47.81
p [rad/s]	0.00	0.00	0.00
q [rad/s]	0.00	0.00	0.00
r [rad/s]	0.00	0.00	0.00
x_f [ft]	3499.6	3498.7	3495.5
y_f [ft]	-0.047	0.090	1.169
z_f [ft]	-2998.0	-2991.9	-2974.7

Table 8.6.2. Approximate standard deviation in state values after 10 seconds over 100,000 simulations with varying turbulence levels.

State	Light	Moderate	Severe
u [ft/s]	2.48	5.00	9.01
v [ft/s]	5.81	11.70	20.91
w [ft/s]	4.84	9.75	17.49
p [rad/s]	0.040	0.080	0.145
q [rad/s]	0.008	0.016	0.029
r [rad/s]	0.019	0.038	0.068
x_f [ft]	8.28	16.64	29.88
y_f [ft]	40.48	80.19	142.0
z_f [ft]	22.62	45.48	82.35

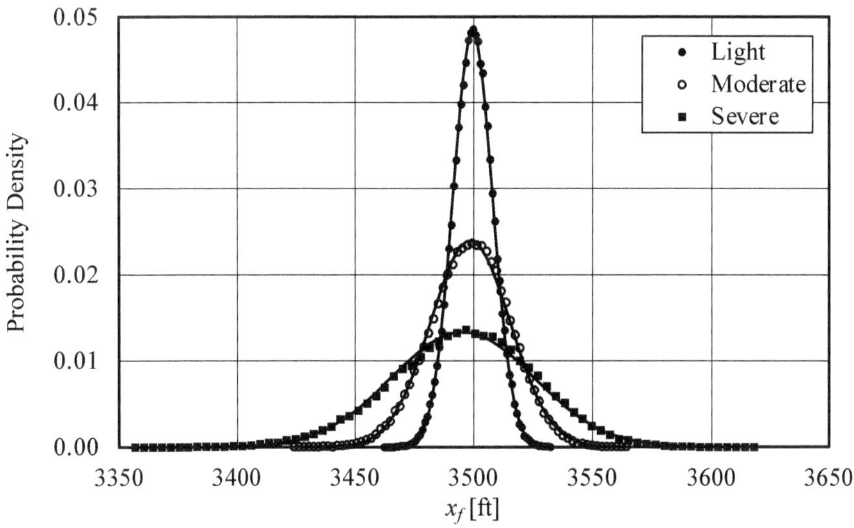

Figure 8.6.1. Probability density results for the x_f coordinate of the vehicle after 10 seconds in varying levels of turbulence with controls fixed.

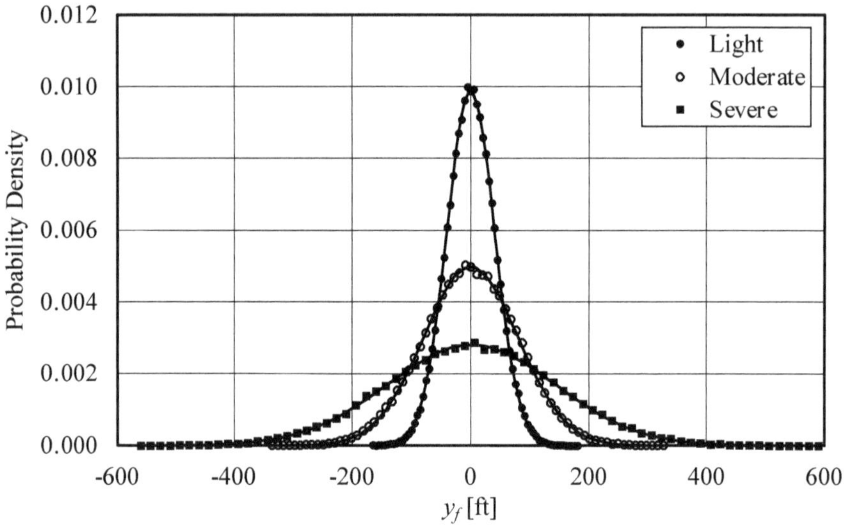

Figure 8.6.2. Probability density results for the y_f coordinate of the vehicle after 10 seconds in varying levels of turbulence with controls fixed.

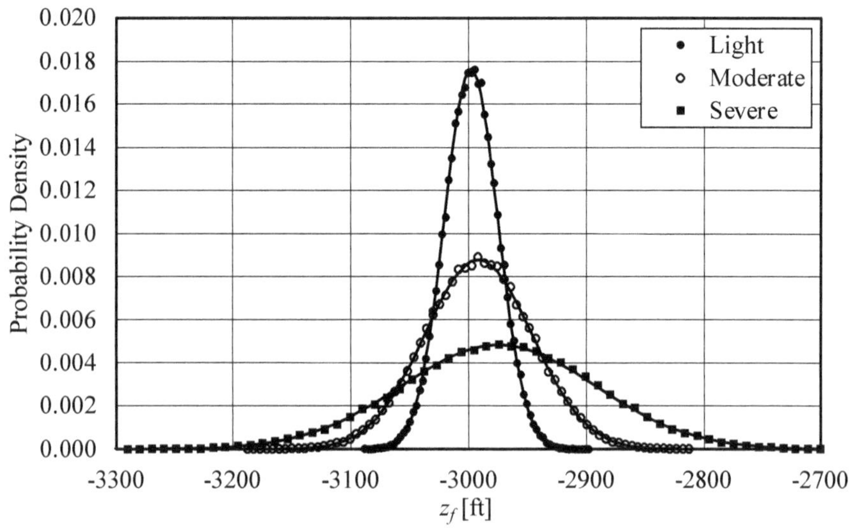

Figure 8.6.3. Probability density results for the z_f coordinate of the vehicle after 10 seconds in varying levels of turbulence with controls fixed.

8.7 Problems

Theory Problems

8.T1 The military specification documents MIL-F-8785C [79] and MIL-STD-1797A [80] include variations for how the Dryden model might be implemented. Although the second document is more recent, some of the equation typesetting is more easily discernible in the older document. Select one variation for implementing the Dryden turbulence model, and implement it in your code. Perform an average periodogram analysis of many distributions generated from your model and compare these to the Dryden turbulence PSD given in Eqs. (8.3.1)–(8.3.3). Comment on the accuracy, advantages, and disadvantages of the implementation.

8.T2 The military specification document MIL-STD-1797A [80] includes a list of many atmospheric models that have been published and used (see Table LVI. A Survey of atmospheric disturbance models). Select one of the models listed in the table and study the associated publication. If time permits, implement the model in your code and compare the spectral output of the model to the Dryden turbulence model.

Coding Problems

8.C1 Repeat Example 8.1.1 in a coding language of your choice.

8.C2 Write a function that computes the normalized periodogram of a signal. The function should accept as input an array of data and a uniform sample rate. The function should return the normalized periodogram of the data with frequency in Hz and the normalized periodogram values in units of 1/Hz.

8.C3 Repeat Example 8.1.2 using your own periodogram code from Problem 8.C2 to compute the estimated normalized PSD of the signals.

8.C4 Write a computer code to generate translational atmospheric turbulence for the Dryden model using the method suggested by Beal [81]. Test your code by repeating Example 8.3.1.

8.C5 Write a computer code to generate translational atmospheric turbulence for the Dryden model using the method suggested in the military standard specification MIL-STD-1797A [80]. Test your code by repeating Example 8.3.2.

8.C6 Write a computer code to generate atmospheric rolling-rate disturbances for the Dryden atmospheric model using the method suggested by Beal [81]. Test your code by repeating Example 8.5.1.

8.C7 Write a computer code to generate atmospheric pitching-rate and yawing-rate disturbances based on a history of $\tilde{v}$ and $\tilde{w}$ and a known distance between the center of gravity and the horizontal and vertical stabilizers. Test your code by repeating Example 8.5.2.

8.C8 Link your atmospheric turbulence model code to your flight simulator such that the aerodynamics of the vehicle are computed using the disturbances from the atmospheric turbulence model as outlined in Section 8.6. Test your code by repeating Example 8.6.1.

Chapter 9

Actuators, Sensors, and Communication

9.1 Introduction

In the previous chapters we considered how to predict the state of a vehicle as a function of time from the rigid-body equations of motion, including geometric, mass, aerodynamic, and altitude effects. We also included the influence of control effectors on the vehicle state as well as algorithms that can be used to obtain appropriate control-effector settings for a variety of common trim conditions.

In the present chapter, we will consider important aspects of flight simulation that must be included to accurately simulate the effects of dynamic pilot interactions with the system. We will use the term *pilot* to mean any system that accepts outputs from the system in order to determine and command new inputs to the system. In this sense, there is no difference whether the physics simulator is interfacing with a human pilot or an auto pilot.

Inputs to the physics simulator from a pilot typically consist of control-effector commands. If a commanded value for a control effector differs from the current value of the control effector, the dynamics of the system requires a finite amount of time for the control effector to assume the new commanded position or value. This is called *actuator dynamics*. In the present chapter, we will consider methods for modeling both first- and second-order actuator dynamics.

Outputs from the physics simulator to the pilot typically consist of a set of commonly used sensors. Accurately modeling sensor outputs, including appropriate sensor error, is imperative to producing realistic simulators with which human pilots and autopilots can be trained. In the present chapter, we will consider how to simulate the outputs of sensors that are commonly used on flight vehicles.

Finally, in order to pass information from a pilot to the simulator or vice versa, communication timing, data-rate limits, and protocols must be agreed upon between the physics simulator and pilot software. These topics will also be addressed in the present chapter.

A computer code capable of predicting actuator dynamics, simulating sensors, and electronically communicating with a human pilot or autopilot, in addition to the capabilities outlined in the previous chapters, constitutes the foundation of a physics simulation package. If properly constructed, it can be linked to many other software packages including control algorithms, graphics software, and human-pilot controls or autopilot systems to accurately predict the dynamic response of a vehicle required in many simulation architectures.

9.2 First-Order Actuator Dynamics

An *actuator* is any device that moves or controls a system or mechanism in response to a commanded input. Actuators are typically mechanical or electromechanical, and are typically used on aircraft to manipulate control surfaces, throttle, or thrust vectoring. Actuators convert the control signal from a pilot or autopilot into mechanical movement such as a control-surface deflection or the fuel-flow rate of an engine. In the following development, we will use the notation δ to represent the state of an actuator, and δ_c to represent the commanded state of the actuator. Both of these will be treated as functions of time.

Actuators typically have physical limits in terms of values that can be attained. These limits are known as *saturation limits*. For example, a control surface typically has a maximum and minimum deflection such that $\delta_{\text{min}} \le \delta \le \delta_{\text{max}}$. Actuators also typically have limits on the rates at which they can move and accelerate such that $\dot{\delta}_{\text{min}} \le \dot{\delta} \le \dot{\delta}_{\text{max}}$ and $\ddot{\delta}_{\text{min}} \le \ddot{\delta} \le \ddot{\delta}_{\text{max}}$. For example, the hydraulic system driving a control surface has a maximum rate at which it can move the control surface, and a jet engine has a maximum rate at which it can accelerate from one throttle setting to another.

Because the state of an actuator can vary with time and depends on the commanded value as well as the actuator dynamics, to generate realistic simulations of vehicle response to control inputs, the dynamic response of the actuators must be included in the simulation. Typically, actuator dynamics are modeled using first-order or second-order dynamic models. In this section we will consider first-order dynamics. Second-order dynamics will be considered in Section 9.3.

The state of an actuator as a function of time $\delta(t)$ can be expressed as a first-order differential equation in the form

$$\frac{d\delta(t)}{dt} = \sigma[\delta_c(t) - \delta(t)] \tag{9.2.1}$$

where $\delta_c(t)$ is the commanded state as a function of time, σ is the damping rate, and $1/\sigma$ is the actuator time constant. The general solution to this differential equation is

$$\delta(t) = \frac{\int \sigma e^{\sigma t} \delta_c(t) dt}{e^{\sigma t}} \tag{9.2.2}$$

To gain some insight into first-order actuator dynamics, let us consider the closed-form solution for the case of constant input.

Closed-Form Solution for Constant Input

Consider an actuator system where the commanded value is a constant, independent of time. Using $\delta_c(t) = \delta_c$ in Eq. (9.2.2) yields the general solution for the case when the control input is independent of time

$$\delta(t) = \frac{\int \sigma e^{\sigma t} \delta_c dt}{e^{\sigma t}} = \frac{\delta_c e^{\sigma t} + C}{e^{\sigma t}} = \delta_c + \frac{C}{e^{\sigma t}} \tag{9.2.3}$$

where C is a constant of integration. The value for C can be obtained by considering the solution at the initial condition $t = t_0$ and $\delta(t_0) = \delta_0$

$$\delta_0 = \delta_c + \frac{C}{e^{\sigma t_0}} \tag{9.2.4}$$

Solving for C gives

$$C = e^{\sigma t_0}(\delta_0 - \delta_c) \tag{9.2.5}$$

Using this in Eq. (9.2.3) yields the closed-form solution for constant input

$$\delta(t) = \delta_c + (\delta_0 - \delta_c)e^{-\sigma(t-t_0)} \tag{9.2.6}$$

For the special case when $t_0 = \delta_0 = 0$, Eq. (9.2.6) simplifies to

$$\delta(t) = \delta_c(1 - e^{-\sigma t}) \tag{9.2.7}$$

EXAMPLE 9.2.1 Evaluate the first-order actuator dynamics for the case of $\delta_0 = 0$ at $t_0 = 0$ and a constant commanded actuator state of $\delta_c = 1$. Consider how the solution changes for various values of the time constant $1/\sigma$.

Solution: Equation (9.2.7) gives the closed-form solution for this case. Using $\delta_c = 1$ in Eq. (9.2.7) gives the analytic solution

$$\delta(t) = 1 - e^{-\sigma t}$$

Figure 9.2.1 shows time-history solutions for this case with varying values of the time constant $1/\sigma$. Note that the time-constant denotes the time at which the solution passes the actuator value of about 0.632, or when the actuator has moved about 63.2% of the distance from the initial value to the commanded value. Larger time constants converge more slowly than small time constants.

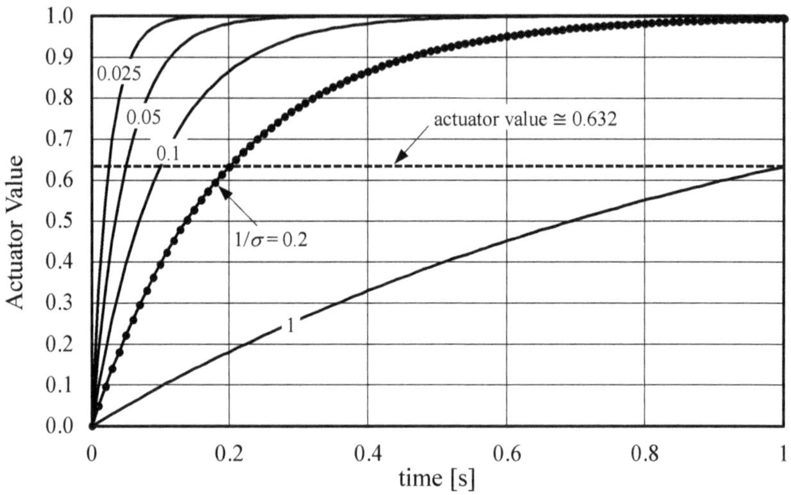

Figure 9.2.1. Actuator response as a function of time for different time constants $1/\sigma$. Solid lines represent exact solutions and the data points represents solutions using fourth-order Runge-Kutta integration for the case of $1/\sigma = 0.2$ and $\Delta t = 0.01$.

The example above is useful for gaining insight into the nature of first-order actuator dynamics and how the time constant is related to the actuator response. However, the example considers only the case when the commanded actuator value is constant and an analytic solution exists. In the case of flight simulation, the commanded actuator value is typically a function of time, and actuator response is obtained using numerical integration.

Numerical Integration

Equation (9.2.1) is typically solved numerically in flight simulation software. Given an initial condition, Eq. (9.2.1) can be numerically integrated using any number of integration schemes. For example, Fig. 9.2.1 shows example solutions using fourth-order Runge-Kutta integration for the case of $\delta_0 = 0$, $\delta_c(t) = \delta_c = 1$, and $1/\sigma = 0.2$, with $\Delta t = 0.01$. As the time step size is decreased, the solutions approach the analytic solutions out to machine precision.

For application in a flight simulator, Eq. (9.2.1) can be added directly to the state vector developed in Chapter 5 and integrated forward in time along with the equations of motion within the function `diff_eq`. For each actuator, an additional state can be added to the array that contains the entire state of the aircraft. For example, if the quaternion formulation is used along with an aircraft with the four traditional controls including aileron deflection δ_a, elevator deflection δ_e, rudder deflection δ_r, and throttle

τ, the differential equations would include those given in Eqs. (5.6.1)–(5.6.4) with the additional four equations

$$\begin{bmatrix} \dot{\delta}_a \\ \dot{\delta}_e \\ \dot{\delta}_r \\ \dot{\tau} \end{bmatrix} = \begin{bmatrix} \sigma_{\delta_a}(\delta_{a_c} - \delta_a) \\ \sigma_{\delta_e}(\delta_{e_c} - \delta_e) \\ \sigma_{\delta_r}(\delta_{r_c} - \delta_r) \\ \sigma_\tau(\tau_c - \tau) \end{bmatrix} \tag{9.2.8}$$

The symbols δ_a, δ_e, δ_r, and τ are the actual values of the actuators as a function of time, and those with the subscript c denote the commanded values as a function of time from a pilot or control algorithm. The time constant or damping rate for each actuator must be known and can be an input to the code.

Using a first-order model for the actuator dynamics yields the value of the actuator δ and the first derivative of the actuator as a function of time $\dot{\delta}$. Because this information is known within the integration routine, saturation limits on δ and $\dot{\delta}$ can be directly applied during integration. For example, any value for the commanded value of an actuator δ_c can be accepted as input, and need not be limited. However, within the `diff_eq` function prior to computing Eq. (9.2.8), the value of the actuator should be limited using

$$\delta = \max(\min(\delta, \delta_{\max}), \delta_{\min}) \tag{9.2.9}$$

This same limiter should also be applied within the `pseudo_aero` function and at the end of each full time step. Similarly, within the `diff_eq` function, the actuator rates should be limited by applying the following after computing the rate $\dot{\delta}$ from Eq. (9.2.8)

$$\dot{\delta} = \max(\min(\dot{\delta}, \dot{\delta}_{\max}), \dot{\delta}_{\min}) \tag{9.2.10}$$

A final limiter should be applied before exiting the `diff_eq` function, which sets the rate to zero if the actuator is already saturated in the direction of the rate. Algorithm 9.2.10 gives an outline of how the first-order dynamics can be implemented, including limits.

Algorithm 9.2.10 First-order actuator dynamics with limiting.

Input
- δ — (real) Actuator value from state array
- δ_c — (real) Commanded actuator value
- σ — (real) Actuator damping rate
- $\delta_{\min}, \dot{\delta}_{\min}$ — (real) Minimum actuator value and rate limits
- $\delta_{\max}, \dot{\delta}_{\max}$ — (real) Maximum actuator value and rate limits

$\delta = \max(\min(\delta, \delta_{\max}), \delta_{\min})$ — Limit value

$\dot{\delta} = \sigma(\delta_c - \delta)$

$\dot{\delta} = \max(\min(\dot{\delta}, \dot{\delta}_{\max}), \dot{\delta}_{\min})$ — Limit rate

if$(\delta <= \delta_{\min}$ and $\dot{\delta} < 0)$ $\dot{\delta} = 0$ — Limit rate if moving against a magnitude limit

if$(\delta >= \delta_{\max}$ and $\dot{\delta} > 0)$ $\dot{\delta} = 0$

EXAMPLE 9.2.2 Consider the F-16 aircraft specified in Appendix C.3 with the aerodynamic model F16-E-1 given in Table C.3.15 at an altitude of 1,000 ft and a velocity of 330 ft/s. Beginning from the trim solution for the vehicle in steady-level flight, compute the time-history of the vehicle states, including actuator states, for a total of 2 seconds, if at the beginning of the simulation the commanded value for the elevator is exactly $\delta_e = 0$. Use a time step of $\Delta t = 0.01$ to compute the states of the vehicle as a function of time for elevator-actuator time constants of $1/\sigma = 0.05$, $1/\sigma = 0.2$, and $1/\sigma = 0.5$. Plot the elevator deflection δ_e as a function of time for each case. Comment on the results. Be sure to enforce the actuator rate limit for the elevator of $-60.0 \leq \dot{\delta}_e \leq 60.0$ deg/s as specified in Appendix C.3.

Solution: The trim solution for this case was obtained using Newton's method along with the specified climb angle of $\gamma = 0$ as outlined in Chapters 6 and 7. The trim solution for steady-level flight is $\theta = \alpha = 8.442080225436$ deg, $\delta_e = -9.439213266902$ deg, $\tau = 0.074041048286$ where all other states and control-surface deflections are zero. Using this as the initial condition, setting the commanded value of the elevator to $\delta_e = 0$ immediately after initialization of the simulation, and employing first-order actuator dynamics with rate limiting as outlined in this section, yields the time histories of the elevator deflection shown in Fig. 9.2.2. Solid lines represent the analytic solution from Eq. (9.2.6), and the data points represent numerical results from the fourth-order integration scheme.

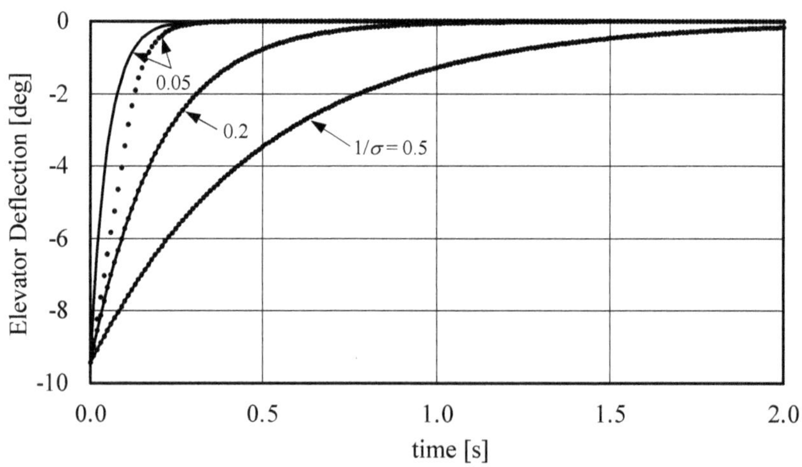

Figure 9.2.2. Elevator deflection as a function of time for different time constants $1/\sigma$. Solid lines represent analytic solutions and the data points represents solutions using fourth-order Runge-Kutta integration with rate limiting.

Table 9.2.1. Sample states of the vehicle after 2.0 seconds for Example 9.2.2.

Time Constant	$1/\sigma = 0.05$ [s]	$1/\sigma = 0.2$ [s]	$1/\sigma = 0.5$ [s]
u [ft/s]	3.243602299E+02	3.257435970E+02	3.295489569E+02
v [ft/s]	1.477195692E−01	1.378582223E−01	1.060075259E−01
w [ft/s]	-5.856150021E+01	-5.511043707E+01	-4.131270239E+01
p [rad/s]	-1.884653037E−03	-1.673431046E−03	-1.216976517E−03
q [rad/s]	-2.969530721E−01	-3.120580584E−01	-3.235500112E−01
r [rad/s]	1.111469790E−05	-8.231771792E−05	-2.018724614E−04
δ_e [deg]	-1.091300422E−16	-4.285398520E−04	-1.728852225E−01
τ	7.404104829E−02	7.404104829E−02	7.404104829E−02

Note that for time constants of 0.2 and 0.5, the difference between the numerical and analytic solutions is nearly imperceptible. However, for the time constant of 0.05, the response of the elevator is slower than that of the analytic solution. This is due to the rate limiter applied in the numerical integration method, which does not allow a rate greater than 60 deg/s. Table 9.2.1 includes the translational and rotational velocity states of the vehicle at a time of $t = 2.0$ seconds for each case. The states of the vehicle as a function of time for these examples are included on the companion website.

EXAMPLE 9.2.3 Consider the F-16 aircraft specified in Appendix C.3 with the aerodynamic model F16-E-1 given in Table C.3.15 at an altitude of 10,000 ft and a velocity of 500 ft/s. Beginning from the trim solution for the vehicle in a steady-heading sideslip with a bank angle of $\phi = 40$ deg and a climb angle of $\gamma = 45.0$ deg, compute the time-history of the vehicle states, including actuator states, over a period of 1.0 second, if at the beginning of the simulation the commanded value for the aileron, elevator, rudder, and throttle is exactly $\delta_a = \delta_e = \delta_r = \tau = 0$. Use a time step of $\Delta t = 0.01$ along with the rate limits and time constants for each actuator given in Appendix C.3. Plot the actual control values as a function of time.

Solution: The trim solution for this case was obtained using Newton's method along with the specified climb angle of $\gamma = 0$ as outlined in Chapters 6 and 7. The trim solution for this case is $\theta = 53.636741877497$ deg, $\alpha = 2.241804969945$ deg, $\beta = 10.184336281017$ deg, $\delta_a = -1.960548979502$ deg, $\delta_e = -7.052530081649$ deg, $\delta_r = 29.602685429820$ deg, $\tau = 0.764708125031$, and $p = q = r = 0.0$. Using this as the initial condition, setting the commanded values of all controls to zero immediately after initialization of the simulation, and employing first-order actuator dynamics with rate limiting as outlined in this section, yields the time histories of the controls shown in Fig. 9.2.3. Table 9.2.2 includes the velocity states of the vehicle at a few time steps. The full states of the vehicle as a function of time for these examples are included on the companion website. Note that due to the throttle time constant of 3 seconds, the throttle setting is still above 50% after 1 second. Additionally, the linear change in rudder deflection during the first 0.2 seconds indicates that the rudder deflection is rate limited during this portion of the transient.

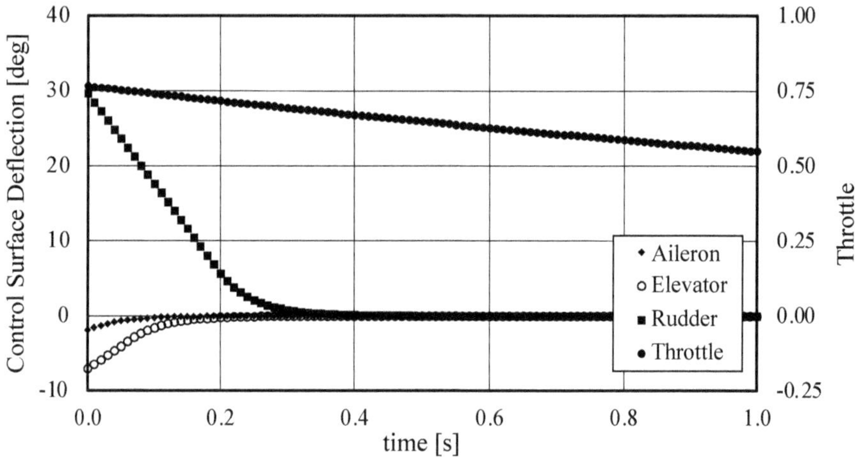

Figure 9.2.3. Example control-surface deflections and throttle setting as a function of time using first-order actuator dynamics and a step change in commanded values.

Table 9.2.2. Sample states of the vehicle at specified time steps for Example 9.2.3.

Time	0.0 [s]	0.5 [s]	1.0 [s]
u [ft/s]	4.9174534180E+02	4.9624536563E+02	4.8445441324E+02
v [ft/s]	8.8407835448E+01	4.6440263720E+01	-2.1055372305E+01
w [ft/s]	1.9250284456E+01	-5.6599421272E+00	-8.2264300609E+01
p [rad/s]	0.0000000000E+00	-7.2187281245E−01	-2.9527283094E−01
q [rad/s]	0.0000000000E+00	-3.6547166305E−01	-4.8741901431E−01
r [rad/s]	0.0000000000E+00	3.4353227848E−01	2.4801901784E−01
δ_a [deg]	-1.9605489795E+00	-8.9022812746E−05	-4.0422663611E−09
δ_e [deg]	-7.0525300816E+00	-5.2614361270E−04	-2.3890647365E−08
δ_r [deg]	2.9602685430E+01	1.3921813036E−02	6.3214893784E−07
τ	7.6470812503E−01	6.4731145271E−01	5.4793731504E−01

As has been shown here, first-order actuator dynamics are relatively simple to model in a flight-simulation code by adding a first-order differential equation and state for each actuator in the system. Additionally, actuation rate limits can be easily applied. This approach for modeling actuator dynamics provides relatively accurate results for a wide range of actuation systems. However, in many cases first-order actuator dynamics are not sufficiently accurate, and second-order actuator dynamics should be modeled.

9.3 Second-Order Actuator Dynamics

Whereas first-order dynamics are sufficient for modeling many actuation systems, in some cases second-order models are needed to capture the main time-dependent dynamics of an actuator. For example, second-order dynamics are required when there is a significant amount of mass, inertia, or mechanical compliance in an actuation system.

In general, a second-order spring-mass-damper system for an actuator can be expressed as

$$m\frac{d^2\delta(t)}{dt^2} + c\frac{d\delta(t)}{dt} + k[\delta(t) - \delta_c(t)] = 0 \tag{9.3.1}$$

where m is the mass, c is the damping coefficient, and k is the spring constant for the system. In this form, $\delta_c(t)$ is considered the forcing function, which is the commanded input to the actuator. Note that the first-order system given in Eq. (9.2.1) is a special case of the second-order system where $m = 0$ and $\sigma = k/c$.

In the absence of the forcing function, i.e. $\delta_c(t) = 0$, Eq. (9.3.1) is a homogeneous second-order differential equation. However, in general, the commanded value $\delta_c(t)$ will be nonzero. Hence, Eq. (9.3.1) is a nonhomogeneous second-order differential equation. The general solution to Eq. (9.3.1) is the sum of a homogeneous solution $\delta_h(t)$ and a particular solution $\delta_p(t)$

$$\delta(t) = \delta_h(t) + \delta_p(t) \tag{9.3.2}$$

Solutions to the homogeneous equation depend on the nature of the eigenvalues of the system, which can be found from solving the characteristic equation

$$m\lambda^2 + c\lambda + k = 0 \tag{9.3.3}$$

This yields two solutions, which are

$$\lambda_{1,2} = \frac{-c \pm \sqrt{c^2 - 4mk}}{2m} \tag{9.3.4}$$

From Eq. (9.3.4) it can be seen that there are three possible forms of the eigenvalues. The eigenvalues will either be distinct and real, identical and real, or complex conjugates. The sign of the real part of the eigenvalues indicates the stability of the dynamic system. If the real parts of both eigenvalues are positive, the system is divergent. If the real parts of both eigenvalues are negative, the system is convergent. **To simplify the following analysis, we will assume that the second-order actuation system is convergent, meaning that the real part of both eigenvalues are negative. This also means that if the forcing function is held constant, the system will eventually converge to the forcing-function value**. For this case, the eigenvalues can alternately be written as

$$\lambda_{1,2} = -\zeta\omega_n \pm \omega_n\sqrt{\zeta^2 - 1} \tag{9.3.5}$$

where the undamped natural frequency ω_n of the system is

$$\omega_n = \sqrt{\lambda_1\lambda_2} = \sqrt{k/m} \tag{9.3.6}$$

and the damping ratio ζ is

$$\zeta = -\frac{\lambda_1 + \lambda_2}{2\sqrt{\lambda_1 \lambda_2}} = \frac{c/m}{2\sqrt{k/m}} \tag{9.3.7}$$

The damping rate is then

$$\sigma = -\text{real}(\lambda) = \zeta\omega_n = \frac{c}{2m} \tag{9.3.8}$$

The eigenvalues obtained from the homogeneous solution describe the dynamics of the system in the absence of a forcing function $\delta_c(t)$. This is called the *free response* of the system. Using the relations between the undamped natural frequency, the damping ratio, and the constants m, c, and k, Eq. (9.3.1) can be expressed as

$$\ddot{\delta}(t) + 2\zeta\omega_n\dot{\delta}(t) + \omega_n^2[\delta(t) - \delta_c(t)] = 0 \tag{9.3.9}$$

Equation (9.3.9) is an attractive form of the second-order system, since the free-response characteristics of a second-order actuator are often known in terms of the undamped natural frequency and damping ratio. The undamped natural frequency ω_n is related to the responsiveness of the free-response system, and the damping ratio ζ is related to the stability of the free-response system. If $\zeta < 1$, the free-response system is *underdamped* and may overshoot the commanded value and oscillate. If $\zeta > 1$, the free-response system is *overdamped* and will not oscillate. If $\zeta = 1$, the free-response system is *critically damped*, and will converge as rapidly as possible without overshoot.

For the case when the eigenvalues are distinct (i.e. $\zeta \neq 1$), solutions to the homogeneous equation have the form

$$\delta_h(t) = C_1 e^{\lambda_1 t} + C_2 e^{\lambda_2 t} \tag{9.3.10}$$

where C_1 and C_2 are constants. For the case when the eigenvalues are identical (i.e. $\zeta = 1$), the system is critically damped. For this case, the eigenvalues have the form

$$\lambda_1 = \lambda_2 = \lambda = -\zeta\omega_n \tag{9.3.11}$$

and solutions to the homogeneous equation have the form

$$\delta_h(t) = (C_1 + C_2 t)e^{\lambda t} \tag{9.3.12}$$

To gain some insight into second-order actuator dynamics, let us consider the closed-form solution for the case of constant input.

Closed-Form Solution for Constant Input

Consider an actuator system where the commanded value is a constant, independent of time. In this case, the particular solution is also a constant, $\delta_p(t) = \delta_c$. The closed-form solution for the homogeneous solution depends on whether the eigenvalues are distinct ($\zeta \neq 1$) or identical ($\zeta = 1$) .

Distinct Eigenvalues

For the case when the eigenvalues are distinct, using the particular solution and Eq. (9.3.10) in Eq. (9.3.2) yields the general solution

$$\delta(t) = C_1 e^{\lambda_1 t} + C_2 e^{\lambda_2 t} + \delta_c \tag{9.3.13}$$

The values for C_1 and C_2 can be obtained by considering the general solution and the first derivative of the general solution at the initial condition. At the initial condition, the initial time is $t = t_0$ and the initial actuator state and rate are $\delta(t_0) = \delta_0$ and $\dot{\delta}(t_0) = \dot{\delta}_0$, respectively. The solution for Eq. (9.3.13) at the initial condition is

$$\delta_0 = C_1 e^{\lambda_1 t_0} + C_2 e^{\lambda_2 t_0} + \delta_c \tag{9.3.14}$$

The first derivative of Eq. (9.3.13) evaluated at the initial condition yields an expression for the actuator rate at the initial condition, which is

$$\dot{\delta}_0 = C_1 \lambda_1 e^{\lambda_1 t_0} + C_2 \lambda_2 e^{\lambda_2 t_0} \tag{9.3.15}$$

Equations (9.3.14) and (9.3.15) form a system of equations that can be used to solve for the constants C_1 and C_2. To simplify this analysis, we will consider the case with the initial time of $t_0 = 0$. For this case, Eqs. (9.3.14) and (9.3.15) simplify to the system of equations

$$C_1 + C_2 = \delta_0 - \delta_c \tag{9.3.16}$$

$$C_1 \lambda_1 + C_2 \lambda_2 = \dot{\delta}_0 \tag{9.3.17}$$

Solving this system of equations for the constants gives

$$C_1 = \frac{\dot{\delta}_0 - \lambda_2(\delta_0 - \delta_c)}{\lambda_1 - \lambda_2} \tag{9.3.18}$$

$$C_2 = \frac{\lambda_1(\delta_0 - \delta_c) - \dot{\delta}_0}{\lambda_1 - \lambda_2} = \delta_0 - \delta_c - C_1 \tag{9.3.19}$$

Using these constants in Eq. (9.3.13), the closed-form solution for constant input $\delta_c(t) = \delta_c$ where $t_0 = 0$ in the case of distinct eigenvalues is

$$\delta(t) = \frac{\left[\dot{\delta}_0 - \lambda_2(\delta_0 - \delta_c)\right] e^{\lambda_1 t} + \left[\lambda_1(\delta_0 - \delta_c) - \dot{\delta}_0\right] e^{\lambda_2 t}}{\lambda_1 - \lambda_2} + \delta_c \tag{9.3.20}$$

The method for numerically computing the solution to this equation depends on the damping ratio ζ. If $\zeta < 1$, the eigenvalues are complex conjugates, as can be seen from Eq. (9.3.5). Hence, numerically evaluating Eq. (9.3.20) must be done using complex arithmetic. If $\zeta > 1$, the eigenvalues are real and distinct, as can be seen from Eq. (9.3.5). For this case, numerically evaluating Eq. (9.3.20) can be done using real arithmetic.

Identical Eigenvalues

For the case when the eigenvalues are identical, using the particular solution and Eq. (9.3.12) in Eq. (9.3.2) yields the general solution

$$\delta(t) = (C_1 + C_2 t)e^{\lambda t} + \delta_c \tag{9.3.21}$$

The values for C_1 and C_2 can be obtained by considering the general solution and the first derivative of the general solution at the initial condition. The solution for Eq. (9.3.21) at the initial condition is

$$\delta_0 = (C_1 + C_2 t_0)e^{\lambda t_0} + \delta_c \tag{9.3.22}$$

The first derivative of Eq. (9.3.21) evaluated at the initial condition yields an expression for the actuator rate at the initial condition, which is

$$\dot{\delta}_0 = (C_1\lambda + C_2 + C_2\lambda t_0)e^{\lambda t_0} \tag{9.3.23}$$

Equations (9.3.22) and (9.3.23) form a system of equations that can be used to solve for the constants C_1 and C_2. This yields

$$C_1 = \frac{(\delta_0 - \delta_c) - t_0\left(\dot{\delta}_0 - \lambda(\delta_0 - \delta_c)\right)}{e^{\lambda t_0}} \tag{9.3.24}$$

$$C_2 = \frac{\dot{\delta}_0 - \lambda(\delta_0 - \delta_c)}{e^{\lambda t_0}} \tag{9.3.25}$$

Using these constants in Eq. (9.3.21), the closed-form solution for constant input $\delta_c(t) = \delta_c$ in the case of identical eigenvalues is

$$\delta(t) = \{(\delta_0 - \delta_c) + [\dot{\delta}_0 - \lambda(\delta_0 - \delta_c)](t - t_0)\}e^{\lambda(t-t_0)} + \delta_c \tag{9.3.26}$$

For the special case when $t_0 = 0$, this simplifies to

$$\delta(t) = [(\delta_0 - \delta_c) + \dot{\delta}_0 t - \lambda t(\delta_0 - \delta_c)]e^{\lambda t} + \delta_c \tag{9.3.27}$$

EXAMPLE 9.3.1 Evaluate the second-order actuator dynamics for the case of $\delta_0 = \dot{\delta}_0 = 0$, an undamped natural frequency of $\omega_n = 20$ rad/s and a constant commanded actuator state of $\delta_c = 1$. Consider how the solution changes for various values of the damping ratio ζ.

Solution: Equation (9.3.20) gives the closed-form solution for this case. Using $\delta_c = 1$ along with $\delta_0 = \dot{\delta}_0 = 0$ in Eq. (9.3.20) gives the analytic solution for the case of $\zeta \neq 1$

$$\delta(t) = \frac{\lambda_2 e^{\lambda_1 t} - \lambda_1 e^{\lambda_2 t}}{\lambda_1 - \lambda_2} + 1 \tag{9.3.28}$$

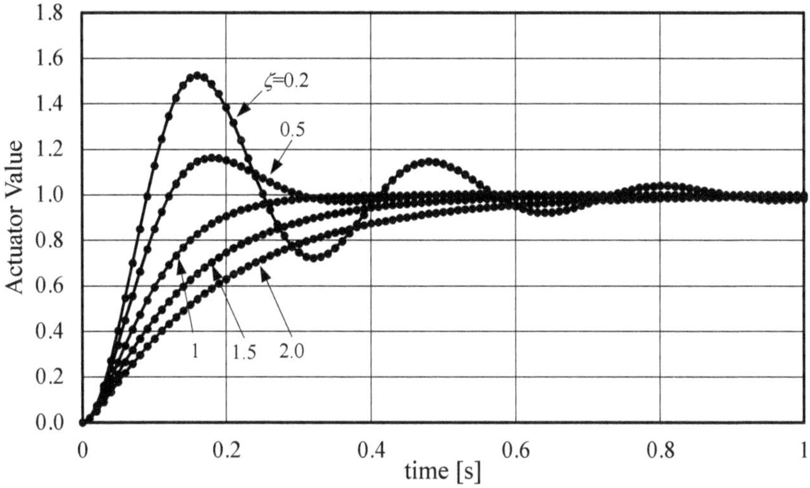

Figure 9.3.1. Actuator response as a function of time for second-order actuator dynamics with $\delta_0 = \dot{\delta}_0 = 0$, $\delta_c(t) = \delta_c = 1$, and $\omega_n = 20$ rad/s. Solutions are shown for various damping ratios ζ. Solid lines represent closed-form analytic solutions and the data points represent solutions using fourth-order Runge-Kutta integration using $\Delta t = 0.01$.

Figure 9.3.1 shows time-history solutions for this case with varying values of ζ. Note that for $\zeta < 1$ the eigenvalues are complex and the solutions are oscillatory. For $\zeta > 1$ the eigenvalues are real and the solutions are non-oscillatory. Using $\delta_c = 1$ along with $\delta_0 = \dot{\delta}_0 = 0$ in Eq. (9.3.27) gives the analytic solution for the case of $\zeta = 1$

$$\delta(t) = (\lambda t - 1)e^{\lambda t} + 1 \tag{9.3.29}$$

This solution is also included in Fig. 9.3.1.

Numerical Integration

In general, the second-order nonhomogeneous equation given in Eq. (9.3.9) can be solved by numerical integration. In order to use traditional numerical integration methods, the second-order equation is typically reformatted as two first-order equations, in what is called the *state-space form.* This can be done by treating the time rate of change of the actuator $\dot{\delta}$ a second dependent variable. With this definition, Eq. (9.3.9) can be rewritten as the system of first-order equations

$$\frac{d}{dt}\begin{bmatrix}\delta\\ \dot{\delta}\end{bmatrix} = \begin{bmatrix}\dot{\delta}\\ \ddot{\delta}\end{bmatrix} = \begin{bmatrix}\dot{\delta}\\ \omega_n^2[\delta_c(t) - \delta] - 2\zeta\omega_n\dot{\delta}\end{bmatrix} \tag{9.3.30}$$

These can be directly integrated using traditional integration schemes such as Runge-Kutta integration. For example, the data points in Fig. 9.3.1 were obtained using fourth-order Runge-Kutta integration with $\Delta t = 0.01$. As the time step size is decreased, the solutions approach the analytic solutions out to machine precision.

For application in a flight simulator, Eq. (9.3.30) can be added directly to the state vector developed in Chapter 5 and integrated forward in time along with the equations of motion within the function `diff_eq`. For each actuator, two states can be added to the array that contains the entire state of the aircraft. For example, if the quaternion formulation is used along with the four traditional fixed-wing controls including aileron δ_a, elevator δ_e, rudder δ_r, and throttle τ, the differential equations would include those given in Eqs. (5.4.5)–(5.4.8) with the additional eight equations

$$\begin{bmatrix} \dot{\delta}_a \\ \dot{\delta}_e \\ \dot{\delta}_r \\ \dot{\tau} \\ \ddot{\delta}_a \\ \ddot{\delta}_e \\ \ddot{\delta}_r \\ \ddot{\tau} \end{bmatrix} = \begin{bmatrix} \dot{\delta}_a \\ \dot{\delta}_e \\ \dot{\delta}_r \\ \dot{\tau} \\ \omega_{n_a}^2 [\delta_{a_c} - \delta_a] - 2\zeta_a \omega_{n_a} \dot{\delta}_a \\ \omega_{n_e}^2 [\delta_{e_c} - \delta_e] - 2\zeta_e \omega_{n_e} \dot{\delta}_e \\ \omega_{n_r}^2 [\delta_{r_c} - \delta_r] - 2\zeta_r \omega_{n_r} \dot{\delta}_r \\ \omega_{n_\tau}^2 [\tau_c - \tau] - 2\zeta_\tau \omega_{n_\tau} \dot{\tau} \end{bmatrix} \tag{9.3.31}$$

where the subscript c denotes the commanded value, which can be a function of time.

Using a second-order model for the actuator dynamics yields the value of the actuator δ, the value of the first derivative of the actuator as a function of time $\dot{\delta}$, and the value of the second derivative of the actuator as a function of time $\ddot{\delta}$. Because this information is known within the integration routine, saturation limits on δ, $\dot{\delta}$, and $\ddot{\delta}$ can be directly applied during integration.

For example, any value for the commanded value of an actuator δ_c can be accepted as input, and need not be limited. Immediately preceding the calculation of Eq. (9.3.30), δ and $\dot{\delta}$ should be limited using

$$\delta = \max(\min(\delta, \delta_{\max}), \delta_{\min}) \tag{9.3.32}$$

$$\dot{\delta} = \max(\min(\dot{\delta}, \dot{\delta}_{\max}), \dot{\delta}_{\min}) \tag{9.3.33}$$

These limiters should also be applied at the end of each full time step. Additionally, the limit on actuator magnitude should be applied within the `pseudo_aero` function. Before computing Eq. (9.3.30), the actuator rate should be set to zero if the actuator position is at a limit and the rate would push the actuator further into that limit. After computing Eq. (9.3.30) $\ddot{\delta}$ should be limited using

$$\ddot{\delta} = \max(\min(\ddot{\delta}, \ddot{\delta}_{\max}), \ddot{\delta}_{\min}) \tag{9.3.34}$$

This should be followed by setting the acceleration to zero if the value for acceleration would push the actuator further into a rate or position limit. Algorithm 9.3.11 outlines the process for computing second-order actuator dynamics within the `diff_eq` function and propertly applying limits.

Algorithm 9.3.11 Second-order actuator dynamics with limiting.

Input
δ (real) Actuator value from state array
$\dot{\delta}$ (real) Actuator rate from state array
δ_c (real) Commanded actuator value
ω_n (real) Actuator undamped natural frequency
ζ (real) Actuator damping ratio
$\delta_{\min}, \dot{\delta}_{\min}, \ddot{\delta}_{\min}$ (real) Minimum actuator value, rate, and acceleration limits
$\delta_{\max}, \dot{\delta}_{\max}, \ddot{\delta}_{\max}$ (real) Maximum actuator value, rate, and acceleration limits
$\delta = \max(\min(\delta, \delta_{\max}), \delta_{\min})$ Limit value
$\dot{\delta} = \max(\min(\dot{\delta}, \dot{\delta}_{\max}), \dot{\delta}_{\min})$ Limit rate
if$(\delta <= \delta_{\min}$ and $\dot{\delta} < 0)$ $\dot{\delta} = 0$ Limit rate if moving against a magnitude limit
if$(\delta >= \delta_{\max}$ and $\dot{\delta} > 0)$ $\dot{\delta} = 0$
$\ddot{\delta} = \omega_n^2[\delta_c - \delta] - 2\zeta\omega_n\dot{\delta}$
$\ddot{\delta} = \max(\min(\ddot{\delta}, \ddot{\delta}_{\max}), \ddot{\delta}_{\min})$ Limit acceleration
if$(\dot{\delta} <= \dot{\delta}_{\min}$ and $\ddot{\delta} < 0)$ $\ddot{\delta} = 0$ Limit accel. if moving against a rate limit
if$(\dot{\delta} >= \dot{\delta}_{\max}$ and $\ddot{\delta} > 0)$ $\ddot{\delta} = 0$
if$(\delta <= \delta_{\min}$ and $\delta_c < \delta_{\min})$ $\ddot{\delta} = 0$ Limit accel. if moving against a mag. limit
if$(\delta >= \delta_{\max}$ and $\delta_c > \delta_{\max})$ $\ddot{\delta} = 0$

EXAMPLE 9.3.2 Consider the F-16 aircraft specified in Appendix C.3 with the aerodynamic model F16-E-1 given in Table C.3.15 at an altitude of 10,000 ft and a velocity of 500 ft/s. Beginning from the trim solution given in Example 9.2.3, compute the time-history of the vehicle states, including actuator states, over a period of 1.0 second, if at the beginning of the simulation the commanded value for the aileron, elevator, rudder, and throttle is exactly $\delta_a = \delta_e = \delta_r = \tau = 0$. Use a time step of $\Delta t = 0.01$ along with the second-order dynamic properties for each actuator given in Appendix C.3. Plot the aileron, elevator, rudder, and throttle values as a function of time. Comment on the differences between the solutions in this example and those in Example 9.2.3.

Solution: Using the trim solution from Example 9.2.3 as the initial condition, setting the commanded values of all controls to zero immediately after initialization of the simulation, and employing second-order actuator dynamics with rate limiting as outlined in this section, yields the time histories of the controls shown in Fig. 9.3.2. Table 9.3.1 includes the translational and rotational velocity states of the vehicle, along with the actuator states, at a few time steps. The full states of the vehicle as a function of time for these examples are included on the companion website.

Note the second-order actuator dynamics apparent in Figure 9.3.2 in comparison to the first-order actuator dynamics shown in Figure 9.2.3. Note that the second-order dynamics include a ramp up in actuator rate at the beginning of the simulation,

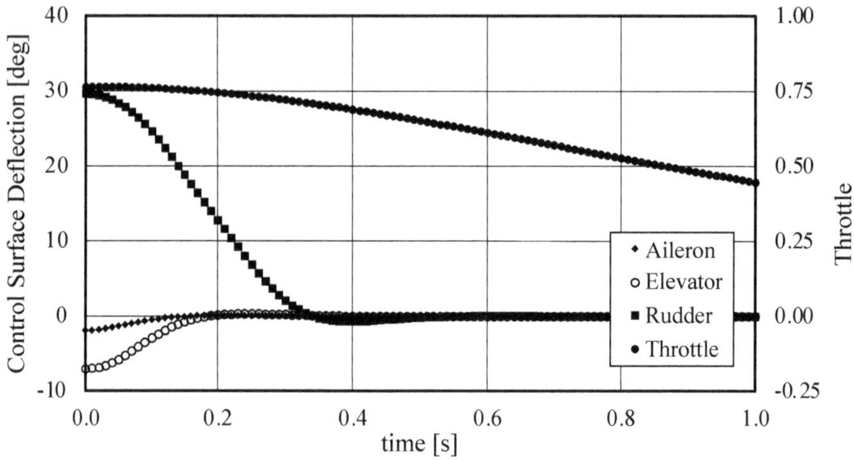

Figure 9.3.2. Example control-surface deflections and throttle setting as a function of time using second-order actuator dynamics and a step change in commanded values.

whereas the first-order dynamics do not. Additionally, note that the damping ratio of the aileron, elevator, and rudder are all less than 1. This results in a slight overshoot of the commanded value, which can be seen in the time-history results for the elevator and rudder shown in Figure 9.3.2. Note that overshoot is a possible characteristic of second-order dynamics, but not first-order dynamics.

Table 9.3.1. Sample states of the vehicle at specified time steps for Example 9.3.2.

Time	0.0 [s]	0.5 [s]	1.0 [s]
u [ft/s]	4.9174534180E+02	4.9562715203E+02	4.8526814150E+02
v [ft/s]	8.8407835448E+01	5.6012261631E+01	-1.4263426130E+01
w [ft/s]	1.9250284456E+01	-3.2909791368E+00	-7.8153896528E+01
p [rad/s]	0.0000000000E+00	-7.2328593871E−01	-3.9902622719E−01
q [rad/s]	0.0000000000E+00	-3.4818075927E−01	-5.0011980416E−01
r [rad/s]	0.0000000000E+00	3.1732431777E−01	2.8001833001E−01
δ_a [deg]	-1.9605489795E+00	-2.4945783604E−03	-1.3436824404E−06
δ_e [deg]	-7.0525300816E+00	-1.1090432528E−02	-8.4452288708E−06
δ_r [deg]	2.9602685430E+01	-1.7740577812E−01	-2.6615807398E−06
τ	7.6470812503E−01	6.5042668154E−01	4.4547296591E−01
$\dot{\delta}_a$ [deg/s]	0.0000000000E+00	3.7887021309E−02	4.5163084966E−05
$\dot{\delta}_e$ [deg/s]	0.0000000000E+00	1.1728204292E−01	2.0489668756E−04
$\dot{\delta}_r$ [deg/s]	0.0000000000E+00	4.6172482355E+00	3.0573575726E−03
$\dot{\tau}$ [/s]	0.0000000000E+00	-3.9036083230E−01	-3.9174313539E−01

9.4 Simulating Sensors

Most flight simulation software is written to mimic real-life flight scenarios. This can include testing vehicle performance and handling qualities, as well as testing a control algorithm in simulation before putting the algorithm on actual flight hardware. In many cases, flight simulation software is used to model the sensors that real aircraft carry on-board. The information from these simulated sensors is then passed to controls software to test control algorithms, or to other instrumentation for visual depiction of the vehicle state. The simulation software can be constructed to provide only the limited sensor data at the sensor refresh rate that the pilot will have access to rather than providing full-state, high-accuracy, high-time-resolution feedback of the vehicle state. If constructed properly, the flight simulation software can mimic real-life flight testing scenarios, including lag in communication, bias and random errors in sensor readings, missing data, and system failures.

The most fundamental sensors used on most aircraft include accelerometers, gyroscopes, a magnetometer, a barometer, a Pitot tube, a global positioning system (GPS), and in some cases, angle of attack and sideslip sensors. The physics simulator has very accurate and precise information that can be used to mimic the output of these sensors. Here we will discuss how the vehicle state information can be used to emulate the output of typical sensors. Additionally, it will be shown how bias and random error, sensor limits, resolution limits, and communication limits can be added to the sensor information to mimic real-world sensor limitations.

Inertial Measurement Unit

The translational and rotational acceleration of a vehicle is typically measured using a suite of sensors called an *inertial measurement unit* (IMU). A typical IMU includes gyroscopes, accelerometers, and a magnetometer. Gyroscopes measure angular velocities ω_x, ω_y, and ω_z about the axes of the IMU. Accelerometers measure translational accelerations a_x, a_y, and a_z along the axes of the IMU. The magnetometer measures the orientation of the IMU relative to the Earth's magnetic field.

In the case of flight testing a real vehicle, it is often not practical to place the IMU exactly at the center of gravity of the vehicle. This is especially true in cases when the vehicle center of gravity changes as a function of time. Additionally, the coordinate system of the IMU may not be aligned exactly with the coordinate system of the vehicle. Hence, in simulation it is often necessary to be able to report simulated measurements for an IMU with an orientation that does not align with the body-fixed coordinate system, and is not located at the vehicle center of gravity. This is addressed in each of the discussions related to the specific sensors.

Gyroscope

A gyroscope measures rotational velocity. Hence, the outputs of gyroscopes can be simulated by reporting the rotational velocities of the vehicle. The rotational velocities along the x_b, y_b, and z_b body-fixed axes of the vehicle are given by

$$\begin{bmatrix} \omega_{x_b} \\ \omega_{y_b} \\ \omega_{z_b} \end{bmatrix} = \begin{bmatrix} p \\ q \\ r \end{bmatrix} \tag{9.4.1}$$

These values are three of the vehicle states and are updated at the end of each simulation time step.

For the special case when the IMU is mounted on the vehicle such that the coordinate system of the IMU is aligned with the vehicle body-fixed coordinate system, the angular velocities reported by the IMU are simply those given in Eq. (9.4.1), which yields

$$\begin{bmatrix} \omega_x \\ \omega_y \\ \omega_z \end{bmatrix}_{\text{IMU}} = \begin{bmatrix} \omega_{x_b} \\ \omega_{y_b} \\ \omega_{z_b} \end{bmatrix} \tag{9.4.2}$$

For the case of an IMU oriented by the Euler angles ϕ, θ, and ψ relative to the vehicle coordinate system, the angular velocities reported by the IMU are related to those in the body-fixed coordinate system by the transformation given in Eq. (1.2.5). This yields

$$\begin{bmatrix} \omega_x \\ \omega_y \\ \omega_z \end{bmatrix}_{\text{IMU}} = \begin{bmatrix} C_\theta C_\psi & C_\theta S_\psi & -S_\theta \\ S_\phi S_\theta C_\psi - C_\phi S_\psi & S_\phi S_\theta S_\psi + C_\phi C_\psi & S_\phi C_\theta \\ C_\phi S_\theta C_\psi + S_\phi S_\psi & C_\phi S_\theta S_\psi - S_\phi C_\psi & C_\phi C_\theta \end{bmatrix} \begin{bmatrix} \omega_{x_b} \\ \omega_{y_b} \\ \omega_{z_b} \end{bmatrix} \tag{9.4.3}$$

The reverse transformation can be used to compute the vehicle body-fixed rotational rates from data reported by an IMU, along with a known orientation of the IMU relative to the vehicle body-fixed coordinate system.

Accelerometer

An accelerometer measures translational acceleration relative to an inertial coordinate system. The translational acceleration measured by an accelerometer is best understood by first considering the equations of motion developed in Chapter 5. Recall that in the development of the equations of motion, the constant-wind coordinate system is used as an inertial coordinate system. Hence, the translational acceleration including gravity relief along the x_b, y_b, and z_b body-fixed axes of the vehicle at the vehicle center of gravity can be expressed as given in Eq. (5.5.4) for the Euler-angle formulation as

$$\begin{bmatrix} \dot{u} \\ \dot{v} \\ \dot{w} \end{bmatrix} = \frac{1}{\mathcal{M}} \begin{bmatrix} \overline{F}_{x_b} \\ \overline{F}_{y_b} \\ \overline{F}_{z_b} \end{bmatrix} + (g - a_c) \begin{bmatrix} -S_\theta \\ S_\phi C_\theta \\ C_\phi C_\theta \end{bmatrix} + \begin{bmatrix} rv - qw \\ pw - ru \\ qu - pv \end{bmatrix} \tag{9.4.4}$$

Each of the terms in this equation has units of acceleration. However, not all the terms in this equation can be sensed by an accelerometer. Rearranging Eq. (9.4.4) such that all terms that can be sensed by an accelerometer are on the right-hand side yields

$$\frac{1}{\mathcal{M}}\begin{bmatrix}\overline{F}_{x_b}\\ \overline{F}_{y_b}\\ \overline{F}_{z_b}\end{bmatrix} = \begin{bmatrix}\dot{u}\\ \dot{v}\\ \dot{w}\end{bmatrix} - (g - a_c)\begin{bmatrix}-S_\theta\\ S_\phi C_\theta\\ C_\phi C_\theta\end{bmatrix} - \begin{bmatrix}rv - qw\\ pw - ru\\ qu - pv\end{bmatrix} \tag{9.4.5}$$

In this form, the three terms on the right-hand side are the translational acceleration of the vehicle relative to the inertial coordinate system, the acceleration of gravity, and Coriolis accelerations. The influence of these three accelerations are directly measured by an accelerometer attached to the vehicle. Because the summation of these three components of acceleration is equal to the pseudo aerodynamic force vector divided by the vehicle mass, the outputs of accelerometers can be simulated from the results of either the left-hand side of Eq. (9.4.5) or the right-hand side of Eq. (9.4.5). Both will yield identical results. Hence, sensor readings from an accelerometer placed at the vehicle CG can be computed from

$$\begin{bmatrix}a_{x_b}\\ a_{y_b}\\ a_{z_b}\end{bmatrix}_{CG} = \frac{1}{\mathcal{M}}\begin{bmatrix}\overline{F}_{x_b}\\ \overline{F}_{y_b}\\ \overline{F}_{z_b}\end{bmatrix} = \begin{bmatrix}\dot{u}\\ \dot{v}\\ \dot{w}\end{bmatrix} - (g - a_c)\begin{bmatrix}-S_\theta\\ S_\phi C_\theta\\ C_\phi C_\theta\end{bmatrix} - \begin{bmatrix}rv - qw\\ pw - ru\\ qu - pv\end{bmatrix} \tag{9.4.6}$$

The components of this vector can easily be computed for any state of the vehicle within the `diff_eq` function.

For the special case when the IMU is mounted on the vehicle at the vehicle center of gravity such that the coordinate system of the IMU is aligned with the body-fixed coordinate system, the translational accelerations reported by the IMU are simply those given in Eq. (9.4.6), which yields

$$\begin{bmatrix}a_x\\ a_y\\ a_z\end{bmatrix}_{\text{IMU}} = \begin{bmatrix}a_{x_b}\\ a_{y_b}\\ a_{z_b}\end{bmatrix}_{CG} \tag{9.4.7}$$

For the case of an IMU located at a distance $\mathbf{r}_{\text{IMU}} = \mathbf{P}_{\text{IMU}} - \mathbf{P}_{CG}$ relative to the vehicle center of gravity in the body-fixed coordinate system, the translational acceleration in the body-fixed coordinate system is related to the translational and rotational acceleration at the center of gravity through the transformation

$$\begin{bmatrix}a_{x_b}\\ a_{y_b}\\ a_{z_b}\end{bmatrix}_{\mathbf{r}} = \begin{bmatrix}a_{x_b}\\ a_{y_b}\\ a_{z_b}\end{bmatrix}_{CG} + \begin{bmatrix}\dot{\omega}_{x_b}\\ \dot{\omega}_{y_b}\\ \dot{\omega}_{z_b}\end{bmatrix} \times \mathbf{r}_{\text{IMU}} + \begin{bmatrix}\omega_{x_b}\\ \omega_{y_b}\\ \omega_{z_b}\end{bmatrix} \times \left(\begin{bmatrix}\omega_{x_b}\\ \omega_{y_b}\\ \omega_{z_b}\end{bmatrix} \times \mathbf{r}_{\text{IMU}}\right) \tag{9.4.8}$$

If the IMU coordinate system is oriented by the Euler angles ϕ, θ, and ψ relative to the body-fixed coordinate system, the translational accelerations reported by the IMU are related to those in the body-fixed coordinate system at the location of the IMU by the transformation given in Eq. (1.2.5). This yields

$$\begin{bmatrix} a_x \\ a_y \\ a_z \end{bmatrix}_{\text{IMU}} = \begin{bmatrix} C_\theta C_\psi & C_\theta S_\psi & -S_\theta \\ S_\phi S_\theta C_\psi - C_\phi S_\psi & S_\phi S_\theta S_\psi + C_\phi C_\psi & S_\phi C_\theta \\ C_\phi S_\theta C_\psi + S_\phi S_\psi & C_\phi S_\theta S_\psi - S_\phi C_\psi & C_\phi C_\theta \end{bmatrix} \begin{bmatrix} a_{x_b} \\ a_{y_b} \\ a_{z_b} \end{bmatrix}_{\mathbf{r}} \tag{9.4.9}$$

Again, the reverse transformation can be used to compute the vehicle translational accelerations from IMU measurements if the orientation of the IMU is known relative to the vehicle body-fixed coordinate system.

Magnetometer

A magnetometer measures the local three-dimensional magnetic field vector. The local magnetic field is a function of Earth's magnetic field, as well as influences from surrounding objects including permanent magnets and ferromagnetic materials. A magnetometer is typically used in conjunction with other sensors to estimate the Euler azimuth angle ψ or heading ψ_h of the vehicle.

Earth's magnetic field is a complex function of time and position on the Earth, and its poles are not aligned with the geographic North and South poles. The magnetic field is affected by currents in the Earth's molten outer core, and is neither symmetric, nor a simple dipole, nor composed of uniformly concentric field lines. The North pole of the magnetic field, known as magnetic North, is not stationary and has moved about 1,200 miles in the past century. In the year 2025, it is located north of Canada in the Arctic Ocean, and is drifting towards Siberia at a rate of approximately 25–35 miles per year.

The difference in the Euler azimuth angle ψ and the direction of magnetic north is called the *declination*, and the angle between the local magnetic field vector and the horizontal surface of the Earth is called the *inclination*.

The most common method for modeling Earth's magnetic field is through the use of the World Magnetic Model (WMM). This model represents the entire three-dimensional magnetic field originating from the Earth's core as a 12th-degree spherical harmonic expansion, which includes dependence on time. This model is updated every five years, and code libraries for this model are available for simple incorporation into other software. Given the geographic coordinates of latitude, longitude, and altitude, as well as the time of interest, the WMM code libraries will compute and return smooth and differentiable values for the declination, inclination, total field strength, and horizontal and vertical magnetic-field components. That information must be rotated from the Earth-fixed coordinate system to the sensor coordinate system. This can be done by first rotating the results into the vehicle body-fixed coordinate system based on the vehicle orientation relative to the Earth-fixed coordinate system, and then to the sensor coordinate system based on the sensor orientation relative to the vehicle body-fixed coordinate system.

Barometer

A barometer measures atmospheric pressure and temperature. In the simulation framework, these are known directly due to the atmospheric model being employed. For most examples in this textbook, the standard atmospheric model outlined in section 3.2 is used. The resulting atmospheric pressure and temperature from the atmospheric model can be directly reported to simulate the output from a barometer.

Pitot tube

In a flight simulation code, there is direct access to the full state of the vehicle, including the speed at which the vehicle is traveling. The speed of the vehicle relative to the surrounding air is called the *true airspeed* (TAS) and will be denoted here as V_∞. It is formally defined in the Federal Aviation Regulations (FAR) Part 1 as *the airspeed of an aircraft relative to undisturbed air.* The true airspeed can be computed from the body-fixed components of velocity as

$$V_\infty = \sqrt{u^2 + v^2 + w^2} \tag{9.4.10}$$

The true airspeed of the vehicle is a very important property. However, it is not easily directly measured from typical instrumentation on most aircraft. In real-life scenarios, airspeed is estimated using an air data system (ADS), which uses measurements from pneumatic sensors to measure pressure in the flow around the vehicle. These measurements are used to estimate the speed at which the vehicle is moving through the fluid. The most common pressure instrument used for this purpose on fixed-wing aircraft is the Pitot tube.

A Pitot tube is a tube with at least two pressure ports oriented to report a differential pressure. For most fixed-wing aircraft, this tube is aligned with the vehicle x_b axis with the tip pointed into the flow. An orifice on the tip of the tube measures stagnation pressure $\mathcal{P}_0$ of the flow, and an orifice on the side of the tube measures the freestream static pressure $\mathcal{P}_\infty$. Since the pressure can vary significantly over the surface of the vehicle, and can be a function of angle of attack, the specific location of this port must be carefully chosen to ensure it yields a value as close to the freestream static pressure as possible. The tubing from the ports is connected to a differential pressure sensor that reports the difference between the two pressure measurements. The combination of ports that yield the difference between the stagnation pressure and static pressure is called the *Pitot-static system*, which yields a value for

$$\Delta\mathcal{P}_I = \mathcal{P}_0 - \mathcal{P}_\infty \tag{9.4.11}$$

where $\Delta\mathcal{P}_I$ represents an indicated pressure difference. This pressure difference is reported to the autopilot, which uses this information to estimate the airspeed of the vehicle. Hence, in order to model this pressure difference, the physics simulator must compute both the freestream static pressure $\mathcal{P}_\infty$ and the stagnation pressure $\mathcal{P}_0$.

The freestream static pressure $\mathcal{P}_\infty$ can be obtained directly from the atmospheric model, as was done with the barometer. For purposes of estimating the stagnation pressure, atmospheric air flow around the vehicle is typically treated as inviscid adiabatic flow of an ideal gas. For this case, the stagnation pressure $\mathcal{P}_0$ is related to the freestream static pressure $\mathcal{P}_\infty$, the specific heat ratio γ, and the freestream Mach number M_∞ as shown in Eq. (3.3.22). Applying the definition of Mach number given in Eq. (3.3.4) yields

$$\mathcal{P}_0 = \mathcal{P}_\infty \left(1 + \frac{\gamma - 1}{2} \frac{V_\infty^2}{\gamma \mathcal{R} \mathcal{T}_\infty}\right)^{\gamma/(\gamma-1)} \tag{9.4.12}$$

The atmospheric temperature $\mathcal{T}_\infty$ along with the gas properties γ and $\mathcal{R}$ are known from the atmospheric model and can be used directly. Hence, the static pressure from the atmospheric model along with results from Eq. (9.4.12) can be used in Eq. (9.4.11) to compute and report the indicated pressure difference.

Airspeed Definitions

At this point it may be helpful to understand common airspeed definitions used in various cockpit displays and autopilot software that result from the relationship shown in Eq. (9.4.12). Equation (9.4.12) can be solved to yield an expression for the freestream velocity in terms of the static and stagnation pressures

$$V_{\text{TAS}} = \sqrt{\frac{2\gamma \mathcal{R} \mathcal{T}_\infty}{\gamma - 1}\left[\left(\frac{\mathcal{P}_0 - \mathcal{P}_\infty}{\mathcal{P}_\infty} + 1\right)^{(\gamma-1)/\gamma} - 1\right]} \tag{9.4.13}$$

If all the properties in Eq. (9.4.13) could be sensed and reported out to high precision, results from this relation would yield a solution equivalent to Eq. (9.4.10), which is the true airspeed. However, with today's technology, this information cannot be measured and reported perfectly. This gives rise to three common airspeeds typically reported in autopilot software and on cockpit displays. These are the indicated airspeed, the calibrated airspeed, and the equivalent airspeed.

The simplest airspeed to compute is the *indicated airspeed* (IAS). The indicated airspeed is formally defined in FAR Part 1 as *the speed of an aircraft as shown on its Pitot-static airspeed indicator calibrated to reflect standard atmosphere adiabatic compressible flow at sea level, uncorrected for airspeed system errors.* This requires only the pressure difference given by a Pitot-static system. Using the indicated Pitot-static $\Delta\mathcal{P}_I$ in Eq. (9.4.13) along with the standard atmospheric temperature and static pressure at sea level yields an expression for the indicated airspeed

$$V_{\text{IAS}} \equiv \sqrt{\frac{2\gamma \mathcal{R} \mathcal{T}_{\text{ssl}}}{\gamma - 1}\left[\left(\frac{\Delta\mathcal{P}_I}{\mathcal{P}_{\text{ssl}}} + 1\right)^{(\gamma-1)/\gamma} - 1\right]} \tag{9.4.14}$$

where $\mathcal{T}_{\text{ssl}} = 288.150$ K and $\mathcal{P}_{\text{ssl}} = 101{,}325$ N/m^2 are the standard temperature and pressure at sea level.

Because of position errors and instrument errors, the indicated pressure difference $\Delta\mathcal{P}_i$ from the Pitot-static system is not exactly equal to the difference between the true stagnation pressure and static pressure $\mathcal{P}_0 - \mathcal{P}_\infty$. Hence, the pressure difference reported by a Pitot-static system is typically calibrated to account for these errors and yield values much closer to the actual value $\mathcal{P}_0 - \mathcal{P}_\infty$. The calibrated pressure difference will be referred to here as $\Delta\mathcal{P}_C$. Using this in Eq. (9.4.14) yields the *calibrated airspeed* (CAS)

$$V_{\text{CAS}} \equiv \sqrt{\frac{2\gamma\mathcal{R}\mathcal{T}_{\text{ssl}}}{\gamma - 1}\left[\left(\frac{\Delta\mathcal{P}_C}{\mathcal{P}_{\text{ssl}}} + 1\right)^{(\gamma-1)/\gamma} - 1\right]} \tag{9.4.15}$$

The calibrated airspeed is formally defined in FAR Part 1 as *the indicated airspeed of an aircraft, corrected for position and instrument error. Calibrated airspeed is equal to true airspeed in the standard atmosphere at sea level.* Strictly speaking, the last statement in the FAR is only true in the case when the Pitot-static system is perfectly calibrated to produce zero error, and the true atmospheric conditions at sea level perfectly match those of the standard atmosphere.

Because airspeed indicators based on the Pitot-static system yield the true airspeed only when the vehicle is flying at sea-level altitude, one alternate airspeed is also typically used. This airspeed is the *equivalent airspeed* (EAS) defined as

$$V_{\text{EAS}} \equiv V_\infty \sqrt{\frac{\rho_\infty}{\rho_{\text{ssl}}}} \tag{9.4.16}$$

Equivalent airspeed is the airspeed at sea level and standard atmospheric conditions that yields the equivalent dynamic pressure as the true airspeed at the actual altitude of the vehicle. This airspeed is important, since aerodynamic loading on the vehicle is proportional to the dynamic pressure. FAR Part 1 defines the equivalent airspeed through the statement *true airspeed is equal to equivalent airspeed multiplied by* $(\rho_{\text{ssl}}/\rho_\infty)^{1/2}$. In mathematical terms, this yields the definition

$$V_\infty \equiv V_{\text{EAS}} \sqrt{\frac{\rho_{\text{ssl}}}{\rho_\infty}} \tag{9.4.17}$$

Note that in order to compute V_{EAS} from V_∞ or vice versa, the true density of the air at altitude ρ_∞ must be known. In order to estimate this, ADS typically use a temperature sensor. Because air is commonly treated as an ideal gas for these purposes, the density at altitude ρ_∞ is related to the temperature T_∞ and pressure $\mathcal{P}_\infty$ at altitude according to

$$\rho_\infty = \frac{\mathcal{P}_\infty}{\mathcal{R}\mathcal{T}_\infty} \tag{9.4.18}$$

It should be noted that some computations for the true airspeed assume that the equivalent airspeed is equal to the calibrated airspeed, i.e. $V_{\text{EAS}} = V_{\text{CAS}}$. Using this in Eq. (9.4.17) yields the approximation

$$V_\infty \approx V_{\text{CAS}} \sqrt{\frac{\rho_{\text{ssl}}}{\rho_\infty}} \tag{9.4.19}$$

The relation given in Eq. (9.4.19) is correct when the vehicle is at sea level in a standard atmosphere. However, **this approximation can yield significant errors at higher altitudes.**

In summary, the true airspeed is given by Eqs. (9.4.10) and (9.4.13), the indicated airspeed is given by Eq. (9.4.14), the calibrated airspeed is given by Eq. (9.4.15), and the equivalent airspeed is given by Eq. (9.4.16). Typically the physics simulation environment is used to report only the indicated pressure difference $\Delta\mathcal{P}_I$ along with freestream temperature and pressure. This information is then used in pilot display or autopilot software to compute the various airspeeds discussed here.

Atmospheric Disturbances

Because real airspeed sensors are exposed to the flow around the aircraft, atmospheric turbulence fluctuations should be included in the simulation of ADS data. Methods for generating realistic disturbances within the Earth's atmosphere as well as methods for incorporating those into the physics simulation is covered in Chapter 8. MIL-STD1797A [80] states that *the direct effect [of gust and turbulence velocities] on the aerodynamic sensors shall be included when such sensors are part of the aircraft augmentation system.* This can be done by storing the turbulent fluctuation components $\tilde{u}$, $\tilde{v}$, and $\tilde{w}$ used in the last call to the `diff_eq` function within the `rk4` function. The velocity fluctuation components can then be added to the body-fixed velocity components u, v, and w when computing the true airspeed to yield

$$V_\infty = \sqrt{(u+\tilde{u})^2 + (v+\tilde{v})^2 + (w+\tilde{w})^2} \tag{9.4.20}$$

This alternate form of true airspeed is then used in Eq. (9.4.12) to compute the stagnation pressure. Note that perturbations to the static pressure $\mathcal{P}_\infty$ could also be modeled depending on what the static pressure ports on a vehicle might sense due to velocity disturbances normal to those ports.

Global Positioning System

A global positioning system (GPS) receiver is a sensor that computes its geographic location, including altitude, using trilateration of timing signals from satellites in the Global Positioning System. The geographic coordinates reported are typically based on an ellipsoidal Earth model. The accuracy of GPS systems can vary, but have significantly improved over time as the number and accuracy of GPS satellites have increased. For any given location on Earth, it is currently typical for an unobstructed GPS sensor to register signals from between 6 and 12 satellites.

Modeling the output from a GPS sensor is rather straightforward, and can be accomplished by simply reporting the geographic coordinates of the vehicle as outlined in Section 7.5. Because most GPS systems use an ellipsoidal model of the Earth, the ellipsoidal geographic locations should be used when possible.

Aerodynamic Angles

Some vehicles carry sensors that measure the direction of the incoming flow to the vehicle, which can be described in terms of the aerodynamic angles α and β. There are a range of sensors that can provide this type of information, including pressure tubes with multiple ports spaced around the tube, rotating vanes that align with the local flow direction, or other optical or image-based systems. The style, size, cost, complexity, and accuracy of these sensors varies widely, and therefore the methods for modeling the output from these sensor can vary as well. For example, to accurately model the output from a wind vane, the dynamic response of the vane may be needed, including second-order dynamics from inertia, friction, and aerodynamic loading.

A simple approach for simulating the output of aerodynamic-angle sensors can be to report the aerodynamic angles directly from the known states of the vehicle. This can be done using the definitions of the aerodynamic angles developed in Section 3.4 and repeated here for convenience. The angle of attack is

$$\alpha \equiv \tan^{-1}\left(\frac{w}{u}\right) \tag{9.4.21}$$

the traditional sideslip angle is

$$\beta \equiv \sin^{-1}\left(\frac{v}{V}\right) \tag{9.4.22}$$

where $V = \sqrt{u^2 + v^2 + w^2}$, and the flank angle is

$$\beta_f \equiv \tan^{-1}\left(\frac{v}{u}\right) \tag{9.4.23}$$

The reader is reminded that the difference between the traditional sideslip angle β and flank angle β_f can be significant. Therefore, care should be taken to report the correct form of the sideslip angle, which may depend on the type of sensor being emulated.

Atmospheric Disturbances

Because aerodynamic sensors are exposed to the flow around the aircraft, the effect of atmospheric turbulence fluctuations should be included in the simulation of aerodynamic-angle sensors. This can be done by adding the translational fluctuation components $\tilde{u}$, $\tilde{v}$, and $\tilde{w}$ to the vehicle velocity components u, v, and w and yields the outputs

$$\alpha = \tan^{-1}\left(\frac{w + \tilde{w}}{u + \tilde{u}}\right) \tag{9.4.24}$$

$$\beta = \sin^{-1}\left(\frac{v + \tilde{v}}{\sqrt{(u + \tilde{u})^2 + (v + \tilde{v})^2 + (w + \tilde{w})^2}}\right) \tag{9.4.25}$$

$$\beta_f = \tan^{-1}\left(\frac{v + \tilde{v}}{u + \tilde{u}}\right) \tag{9.4.26}$$

Methods for generating realistic atmospheric turbulence is discussed in Chapter 8.

9.5 Error and Quantization

The analytic and numerical solutions for the actuator states and sensors presented in the previous sections can be computed out to machine precision during a simulation, which for modern computers is typically double precision. However, most actuation systems and sensors do not have double-precision resolution or accuracy due to the physical limitations of the actuator or sensor. The limitations of actuators and sensors in this regard are due to physical error in the signals, actuators, and sensors, as well as a limited resolution in what can be measured or reported. Hence, simulators that are used to depict real-life scenarios and test control algorithms must model these limitations in actuators and sensors.

Error Modeling

Due to manufacturing, installation, and calibration limitations and defects, data reported by sensors includes error. With today's technology, it is virtually impossible to produce a sensor that does not exhibit some source of error in the output. Error from sensor readings is typically quantified in terms of two different types of error. These are bias error and random error.

Bias error is the tendency of a sensor to report values that are consistently offset from the actual value in some direction. For example, due to manufacturing and calibration inconsistencies, a particular thermometer may always read 1.0° less than the true temperature. Bias error can also be a function of operating condition. For example, an accelerometer may have a temperature dependence, and therefore include a bias error that changes with altitude or time of day.

Random error is the variation in a sensor measurement relative to the mean value if the measurement were to be repeated under near identical conditions. For example, due to the time history or random noise in the sensor state, a temperature sensor may sometimes read 0.1° higher than the actual temperature, and sometimes read 0.1° lower than the actual temperature.

Bias and random errors are inherent in virtually every sensor in use today, and significant effort is expended to quantify the error in sensors before they are used in aerospace applications. Many sensor manufactures publish information about their sensors, which can be used to estimate the bias and random error of the sensors. The accuracy and precision needed for any particular application is a driving consideration in sensor selection. Furthermore, post-processing of sensor data using filters such as the Kalman filter or Extended Kalman Filter is commonly performed in an attempt to estimate the error in sensors as well as the resulting estimated state of a vehicle. Hence, in order to use a flight-simulation framework to model and test vehicle control algorithms or post-processing software, sensor error should be modeled in the simulation.

Modeling sensor error can be fairly straightforward, depending on the level of accuracy needed in simulating the sensor error. For example, if the bias error of a sensor is a known constant, it can simply be added to the sensor output. If the bias error is a known function of another property, it can be easily modeled and added. If the random error is normally distributed with a known variance or standard deviation, the random error of a sensor can be modeled by adding normally distributed noise with the correct variance to a signal. Example 9.1.1 demonstrates the process of generating random numbers with a specified mean and standard deviation.

Some sensors require more effort to accurately model the expected error. For example, the accuracy of a GPS sensor reading depends on the number of satellites the sensor can detect, the relative position of the sensor to the detected satellites, the geometry between the detected satellites, uncertainty in satellite location, satellite clock timing errors, and atmospheric effects on the transmitted satellite signals. Similar types of challenges exist in accurately modeling other sensors typically used on aircraft. For an initial implementation, bias and random error can be added to a simulation as discussed here. However, to more accurately model error from a sensor, more detail is typically required.

Quantization

Actuators and sensors are limited in the resolution they can enforce or sense. For example, an actuator driving a control surface may only be able to ensure that the controls surface is correct to within 0.1 deg, or a GPS sensor may only be able to provide location measurements at a resolution of 0.1 ft. Therefore, one way to increase the accuracy of a simulation to match real-life behavior is to round or truncate the commanded values used in the actuator dynamics, the actuator states computed from the actuator dynamics, and sensor outputs to appropriate resolutions. This is called *quantization.*

Signals used to pass information to actuators or from sensors are typically digitally encoded using a binary format. The number of values N_s that can be specified or represented is related to the bit count N_b of the binary signal in the form

$$N_s = 2^{N_b} \tag{9.5.1}$$

For example, an 8-bit signal can represent $2^8 = 256$ discrete values, and a 16-bit signal can represent $2^{16} = 65,536$ discrete values. Actuators and sensors vary widely in their precision and resolution. For example, some actuators have 1-bit resolution, such as on/off or open/close switches and valves. Low-cost servos were historically 8-bit, with modern digital servos typically having 10-bit resolution. Precision military actuators can have as high as 16- or 20-bit resolution.

In the following development, we will use the symbol s to represent the value of a signal. The resolution or step size of a signal Δs is related to the range of values possible, as well as the bit-count of the signal

$$\Delta s = \frac{s_{\max} - s_{\min}}{2^{N_b}} = \frac{s_{\max} - s_{\min}}{N_s} \tag{9.5.2}$$

In order to better model the real-life actuator and sensor signals, the signals can be rounded using

$$s_q = \text{round}\left(\frac{s}{\Delta s}\right) \Delta s \tag{9.5.3}$$

where s_q represents the quantized value.

Quantizing signals in a flight simulator is sometimes necessary for accurately modeling some dynamic behaviors that result from interactions between the controller and the actuator. Exact methods for implementation in a flight simulation framework depend on the actuator properties and communication protocols, as well as the accuracy needed for capturing these effects in the aerodynamics, propulsion modeling, or vehicle dynamics. In general, quantization should be applied to any input signals to the simulation, as well as any output signals from the simulation that are used to emulate real-life scenarios. However, quantization should not typically be applied in the integration of the equations of motion, since those require high precision to capture the relevant physics, which physics is not generally quantized.

9.6 Communication

Various frameworks, coding structures, and programming languages have been used for flight simulation in the past several decades, and best practices for software development continue to evolve. One modern approach for simulation software development is to separate the computation required for physics, controls, and graphics by implementing each of these processes in distinct and separate executables or codes. These codes can run independently on separate computers, or on different threads or processors of a single computer. In order to support this type of architecture for a fully-featured flight simulator, the computer programs must be able to communicate and share information in real-time.

There are many ways to pass information within computer architectures, including copying information from a hard disk or long-term memory to the computer random access memory (RAM), passing pointers to RAM addresses, accepting input from a keyboard or mouse, sending information to a computer screen, and sending or receiving information over networks and ports. Each of these forms of communication for passing information involves specific protocols. For example, addresses to RAM are typically defined in hexadecimal format, while networking information is often sent and received using hypertext transfer protocol (HTTP). New communication protocols that are more efficient or can transfer larger amounts of data are constantly being developed, and computer storage and communication protocols will continue to evolve as technology improves.

Rather than presenting a modern protocol that will likely become outdated, this section focuses on one of the fundamental communication options that has been around since the birth of the computer, and will likely persist as an option for many years to come. The messaging protocol shown here does not comply with modern security standards, and can introduce insecurities or vulnerabilities to a code if used as shown here. By presenting this method, the author is not advocating that it is the best method for flight simulation. However, for students just beginning to understand messaging protocols, it is an excellent starting point.

User Datagram Protocol (UDP)

User datagram protocol (UDP) is a widely used communication system in which packets (datagrams) of information are passed through computer ports. It is a connectionless protocol, meaning that the data is sent and received without any "handshake" or testing of whether the full packet was successfully and accurately received. It is akin to a radio broadcast, in that there is no way for the sender to verify whether the message was actually received, or received in its entirety. If information was lost in the transmission, it is not re-sent.

UDP is similar to transmission control protocol (TCP), but differs in that TCP performs a handshake between sender and receiver to ensure that all information is correctly communicated. UDP is often used in real-time applications where speed is more important than reliability, such as video conferencing and online gaming. In such situations, information that was not correctly received in the past is not as important as receiving the current real-time information. TCP on the other hand is typically used in scenarios where reliability is more important than speed, such as email, file transfers, and website traffic.

UDP passes datagrams of information, called packets, in binary format. Each packet includes an 8-byte header, which contains information about the source port, destination port, length of the packet, and a checksum. The remainder of the packet includes the data to be transmitted in binary form. Note that the source port and checksum are optional for internet protocol version 4 (IPv4), and the source port is optional for IPv6.

It can be helpful to generate a communication class that handles all UDP communication for sending and receiving. A class of this sort can vary widely in complexity, but a simple class constructor could accept the target IP address, port ID, and a boolean flag denoting whether the port should be used for sending or receiving. The class can have public functions for opening and closing the port, testing whether the port is open, sending an array of real numbers, or receiving an array of real numbers. The following is a basic C++ class definition for a UDP interface. It does not include the required header dependencies and implementation details, but defines a usable interface.

```
class UDP_Port {
    public:
        UDP_Port(const char * targetIP, int portID, bool read);
        ~UDP_Port(){close(mSockfd);};
        void start(){};
        void stop(){close(mSockfd);}
        bool is_running(){return is_open;}
        int receive_array(std::vector<double> &doubleArray);
        int send_array(const std::vector<double> &doubleArray);

    private:
        bool mRead;
        int mPortID;
        const char *mIP;
        int mSockfd;
        struct sockaddr_in mAddress;
        std::mutex mMutex;
        bool is_open;
};
```

With modern artificial intelligence code generation tools, generating the code required for the remaining undefined functions is straightforward.

Packet Contents

As discussed previously, information is typically transferred in packets. The type of information to be included in each packet varies by application and use. For example, information about the location and orientation of the vehicle is needed by a graphics engine. The location of the vehicle can be specified using the three components of the Earth-fixed coordinate system or the three components of the geographic coordinates. The orientation of the vehicle can be specified using the three components of the Euler angles or the four components of the quaternion representation. It may be easiest to package all of this information into a single array that includes all 13 values representing the state of the vehicle. With this packet of information, the graphics software should have access to all information needed to represent the location and orientation of the vehicle at any point in time. Of course, other information could be added, such as the current time, airspeed, etc. which may need to be displayed graphically.

Transmission Rate

The timing of communication is also very important for at least two reasons. These include computational cost and fidelity of modeling real-world sensor update rates. Let us briefly consider each of these.

Any form of communication requires computational time and effort. Hence, communication can impact the speed and accuracy of a dynamic simulation if used too often. The rate at which communication is needed depends on the application. For example, the human eye can detect motion at a rate of about 24–30 Hz. Therefore, a reasonable rate for a graphics package to update is around 30 Hz. If the graphics update at a much slower pace than this, the motion will not appear to be smooth to a human. On the other hand, updating the graphics much faster than 30 Hz does not make sense, since it requires computational effort to communicate and refresh the screen, but the changes at this refresh rate won't be perceptible by a human. Hence, if the physics simulation is running at a very high rate, such as 3,000 Hz, there is no need to send the graphics information at every time step. Rather, it can be sent at a rate of 30 Hz, or only once every 100 steps of the physics simulation. This reduces the computational cost of the physics simulation, since it only needs to send information to the graphics periodically, rather than at every time step.

The best timing of transmission can also be affected by update rates of controls and sensors. For example, if the simulation is being used to emulate the actual characteristics of a particular sensor, such as a Pitot tube, the actual refresh rate of the sensor should be used to set the frequency of the signal being sent by the physics simulator. Depending on sensor quality, the refresh rate of a Pitot tube can vary from 100 to 10,000 Hz. For a hardware-in-the-loop simulation, where the physics simulation is being used to emulate sensor readings from a Pitot tube, the rate at which information is sent should reflect the refresh rate of the Pitot tube that will be used in the actual flight-test article.

In some cases, the transmission and reception rates may differ. For example, a sensor such as a Pitot tube may transmit at 1,000 Hz, while a flight controller may only process updates at 100 Hz. In such cases, the receiving code can be written to purge the port buffer and retain only the most recent packet. For a first-in-first-out (FIFO) buffer, this requires reading all available data and storing only the very last packet. This behavior can be implemented in the `receive_array` function of the UDP class to enable synchronization across asynchronous processes.

9.7 Problems

Theory Problems

9.T1 Repeat Example 9.2.1 using the analytic solution and a spreadsheet to obtain the time history of the actuator.

9.T2 Repeat Example 9.2.1 using fourth-order Runge-Kutta integration to obtain the time history of the actuator.

9.T3 Repeat Example 9.3.1 using the analytic solution and a spreadsheet to obtain the time history of the actuator.

9.T4 Repeat Example 9.3.1 using fourth-order Runge-Kutta integration to obtain the time history of the actuator.

Coding Problems

9.C1 Modify your code so that it can incorporate first-order actuator dynamics including deflection and rate limiting for an actuator with a given time constant. Consider constructing your code to accept actuator information from a JSON dictionary within the `vehicle` dictionary in the form

```
"vehicle" : {
    "control_effectors" : {
        "1" : {
            "name" : "aileron",
            "units" : "deg",
            "dynamics_order" : 1,
            "magnitude_limits" : [-21.5, 21.5],
            "rate_limits[/s]" : [-80.0, 80.0],
            "time_constant[s]" : 0.05
        },
        "2" : {
            "name" : "elevator",
            "units" : "deg",
            "dynamics_order" : 1,
            "magnitude_limits" : [-25.0, 25.0],
            "rate_limits[/s]" : [-60.0, 60.0],
            "time_constant[s]" : 0.05
        },
```

```
            "3" : {
                "name" : "rudder",
                "units" : "deg",
                "dynamics_order" : 1,
                "magnitude_limits" : [-30.0, 30.0],
                "rate_limits[/s]" : [-120.0, 120.0],
                "time_constant[s]" : 0.05
            },
            "4" : {
                "name" : "throttle",
                "dynamics_order" : 1,
                "magnitude_limits" : [0.0, 1.0],
                "rate_limits[/s]" : [-1.0, 1.0],
                "time_constant[s]" : 3.0
            }
        }
}
```

Allow the user to specify the dynamics order using the `dynamics_order` setting. If this is set to 1, the code should use first-order dynamics. If this is set to 0, the code should not model any actuator dynamics and the code should use the commanded value for the actuator as the current value for the actuator. Test your code by repeating Examples 9.2.2 and 9.2.3.

9.C2 Modify your code so that it can incorporate second-order actuator dynamics including deflection limiting, rate limiting, and acceleration limiting for an actuator with a given natural frequency and damping ratio. Consider constructing your code to accept actuator information from a JSON dictionary within the `vehicle` dictionary in the form

```
"vehicle" : {
    "control_effectors" : {
        "aileron[deg]" : {
            "magnitude_limits" : [-21.5, 21.5],
            "rate_limits[/s]" : [-80.0, 80.0],
            "acceleration_limits[/s^2]" : [-1000.0, 1000.0],
            "natural_frequency[rad/s]" : 20.0,
            "damping_ratio" : 0.7
        },
        .
        .
        .
    }
}
```

Allow the user to specify the use of second-order actuator dynamics by setting the `dynamics_order` flag to a value of 2. Test your code by repeating Example 9.3.2.

9.C3 Modify your code to output information that would be reported by an IMU, including information reported by gyroscopes, accelerometers, and a magnetometer. Construct your code to accept the location and orientation of the IMU relative to the vehicle coordinate system as an input in the JSON file.

9.C4 Modify your code to output information that would be reported by a barometer.

9.C5 Modify your code to output information that would be reported by an airspeed sensor. Also output indicated airspeed, calibrated airspeed, and equivalent airspeed. Be sure to include the effects of atmospheric disturbances.

9.C6 Modify your code to output information that would be reported by a GPS sensor.

9.C7 Modify your code to output information that would be reported by an aerodynamic angle sensor. Be sure to include the effects of atmospheric disturbances.

9.C8 Modify your code to apply bias error and normally distributed error for an output signal. Construct your code to allow the user to specify the bias and standard deviation of the sensor error for any sensor.

9.C9 Modify your code to apply quantization to input and output signals to and from the physics simulator. Construct your code to allow the user to specify the bit count or number of steps, as well as the maximum and minimum values for any input or output signal.

9.C10 Develop a communication class that can send and receive information using the UDP transmission protocol. Construct your code to allow the same class to be instantiated and used for any input coming from a controller, or any output from a sensor. Write a separate dummy code that simply receives or sends information to communicate with the physics simulator to test your code.

Chapter 10

Graphics and Visualization

10.1 Introduction

The previous chapters have considered how to mathematically predict the location and orientation of a vehicle as a function of time in the Earth-fixed coordinate system, and how to map that information to geographic coordinates. The numerical results that come from these calculations and simulations in real time are inherently in a digital format that is difficult for the human mind to rapidly process. For example, the states of the vehicle could be printed out to high precision in columns on a screen in real time. However, this form of information is very difficult, if not impossible, for the human mind to read, let alone process, in real time. Furthermore, that information is challenging to assimilate into the human brain's understanding of the state of a vehicle relative to the Earth. Rather than being able to process large amounts of digital information, the human mind is much more capable of rapidly processing imagery. Because of this, in order for a simulation to be useful to a human such as a pilot or analyst in real time, vehicle state information must be portrayed in graphical form. Hence, the ability to visualize the state of a vehicle in three dimensions is integral to flight simulation.

The challenge of visualizing the state of an object in three dimensions is compounded by the fact that most modern screens are inherently two-dimensional. A computer screen is typically made of a large two-dimensional matrix of pixels, each able to turn on and off or change color rapidly. Even virtual-reality goggles are simply two-dimensional screens that are rapidly updated to represent a three-dimensional simulated environment. Hence, in light of current technology trends and capabilities, visualization of flight simulation involves portraying the three-dimensional motion of the vehicle on a two-dimensional screen.

Computer graphics and three-dimensional visualization have been studied in detail in the past 50 years, and are covered in many textbooks. There are also many online resources available for students who wish to study this topic in more detail. Modern flight simulators typically use sophisticated graphics libraries such as OpenGL, which can be used to generate realistic visualization of objects in a three-dimensional space, with lighting, shading, texturing, etc. Some software packages, such as Unreal Engine®, are so advanced that users need not understand the inner-workings of computer algorithms to make very realistic graphics. However, since graphics is so integral to flight simulation, any flight-simulation expert should have some understanding of the underlying principles and practices commonly used in computer graphics. Hence, this chapter gives a basic introduction to computer graphics and how three-dimensional imagery can be generated on a two-dimensional screen.

In the following development, it will be assumed that the state of the vehicle is known. In an integrated physics and graphics framework, this information is readily available during run time of the program executable. However, it is sometimes more convenient and flexible to have separate and distinct codes or executables for the physics and graphics. In this case, information can be passed from the physics engine to the graphics engine using the methods discussed in Chapter 9.

10.2 Center of Projection, View Plane, and Camera

To understand how three-dimensional images are rendered on a two-dimensional display, we must first understand two foundational concepts in computer graphics: the center of projection and the view plane. The *center of projection* represents the location from which an observer is viewing the three-dimensional space. This location is also known as the *observer location*. The *view plane* is a two-dimensional plane in three-dimensional space through which the three-dimensional environment is viewed by the observer. The view plane can be thought of as a window through which the observer views the three-dimensional world. The combination of an observer location and a view plane is often referred to as a *camera* in graphics systems.

Here we will define the center of projection as a single point in space with a local coordinate system. The subscript c will be used to denote the coordinate system used here, as shown in Fig. 10.2.1, and will be referred to as the *camera coordinate system*. In this coordinate system, the x_c axis points towards the view plane. In other words, x_c is the direction the camera is pointing. Note that it is customary for graphics systems to define this coordinate system with the z_c axis orthogonal to the view plane. The choice to orient the coordinate system with the x axis pointed towards the view plane is purely for convenience, such that the rotational matrices developed in Chapter 1 and applied throughout the text for other components can be directly applied. For the purposes of this development, the view plane will be defined as a rectangle at a distance d_{vp} from the observer along the x_c axis and perpendicular to the x_c axis as shown in Fig. 10.2.1. The view plane width w_{vp} is related to the view-plane angle θ_{vp} according to

$$w_{vp} = 2d_{vp} \tan \theta_{vp} \tag{10.2.1}$$

The height h_{vp} of the view plane can be found from a user-specified view-plane aspect ratio defined as

$$R_{A_{vp}} \equiv \frac{w_{vp}}{h_{vp}} \tag{10.2.2}$$

Because the view plane represents the "window" through which the observer views the three-dimensional world, the dimensions of the view plane defined above are directly related to the two-dimensional plot on which information will be viewed on a computer screen. In order to provide a geometrically consistent image, the two-dimensional plot must have the same aspect ratio as the view plane, and have limits specified by the

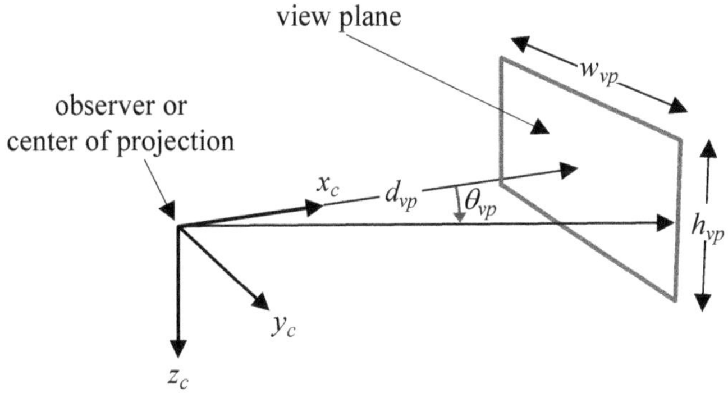

Figure 10.2.1. Definition of a camera, consisting of an observer and a view plane.

view-plane dimensions. Here we will use the following arrays to store the corner points of the view plane

$$\mathbf{x}_{c_{vp}} = \begin{bmatrix} d_{vp} \\ d_{vp} \\ d_{vp} \\ d_{vp} \end{bmatrix}, \qquad \mathbf{y}_{c_{vp}} = \frac{1}{2}\begin{bmatrix} -w_{vp} \\ -w_{vp} \\ w_{vp} \\ w_{vp} \end{bmatrix}, \qquad \mathbf{z}_{c_{vp}} = \frac{1}{2}\begin{bmatrix} -h_{vp} \\ h_{vp} \\ h_{vp} \\ -h_{vp} \end{bmatrix} \tag{10.2.3}$$

These arrays include the points ordered as top-left, bottom-left, bottom-right, and top-right when viewed from the point of the observer. Note that these arrays contain information in the camera coordinate system, denoted by the subscript c. The y_c components in the camera coordinate system correlate to x coordinates on a two-dimensional plot, and the z_c components in the camera coordinate system correlate to $-y$ coordinates on a two-dimensional plot. Hence, the corner points given in Eq. (10.2.3) can be used to generate the limits of a two-dimensional x-y plot to be viewed on a computer screen. The exact syntax for doing this differs by computer language. The following example demonstrates how this can be done in the Python computer language.

EXAMPLE 10.2.1 Consider a camera with a view plane at a distance of 1 ft from the center of projection, a view-plane angle of 60 deg, and a view-plane aspect ratio of 2. Compute the coordinates of the corner points of the view plane in the camera coordinate system. Write a computer code to generate a two-dimensional plot to represent the view plane.

Solution: The width of the view plane is related to the view-plane distance and angle according to Eq. (10.2.1), which yields

$$w_{vp} = 2d_{vp}\tan\theta_{vp} = 2(1)\tan(60\times\pi/180) = 3.46410161514 \text{ ft}$$

The height of the view plane is related to the width and aspect ratio according to Eq. (10.2.2). Solving for the height yields

$$h_{vp} = \frac{w_{vp}}{R_{A_{vp}}} = \frac{3.46410161514}{2} = 1.73205080757 \text{ ft}$$

Using these results in Eq. (10.2.3) yields the corner points of the view plane in the camera coordinate system

$$\mathbf{x}_{C_{vp}} = \begin{bmatrix} d_{vp} \\ d_{vp} \\ d_{vp} \\ d_{vp} \end{bmatrix} = \begin{bmatrix} 1.0 \\ 1.0 \\ 1.0 \\ 1.0 \end{bmatrix} \text{ ft}$$

$$\mathbf{y}_{C_{vp}} = \frac{1}{2} \begin{bmatrix} -w_{vp} \\ -w_{vp} \\ w_{vp} \\ w_{vp} \end{bmatrix} = \begin{bmatrix} -1.73205080757 \\ -1.73205080757 \\ 1.73205080757 \\ 1.73205080757 \end{bmatrix} \text{ ft}$$

$$\mathbf{z}_{C_{vp}} = \frac{1}{2} \begin{bmatrix} -h_{vp} \\ h_{vp} \\ h_{vp} \\ -h_{vp} \end{bmatrix} = \begin{bmatrix} -0.866025403784 \\ 0.866025403784 \\ 0.866025403784 \\ -0.866025403784 \end{bmatrix} \text{ ft}$$

The following Python code generates a two-dimensional plot with the correct limits, which represents the view plane as viewed by the observer or center of projection. Note that in this code, the commands `subplots_adjust` are used to remove white space around the plot, so the plot limits are equal to the limits of the image. Additionally, all axes, labels, and tick marks are removed from the figure. The student may wish to retain axes, labels, and tick marks in the early stages of code development. The resulting limits of the plot are shown in Fig. 10.2.2.

```
import matplotlib.pyplot as plt

fig = plt.figure(figsize=(aspect_ratio_vp*5.0,5.0))
ax = fig.add_subplot(111)
plt.subplots_adjust(top=1.0, bottom=0.0, left=0.0, right=1.0)
plt.axis('off')
ax.axes.set_xlim( y_c_vp[0], y_c_vp[2])
ax.axes.set_ylim(-z_c_vp[1],-z_c_vp[0])
ax.axes.xaxis.set_ticklabels([])
ax.axes.yaxis.set_ticklabels([])
ax.set_xticks([])
ax.set_yticks([])
ax.axes.set_aspect('equal')
fig.canvas.draw()
plt.show()
```

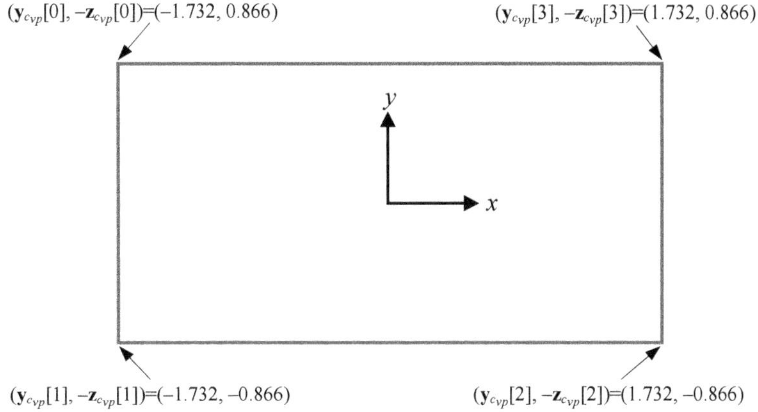

Figure 10.2.2. Example two-dimensional plot representing the view plane to an observer.

Camera Position in Earth-Fixed Coordinates

The position of the camera, or center of projection, in Earth-fixed coordinates can be expressed as

$$\mathbf{P}_{f_c} = \begin{bmatrix} x_{f_c} \\ y_{f_c} \\ z_{f_c} \end{bmatrix} \tag{10.2.4}$$

where the subscript f represents the Earth-fixed coordinate system. The orientation of the camera relative to the Earth-fixed coordinate system can be represented by any of the attitude descriptors outlined in Chapter 1. In general, the components of a vector in the camera coordinate system can be expressed in the Earth-fixed coordinate system using the transformation

$$\begin{bmatrix} v_{x_f} \\ v_{y_f} \\ v_{z_f} \end{bmatrix} = [\mathbf{R}]^{-1} \begin{bmatrix} v_{x_c} \\ v_{y_c} \\ v_{z_c} \end{bmatrix} \tag{10.2.5}$$

where $[\mathbf{R}]$ is a rotation matrix describing the orientation of the camera relative to the Earth-fixed coordinate system. For example, the rotation matrix $[\mathbf{R}]$ can be expressed in Euler angles from Eq. (1.2.5) or in quaternions from Eq. (1.5.2) as

$$\begin{aligned} [\mathbf{R}] &= \begin{bmatrix} C_\theta C_\psi & C_\theta S_\psi & -S_\theta \\ S_\phi S_\theta C_\psi - C_\phi S_\psi & S_\phi S_\theta S_\psi + C_\phi C_\psi & S_\phi C_\theta \\ C_\phi S_\theta C_\psi + S_\phi S_\psi & C_\phi S_\theta S_\psi - S_\phi C_\psi & C_\phi C_\theta \end{bmatrix} \\ &= \begin{bmatrix} e_x^2 + e_0^2 - e_y^2 - e_z^2 & 2(e_x e_y + e_z e_0) & 2(e_x e_z - e_y e_0) \\ 2(e_x e_y - e_z e_0) & e_y^2 + e_0^2 - e_x^2 - e_z^2 & 2(e_y e_z + e_x e_0) \\ 2(e_x e_z + e_y e_0) & 2(e_y e_z - e_x e_0) & e_z^2 + e_0^2 - e_x^2 - e_y^2 \end{bmatrix} \end{aligned} \tag{10.2.6}$$

In computer code, it is much more efficient to store and compute orientation information using the quaternion formulation rather than the Euler-angle formulation. Therefore, although Euler angles may be used as the formulation for the user to specify the orientation of the camera, this information should be converted to quaternions for efficient computations required for visualization. This can be done using Eq. (1.6.2).

With these definitions, the corner points of the view plane in the Earth-fixed coordinate system can be stored in arrays defined as

$$\begin{bmatrix} \mathbf{x}_{f_{vp}} \\ \mathbf{y}_{f_{vp}} \\ \mathbf{z}_{f_{vp}} \end{bmatrix} = [\mathbf{R}]^{-1} \begin{bmatrix} \mathbf{x}_{c_{vp}} \\ \mathbf{y}_{c_{vp}} \\ \mathbf{z}_{c_{vp}} \end{bmatrix} + \begin{bmatrix} x_{f_c} \\ y_{f_c} \\ z_{f_c} \end{bmatrix} \tag{10.2.7}$$

There are two common methods for positioning and orienting the camera in a three-dimensional space for flight simulation. Each of these will be discussed here.

Camera Attached to Vehicle

It is often useful to attach a camera to the vehicle, such that the three-dimensional space can be viewed from the perspective of a pilot or a point just behind the vehicle. This setup is shown in Fig. 10.2.3. In this case, the position of the camera is specified in the body-fixed coordinate system as

$$\mathbf{P}_{b_c} = \begin{bmatrix} x_{b_c} \\ y_{b_c} \\ z_{b_c} \end{bmatrix} \tag{10.2.8}$$

The orientation of the camera is equal to the orientation of the vehicle, which can be obtained from the vehicle states discussed in Chapter 5. Typically this is stored using

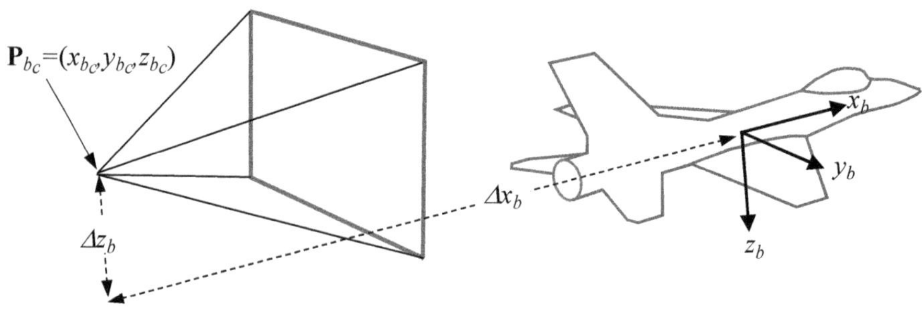

Figure 10.2.3. Camera attached to vehicle.

the quaternion formulation. With these definitions, the position of the camera center of projection in the Earth-fixed coordinate system is

$$\mathbf{P}_{f_c} = [\mathbf{R}]^{-1} \begin{bmatrix} x_{b_c} \\ y_{b_c} \\ z_{b_c} \end{bmatrix} + \begin{bmatrix} x_f \\ y_f \\ z_f \end{bmatrix} \tag{10.2.9}$$

where (x_f, y_f, z_f) is the position of the vehicle in the Earth-fixed coordinate system, and $[\mathbf{R}]$ is the orientation matrix of the vehicle relative to the Earth-fixed coordinate system as computed from Eq. (10.2.6) using the vehicle states. The position of the view plane in the Earth-fixed coordinate system is related to the position of the vehicle according to

$$\begin{aligned} \begin{bmatrix} \mathbf{x}_{f_{vp}} \\ \mathbf{y}_{f_{vp}} \\ \mathbf{z}_{f_{vp}} \end{bmatrix} &= [\mathbf{R}]^{-1} \begin{bmatrix} \mathbf{x}_{c_{vp}} \\ \mathbf{y}_{c_{vp}} \\ \mathbf{z}_{c_{vp}} \end{bmatrix} + \begin{bmatrix} x_f \\ y_f \\ z_f \end{bmatrix} + [\mathbf{R}]^{-1} \begin{bmatrix} x_{b_c} \\ y_{b_c} \\ z_{b_c} \end{bmatrix} \\ &= [\mathbf{R}]^{-1} \begin{bmatrix} \mathbf{x}_{c_{vp}} + x_{b_c} \\ \mathbf{y}_{c_{vp}} + y_{b_c} \\ \mathbf{z}_{c_{vp}} + z_{b_c} \end{bmatrix} + \begin{bmatrix} x_f \\ y_f \\ z_f \end{bmatrix} \end{aligned} \tag{10.2.10}$$

Camera Attached to Earth

It is sometimes helpful to view the vehicle from a stationary point in the Earth-fixed coordinate system. Such an approach allows the graphics to represent the view from a tower or the view from the ground for a pilot flying a remote-controlled vehicle. In this case, the center of projection is set at a fixed point in the Earth-fixed coordinate system and, because the vehicle is moving, the orientation of the camera must change with time such that the x_c axis points towards the vehicle at all times. This is shown in Fig. 10.2.4.

In this case, a camera location in the Earth-fixed coordinate system is specified by the user

$$\mathbf{P}_{f_c} = \begin{bmatrix} x_{f_c} \\ y_{f_c} \\ z_{f_c} \end{bmatrix} \tag{10.2.11}$$

The orientation of the camera coordinate system is most easily visualized using the Euler-angle formulation. The Euler angles for the rotation are related to the difference in the position of the vehicle relative to the camera

$$\begin{bmatrix} \Delta x_f \\ \Delta y_f \\ \Delta z_f \end{bmatrix} = \begin{bmatrix} x_f - x_{f_c} \\ y_f - y_{f_c} \\ z_f - z_{f_c} \end{bmatrix} \tag{10.2.12}$$

The associated Euler angles that describe the orientation of the camera relative to the Earth-fixed coordinate system are

$$\begin{bmatrix} \phi \\ \theta \\ \psi \end{bmatrix} = \begin{bmatrix} 0 \\ \tan^{-1}(-\Delta z_f/d) \\ \tan^{-1}(\Delta x_f/\Delta y_f) \end{bmatrix} \tag{10.2.13}$$

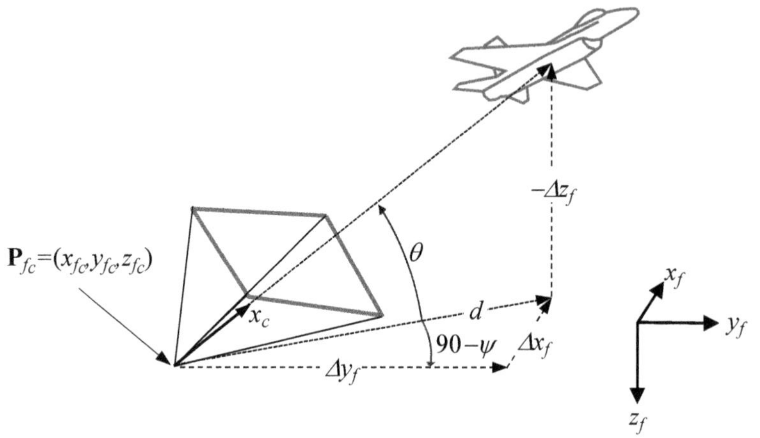

Figure 10.2.4. Camera attached to Earth.

where

$$d = \sqrt{\Delta x^2 + \Delta y^2} \tag{10.2.14}$$

In practice, the `atan2` function should be used to compute angles in Eq. (10.2.13). The Euler angles shown in Eq. (10.2.13) can be used in Eq. (10.2.6) to compute the rotation matrix, and the result can then be used in Eq. (10.2.7) to compute the corner points of the view plane in the Earth-fixed coordinate system.

EXAMPLE 10.2.2 Consider the camera given in Example 10.2.1 located at the Earth-fixed coordinates

$$\mathbf{P}_{f_c} = \begin{bmatrix} x_{f_c} \\ y_{f_c} \\ z_{f_c} \end{bmatrix} = \begin{bmatrix} -20.0 \\ 0.0 \\ -5.0 \end{bmatrix}$$

and oriented relative to the Earth-fixed coordinate system with Euler angles $(\phi, \theta, \psi) = (5, -10, 0)$ deg. Compute the coordinates of the corner points of the view plane in the Earth-fixed coordinate system.

Solution: Using the Euler angles $(\phi, \theta, \psi) = (5, -10, 0)$ deg in Eq. (10.2.6) to compute the rotation matrix $[\mathbf{R}]$, and using this rotation matrix along with the results of Example 10.2.1 in Eq. (10.2.7) along with the camera position $\mathbf{P}_{f_c} = (x_{f_c}, y_{f_c}, z_{f_c}) = (-20.0, 0.0, -5.0)$ ft, yields the corner points of the view plane in the Earth-fixed coordinate system in units of ft

$$\mathbf{x}_{f_{vp}} = \begin{bmatrix} -18.8391671574 \\ -19.1387901127 \\ -19.1912173366 \\ -18.8915943812 \end{bmatrix}, \quad \mathbf{y}_{f_{vp}} = \begin{bmatrix} -1.64998074402 \\ -1.80093891863 \\ 1.64998074402 \\ 1.80093891863 \end{bmatrix}, \quad \mathbf{z}_{f_{vp}} = \begin{bmatrix} -5.82463971277 \\ -4.12539349337 \\ -3.82806393190 \\ -5.52731015130 \end{bmatrix}$$

10.3 Projection of a Point onto the View Plane

With the definition of the location and orientation of the observer and view plane as given in the previous section, let us now consider how to project an image of the surrounding three-dimensional environment onto the view plane. Recall that a view plane is similar to a window, through which the observer can view the three-dimensional space. Any point in the three-dimensional space will be visible to the observer if the line passing from the point of interest to the observer intersects the view plane as shown in Fig. 10.3.1. Hence, in order to construct the image that the observer would see as projected on the view plane, we must find the points of intersection of the view plane and the lines passing from the observer to the points of interest.

Let $\mathbf{P}_c$ be the location of the camera, or center of projection, and $\mathbf{P}_a$ be the point of interest. The line from $\mathbf{P}_c$ to $\mathbf{P}_a$ is

$$\mathbf{l}_{ca} = \mathbf{P}_a - \mathbf{P}_c \tag{10.3.1}$$

Any point $\mathbf{P}_b$ along that line can be obtained from

$$\mathbf{P}_b = \mathbf{P}_c + \lambda \mathbf{l}_{ca} \tag{10.3.2}$$

where λ is a real number. We wish to find the point $\mathbf{P}_b$ that lies on the view plane.

The plane of interest, or view plane, can be described by the three points $\mathbf{P}_0$, $\mathbf{P}_1$, and $\mathbf{P}_2$ where $\mathbf{P}_0$ represents the center of the plane and the points $\mathbf{P}_1$ and $\mathbf{P}_2$ are two neighboring corners of the view plane. If the points $\mathbf{P}_1$ and $\mathbf{P}_2$ are ordered counter-clockwise as viewed from the observer, the view-plane normal pointing toward the center of projection can be expressed in vector notation as

$$\mathbf{n}_{vp} = (\mathbf{P}_1 - \mathbf{P}_0) \times (\mathbf{P}_2 - \mathbf{P}_0) \tag{10.3.3}$$

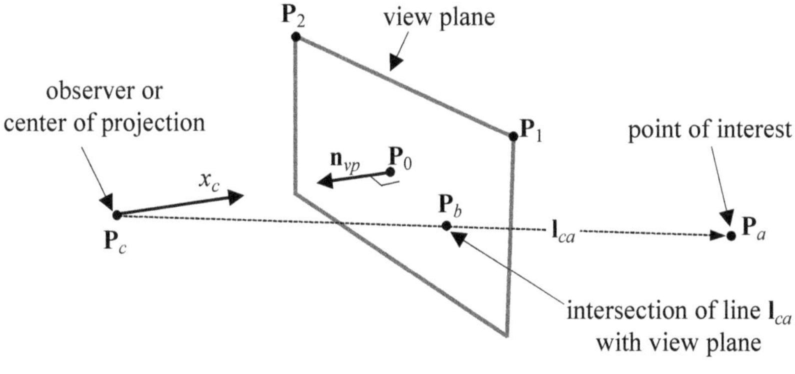

Figure 10.3.1. Projection of a point onto the view plane.

In order for the point $\mathbf{P}_b$ to lie on the view plane, the line connecting $\mathbf{P}_0$ and $\mathbf{P}_b$ must be perpendicular to the plane normal $\mathbf{n}_{vp}$. Hence,

$$(\mathbf{P}_b - \mathbf{P}_0) \cdot \mathbf{n}_{vp} = 0 \tag{10.3.4}$$

Using Eq. (10.3.2) in Eq. (10.3.4) and solving for λ yields

$$\lambda = \frac{(\mathbf{P}_0 - \mathbf{P}_c) \cdot \mathbf{n}_{vp}}{\mathbf{l}_{ca} \cdot \mathbf{n}_{vp}} \tag{10.3.5}$$

This value for λ can then be used in Eq. (10.3.2) to compute the location of the point $\mathbf{P}_b$. If $0 < \lambda < 1$, the view plane lies between points $\mathbf{P}_c$ and $\mathbf{P}_a$ as depicted in Fig. 10.3.1. If $\lambda > 1$, the point $\mathbf{P}_a$ lies between $\mathbf{P}_c$ and the view plane. If $\lambda < 0$ the point need not be included in plotting, since it represents a case where the point is behind the observer. Note that when the denominator $\mathbf{l}_{ca} \cdot \mathbf{n}_{vp} = 0$, the line $\mathbf{l}_{ca}$ is parallel to the plane. The computational cost of computing λ for each point of interest can be reduced by noting that the vector $\mathbf{l}_{ca}$ is the only component of Eq. (10.3.5) that depends on the point of interest.

With the observer and view-plane location specified in the Earth-fixed coordinate system as was presented in the previous section, the coordinates used in Eqs. (10.3.1)–(10.3.5) are all in the Earth-fixed coordinate system. Therefore, the point that intersects the view plane $\mathbf{P}_b$ as computed from Eq. (10.3.2) is also in the Earth-fixed coordinate system. These points can be expressed in the camera coordinate system by applying the translation and rotation

$$\begin{bmatrix} x_{vp} \\ y_{vp} \\ z_{vp} \end{bmatrix} = [\mathbf{R}](\mathbf{P}_b - \mathbf{P}_c) = [\mathbf{R}](\lambda \mathbf{l}_{ca}) \tag{10.3.6}$$

where $[\mathbf{R}]$ is the rotation matrix describing the orientation of the camera relative to the Earth-fixed coordinate system given in Eq. (10.2.6). Note that, for viewing purposes, the points of intersection on the view plane are only needed in the camera coordinate system, and not the Earth-fixed coordinate system. These can be directly computed from the right-hand side of Eq. (10.3.6), and the values for $\mathbf{P}_b$ in the Earth-fixed coordinate system need never be computed.

If the view-plane boundaries are used as the boundaries of an x-y plot on a computer screen, as was presented in the previous section, the corresponding x and y coordinates to be plotted in a figure on a computer screen can be found from

$$\begin{bmatrix} x \\ y \end{bmatrix} = \begin{bmatrix} y_{vp} \\ -z_{vp} \end{bmatrix} \tag{10.3.7}$$

Note that if the point on the plot (x, y) is outside the bounds of the plot, it need not be included when plotting. Additionally, any point for which $\lambda < 0$ need not be included in the plot, since it represents a point behind the observer.

EXAMPLE 10.3.1 Consider the list of points in the Earth-fixed coordinate system

$$\mathbf{x}_f = \begin{bmatrix} -10 \\ 10 \\ -10 \\ 10 \\ -10 \\ 10 \\ -10 \\ 10 \\ -10 \\ 10 \\ -10 \\ -10 \\ -5 \\ -5 \\ 0 \\ 0 \\ 5 \\ 5 \\ 10 \\ 10 \end{bmatrix}, \quad \mathbf{y}_f = \begin{bmatrix} -10 \\ -10 \\ -5 \\ -5 \\ 0 \\ 0 \\ 5 \\ 5 \\ 10 \\ 10 \\ -10 \\ 10 \\ -10 \\ 10 \\ -10 \\ 10 \\ -10 \\ 10 \\ -10 \\ 10 \end{bmatrix}, \quad \mathbf{z}_f = \begin{bmatrix} 0 \\ 0 \\ 0 \\ 0 \\ 0 \\ 0 \\ 0 \\ 0 \\ 0 \\ 0 \\ 0 \\ 0 \\ 0 \\ 0 \\ 0 \\ 0 \\ 0 \\ 0 \\ 0 \\ 0 \end{bmatrix}$$

For each point in this list, compute the point of intersection in the camera coordinate system on the view plane given in Examples 10.2.1 and camera location and orientation given in Example 10.2.2. Use the results to generate a plot of the points as viewed by the observer on a view plane.

Solution: The center of the view plane in Earth-fixed coordinates can be found by averaging the limits of the view plane, which yields

$$\mathbf{P}_0 = \begin{bmatrix} -19.01519224699 \\ 0.0 \\ -4.826351822333 \end{bmatrix} \text{ ft}$$

Using any two consecutive points $\mathbf{P}_1$ and $\mathbf{P}_2$ ordered counter-clockwise around the boundary of the view plane in Earth-fixed coordinates in Eq. (10.3.3) yields

$$\mathbf{n}_{vp} = (\mathbf{P}_1 - \mathbf{P}_0) \times (\mathbf{P}_2 - \mathbf{P}_0) = \begin{bmatrix} -2.95442325904 \\ 0.0 \\ -0.520944533001 \end{bmatrix} \text{ ft}$$

The vector from the center of projection to the center of the view plane is

$$\mathbf{P}_0 - \mathbf{P}_c = \begin{bmatrix} 0.984807753012 \\ 0.0 \\ 0.173648177667 \end{bmatrix} \text{ ft}$$

The vector $\mathbf{l}_{ca}$ which extends from the center of projection to each point can be computed, which yields

$$\mathbf{l}_{ca_x} = \begin{bmatrix} 10.0 \\ 30.0 \\ 10.0 \\ 30.0 \\ 10.0 \\ 30.0 \\ 10.0 \\ 30.0 \\ 10.0 \\ 30.0 \\ 10.0 \\ 10.0 \\ 15.0 \\ 15.0 \\ 20.0 \\ 20.0 \\ 25.0 \\ 25.0 \\ 30.0 \\ 30.0 \end{bmatrix}, \quad \mathbf{l}_{ca_y} = \begin{bmatrix} -10.0 \\ -10.0 \\ -5.0 \\ -5.0 \\ 0.0 \\ 0.0 \\ 5.0 \\ 5.0 \\ 10.0 \\ 10.0 \\ -10.0 \\ 10.0 \\ -10.0 \\ 10.0 \\ -10.0 \\ 10.0 \\ -10.0 \\ 10.0 \\ -10.0 \\ 10.0 \end{bmatrix}, \quad \mathbf{l}_{ca_z} = \begin{bmatrix} 5.0 \\ 5.0 \\ 5.0 \\ 5.0 \\ 5.0 \\ 5.0 \\ 5.0 \\ 5.0 \\ 5.0 \\ 5.0 \\ 5.0 \\ 5.0 \\ 5.0 \\ 5.0 \\ 5.0 \\ 5.0 \\ 5.0 \\ 5.0 \\ 5.0 \\ 5.0 \end{bmatrix}$$

which have units of ft. Using these results in Eq. (10.3.5) yields

$$\lambda = \begin{bmatrix} 0.09331562958 \\ 0.03288124528 \\ 0.09331562958 \\ 0.03288124528 \\ 0.09331562958 \\ 0.03288124528 \\ 0.09331562958 \\ 0.03288124528 \\ 0.09331562958 \\ 0.03288124528 \\ 0.09331562958 \\ 0.09331562958 \\ 0.06393715874 \\ 0.06393715874 \\ 0.04862773516 \\ 0.04862773516 \\ 0.03923348025 \\ 0.03923348025 \\ 0.03288124528 \\ 0.03288124528 \end{bmatrix}$$

Because all λ values are in the range $0 < \lambda < 1$, all points in this list are located on the opposite side of the view plane as the center of projection. The intersection point in the camera coordinate system on the view plane for each line $\mathbf{l}_{ca}$ can be computed from Eq. (10.3.6), and the results used in Eq. (10.3.7) to yield the final set of x and y values to be displayed on the plot

$$x = \begin{bmatrix} -0.903680975676 \\ -0.328379137233 \\ -0.438878298491 \\ -0.164598526165 \\ 0.025924378693 \\ -0.000817915096 \\ 0.490727055877 \\ 0.162962695973 \\ 0.955529733062 \\ 0.326743307042 \\ -0.903680975676 \\ 0.955529733062 \\ -0.624014218412 \\ 0.649862952554 \\ -0.478276992315 \\ 0.490576846702 \\ -0.388848892995 \\ 0.392834807306 \\ -0.328379137233 \\ 0.326743307042 \end{bmatrix} \text{ft}, \quad y = \begin{bmatrix} -0.377646934437 \\ -0.019309081226 \\ -0.336981969407 \\ -0.004980134452 \\ -0.296317004376 \\ 0.009348812322 \\ -0.255652039346 \\ 0.023677759096 \\ -0.214987074316 \\ 0.038006705870 \\ -0.377646934437 \\ -0.214987074316 \\ -0.203451097198 \\ -0.092001286014 \\ -0.112675853243 \\ -0.027912125715 \\ -0.056973835648 \\ 0.011414626591 \\ -0.019309081226 \\ 0.038006705870 \end{bmatrix} \text{ft}$$

Figure 10.3.2 shows the points plotted on the final x-y plot representing the view of the observer.

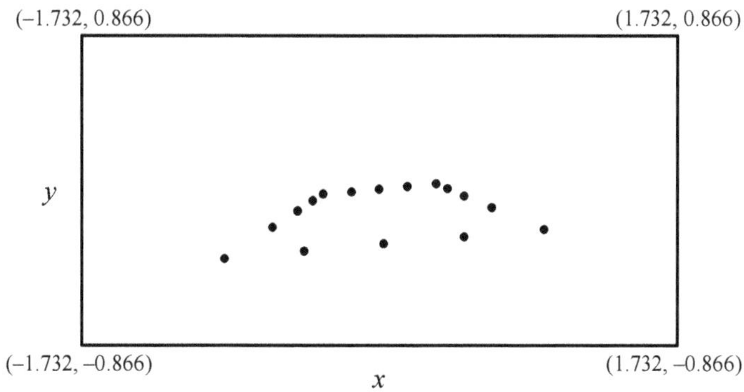

Figure 10.3.2. Points projected onto the camera view plane.

Clipping

Not all points in the line-of-sight of a camera need be included when plotting. For example, points that are sufficiently close to an observer or camera are sometimes not included, and points that are sufficiently far away may not be included. The process of removing points within the line-of-sight of a camera is called *clipping*. In most graphical software packages, both near and far clipping planes can be defined. If points are closer to the observer than the near clipping plane, the point is not included in the graphical representation. Likewise, if a point is further from the observer than the far clipping plane, the point is not included in the image. As discussed above, the value of λ for any point yields information about how that point is positioned relative to the observer and the view plane. A similar process can be used to identify where points are located relative to near or far clipping planes. Hence, with the development included above, near and far clipping planes can easily be defined and implemented.

The concepts of an observer and view plane presented in Section 10.2, and the mathematics for finding the point of intersection of a line with a plane as shown here, form the basis for computer graphics and visualization. With this fundamental understanding, we can now turn to the application of these principles for representing more complex geometric entities.

10.4 Representing Geometries

The definition of the observer and view plane presented in Section 10.2, and a method for projecting a point onto a plane by computing the point at which a line intersects a plane as presented in Section 10.3 form the foundation for visually representing complex three-dimensional geometries on a two-dimensional view plane.

Lines

The principles discussed in Section 10.3 can be used to project the view of a line segment in three-dimensional space onto a two-dimensional view plane. A line segment represents a straight line along the shortest distance between two points in space. A line segment in three-dimensional space is therefore fully defined by two end points. Following the process outlined in Section 10.3, each end point can be projected onto the view plane. Once the end-point coordinates of the line in the view plane are known, they can simply be connected by a line in the two-dimensional view plane or plot on the computer screen to represent the observer perspective of the line in three-dimensional space.

EXAMPLE 10.4.1 The points given in Example 10.3.1 can be used to generate a grid that represents a ground plane by connecting each set of two points with a line. Note that all the $\mathbf{z}_f$ values of the original points in the Earth-fixed coordinate system are zero, representing points at an altitude of zero. The ground plane is centered at the origin of the Earth-fixed coordinate system. Example 10.3.1 demonstrated how to project those points onto a view plane of a camera specified in Example 10.2.1,

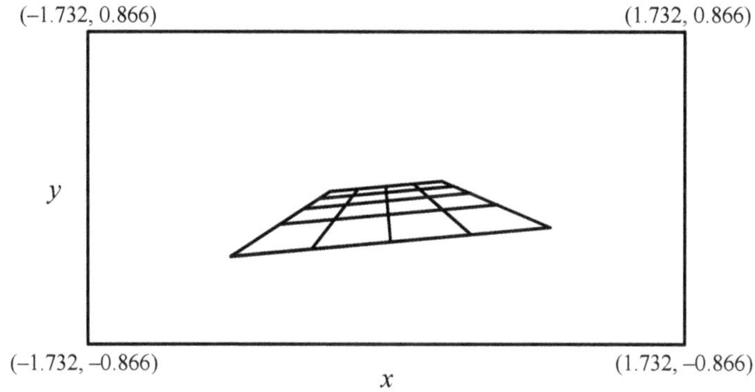

Figure 10.4.1. Observer perspective of a ground plane.

located at $\mathbf{P}_{f_c} = (-20.0, 0.0, -5.0)$ ft, and oriented relative to the Earth-fixed coordinate system with Euler angles $(\phi, \theta, \psi) = (5, -10, 0)$. Use plotting software to draw a line between each set of two points to generate the view of the ground plane to the observer.

Solution: The final (x, y) coordinates given in Example 10.3.1 represent the original points projected onto the camera view plane in a coordinate system that represents the view of the observer and can be used directly for plotting. Connecting each set of two points with a line in the two-dimensional plot generates the results shown in Figure 10.4.1. The reader is encouraged to pause for a moment and ensure that the view of the ground plane matches the reader's intuition for how it would appear given the camera location and orientation relative to the Earth-fixed coordinate system.

Polygons

A polygon is a set of connected line segments that form a closed surface. The method discussed above can be used to find the observer perspective of each line segment. In the view plane, the collection of two-dimensional line segments will also represent a closed surface with the outline defined by the line segments. The surface can be filled in with a specific color, or shaded, using common two-dimensional plotting software.

Three-Dimensional Geometries

Any three-dimensional geometry can be represented by a collection of polygons. For example, Fig. 10.4.2 shows a coarse polygon representation of an F-16. If the polygons all connect to represent a closed volume, the geometry is said to be *watertight*. To represent a three-dimensional geometry, each polygon can be projected onto the view plane as described above.

Figure 10.4.2. Coarse polygon model of an F-16.

Clipping

The previous section included a discussion on implementing both near and far plane clipping. Applying clipping to a set of points is straightforward, since the λ value for an individual point can be computed, and based on the value of λ, that point can be included or excluded from the plot. However, for lines, polygons, and more complex geometries, the process of clipping requires additional consideration.

For example, consider the case of plotting lines. If both points representing the end of a line are to be included in the image, the line should also be included. Likewise, if both points are to be clipped, the entire line should be clipped. However, there are often scenarios in which one point should be clipped while the other point should not. If the entire line connecting the two points is clipped whenever one of the points should be clipped, lines will appear to the user to sporadically disappear. When one end point remains in view when the other end point should be clipped, only a portion of the line should be clipped to allow the portion of the line that is in view to remain in view. This can be done by computing the intersection point of the line of interest on the view plane or the clipping plane. This point is then used as the end point to the line instead of the end point that is to be clipped. This method can be used in a similar manner for handling the clipping of polygons.

10.5 Common Geometry File Types

Many different file types are used in various computer-graphics software packages for storing geometric information for three-dimensional objects. Some of these file types preserve the actual geometric shape, including the contour of the geometry, and some only approximate the geometric shape using a collection of polygons to represent the surface.

The difference between these two main types of geometric files can be understood by considering two methods for storing two-dimensional images. Two-dimensional images can either be stored in a vector format or raster format. *Vector* files store the image information as a collection of mathematical lines and curves with colors filling the regions that make up the image. This allows the images to be enlarged without any loss of resolution, since the mathematical lines and curves that make up the image can be generated at any resolution or size. *Rasterized* files store the image information in a format similar to a two-dimensional matrix or grid. Each element in the matrix or grid contains a different color. As the image is enlarged, the discretized grid used to store the image information can be seen, which does not preserve the true contours within the image. Whereas vector format preserves the actual contour of the image contents, a rasterized image does not. For example, Fig. 10.5.1 shows an enlarged vector image and rasterized image of the letter x. The vector image can be enlarged indefinitely without loss of clarity or resolution on a computer screen, whereas the rasterized image cannot.

Two methods for storing three-dimensional information about an object can be taken, which are analogous to vector images and rasterized images. The three-dimensional analog to a two-dimensional vector image is any file type that stores the three-dimensional information about the object in a mathematical form from which the curvature and connectivity of the object can be obtained at any resolution. Such file types preserve the actual geometric shape of the object, and store the surface properties as three-dimensional equations in space. Although each computer-aided design (CAD) package typically has a unique format for storing this information, perhaps the most common cross-package CAD files of this type are IGES and STEP files. These two file types are so prevalent that nearly every CAD software package can read and write these file types. Fig. 10.5.2 shows the outer geometry of an F-16 stored as a STEP file, which preserves curvature.

The three-dimensional analog to a two-dimensional rasterized image is any file type that stores the three-dimensional information about an object in a *mesh* format. Such is the case when a collection of planar polygons are used to represent the surface of an object. In this case, the actual curvature of the surface is not stored, but only represented by a finite number of planar elements along the surface, very similar to the matrix or grid elements used to represent a two-dimensional rasterized image. As the number of polygons used to represent a geometry increases, the actual surface is better approximated. The most prevalent file type for storing geometric information in this

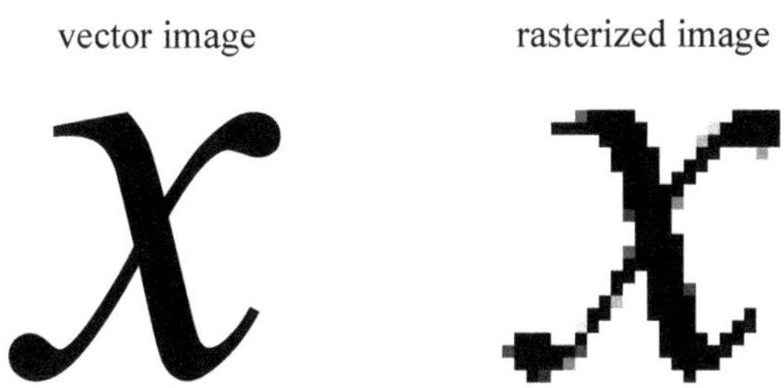

Figure 10.5.1. Vector and rasterized images of the letter x.

Figure 10.5.2. Image of a CAD model of an F-16 that preserves curvature.

format is the stereolithography (STL) format. Fig. 10.4.2 shows the outer geometry of an F-16 represented by a surface mesh of elements.

When high-precision design work must be performed, files that preserve the three-dimensional curvature of the geometry must be used. However, in simulation frameworks when the exact curvature of the geometry is not needed, mesh representations of geometric objects can be used. Because using a mesh, or collection of polygons, to represent a complex geometry is such a common practice in simulation, many file types exist for storing surface meshes. Here we will discuss two of the most commonly used file types.

Stereolithography (STL) File Format

The stereolithography (STL) file type is perhaps the most widely used format for storing geometric shape information in a mesh format. Nearly all CAD and graphics software can read and write the STL file format, and the STL format is the most prevalent format used in 3D printing. The basic premise of the STL file type is that every complex geometry can be represented by a collection of polygons. Although the file format can allow other polygons to be represented, in practice triangles are used exclusively. All polygon shapes can be subdivided into triangular elements, which makes triangular elements an attractive and versatile basic shape.

The STL file format is simple to read and write in ASCII format. The first line of the file must be

```
solid name
```

where `name` is an arbitrary string, which can be used to store not only the name of the solid, but also other metadata. The contents of the first line following the word `solid` is typically ignored in software packages. The remainder of the file consists of any number of triangular elements, each represented by the form

```
facet normal nx ny nz
   outer loop
      vertex x1 y1 z1
      vertex x2 y2 z2
      vertex x3 y3 z3
   endloop
endfacet
```

where n_x, n_y, and n_z are the three components of the triangle normal vector **n**, and the three triangle corners are defined by the three points $\mathbf{P}_1 = (x_1, y_1, z_1)$, $\mathbf{P}_2 = (x_2, y_2, z_2)$, and $\mathbf{P}_3 = (x_3, y_3, z_3)$. Each of these entries must be formatted as a floating-point number in the sign-mantissa-e-sign-exponent format. For example,

```
facet normal 1.234567e-000 -1.234567e-000 1.234567e+000
```

The final line of the file is

```
endsolid name
```

Although the tabbed format shown here is helpful for visualizing individual triangular components in the ASCII file, the file format for reading and writing in software is agnostic to the use of whitespace. Whitespace (tabs, spaces, newlines) can be used at any point in the file except within a word or number. Note that the space between `facet` and `normal` as well as that between `outer` and `loop` are required.

EXAMPLE 10.5.1 A cuboid is defined as shown in Fig. 2.3.1 with lengths $l_x = l_y = l_z = 2.0$. Generate an ASCII STL file to store the faces of the cuboid geometry that are normal to the x axis, with the face normals pointing outward from the center of the geometry.

Solution: Figure 10.5.3 shows the cuboid with the proper dimensions and the values of the corner points labeled. Notice that each face of the cuboid was split into two triangular elements so that it could be represented in STL format. The STL ASCII file for storing just the two faces of the cuboid that are normal to the x axis is shown below the figure. Note the ordering of the vertices to produce an outward normal vector consistent with the right-hand rule. In order to specify two faces of the cuboid, four triangular elements are needed. This process could be repeated for the faces normal to the y and z axes, which would result in a total of 12 triangular elements to represent all 6 sides of the cuboid.

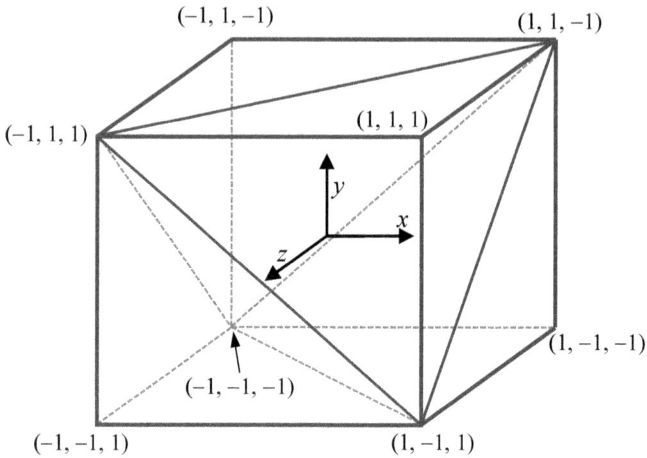

Figure 10.5.3. Triangulated tesselation of a cuboid.

```
solid cuboid example (just the faces normal to the x axis)
facet normal  1.000000e+000  0.000000e+000  0.000000e+000
   outer loop
      vertex  1.000000e+000  1.000000e+000 -1.000000e+000
      vertex  1.000000e+000  1.000000e+000  1.000000e+000
      vertex  1.000000e+000 -1.000000e+000  1.000000e+000
   endloop
endfacet
facet normal  1.000000e+000  0.000000e+000  0.000000e+000
   outer loop
      vertex  1.000000e+000  1.000000e+000 -1.000000e+000
      vertex  1.000000e+000 -1.000000e+000  1.000000e+000
      vertex  1.000000e+000 -1.000000e+000 -1.000000e+000
   endloop
endfacet
facet normal -1.000000e+000  0.000000e+000  0.000000e+000
   outer loop
      vertex -1.000000e+000  1.000000e+000  1.000000e+000
      vertex -1.000000e+000  1.000000e+000 -1.000000e+000
      vertex -1.000000e+000 -1.000000e+000 -1.000000e+000
   endloop
endfacet
facet normal -1.000000e+000  0.000000e+000  0.000000e+000
   outer loop
      vertex -1.000000e+000  1.000000e+000  1.000000e+000
      vertex -1.000000e+000 -1.000000e+000 -1.000000e+000
      vertex -1.000000e+000 -1.000000e+000  1.000000e+000
```

```
    endloop
  endfacet
  endsolid cuboid example (just the faces normal to the x axis)
```

As stated previously, the format of the STL file clearly could be used for polygons of any size. However, in practice, it is exclusively used only for triangles. Because it is used only for triangles, the ASCII format of STL files is very verbose, and requires a significant amount of memory in comparison to the amount of unique information it contains. For example, if the vertices are properly ordered, the normal vector could be computed from the vertex information, and need not be stored in the file. Furthermore, because STL files are traditionally used only for triangular elements, each loop always contains only three vertices, which need not be labeled. Hence, the file could be reduced simply to a list of numbers. Each set of nine numbers would contain the vertex information for three corner points, and be ordered such that the normal of each panel could be directly computed. White space could still be used to separate each triangular element to yield the file format

```
solid name
x1 y1 z1
x2 y2 z2
x3 y3 z3

x1 y1 z1
x2 y2 z2
x3 y3 z3
...
endsolid name
```

A typical STL ASCII format stores approximately 230 ASCII characters for each triangular element. The modified format suggested here would require storing only 168 ASCII characters for each triangular element. Hence, without any loss of information, the modified file format would require about 27% less memory.

ASCII files of any type are inherently large due to the need to store the binary representation of every ASCII character in the file. Hence, rather than storing data in ASCII format, it is common for data files to use a pure binary form. In this form, rather than storing the binary form of each ASCII character for a floating-point number, the binary value of the floating-point number is stored directly. Binary files constructed in this way are typically much faster to read and write and require less memory for storage. STL files can be stored in a binary format as well[2], which significantly reduces the file size. Free and open-source libraries are available for most major coding languages for reading binary STL files and storing the data to arrays within the code, as well as writing

[2]https://en.wikipedia.org/wiki/STL_(file_format)

binary STL files from information stored in arrays. For example, in Python, an STL file can be read, and the number of points and faces obtained from the code

```
from stl import mesh
myMesh = mesh.Mesh.from_file('example.stl')
npoints = myMesh.vectors
nfaces = len(myMesh.vectors)
```

An additional inefficiency of the STL file format is perhaps less obvious. Notice that all three corner points of each triangular element are specified out to a set precision within the loop for any given triangular element. For any water-tight triangular mesh, triangular elements share corner points with neighboring triangular elements. This can clearly be seen in Example 10.5.1. The triangular mesh used to represent the cube in that example required 12 triangular elements. To represent the cuboid in STL file format, each of the three corner points of each triangular element was stored separately in the file. This required 36 points, each with x, y, and z components for a total of 108 floating-point numbers. However, in reality, only 8 unique vertices define the cuboid, each with x, y, and z components. Hence, there are only a total of 24 potential unique floating-point numbers required to represent a cuboid.

Although STL files are widely used, their format is not particularly useful or efficient for direct use in computer graphics. Hence, once the information from an STL file is read into memory for generating computer graphics, it is typically manipulated to store the information in a more efficient manner. Other file formats can also be used to represent three-dimensional geometries in the form of surface meshes, which are much more efficient to store and process than the STL file format. One commonly used format is the VTK file format.

Visualization Toolkit (VTK) File Format

The visualization toolkit (VTK) file format is highly versatile, and can be used to store all types of scalar and vector information, including representations for geometric shapes. Here we will not attempt to give a comprehensive overview of the VTK file format, but will only present a few key elements of the legacy file format that can be useful for basic representation of geometric shapes. More information about the VTK file formats currently supported and in use can be found on the VTK website[3].

A basic legacy ASCII VTK file contains five main components. These include the header, title, data type, geometry/topology, and dataset attributes. We will give a brief overview of each.

Header

The first line of the file is the header line, which for this legacy version must be

```
# vtk DataFile Version 3.0
```

[3]https://docs.vtk.org/

Title

The second line of the file is the title line. The title line can contain a maximum of 256 characters, which can be used to store any metadata, such as a title or name for the dataset. For example, the title line can be

```
my new VTK file
```

Data Type

The third line of the file is the data type, which must be either ASCII or BINARY. In this development, we will consider only ASCII data types. Hence, line three of the file is

```
ASCII
```

Geometry/Topology

The fourth line of the file is the beginning of the geometry/topology section. This section begins with a type, which must be one of the following: `STRUCTURED_POINTS`, `STRUCTURED_GRID`, `UNSTRUCTURED_GRID`, `POLYDATA`, `STRUCTURED_POINTS`, `RECTILINEAR_GRID`, or `FIELD`. Here we will consider only the `POLYDATA` type. For this type, the next line must specify the number of points as well as the data type of those points. This is followed by a listing of all points used for the corners of the polygons in the format

```
DATASET POLYDATA
POINTS N float
x0 y0 z0
x1 y1 z1
x2 y2 z2
x3 y3 z3
...
xN-1 yN-1 zN-1
```

where N is the number of points. Once all the points have been listed, the following line specifies how many polygons will be used to generate the geometry in the form

```
POLYGONS Ng Ni
```

where N_g is the number of polygons, and N_i is the number of integers in the following list of polygon descriptions. This is followed by a description for each polygon of the form

$$N_p \; P_0 \; P_1 \; P_2 \; P_3 \; \cdots \; P_{N_p-1}$$

where N_p is the number of vertices of the polygon and P_0–P_{N_p-1} are integers referring to indices of the points in the array of points listed above that correspond to each vertex of the polygon. Note that the array of points is indexed from zero.

A set of lines can also be specified by using a similar format of the form

---

`LINES` N_l N_i

---

where N_l is the number of lines, N_i is the number of integers in the following list of line descriptions. This is followed by a description for each line of the form

---

$N_p \; P_0 \; P_1 \; P_2 \; P_3 \; \cdots \; P_{N_p-1}$

---

where N_p is the number of points used to define the line and P_0–P_{N_p-1} are integers referring to indices of the points in the array of points listed above that correspond to each point on the line. Note that the VTK file format can handle lines defined by more than two points. Such lines are referred to as *polylines*, which are defined as straight line segments connecting an array of points.

Dataset Attributes

The final section of the VTK file is the dataset attributes. This section can be used to store scalar and vector quantities for each polygon, such as color, temperature, normal vector, etc. In the examples shown here, we will not use this section of the VTK file.

EXAMPLE 10.5.2 Generate a VTK file that uses quadrilaterals to store all six sides of the cuboid geometry used in Example 10.5.1. Order the points representing each polygon such that the normal vector of each polygon points outward.

Solution: Figure 10.5.3 shows the cuboid with the proper dimensions and the values of the corner points labeled. Figure 10.5.4 shows the cuboid with the corner points numbered, indexed from zero. Here we will specify the cuboid geometry using a four-sided polygon, or quadrilateral, for each face of the cuboid. We will list the points in the VTK file in the order numbered on the image. Each polygon is specified with points using the right-hand rule to produce a surface normal pointing outward. The resulting VTK file is shown below Figure 10.5.4.

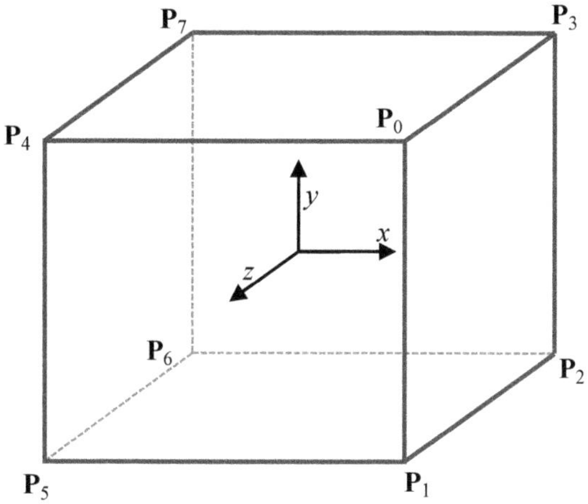

Figure 10.5.4. Representation of a cuboid as specified by points for VTK file format.

```
# vtk DataFile Version 3.0
cuboid example using quadrilaterals
ASCII
DATASET POLYDATA
POINTS 8 float
 1.0  1.0  1.0
 1.0 -1.0  1.0
 1.0 -1.0 -1.0
 1.0  1.0 -1.0
-1.0  1.0  1.0
-1.0 -1.0  1.0
-1.0 -1.0 -1.0
-1.0  1.0 -1.0
POLYGONS 6 30
4 0 1 2 3
4 4 7 6 5
4 0 3 7 4
4 1 5 6 2
4 0 4 5 1
4 3 2 6 7
```

If surface information is not needed and only the outline of the geometry is needed, the geometry could alternately be specified by a collection of line segments rather than a collection of polygons. If we use line segments specified by two end points, a VTK file that contains sufficient information to generate a visualization of the cuboid is given by

```
# vtk DataFile Version 3.0
cuboid example using line segments
ASCII
DATASET POLYDATA
POINTS 8 float
 1.0  1.0  1.0
 1.0 -1.0  1.0
 1.0 -1.0 -1.0
 1.0  1.0 -1.0
-1.0  1.0  1.0
-1.0 -1.0  1.0
-1.0 -1.0 -1.0
-1.0  1.0 -1.0
LINES 12 36
2 0 1
2 1 2
2 2 3
2 3 0
2 4 5
2 5 6
2 6 7
2 7 4
2 0 4
2 3 7
2 1 5
2 2 6
```

Use in Flight Simulation

The file types outlined above can be used to read in geometric information of objects to be displayed in a simulation. When this information is first read into a computer program for generating graphics, it is good practice to reorganize the information such that it can be efficiently processed and used in real time during the simulation. As discussed previously, STL file formats are extremely inefficient in the storage of data, since multiple instances of each point representing the corners of triangular elements is typically stored. An additional, and perhaps more subtle, inefficiency that can be present in STL and VTK files is the existence of collocated lines. If a polygonal face is not needed, and only the outline of the polygon is needed, storage space and processing time can be reduced by drawing a single line connecting two points only once, rather than drawing the line each time it is included in a different polygon. In any case, the amount of information required for viewing graphics should be minimized to allow the graphics to update as rapidly as possible.

For example, consider how the information about a cuboid would be used to generate the graphical image of the cuboid on a screen during simulation. In a flight simulation framework, the cuboid could change location and orientation relative to the observer and view plane at each time step. In this case, all the corner points of the cuboid must be rotated and translated at every time step, and the location of each point on the view plane must be found by computing the intersection of the line connecting the point of interest and observer, as shown in Fig. 10.3.1. If the geometric format of an STL is used without modification, 36 points must be rotated, translated, and then projected onto the view plane at each time step. However, if only the unique corner points of the cuboid are retained, only 8 points must be rotated, translated, and projected onto the view plane at each time step. A similar thought process can be followed that motivates removing collocated lines.

One method for reducing the graphics computational time is to pre-process the geometry information as it is read in from an STL or VTK file. This pre-processing step can be used to remove duplicate points and lines. If polygonal representation for the surface of an object is needed, each polygon can contain an array of pointers to elements within the unique array of points and/or lines. For example, the following Python code can be used to open an STL file, remove duplicate line and point entries, and store the geometry in a VTK file with only unique lines.

```
import numpy as np
from stl import mesh
import math as m

PRECISION = 1.0e-5 # points within this distance will be treated as non-unique

stlfile = 'example.stl' # input file
vtkfile = 'example.vtk' # output file

myMesh = mesh.Mesh.from_file(stlfile)
meshpoints_orig = myMesh.vectors
nfaces = len(myMesh.vectors)

# Get unique lines
templines = np.zeros((3*nfaces,2,3)) # initialize first two lines
print("Removing duplicate lines for ",nfaces," faces.")
num_unique = 0
for f in range(nfaces): # loop over all the faces
    print("Checking face ",f," of ",nfaces," faces")
    for i in range(3): # loop over all the lines on each face
        if(i+1 < 3): # if looking at last vertex, use vertex 2 and 0 for comparison
            ip1 = i+1
        else:
            ip1 = 0
        pa = myMesh.vectors[f,i,:]
        pb = myMesh.vectors[f,ip1,:]
        found = False
        for j in range(num_unique): # loop through the unique lines in lines3D
            p1 = templines[j,0,:]
            p2 = templines[j,1,:]

            dx1 = pa[0] - p1[0]
            dy1 = pa[1] - p1[1]
            dz1 = pa[2] - p1[2]

            dx2 = pb[0] - p2[0]
            dy2 = pb[1] - p2[1]
            dz2 = pb[2] - p2[2]
```

```
            if(abs(dx1)+abs(dy1)+abs(dz1)+abs(dx2)+abs(dy2)+abs(dz2) < PRECISION):
                found = True #lines match
                break
            else: # check to see if other direction matches
                dx1 = pa[0] - p2[0]
                dy1 = pa[1] - p2[1]
                dz1 = pa[2] - p2[2]

                dx2 = pb[0] - p1[0]
                dy2 = pb[1] - p1[1]
                dz2 = pb[2] - p1[2]

                if(abs(dx1)+abs(dy1)+abs(dz1)+abs(dx2)+abs(dy2)+abs(dz2) <
    PRECISION):
                    found = True  #lines match
                    break
        if(not found): #lines don't match, so add lines to unique list
            templines[num_unique,0,:] = pa
            templines[num_unique,1,:] = pb
            num_unique += 1

nlines = num_unique
print("Number of unique lines = ",nlines)
lines = np.zeros((nlines,2,3))
lines = templines[0:nlines,:,:]

# Get unique points
points = np.zeros((1,3))
points[0,:] = lines[0,0,:]

lineid = np.zeros((nlines,2), dtype=int)

print("Removing duplicate points for ",nlines," lines.")
num_unique = 1
for f in range(nlines): # loop over all the lines
    print("Checking line ",f," of ",nlines," lines")
    for i in range(2): # loop over all the points on each line
        pa = lines[f,i,:]
        found = False
        for j in range(num_unique): # loop through the unique points
            p1 = points[j,:]
            dx = pa[0] - p1[0]
            dy = pa[1] - p1[1]
            dz = pa[2] - p1[2]

            if(abs(dx) + abs(dy) + abs(dz) < PRECISION): #points match
                found = True
                lineid[f,i] = j
                break
        if(not found): #points don't match, so add point to unique list
            points = np.append(points,[pa],axis=0)
            lineid[f,i] = num_unique
            num_unique += 1

npoints = len(points)

# Create VTK file and write only unique points
with open(vtkfile, 'w') as writer:
    writer.write('# vtk DataFile Version 3.0\n')
    writer.write(f'{vtkfile}\n')
    writer.write('ASCII\n')
    writer.write('DATASET POLYDATA\n')
    writer.write(f'POINTS {npoints} float\n')
    for i in range(npoints):
        writer.write(f'{points[i,0]:.6e} {points[i,1]:.6e} {points[i,2]:.6e}\n')
```

```
    writer.write(f'LINES {nlines} {3*nlines}\n')
    for i in range(nlines):
        writer.write(f'2 {lineid[i,0]} {lineid[i,1]}\n')

print('')
print("Finished writing ",vtkfile)
print('')
print("File Statistics:")
print("--> Original number of points    = ",nfaces*3)
print("--> Original number of lines     = ",nfaces*3)
print("--> VTK number of unique points = ",npoints)
print("--> VTK number of unique lines  = ",nlines)
```

EXAMPLE 10.5.3 The STL file for the F-16 geometry shown in Fig. 10.4.2 is available in binary format on the book website and titled `F16_coarse.stl`. Download this file and use the Python code presented above to convert the STL file to a VTK file with all duplicate points and lines removed. Compare the resulting file sizes and number of points and lines within the files. Use the resulting VTK file to plot a wireframe image of the F-16 including all unique lines.

Solution: The `F16_coarse.stl` file available on the book website was copied into the working directory where the Python conversion script listed above was also stored. The Python script was run with `stlfile='F16_coarse.stl'` and `vtkfile='F16_coarse.vtk'`. The STL file is in binary format, and the resulting VTK file is in ASCII format. Table 10.5.1 shows the results of the script. Note that of the 5,448 points in the original STL file, only 909 of those points are unique. This is a reduction of roughly 80% in the number of points that need be stored. Additionally, there are exactly half as many unique lines as original lines in the STL file. The original STL file was stored in binary format, which is much more efficient than ASCII format. Even storing the resulting VTK file in ASCII format, the VTK file size is about 30% smaller than the original binary STL file.

Figure 10.5.5 shows the wireframe representation of the F-16 with only the unique lines contained in the ASCII VTK file. Because none of the polygons are shaded, all lines are visible and the tessellation looks much more dense than that shown in Fig. 10.4.2.

Table 10.5.1. Results of STL to VTK converter, storing only unique points and lines.

Property	Original STL File	Resulting VTK File
Number of points	5,448	909
Number of lines	5,448	2,724
File Type	Binary	ASCII
File Size	91 KB	64 KB

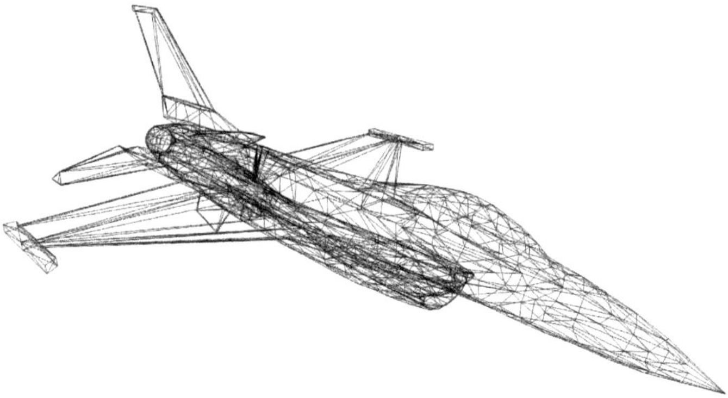

Figure 10.5.5. Unique lines included in the VTK file to represent an example F-16 geometry.

10.6 Instrumentation

The position and orientation of the vehicle relative to the Earth-fixed coordinate system can be visualized using the methods presented in the previous sections. However, the position and orientation of the vehicle are not the only states of the vehicle that are important to pilots and autopilot systems. Additionally, due to limited visibility from clouds or lighting, the pilot does not always have a visual indicator from their window of the vehicle orientation and position relative to the Earth. Hence, in traditional aircraft, instrumentation is included in the cockpit that provides feedback to the pilot for many states of the aircraft. Because the human brain can rapidly process graphical information, the instruments typically provide the most important information about the vehicle in a graphical form. However, since the actual numerical values are also of great importance, cockpit instruments often include the numerical values as well.

Traditional Instruments

Instrumentation within the cockpit of an aircraft has changed over time as technology and aircraft designs and capabilities have evolved. The most common instruments that are nearly universal among all types of aircraft include the airspeed indicator, attitude indicator, altimeter, turn coordinator, heading indicator, and vertical speed indicator. These six instruments are commonly referred to as the *six-pack* and are present in some form on nearly every aircraft. Each of these instruments has a traditional standard form, which should be adhered to in order for the flight simulator to be intuitive to pilots.

Instrumentation within cockpits generally falls into three categories. *Traditional instrumentation* uses mechanical, analog gauges with physical dials and needles. Cockpits containing traditional instrumentation have been around for many years. Figure 10.6.1 shows an example of a cockpit with traditional instrumentation. Most modern aircraft

Figure 10.6.1. Example of traditional instrumentation in a cockpit.

use *glass cockpits* which include modern digital LCD and/or touchscreen displays for the main instrumentation. Glass cockpits typically have large displays that provide most of the relevant information about the vehicle state to the pilot on configurable, color-coded screens. An example of a glass cockpit is shown in Fig. 10.6.2. A third very common form of instrumentation falls under a category called a *heads-up display* (HUD). A HUD includes instrumentation that is projected onto the windshield or a combiner glass in the field of view of the pilot. This allows the pilot to view critical information without needing to look down at an instrument panel and break their view of the surrounding environment. HUDs allow more accurate targeting and maneuvering in high workload environments and are often used on military aircraft. HUDs are also becoming more common on modern commercial and business aircraft. Figure 10.6.3 shows an example of a HUD in a modern cockpit.

Graphical Representation

In a simulation environment, the instrumentation information must be overlaid on the screen to block any graphical imagery outside the cockpit or further away from the pilot. This can be done by controlling either the order in which information is drawn on the screen, or the depth at which information is drawn on the screen. For example, one approach is to add the graphical information for the instrumentation to the graphics plot or screen after the three-dimensional imagery has been drawn. If the instrumentation information is the last information to be displayed on each frame, it will, in some plotting

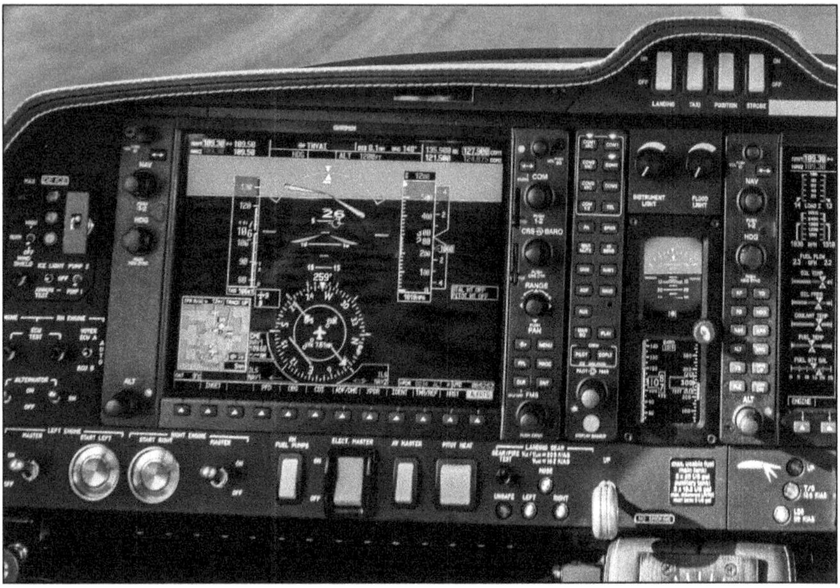

Figure 10.6.2. Example of a glass cockpit.

Figure 10.6.3. Example of a heads-up display in a cockpit.

software, naturally overlay any other information on the screen. Additionally, in some plotting software, the depth of anything drawn to the two-dimensional screen can be controlled. For example, in the main Python plotting package Matplotlib, a *z-order* value can be set, which controls the depth at which any component is drawn. Setting the z-order value of components drawn to the screen gives the user direct control over how information or graphics are layered on the screen.

Plain-Text Format

In the early stages of developing flight simulation code, it is helpful to have numerical information about the state of the vehicle displayed in a plain-text format directly in the graphical environment. Additionally, even in very mature flight simulation frameworks, plain-text format is commonly used for some instruments. Plain text is quite straightforward to represent on a two-dimensional screen. For example, plain text for displaying the simulation time and graphics frames per second in Python using the Matplotlib library can be initialized with the code

```
t_time = ax.text(camera.vp_yb[0],camera.vp_zb[0]+0.05,' time [s] =
     '+str("{:.2f}".format(time)))
t_fps  = ax.text(camera.vp_yb[0],camera.vp_zb[0]+0.02,'      fps =
     '+str("{:.2f}".format(fps)))
```

where ax is the axis system for plotting, and the first two terms in each function call represent the x and y coordinates for the location within the plot where the text should be displayed. Notice in this example the text is being displayed very near the bottom-left edge of the screen, which is tied to the view plane limits. Once this is initialized, the information can be updated within a loop with the commands

```
t_time.set_text(' time[s] = '+str("{:.2f}".format(time)))
t_fps.set_text('      fps = '+str("{:.2f}".format(fps)))
```

Heads-Up Display

Heads-up displays are visually appealing and relatively simple to generate. Figure 10.6.4 shows a simple HUD overlayed in a two-dimensional Python plot, which includes airspeed on the left, altitude on the right, and heading at the bottom. Actual values at any given time are boxed, and ticker tapes behind the boxed values move in real-time with major and minor tick marks and labels. To generate the HUD shown in Fig. 10.6.4, a general ticker-tape class was written in Python, which accepts information about a particular ticker tape such as a location on the screen where the tape should be displayed, a delta value to be used between tick marks, a length to be used for the tick marks, and a flag denoting whether the ticker tape should be vertical or horizontal. A HUD class was also generated, which contains all the ticker tapes for the HUD. The following code shows the HUD class constructor including the generation of the altitude and heading ticker tapes.

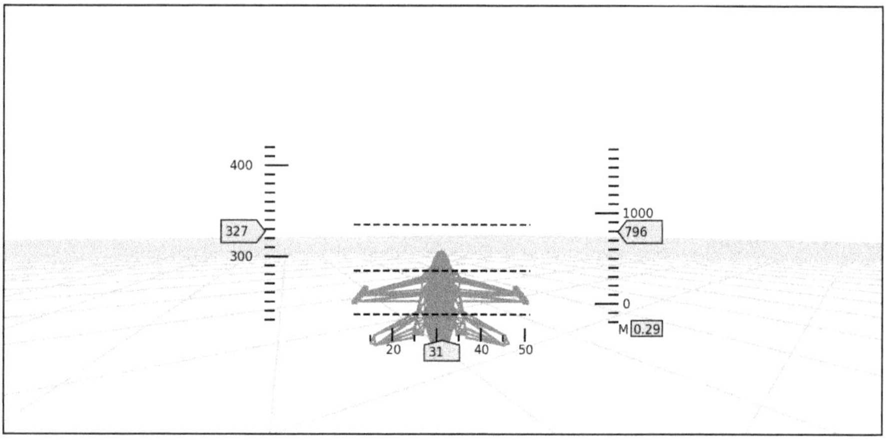

Figure 10.6.4. Example of a simple HUD overlayed in a simple graphics rendering.

```
class HUD:
    def __init__(self,json_data,ax,camera):
        color = json_data["color"]
        box_background_color = 'lightgrey'
        dx = camera.dx
        dy = camera.dy

        # Altitude
        self.altitude_minor = TickerTape(ax,color,'vertical',0.4*dy,10, 100,
    0.40*dx, -0.02*dx)
        self.altitude_major = TickerTape(ax,color,'vertical',0.4*dy,1 ,
    1000,0.40*dx, -0.05*dx,True,0.01*dx,-0.02*dy)
        ax.fill([0.4*dx, 0.42*dx, 0.5*dx, 0.5*dx, 0.42*dx, 0.4*dx],[0.0, 0.05*dy,
    0.05*dy, -0.05*dy, -0.05*dy,
    0.0],facecolor=box_background_color,edgecolor=color,linewidth=1,zorder=100)
        self.altitude_box = ax.text(0.415*dx, -0.02*dy,
    str("{:0.0f}".format(0.0)), color=color,zorder=101)

        # Heading
        self.heading_minor = TickerTape(ax,color,'horizontal',0.2*dx,4, 5,
    -0.48*dy, 0.02*dy)
        self.heading_major = TickerTape(ax,color,'horizontal',0.2*dx,2,
    10,-0.48*dy, 0.05*dy,True,-0.03*dx,-0.03*dy)
        ax.fill([0.0, -0.04*dx, -0.04*dx, 0.04*dx, 0.04*dx, 0.0],[-0.48*dy,
    -0.5*dy, -0.57*dy, -0.57*dy, -0.5*dy,
    -0.48*dy],facecolor=box_background_color,edgecolor=color,linewidth=1,zorder=
    100)
        self.heading_box = ax.text(-0.03*dx, -0.55*dy, str("{:0.0f}".format(0.0)),
    color=color,zorder=101)
```

During the main graphics loop, the values for the ticker tapes are then updated in a `draw` function within the HUD class. The following shows sample code from this function to update the altitude and heading ticker tapes.

```
def draw(self, camera, state):
    dx = camera.dx
```

```
dy = camera.dy

# Altitude Ticker
self.altitude_minor.update(-state.location[2])
self.altitude_major.update(-state.location[2])
self.altitude_box.set_text(str("{:0.0f}".format(-state.location[2])))

# Heading Ticker
self.heading_minor.update(state.eul[2]*180.0/pi)
self.heading_major.update(state.eul[2]*180.0/pi,True)
self.heading_box.set_text(str("{:0.0f}".format(state.eul[2]*180.0/pi)))
```

The code above demonstrates how a class structure can be used to represent ticker tapes on a HUD. Developing similar classes for dials or other common instrumentation types can be a very useful way to rapidly generate and position instrumentation in a HUD or cockpit.

Unreal Engine® is a modern graphics package that is widely used in the computer graphics and gaming industry. It can be used to generate realistic-looking images including textures, lighting, and shading. Some plugins allow satellite imagery and Earth topology to be used, along with atmospheric lighting and effects. Such combinations produce very realistic imagery, which can greatly enhance the utility of flight simulators. Figure 10.6.5 shows a grayscale image of a full-color screenshot generated in Unreal Engine®, which includes a representative HUD for an F-16. This HUD includes information about the airspeed, altitude, heading, bank, direction indicator, Mach number, RADAR altitude, control-surface deflections, and throttle setting.

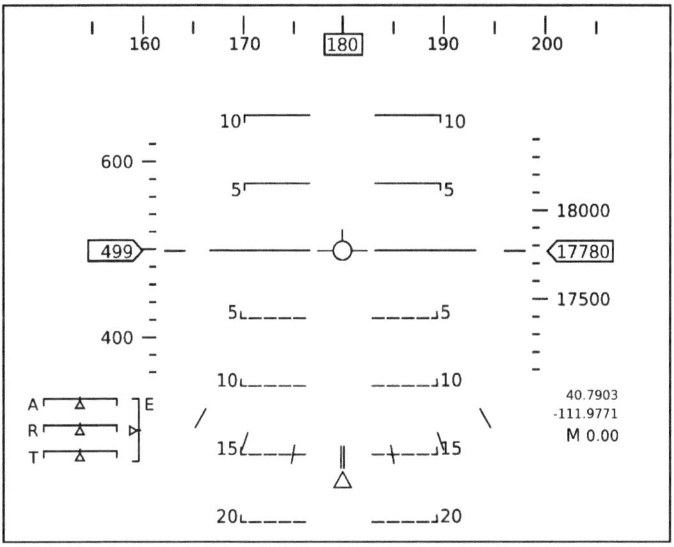

Figure 10.6.5. Example of a HUD generated in UnrealEngine®. (Created by Austin Kohler)

10.7 Code Structure

Best practices for modern code architectures recommend organizing functionality into classes. The following is a listing of example Python code, which employs a `LinesObject` class for all geometries to be represented in the scene including the ground, vehicle, and other scenery. A `HUD` class is used for all the HUD information and display, and a `UDP` class is used to manage reading data from the physics simulator over a computer port. The information within these objects is updated each time through the main loop, and the time required to update each object is recorded and printed to the screen. Such a setup can help profile code to identify methods and functions that require the most time. Such information can be used to direct efforts for increasing the computational speed of the code.

```
# Read in settings
with open('input.json', 'r') as file:
    data = json.load(file)

# Create view plane in local coordinate system
camera = Camera(data["camera"])

# Create Figure
fig = plt.figure(figsize=(camera.vp_aspect_ratio*5.0,5.0))
ax = fig.add_subplot(111)
plt.subplots_adjust(top=1.0, bottom=0.0, left=0.0, right=1.0)
plt.axis('off')
ax.axes.set_xlim( camera.vp_yb[0], camera.vp_yb[2])
ax.axes.set_ylim(-camera.vp_zb[1],-camera.vp_zb[0])
ax.axes.xaxis.set_ticklabels([])
ax.axes.yaxis.set_ticklabels([])
ax.set_xticks([])
ax.set_yticks([])
ax.axes.set_aspect('equal')
fig.canvas.draw()
plt.show(block=False)


# Create Ground Plane
show_ground = False
if("ground" in data["scene"]):
    show_ground = True
    ground = LinesObject(data["scene"]["ground"],ax)

# Create Scenery
show_scenery = False
if("scenery" in data["scene"]):
    show_scenery = True
    scenery = LinesObject(data["scene"]["scenery"],ax)

# Create Vehicle
show_vehicle = False
if("vehicle" in data):
    show_vehicle = True
    vehicle = LinesObject(data["vehicle"],ax)

# Create HUD
show_hud = False
if("hud" in data):
    show_hud = True
    hud = HUD(data["hud"],ax,camera)
```

```
# Create connections
connection = UDP(data["connections"]["receive_states"])

fps = 0.0

# Create HUD text
t_time = ax.text(camera.vp_yb[0],camera.vp_zb[0]+0.05,' time [s] =
     '+str("{:.2f}".format(connection.time)))
t_fps  = ax.text(camera.vp_yb[0],camera.vp_zb[0]+0.02,'      fps =
     '+str("{:.2f}".format(fps)))

while(True):
    timebegin = time.time()
    camera.set_state(connection.location,connection.quat)

    if(show_ground):
        time1 = time.time()
        ground.draw(camera)
        time2 = time.time()
        print(" time to draw ground = ",time2-time1)

    if(show_scenery):
        time1 = time.time()
        scenery.draw(camera)
        time2 = time.time()
        print(" time to draw scenery = ",time2-time1)

    if(show_vehicle):
        time1 = time.time()
        vehicle.set_state(connection.location,connection.quat)
        time2 = time.time()
        print(" time to set state = ",time2-time1)

        time1 = time.time()
        vehicle.draw(camera)
        time2 = time.time()
        print(" time to draw vehicle = ",time2-time1)

    if(show_hud):
        time1 = time.time()
        hud.draw(camera,connection)
        time2 = time.time()
        print(" time to draw HUD = ",time2-time1)

    t_time.set_text(' time[s] = '+str("{:.2f}".format(connection.time)))
    t_fps.set_text('      fps = '+str("{:.2f}".format(fps)))

    time1 = time.time()
    fig.canvas.draw()
    time2 = time.time()
    print(" time to draw canvas = ",time2-time1)

    time1 = time.time()
    fig.canvas.flush_events()
    time2 = time.time()
    print(" time to flush canvas = ",time2-time1)

    connection.update()
    timeend = time.time()
    fps = 1/(timeend-timebegin)
    print(" graphics framerate [hz] = ",fps)
```

10.8 Problems

Theory Problems

10.T1 Develop the mathematics to describe an observer location and orientation that could be used as a camera out the right window of an aircraft.

10.T2 Develop the mathematics to describe an observer location and orientation that could be used as a camera looking out the bottom of an aircraft.

10.T3 Develop the mathematics to describe an observer location, orientation, and view plane that could be used as a camera mounted to the ground, but that can zoom to maintain a zoomed image of the aircraft during flight, even at great distances.

Coding Problems

10.C1 Using a scripted language such as Python or Matlab®, generate a graphics code to create an x-y plot representing the view plane of an observer. Construct the code to read in the view-plane properties from a JSON input file with the options

```
{
    "camera" : {
        "view_plane": {
            "distance[ft]": 1.0,
            "angle[deg]" : 60,
            "aspect_ratio" : 2
        }
    }
}
```

Test your graphics code by comparing your solutions for these input parameters to the solution given in Example 10.2.1.

10.C2 Modify your graphics code to compute the view-plane corner points in the Earth-fixed coordinate system based on camera location and orientation as specified in the JSON file with the additional options

```
{
    "camera" : {
        "location[ft]" : [-20.0, 0.0, -5.0],
        "orientation[deg]" : [5.0, -10.0, 0.0],
    }
}
```

Test your code by comparing your solutions for these input parameters to the solutions given in Example 10.2.2.

10.C3 Modify your graphics code to be able to generate a rectangular gridded ground plane specified by the number of grid lines in each direction, and a distance between each parallel line. Your code should store just the end points of each full-length line as an array of points, and store an array of lines defined by pointers to the array of points. Construct your code to generate this ground plane by reading the properties from a JSON file of the form

```
{
    "ground" : {
        "altitude[ft]" : 0.0,
        "grid_number" : 2,
        "grid_scale[ft]" : 5.0,
        "color" : "black"
    }
}
```

Test your code by using these given input parameters to reproduce the grid points in the Earth-fixed coordinate system given in Example 10.3.1.

10.C4 Modify your graphics code to compute the intersection of each ground-plane grid point on the view plane of the camera. Store these points in an array. Use the array of lines, which include pointers to the point data, to draw lines between each pair of consecutive points. Test your code by viewing the ground plane shown in Example 10.4.1.

10.C5 Modify your graphics code to handle clipping as outlined in Sections 10.3 and 10.4.

10.C6 Modify your graphics code to read in a VTK file representing a three-dimensional geometry, such as the cuboid described in Example 10.5.2, or the F-16 vehicle used in Example 10.5.3. Make the code general to store an array of unique points, and an array of unique lines. Each line is defined by two pointers, which point to the end points of each line. Construct your code to read in the information required from a JSON file of the form

```
{
    "vehicle" : {
        "filename" : "f16_coarse.vtk",
        "color" : "gray"
    }
}
```

Test your code by placing the geometry read in from the file at the origin of the Earth-fixed coordinate system, and changing the camera position and orientation to view the geometry from multiple angles.

10.C7 Modify your graphics code so that each object to be drawn (ground plane, vehicle, scenery, etc.) is an instance of a class that holds the collection of points and lines required to draw the object. The class should have a constructor, which generates the initial point and line arrays from the VTK file or the definition of the grid given in the JSON input file, and two functions. A function called `set_state` should accept a location and orientation of the object relative to Earth-fixed coordinates, and compute all the locations of the points that make up the object in Earth-fixed coordinates. A function called `draw` should accept the camera location and orientation and generate the (x, y) points needed to plot the points and lines on the view-plane plot.

10.C8 Modify your graphics code to run in a loop. Allow the position and orientation of the camera to be updated each time through the loop, and draw the scene as it would appear to the observer each time. In Python, you may choose to first set up the data plotting structure using Pyplot with the code

```
ax = fig.add_subplot(111)
ax.plot([],[],marker='',ls='-',color=color)
```

and then update the data on the plot each time through the loop using the code

```
ax.set_data([xvals], [yvals])
fig.canvas.draw()
fig.canvas.flush_events()
```

10.C9 Modify your graphics code to read the state of the vehicle through a UDP connection in real-time as you run your physics simulator. Each time through the loop, overwrite the position and orientation of the camera to match the position and orientation of the vehicle as read in from the physics simulator.

10.C10 Modify your graphics code to use the camera location information specified by the user to generate an offset of the camera from the vehicle in the vehicle-fixed coordinate system.

10.C11 Modify your graphics code to include instrumentation information that is printed to the screen in text format. Example code for the Python language can be found in Section 10.6.

10.C12 Modify your graphics code to include instrumentation in graphical format. The instrumentation can be in dial or ticker-tape format, and can emulate traditional instrumentation, a glass cockpit, or a heads-up display.

10.C13 Modify your graphics code to be in object-oriented format and follow the structure given in Section 10.7.

10.C14 Profile your graphics code and make adjustments to your code functions to run as quickly as possible.

Appendices

Appendix A

Constants and Conversion Factors

A.1 Constants

The following constants are used in this textbook out to the precision shown.

Table A.1.1. Constants used throughout the textbook.

Name	Symbol	Value
Pi = $\cos^{-1}(-1)$	π	3.141592653589793
Gravitational Acceleration at Sea Level	g_{ssl}	9.80665 m/s^2
Earth Radius for Geopotential Altitude	R_{E_Z}	$6,356,766$ m
Mean Radius of the Earth at Sea Level	R_E	$6,366,707.01949371$ m
Gas Constant for Dry Air	R	287.0528 J/(kg·K)
Heat Capacity Ratio	γ	1.4

A.2 Conversion Factors

The conversion factors shown in the following tables come from a 1973 NASA publication Physical Constants and Conversion Factors by Mechtley [83], and are those used in this book. In order to match the solutions to the precision shown in the example problems within this book, it is important to use the same conversion factors included here. Doing so will help the student in the code debugging process.

Some conversions between the SI and English systems are exact, while others are approximate. In the following table, the exact solutions are denoted with an = sign, and the approximate solutions are denoted with an $\approx$ sign. The exact solutions are included at the top of the table and separated by a horizontal line from the approximate solutions.

Table A.2.1. Conversion factors between English and SI units [83].

Property	English		SI	
Length	1 ft	=	0.3048	m
Mass	1 lbm	=	0.45359237	kg
Force	1 lbf	=	4.4482216152605	N
Temperature	1 °F	=	5/9	°K
Mass	1 slug	≈	14.5939029	kg
Pressure	1 lbf/ft^2	≈	47.880258	N/m^2
Density	1 $slug/ft^3$	≈	515.379	kg/m^3
Viscosity	1 slug/(ft·s)	≈	47.880258	kg/(m·s)

Table A.2.2. Helpful exact relations.

Property			
Length	1 mile	=	5280 ft
Length	1 nautical mile	=	1852 m
Frequency	1 Hz (cycles/s)	=	2π rad/s

A.3 Temperature Relations

Relationships between units of temperature are best expressed in terms of equations rather than conversion factors. Common relationships include

$$°K = °C + 273.15 \tag{A.3.1}$$

$$°F = 1.8(°C) + 32 \tag{A.3.2}$$

$$°R = °F + 459.670 \tag{A.3.3}$$

$$°R = 1.8(°K) \tag{A.3.4}$$

Appendix B

Earth and Atmospheric Properties

B.1 Gravity and Atmosphere

Table B.1.1. Gravity as a function of geometric altitude.

Geometric Altitude [m]	Gravity [m/s²]	Geometric Altitude [ft]	Gravity [ft/s²]
0	9.806650000000	0	32.17404855643
5,000	9.791241076983	10,000	32.14321652255
10,000	9.775868442887	20,000	32.11242878644
15,000	9.760531983854	30,000	32.08168526328
20,000	9.745231586467	40,000	32.05098586845
25,000	9.729967137757	50,000	32.02033051754
30,000	9.714738525195	60,000	31.98971912634
35,000	9.699545636693	70,000	31.95915161084
40,000	9.684388360600	80,000	31.92862788722
45,000	9.669266585702	90,000	31.89814787188
50,000	9.654180201219	100,000	31.86771148141
55,000	9.639129096800	110,000	31.83731863259
60,000	9.624113162527	120,000	31.80696924242
65,000	9.609132288908	130,000	31.77666322807
70,000	9.594186366878	140,000	31.74640050692
75,000	9.579275287794	150,000	31.71618099656
80,000	9.564398943436	160,000	31.68600461477
85,000	9.549557226004	170,000	31.65587127950
90,000	9.534750028115	180,000	31.62578090892
95,000	9.519977242803	190,000	31.59573342140
100,000	9.505238763515	200,000	31.56572873548

Table B.1.2. Properties of the Standard Atmosphere in SI units.

Geometric Altitude [m]	Geopotential Altitude [m]	Temperature [K]	Pressure [N/m²]	Density [kg/m³]	Speed of Sound [m/s]	Dynamic Viscosity [kg/(m·s)]
0	0	288.150000	1.01325000E+05	1.22500032E+00	340.293947	1.78929763E−05
5,000	4,996.0703	255.675543	5.40482540E+04	7.36428680E−01	320.545368	1.62817293E−05
10,000	9,984.2934	223.252093	2.64998645E+04	4.13510295E−01	299.531624	1.45759516E−05
15,000	14,964.6880	216.650000	1.21118013E+04	1.94754839E−01	295.069458	1.42154741E−05
20,000	19,937.2723	216.650000	5.52929665E+03	8.89097543E−02	295.069458	1.42154741E−05
25,000	24,902.0647	221.552065	2.54921410E+03	4.00837848E−02	298.389002	1.44835756E−05
30,000	29,859.0836	226.509084	1.19702661E+03	1.84101104E−02	301.708623	1.47520772E−05
35,000	34,808.3472	236.513372	5.74591714E+02	8.46334162E−03	308.299458	1.52862153E−05
40,000	39,749.8736	250.349646	2.87142359E+02	3.99565971E−03	317.189208	1.60085509E−05
45,000	44,683.6810	264.164307	1.49100496E+02	1.96627006E−03	325.823181	1.67117634E−05
50,000	49,609.7875	270.650000	7.97785526E+01	1.02687205E−03	329.798691	1.70359966E−05
55,000	54,528.2111	265.593578	4.27513233E+01	5.60751192E−04	326.703432	1.67835331E−05
60,000	59,438.9697	255.772061	2.24604297E+01	3.05916689E−04	320.605865	1.62866556E−05
65,000	64,342.0813	239.281675	1.14461881E+01	1.66643989E−04	310.098476	1.54322322E−05
70,000	69,237.5637	219.699745	5.52037697E+00	8.75341249E−05	297.139021	1.43825736E−05
75,000	74,125.4346	200.148262	2.49032629E+00	4.33453630E−05	283.609532	1.32934007E−05
80,000	79,005.7119	180.650000	1.03658015E+00	1.99895554E−05	269.441132	1.21626108E−05
85,000	83,878.4132	180.650000	4.12481797E−01	7.95435619E−06	269.441132	1.21626108E−05
90,000	88,743.5561	180.650000	1.64371857E−01	3.16976970E−06	269.441132	1.21626108E−05

Table B.1.3. Properties of the Standard Atmosphere in English units.

Geometric Altitude [ft]	Geopotential Altitude [ft]	Temperature [R]	Pressure [lbf/ft^2]	Density [slug/ft^3]	Speed of Sound [ft/s]	Dynamic Viscosity [slug/(ft·s)]
0	0	518.670000	2.11621667E+03	2.37689218E−03	1,116.44996	3.73702587E−07
10,000	9,995.207	483.025491	1.45560192E+03	1.75554938E−03	1,077.40434	3.53408939E−07
20,000	19,980.839	447.415132	9.73274308E+02	1.26725809E−03	1,036.92902	3.32420177E−07
30,000	29,956.908	411.838873	6.29667313E+02	8.90685317E−04	994.84945	3.10676345E−07
40,000	39,923.429	389.970000	3.93127415E+02	5.87276487E−04	968.07565	2.96896357E−07
50,000	49,880.414	389.970000	2.43609478E+02	3.63917938E−04	968.07565	2.96896357E−07
60,000	59,827.879	389.970000	1.51026714E+02	2.25612446E−04	968.07565	2.96896357E−07
70,000	69,765.836	392.246328	9.37267608E+01	1.39201913E−04	970.89696	2.98346071E−07
80,000	79,694.300	397.693481	5.85113440E+01	8.57101161E−05	977.61517	3.01800521E−07
90,000	89,613.283	403.135431	3.67783030E+01	5.31472990E−05	984.28118	3.05231253E−07
100,000	99,522.799	408.572188	2.32721123E+01	3.31823763E−05	990.89605	3.08638596E−07
110,000	109,422.862	418.384526	1.48372428E+01	2.06594174E−05	1,002.72422	3.14738172E−07
120,000	119,313.486	433.578422	9.60135146E+00	1.29004597E−05	1,020.76913	3.24059026E−07
130,000	129,194.683	448.757839	6.30948078E+00	8.19071953E−06	1,038.48379	3.33225242E−07
140,000	139,066.468	463.922796	4.20616469E+00	5.28178893E−06	1,055.88484	3.42242404E−07
150,000	148,928.854	479.073313	2.84186692E+00	3.45574938E−06	1,072.98756	3.51115836E−07
160,000	158,781.853	487.170000	1.94191426E+00	2.32214853E−06	1,082.01670	3.55804194E−07
170,000	168,625.481	487.170000	1.32969797E+00	1.59005794E−06	1,082.01670	3.55804194E−07
180,000	178,459.749	478.549687	9.08358100E−01	1.10578460E−06	1,072.40101	3.50811358E−07
190,000	188,284.671	467.768996	6.15494561E−01	7.66537256E−07	1,060.25276	3.44507670E−07
200,000	198,100.262	456.998545	4.13443436E−01	5.27037801E−07	1,047.97544	3.38142289E−07

Appendix C

Example Vehicle Properties

C.1 Arrow

This model is an adaptation of a model for an arrow published by Phillips [4]. The mass and inertial properties of an arrow can be approximated from the model

$$W = 0.0697 \quad \text{lbf at sea level} \tag{C.1.1}$$

$$I_{xx_b} = 0.0000194 \quad \text{slug·ft}^2 \tag{C.1.2}$$

$$I_{yy_b} = I_{zz_b} = 0.00097 \quad \text{slug·ft}^2 \tag{C.1.3}$$

$$I_{xy_b} = I_{yz_b} = I_{xz_b} = 0 \tag{C.1.4}$$

Reference values for the aerodynamic model are given in Table C.1.1. The aerodynamic model of the arrow depends on whether the fletchings are mounted straight along the axis of the arrow, or mounted at an angle to the arrow axis.

Table C.1.1. Aerodynamic reference values for an arrow.

Reference	Value
Longitudinal Length [ft]	2.3
Lateral Length [ft]	2.3
Area [ft^2]	0.000218

Angled Fletchings

The dimensionless aerodynamic forces and moments at small angles of attack and sideslip angles on a rigid arrow with angled fletchings can be expressed as

$$C_L = C_{L,\alpha}\alpha \tag{C.1.5}$$

$$C_S = -C_{L,\alpha}\beta_f \tag{C.1.6}$$

$$C_D = C_{D_{L0}} + C_{D,L^2}C_L^2 + C_{D,L^2}C_S^2 \tag{C.1.7}$$

$$C_\ell = C_{\ell_0} + C_{\ell,\overline{p}}\overline{p} \tag{C.1.8}$$

$$C_m = C_{m,\alpha}\alpha + C_{m,\overline{q}}\overline{q} \tag{C.1.9}$$

$$C_n = -C_{m,\alpha}\beta_f + C_{m,\overline{q}}\overline{r} \tag{C.1.10}$$

where β_f is the flank angle, and the aerodynamic constants are

$$C_{L,\alpha} = 4.929 \tag{C.1.11}$$
$$C_{D_{L0}} = 5.096 \tag{C.1.12}$$
$$C_{D,L^2} = 48.138 \tag{C.1.13}$$
$$C_{m,\alpha} = -2.605 \tag{C.1.14}$$
$$C_{m,\overline{q}} = -9.06 \tag{C.1.15}$$
$$C_{\ell_0} = 3.223 \tag{C.1.16}$$
$$C_{\ell,\overline{p}} = -5.378 \tag{C.1.17}$$

Note that all moments are about the arrow center of gravity.

Straight Fletchings

The dimensionless aerodynamic forces and moments at small angles of attack and sideslip angles on a rigid arrow with straight fletchings can be expressed as a special case of that with angled fletchings where

$$C_{\ell_0} = 0 \tag{C.1.18}$$

C.2 Boeing Insitu RQ-21 Blackjack

A photo of a Boeing Insitu RQ-21 Blackjack is shown in Fig. C.2.1. Properties of the aircraft can be roughly estimated using the components listed in Table C.2.1. Figure

Figure C.2.1. Photo of a Boeing Insitu RQ-21 Blackjack. Image in public domain.

2.6.2 shows a representation of this vehicle as a collection of these simple geometric shapes.

Table C.2.1. Approximate geometric properties of the RQ-21 Blackjack.

Component	Type	Location [ft]	Weight [lbf]
Main Wing	wing	$[0.37, 0.0, 0.0]$	5.0
Winglet	wing	$[-0.35, 7.7, 0.0]$	1.0
Horizontal Stabilizer	wing	$[-3.5, 0.0, -1.2]$	2.0
Vertical Stabilizer	wing	$[-2.9, 1.5, 0.4]$	2.0
Fuselage	cuboid	$[0.1, 0.0, 0.43]$	60.0
Camera	sphere	$[3.0, 0.0, 0.5]$	13.0
Right Boom	cylinder	$[-1.7, 1.5, 0.0]$	1.0
Left Boom	cylinder	$[-1.7, -1.5, 0.0]$	1.0
Propeller	rotor	$[-2.7, 0.0, 0.27]$	5.0

Each wing segment is defined on the right-hand side of the vehicle, but mirrored to the left-hand side as well. Here we will assume that all airfoils used on lifting surfaces and for the propeller blades are of the NACA 4-digit type, and have a thickness of 12%. The additional wing properties are given in Table C.2.2. The aerodynamic properties of the wings are given in Table C.2.3. Additional approximate component properties are given in Table C.2.4. Approximate rotor properties are given in Table C.2.5. For estimating the propulsive properties of the propeller, the propeller model given in Eqs. (4.5.13)–(4.5.24) can be used with an estimated pitch-to-diameter ratio $K_c = 0.8$.

Table C.2.2. Approximate lifting-surface geometric properties of the RQ-21 Blackjack.

	Span [ft]	Chord [ft]	Mounting Angle [deg]	Sweep [deg]	Dihedral [deg]
Main Wing	7.7	$[1.0, 0.7]$	5.0	5.0	0.5
Winglet	1.0	$[0.5, 0.4]$	0	30.0	80.0
Horiz. Stab.	1.5	$[0.6, 0.6]$	0	0	0
Vert. Stab.	1.6	$[1.1, 0.6]$	0	20.0	90.0

Table C.2.3. Approximate lifting-surface aerodynamic properties of the RQ-21 Blackjack.

	α_{L0} [deg]	$C_{L,\alpha}$ [rad^{-1}]	$C_{D_{L0}}$	$C_{D,L}$	e	C_{m_0}	$C_{m,\alpha}$ [rad^{-1}]
Main Wing	−2.13	5.6	0.01	−0.004	0.8	−0.14	−2.2
Winglet	−2.13	5.6	0.01	−0.004	0.6	−0.14	−2.2
Horiz. Stab.	0.0	5.6	0.01	0.0	0.6	0.0	0.0
Vert. Stab.	0.0	5.6	0.01	0.0	0.6	0.0	0.0

Table C.2.4. Approximate component properties of the RQ-21 Blackjack.

Component	length [ft]	radius [ft]
Fuselage	$[5.3, 0.8, 1.0]$	N/A
Camera	N/A	0.4
Right Boom	3.0	0.05
Left Boom	3.0	0.05

Table C.2.5. Approximate rotor properties of the RQ-21 Blackjack.

Property	Value
Diameter [ft]	2.0
Pitch-to-Diameter Ratio	0.8
Blade Chord [ft]	$[0.2, 0.05]$
Rotation	right-hand
Number of Blades	2
Hub Diameter [ft]	0.2
Hub Height [ft]	0.1

C.3 F-16 Fighting Falcon

A photo of an F-16 Fighting Falcon is shown in Fig. C.3.1. The properties of this aircraft can be estimated from drawings and flight-simulation data published by NASA [39], Fox and Forrest [84], and Butcher [85], and summarized by Bolander [62].

Figure C.3.1. Photo of an F-16 Fighting Falcon. Image in public domain.

Geometry and Inertia

The approximate vehicle geometric properties are summarized in Table C.3.1. Approximate inertial properties are summarized in Table C.3.2.

Table C.3.1. Approximate geometric properties of the F-16 Fighting Falcon.

Property	Main Wing	Horizontal Tail	Vertical Tail
Planform Area [ft^2]	300.0	63.675	54.675
Span [ft]	30.0	11.605	8.42
Aspect Ratio	3	2.116	1.44
Taper Ratio	0.2275	0.391	0.52
Root Chord [ft]	16.293	7.980	7.70
Mean Aerodynamic Chord [ft]	11.32	5.906	6.03
Mean Geometric Chord [ft]	10.0	5.550	5.852
Leading-Edge Sweep [deg]	40.0	40.0	47.5
Quarter-Chord Sweep [deg]	32.0	32.0	43.0
Half-Chord Sweep [deg]	23.0	22.0	38.0
Dihedral [deg]	0.0	−10.0	90.0
Quarter-Chord Location [ft]	4.86	−13.13	−8.83
Airfoil	NACA 64A204	NACA 0005	NACA 0004

Table C.3.2. Approximate inertial properties of the F-16 Fighting Falcon.

Property	Value
Weight [lbf]	$20,500$
Inertia, I_{xx} [slug·ft^2]	$9,496$
Inertia, I_{yy} [slug·ft^2]	$55,814$
Inertia, I_{zz} [slug·ft^2]	$63,100$
Inertia, I_{xy} [slug·ft^2]	0
Inertia, I_{xz} [slug·ft^2]	982
Inertia, I_{yz} [slug·ft^2]	0
Engine Angular Momentum, h_x [slug·ft^2/s]	160
Engine Angular Momentum, h_y [slug·ft^2/s]	0
Engine Angular Momentum, h_z [slug·ft^2/s]	0

Aerodynamics

Aerodynamic reference properties are given in Table C.3.3. Approximate aerodynamic properties that can be used in Eqs. (3.7.24)–(3.7.30) for the aerodynamic model with the leading-edge flap extended are summarized in Table C.3.4. Sample aerodynamic data published by Nguyen et. al [39] is also included in tabular format later in this section.

Table C.3.3. Aerodynamic reference values for the F-16 Fighting Falcon.

Reference	Value
Longitudinal Length [ft]	11.32
Lateral Length [ft]	30.0
Area [ft^2]	300
CG location	$0.35\overline{c}$

Table C.3.4. Approximate aerodynamic properties of the F-16 Fighting Falcon below stall with leading-edge flaps extended for use in the model given in Eqs. (3.7.24)–(3.7.30).

Coefficient	Value	Coefficient	Value	Coefficient	Value
C_{L_0}	0.0535	$C_{S,\beta}$	−1.1086	$C_{D_{L0}}$	0.0425
$C_{L,\alpha}$	3.8459	$C_{S,\overline{p}}$	−0.0207	$C_{D,L}$	−0.1154
$C_{L,\hat{\alpha}}$	0.0	$C_{S,\alpha\overline{p}}$	1.1566	C_{D,L^2}	0.2427
$C_{L,\overline{q}}$	29.5	$C_{S,\overline{r}}$	0.9900	C_{D,S^2}	0.8733
C_{L,δ_e}	0.5153	C_{S,δ_a}	0.0601	$C_{D,\overline{q}}$	−1.498
		C_{S,δ_r}	0.1668	$C_{D,\alpha\overline{q}}$	22.99
				C_{D,δ_e}	−0.0006
				$C_{D,\alpha\delta_e}$	0.7084
				C_{D,δ_e^2}	0.3251
$C_{\ell,\beta}$	−0.1048	C_{m_0}	−0.0642	$C_{n,\beta}$	0.2194
$C_{\ell,\overline{p}}$	−0.3766	$C_{m,\alpha}$	0.0955	$C_{n,\overline{p}}$	−0.0074
$C_{\ell,\overline{r}}$	−0.0206	$C_{m,\overline{q}}$	−5.5	$C_{n,\alpha\overline{p}}$	−0.0904
$C_{\ell,\alpha\overline{r}}$	1.0114	$C_{m,\hat{\alpha}}$	0.0	$C_{n,\overline{r}}$	−0.4605
C_{ℓ,δ_a}	−0.1337	C_{m,δ_e}	−0.5926	C_{n,δ_a}	−0.0379
C_{ℓ,δ_r}	0.0272			$C_{n,\alpha\delta_a}$	0.0922
				C_{n,δ_r}	−0.0854

Controls

Approximate control properties are summarized in Table C.3.5.

Table C.3.5. Control properties of the F-16 used in this textbook.

Property	Aileron	Elevator	Rudder	Throttle [4]
Units	deg	deg	deg	none
Value Limits	± 21.5	± 25	± 30	$[0, 1]$
Rate Limits [/s]	± 80	± 60	± 120	± 1.0
Accel. Limits [/s^2]	$\pm 1,000$	$\pm 1,000$	$\pm 1,000$	± 1.0
Time Constant [s]	0.05	0.05	0.05	3.0
Natural Freq. [rad/s]	20	20	20	1.5
Damping Ratio	0.7	0.7	0.7	1.0

Propulsion

Maximum thrust with afterburner data from the NASA report [39] can be fit to the model given in Eq. (4.3.1) to yield the coefficients

$$T_a = 0.84, \quad T_0 = 29,550 \text{ lbf},$$

JSON Input File

A vehicle model for the F-16 can be formatted for input to a computer code using the following JSON format. The mass properties are taken from Table C.3.2. The aerodynamic coefficients are those given in Table C.3.4.

```
"vehicle" : {
    "mass" : {
        "weight[lbf]" : 20500.0,
        "Ixx[slug-ft^2]" : 9496.0,
        "Iyy[slug-ft^2]" : 55814.0,
        "Izz[slug-ft^2]" : 63100.0,
        "Ixy[slug-ft^2]" : 0.0,
        "Ixz[slug-ft^2]" : 982.0,
        "Iyz[slug-ft^2]" : 0.0,
        "h[slug-ft^2/s]" : [160.0, 0.0, 0.0]
    },
    "thrust" : {
        "location[ft]" : [0.0, 0.0, 0.0],
        "orientation[deg]" : [0.0, 0.0, 0.0],
        "T0[lbf]" : 29550,
        "Ta" : 0.84
    },
```

[4] Note that the dynamics of engines varies widely and can be a strong function of operating condition. For example, the dynamics may depend on throttle setting as well as whether the throttle is increasing or decreasing in magnitude. The values indicated here are simply those used for the examples contained in this textbook, and may not represent realistic behavior.

```
    "aerodynamics" : {
        "reference" : {
            "area[ft^2]" : 300.0,
            "longitudinal_length[ft]" : 11.32,
            "lateral_length[ft]" : 30.0,
            "location[ft]" : [0.0, 0.0, 0.0]
        },
        "coefficients" : {
            "CL" : {
                "0" : 0.0535,
                "alpha" : 3.8459,
                "alphahat" : 0.0,
                "qbar" : 29.5,
                "elevator" : 0.5153
            },
            "CS" : {
                "beta" : -1.1086,
                "pbar" : -0.0207,
                "alpha_pbar" : 1.1566,
                "rbar" : 0.9900,
                "aileron" : 0.0601,
                "rudder" : 0.1668
            },
            "CD" : {
                "L0" : 0.0425,
                "CL1" : -0.1154,
                "CL1_CL1" : 0.2427,
                "CS_CS" : 0.8733,
                "qbar" : -1.498,
                "alpha_qbar" : 22.99,
                "elevator" : -0.0006,
                "alpha_elevator" : 0.7084,
                "elevator_elevator" : 0.3251
            },
            "Cl" : {
                "beta" : -0.1048,
                "pbar" : -0.3766,
                "rbar" : -0.0206,
                "alpha_rbar" : 1.0114,
                "aileron" : -0.1337,
                "rudder" : 0.0272
            },
            "Cm" : {
                "0" : -0.0642,
                "alpha" : 0.0955,
                "qbar" : -5.5,
                "alphahat" : 0.0,
                "elevator" : -0.5926
            },
            "Cn" : {
                "beta" : 0.2194,
                "pbar" : -0.0074,
                "alpha_pbar" : -0.0904,
                "rbar" : -0.4605,
                "aileron" : -0.0379,
                "alpha_aileron" : 0.0922,
                "rudder" : -0.0854
            }
        }
    }
}
```

Aerodynamic Datasets

1979 NASA Dataset

In 1979, NASA published a large dataset of aerodynamic properties for a representative F-16 [39]. A link to the complete digitized original dataset can be found on the book website. An alternate form of the original NASA data is provided here, which conforms to the structure of the datasets presented in Section 3.11 and matches the original NASA data over the entire domain. A link to this alternate dataset is also available on the book website and consists of 15 individual datasets named according to the format

```
C(dep.variables),multipliers(indep.variables).csv
```

The datasets have the properties shown in Table C.3.8 and are combined in an additive form as discussed in Section 3.11 to yield the final aerodynamic coefficients given by

$$\begin{aligned} C_{(\mathbf{F},\mathbf{M})} &= C_{(x,y,z,\ell,m,n)_0}(\delta_e,\beta,\alpha)^\dagger + \Delta C_{(m)}(\beta,\alpha) + \Delta C_{(m)}(\delta_e,\alpha) \\ &+ C_{(\ell,n),\beta}(\alpha)\beta + C_{(x,y,z,\ell,m,n),\delta_{lef}}(\beta,\alpha)\delta_{lef} \\ &+ C_{(x,z,m),\overline{q}}(\alpha)\overline{q} + C_{(x,z,m),\overline{q}\delta_{lef}}(\alpha)\overline{q}\delta_{lef} \\ &+ C_{(x,z,m),\delta_{sb}}(\alpha)\delta_{sb} + C_{(y,\ell,n),\overline{p}}(\alpha)\overline{p} \\ &+ C_{(y,\ell,n),\overline{r}}(\alpha)\overline{r} + C_{(y,\ell,n),\delta_a}(\beta,\alpha)\delta_a \\ &+ C_{(y,\ell,n),\delta_r}(\beta,\alpha)\delta_r + C_{(y,\ell,n),\overline{p}\delta_{lef}}(\alpha)\overline{p}\delta_{lef} \\ &+ C_{(y,\ell,n),\overline{r}\delta_{lef}}(\alpha)\overline{r}\delta_{lef} + C_{(y,\ell,n),\delta_a\delta_{lef}}(\beta,\alpha)\delta_a\delta_{lef} \end{aligned} \tag{C.3.1}$$

The units used in these datasets for the independent variables and multipliers are given in Table C.3.6. The limits for the control surfaces are given in Tables C.3.5 and C.3.7.

The † symbol in Eq. (C.3.1) represents one caveat to using the first dataset directly. In the simulation framework published by NASA [39], the pitching moment computed from the dataset $C_{(x,y,z,\ell,m,n)_0}(\delta_e,\beta,\alpha)$ is multiplied by an elevator effectiveness η_h. The elevator effectiveness has a value of $\eta_h = 1.0$ in the range $-25 \le \delta_e \le +10$ deg and decreases linearly to a value of $\eta_h = 0.95$ at +25 deg. Note that this elevator effectiveness is applied to **only the pitching moment** computed from that dataset.

It should also be noted that the 1979 NASA publication [39] includes suggestions for scheduling the leading-edge flap as a function of angle of attack and pitch rate, as well as first-order actuator dynamic properties for the leading-edge flap.

Table C.3.6. Units used in alternate form of the NASA aerodynamic dataset.

Independent Variable	Symbol	Units	Multiplier	Symbol	Units
Angle of Attack	α	deg	Sideslip Angle	β	rad
Sideslip Angle	β	deg	Rolling Rate	$\overline{p}$	none
Elevator Deflection	δ_e	deg	Pitching Rate	$\overline{q}$	none
			Yawing Rate	$\overline{r}$	none
			Aileron Deflection	δ_a	rad
			Rudder Deflection	δ_r	rad
			Leading-Edge Flap	δ_{lef}	rad
			Speed Brake	δ_{sb}	rad

Table C.3.7. Additional control surface limits used in NASA aerodynamic dataset.

Control Surface	Minimum [deg]	Maximum [deg]
Leading-Edge Flap	0	25
Speed Brake	0	60

Table C.3.8. Alternate form of NASA aerodynamic dataset.

Number	Filename	Dataset Format
1	`01_C(x,y,z,l,m,n)0(elevator,beta,alpha).csv`	$C_{(x,y,z,\ell,m,n)_0}(\delta_e,\beta,\alpha)$
2	`02_DC(m)(beta,alpha).csv`	$\Delta C_{(m)}(\beta,\alpha)$
3	`03_DC(m)(elevator,alpha).csv`	$\Delta C_{(m)}(\delta_e,\alpha)$
4	`04_C(l,n),beta(alpha).csv`	$C_{(\ell,n),\beta}(\alpha)$
5	`05_C(x,y,z,l,m,n),lef(beta,alpha).csv`	$C_{(x,y,z,\ell,m,n),\delta_{lef}}(\beta,\alpha)$
6	`06_C(x,z,m),qbar(alpha).csv`	$C_{(x,z,m),\overline{q}}(\alpha)$
7	`07_C(x,z,m),qbar_lef(alpha).csv`	$C_{(x,z,m),\overline{q}\delta_{lef}}(\alpha)$
8	`08_C(x,z,m),sb(alpha).csv`	$C_{(x,z,m),\delta_{sb}}(\alpha)$
9	`09_C(y,l,n),pbar(alpha).csv`	$C_{(y,\ell,n),\overline{p}}(\alpha)$
10	`10_C(y,l,n),rbar(alpha).csv`	$C_{(y,\ell,n),\overline{r}}(\alpha)$
11	`11_C(y,l,n),aileron(beta,alpha).csv`	$C_{(y,\ell,n),\delta_a}(\beta,\alpha)$
12	`12_C(y,l,n),rudder(beta,alpha).csv`	$C_{(y,\ell,n),\delta_r}(\beta,\alpha)$
13	`13_C(y,l,n),pbar_lef(alpha).csv`	$C_{(y,\ell,n),\overline{p}\delta_{lef}}(\alpha)$
14	`14_C(y,l,n),rbar_lef(alpha).csv`	$C_{(y,\ell,n),\overline{r}\delta_{lef}}(\alpha)$
15	`15_C(y,l,n),aileron_lef(beta,alpha).csv`	$C_{(y,\ell,n),\delta_a\delta_{lef}}(\beta,\alpha)$

Subset of NASA Dataset

The following datasets are a subset of data published by Nguyen et al. [39] for a representative F-16 Fighting Falcon with the leading-edge flaps extended. All forces and moments are reported in the body-fixed coordinate system.

Table C.3.9. Dataset $C_{(\mathbf{F},\mathbf{M})}(\alpha)$ at $\beta = 0$.

α [deg]	C_x	C_y	C_z	C_ℓ	C_m	C_n
−20	−0.0933	0.0000	1.1160	0.0000	0.0127	0.0000
−15	−0.0978	0.0000	0.9590	0.0000	−0.0755	0.0000
−10	−0.0982	0.0000	0.6920	0.0000	−0.1025	0.0000
−5	−0.0784	0.0000	0.2870	0.0000	−0.0744	0.0000
0	−0.0489	0.0000	−0.0250	0.0000	−0.0598	0.0000
5	−0.0066	0.0000	−0.3670	0.0000	−0.0498	0.0000
10	0.0490	0.0000	−0.7500	0.0000	−0.0437	0.0000
15	0.1072	0.0000	−1.1120	0.0000	−0.0407	0.0000
20	0.1283	0.0000	−1.4180	0.0000	−0.0342	0.0000
25	0.1300	0.0000	−1.6580	0.0000	−0.0507	0.0000
30	0.1536	0.0000	−2.0080	0.0000	−0.0459	0.0000
35	0.1605	0.0000	−2.2000	0.0000	−0.0605	0.0000
40	0.1552	0.0000	−2.3280	0.0000	−0.0835	0.0000
45	0.1382	0.0000	−2.3110	0.0000	−0.0923	0.0000
50	0.1281	0.0000	−2.3260	0.0000	−0.0826	0.0000
55	0.1200	0.0000	−2.2520	0.0000	−0.0738	0.0000
60	0.1147	0.0000	−2.2080	0.0000	−0.1414	0.0000
70	0.1025	0.0000	−2.1340	0.0000	−0.3216	0.0000
80	0.0821	0.0000	−2.0040	0.0000	−0.4678	0.0000
90	0.0864	0.0000	−2.1400	0.0000	−0.6184	0.0000

Table C.3.10. Dataset $C_{(\mathbf{F},\mathbf{M})}(\beta)$ at $\alpha = 0$.

β [deg]	C_x	C_y	C_z	C_ℓ	C_m	C_n
−30	−0.0355	0.6218	−0.0060	0.0157	−0.0550	−0.1058
−25	−0.0381	0.5258	−0.0300	0.0190	−0.0460	−0.0916
−20	−0.0404	0.4208	−0.0500	0.0199	−0.0590	−0.0749
−15	−0.0416	0.3088	−0.0500	0.0207	−0.0640	−0.0578
−10	−0.0450	0.2014	−0.0360	0.0185	−0.0660	−0.0413
−8	−0.0461	0.1553	−0.0350	0.0153	−0.0660	−0.0317
−6	−0.0469	0.1138	−0.0280	0.0110	−0.0639	−0.0226
−4	−0.0478	0.0726	−0.0270	0.0071	−0.0615	−0.0138
−2	−0.0485	0.0371	−0.0270	0.0033	−0.0605	−0.0066
0	−0.0489	0.0000	−0.0250	0.0000	−0.0598	0.0000
2	−0.0483	−0.0394	−0.0280	−0.0030	−0.0600	0.0061
4	−0.0475	−0.0764	−0.0280	−0.0067	−0.0606	0.0135
6	−0.0467	−0.1191	−0.0290	−0.0107	−0.0608	0.0225
8	−0.0449	−0.1674	−0.0310	−0.0147	−0.0617	0.0324
10	−0.0433	−0.2134	−0.0340	−0.0182	−0.0621	0.0414
15	−0.0399	−0.3198	−0.0480	−0.0204	−0.0606	0.0579
20	−0.0387	−0.4315	−0.0370	−0.0196	−0.0587	0.0746
25	−0.0364	−0.5369	−0.0330	−0.0187	−0.0484	0.0914
30	−0.0338	−0.6400	−0.0060	−0.0154	−0.0517	0.1055

Table C.3.11. Dataset $C_{(x,z,m),\bar{q}}(\alpha)$.

α [deg]	$C_{x,\bar{q}}$	$C_{z,\bar{q}}$	$C_{m,\bar{q}}$
−20	0.9530	−23.9000	−6.8400
−15	0.9530	−23.9000	−6.8400
−10	0.9530	−23.9000	−6.8400
−5	1.5500	−29.5000	−3.4200
0	1.9000	−29.5000	−5.4800
5	2.4600	−30.5000	−5.4500
10	2.9200	−31.3000	−6.0200
15	3.3000	−30.1000	−6.7000
20	2.7600	−27.7000	−5.6900
25	2.0500	−28.2000	−6.0000
30	1.5000	−29.0000	−6.2000
35	1.4900	−29.8000	−6.4000
40	1.8300	−38.3000	−6.6000
45	1.2100	−35.3000	−6.0000
50	1.3300	−32.3000	−5.5000
55	1.6100	−27.3000	−5.0000
60	0.9100	−25.2000	−4.5000
70	3.4300	−27.3000	−3.5000
80	0.6170	−9.3500	−5.6000
90	0.2730	−2.1600	−4.0400

Table C.3.12. Dataset $C_{(y,\ell,n),\overline{p}/\overline{r}}(\alpha)$.

α [deg]	$C_{y,\overline{p}}$	$C_{\ell,\overline{p}}$	$C_{n,\overline{p}}$	$C_{y,\overline{r}}$	$C_{\ell,\overline{r}}$	$C_{n,\overline{r}}$
−20	0.0333	−0.3660	−0.0006	1.4400	−0.1550	−0.5170
−15	0.0333	−0.3660	−0.0006	1.4400	−0.1550	−0.5170
−10	0.0333	−0.3660	−0.0006	1.4400	−0.1550	−0.5170
−5	−0.1770	−0.3770	0.0424	1.0500	−0.2010	−0.4610
0	0.0055	−0.3450	−0.0075	0.9810	−0.0024	−0.4140
5	0.0679	−0.4340	−0.0214	0.9390	0.0880	−0.3970
10	0.3100	−0.4080	−0.0320	0.9990	0.2050	−0.3730
15	0.2340	−0.3880	−0.0320	0.9810	0.2200	−0.4550
20	0.3440	−0.3290	0.0500	0.8190	0.3190	−0.5500
25	0.3620	−0.2940	0.1500	0.4830	0.4370	−0.5820
30	0.6110	−0.2300	0.1300	0.5900	0.6800	−0.5950
35	0.5290	−0.2100	0.1580	1.2100	0.1000	−0.6370
40	0.2980	−0.1200	0.2400	−0.4930	0.4470	−1.0200
45	−2.2700	−0.1000	0.1500	−1.0400	−0.3300	−0.8400
50	0.9710	−0.1000	0.0000	−1.2100	−0.0680	−0.5410
55	1.0200	−0.1200	−0.2000	−1.5800	0.1180	−0.3500
60	2.9000	−0.1400	−0.3000	−1.3700	0.0802	−0.3500
70	0.4510	−0.1000	0.1500	−0.0259	0.0529	−0.0700
80	−0.2940	−0.1500	0.0000	−0.1270	0.0868	−0.1500
90	−0.2610	−0.2000	0.0000	0.1930	−0.0183	−0.1500

Table C.3.13. Dataset $C_{(x,z,m)}(\alpha)$ for $\delta_e = \pm 10$ deg.

	$\delta_e = -10$ deg			$\delta_e = +10$ deg		
α [deg]	C_x	C_z	C_m	C_x	C_z	C_m
−20	−0.1223	1.228	0.0864	−0.0884	1.039	−0.0835
−15	−0.1188	1.059	0.0328	−0.1010	0.849	−0.1719
−10	−0.1147	0.815	0.0041	−0.1092	0.596	−0.2153
−5	−0.0893	0.356	0.0076	−0.0902	0.205	−0.1888
0	−0.0617	0.064	0.0430	−0.0606	−0.114	−0.1610
5	−0.0172	−0.287	0.0501	−0.0200	−0.490	−0.1606
10	0.0399	−0.650	0.0553	0.0313	−0.849	−0.1548
15	0.1027	−0.980	0.0706	0.0829	−1.177	−0.1452
20	0.1322	−1.306	0.0674	0.0971	−1.442	−0.1264
25	0.1407	−1.594	0.0492	0.0949	−1.789	−0.1530
30	0.1651	−1.863	0.0528	0.1104	−2.082	−0.1440
35	0.1795	−2.090	0.0278	0.1201	−2.308	−0.1411
40	0.1798	−2.216	−0.0094	0.1127	−2.411	−0.1450
45	0.1671	−2.263	−0.0411	0.0996	−2.306	−0.1411
50	0.1544	−2.097	−0.0129	0.1071	−2.261	−0.1008
55	0.1488	−2.088	0.0202	0.1030	−2.231	−0.0679
60	0.1383	−2.051	−0.0708	0.0914	−2.185	−0.1556
70	0.1328	−2.027	−0.3137	0.1190	−2.268	−0.1983
80	0.1211	−1.916	−0.4236	0.0519	−1.940	−0.4721
90	0.1247	−1.998	−0.5718	0.0504	−2.057	−0.6083

Table C.3.14. Dataset $C_{(y,\ell,n)}(\alpha)$ for $\delta_a = 20$ deg and $\delta_r = 30$ deg.

	$\delta_a = 20$ deg			$\delta_r = 30$ deg		
α [deg]	C_y	C_ℓ	C_n	C_y	C_ℓ	C_n
−20	0.0190	−0.0226	−0.0089	0.0854	0.0201	−0.0481
−15	0.0157	−0.0327	−0.0098	0.0900	0.0176	−0.0484
−10	0.0160	−0.0328	−0.0091	0.0869	0.0184	−0.0476
−5	0.0152	−0.0401	−0.0111	0.0815	0.0154	−0.0449
0	0.0235	−0.0481	−0.0120	0.0859	0.0146	−0.0451
5	0.0288	−0.0511	−0.0105	0.0923	0.0144	−0.0450
10	0.0284	−0.0499	−0.0090	0.0851	0.0137	−0.0441
15	0.0222	−0.0491	−0.0066	0.0834	0.0135	−0.0446
20	0.0181	−0.0418	0.0001	0.0936	0.0137	−0.0475
25	0.0141	−0.0372	0.0045	0.0994	0.0147	−0.0483
30	0.0143	−0.0308	0.0065	0.1071	0.0126	−0.0494
35	0.0067	−0.0256	0.0099	0.0885	0.0114	−0.0449
40	0.0005	−0.0166	0.0044	0.0749	0.0059	−0.0328
45	−0.0250	−0.0122	0.0097	0.0387	0.0007	−0.0162
50	−0.0412	−0.0076	0.0130	0.0251	0.0026	−0.0081
55	−0.0544	−0.0095	0.0167	0.0122	0.0019	−0.0040
60	−0.0497	−0.0092	0.0182	0.0066	0.0015	−0.0012
70	−0.0208	−0.0075	0.0154	0.0061	0.0008	−0.0015
80	−0.0081	−0.0041	0.0138	0.0175	−0.0023	−0.0061
90	−0.0060	0.0022	0.0125	0.0052	0.0018	−0.0024

Model Variations

The aerodynamic information for the F-16 given in this section can be used several different ways to produce an aerodynamic model of the vehicle. For example, the aerodynamic coefficients given in Table C.3.4 can be used in the model given in Eqs. (3.7.24)–(3.7.30). Alternately, the datasets given in Tables C.3.9 – C.3.14 can be used directly by interpolation. If the aerodynamic coefficients given in Table C.3.4 are used in the model given in Eqs. (3.7.24)–(3.7.30), a stall model must also be incorporated to provide reasonable results at high angles of attack. Additionally, these models can also be altered to account for compressibility and ground effect as outlined in Chapter 3.

The aerodynamic model of the F-16 derived from the NASA data is statically unstable in pitch. Similar pitch instability is common for many fighter aircraft, since a reduction in stability can increase maneuverability. Without an active stability augmentation system, this vehicle can be challenging to fly, especially for an introductory course on flight simulation. Hence, one variation of the model provided above is to assume that the aerodynamic model reference location is aft of the vehicle center of gravity. The aerodynamic moments $\overline{\mathbf{M}}_b$ about the CG location, or origin of the body-fixed coordinate system, can be computed from

$$\overline{\mathbf{M}}_b = \overline{\mathbf{M}}_{\text{ref}} + \mathbf{r}_{\text{ref}} \times \overline{\mathbf{F}}_{\text{ref}} \tag{C.3.2}$$

where $\overline{\mathbf{F}}_{\text{ref}}$ and $\overline{\mathbf{M}}_{\text{ref}}$ are the aerodynamic forces and moments from the aerodynamic model about the aerodynamic reference location, and $\mathbf{r}_{\text{ref}}$ is the vector from the vehicle CG to the aerodynamic reference location. Assuming the aerodynamic reference location lies 1 ft aft of the vehicle CG, i.e. $\mathbf{r}_{\text{ref}} = [-1, 0, 0]$, for the F-16 aerodynamic model given in this section is more than sufficient to ensure the vehicle is statically stable in pitch. This model is used for many examples in this textbook.

The aerodynamic models given above can be modified and applied for various needs for flight simulation. Due to the large number of modifications that can be made to the model, Table C.3.15 includes a list of variations of the model that are used throughout the textbook for various examples.

Table C.3.15. Aerodynamic model variations used in examples problems.

Model ID	Description	Reference Location $\mathbf{r}_{\text{ref}} = [x_b, y_b, z_b]$	Stall	Compressibility	Ground Effect
F16-D	interpolated from Tables C.3.9 – C.3.14	[0, 0, 0]	None	None	None
F16-D-1	Same as F16-D	[-1, 0, 0]	None	None	None
F16-AD	interpolated from datasets described in Eq. (C.3.1) and Table C.3.8	[0, 0, 0]	None	None	None
F16-AD-1	interpolated from datasets described in Eq. (C.3.1) and Table C.3.8	[-1, 0, 0]	None	None	None
F16-E	coefficients given in Table C.3.4 used in Eqs. (3.7.24)–(3.7.30)	[0, 0, 0]	None	None	None
F16-E-1	Same as F16-E	[-1, 0, 0]	None	None	None
F16-ES-1	Same as F16-E	[-1, 0, 0]	Properties from Example 3.8.1	None	None

C.4 Simple Quadrotor

Figure C.4.1 shows a simple generic quadrotor, which consists of a single cuboid with side lengths of $l_x = l_y = 1.0$ ft, and $l_z = 0.5$ ft, and four rotors with the geometric properties given in Table C.4.1. The cuboid is centered at the center of gravity, and the four rotors are located relative to the vehicle center of gravity as given in Table C.4.2.

Table C.4.1. Properties of a single rotor used for the Simple Quadrotor.

Property	Value
Diameter [ft]	1.25
Pitch [ft]	0.5
Pitch-to-Diameter Ratio	0.4
Blade Count	2
Weight [lbf]	0.0567
Axial Inertia, I_{xx} [slug·ft^2]	0.000095231

Table C.4.2. Locations of the rotors used for the Simple Quadrotor.

Rotor	Body-Fixed Location (x_b, y_b, z_b)	Rotation
1	[1.0, 1.0, −0.25]	RH
2	[−1.0, 1.0, −0.25]	LH
3	[−1.0, −1.0, −0.25]	RH
4	[1.0, −1.0, −0.25]	LH

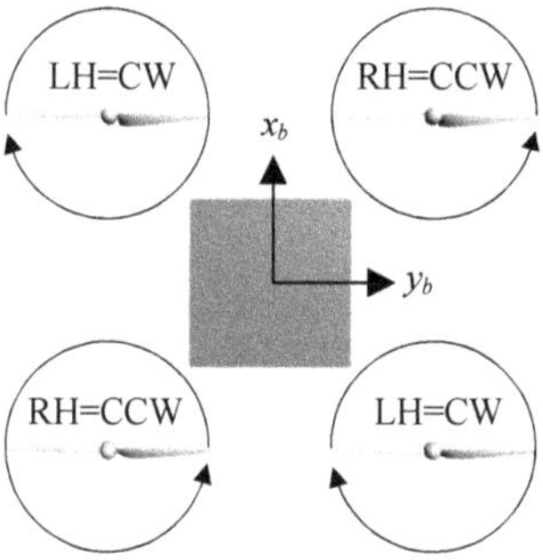

(a) Top view

(b) Side view

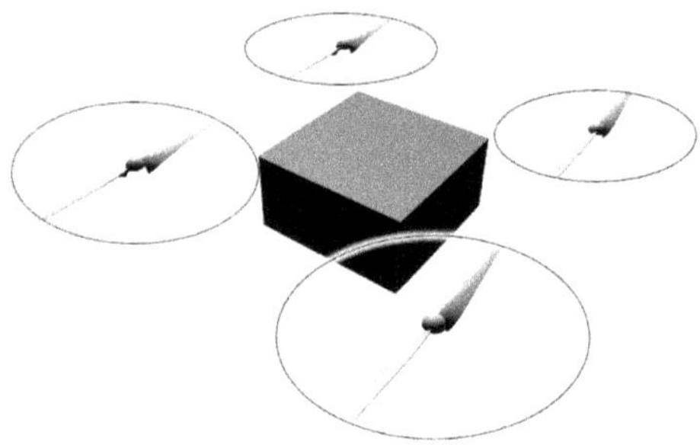

(c) Isometric view

Figure C.4.1. Drawing of a simple generic quadrotor used for examples in this textbook.

The dimensionless thrust, brake power, normal-force, and yawing-moment coefficients of the rotors can be computed as a function of advance ratio from the relations

$$C_T = 0.0560 - 0.0704J - 0.1027J^2 \tag{C.4.1}$$

$$C_{P_b} = 0.0137 + 0.0144J - 0.0850J^2 \tag{C.4.2}$$

$$C_{N,\alpha} = 0.0036J + 0.0160J^2 + 0.0148J^3 \tag{C.4.3}$$

$$C_{n,\alpha} = -0.0112J + 0.0029J^2 + 0.0050J^3 \tag{C.4.4}$$

The total vehicle mass and inertial properties are given in Table C.4.3.

Table C.4.3. Approximate inertial properties of the Simple Quadrotor.

Property	Value
Weight [lbf]	5.0
Inertia, I_{xx} [slug·ft^2]	0.0231
Inertia, I_{yy} [slug·ft^2]	0.0231
Inertia, I_{zz} [slug·ft^2]	0.0391
Inertia, I_{xy} [slug·ft^2]	0
Inertia, I_{xz} [slug·ft^2]	0
Inertia, I_{yz} [slug·ft^2]	0

Appendix D

Numerical Methods

The following is an overview of commonly used numerical methods for various mathematical operations. The content of this appendix does not represent an exhaustive list of numerical methods required for flight simulation, nor does it represent the most efficient coding structure or best practices for the algorithms. The content of this appendix is simply a listing of numerical methods the author has found helpful in the code development of flight-simulation and computational-fluid-dynamics software, and is intended to serve as a quick reference for some helpful methods and algorithms.

D.1 Vector Algebra

Dot Product

Let $\mathbf{A} = (a_1, a_2, a_3)$ and $\mathbf{B} = (b_1, b_2, b_3)$ be two vectors each of length 3. The *dot product*, also referred to as the *scalar product*, is defined as

$$\mathbf{A} \cdot \mathbf{B} \equiv ||\mathbf{A}||\ ||\mathbf{B}|| \cos(\theta) \tag{D.1.1}$$

where $||\mathbf{A}||$ is the magnitude of vector **A**, $||\mathbf{B}||$ is the magnitude of vector **B**, and θ is the angle between the two vectors. The dot product is commutative, which means that

$$\mathbf{A} \cdot \mathbf{B} = \mathbf{B} \cdot \mathbf{A} \tag{D.1.2}$$

The dot product can be computed from

$$\mathbf{A} \cdot \mathbf{B} = a_1 b_1 + a_2 b_2 + a_3 b_3 \tag{D.1.3}$$

The vectors **A** and **B** are orthogonal if and only if $\mathbf{A} \cdot \mathbf{B} = 0$.

Cross Product

Let $\mathbf{A} = (a_1, a_2, a_3)$ and $\mathbf{B} = (b_1, b_2, b_3)$ be two vectors each of length 3. The *cross product*, also referred to as the *vector product*, is defined as

$$\mathbf{A} \times \mathbf{B} \equiv ||\mathbf{A}||\ ||\mathbf{B}|| \sin(\theta)\mathbf{n} \tag{D.1.4}$$

where $||\mathbf{A}||$ is the magnitude of vector **A**, $||\mathbf{B}||$ is the magnitude of vector **B**, θ is the angle between the two vectors, and **n** is a unit vector perpendicular to the plane containing the two vectors. The direction of $\mathbf{n}$ is based on the right-hand rule ordered as $(\mathbf{A}, \mathbf{B}, \mathbf{n})$. In other words, if you point your right-hand fingers in the direction of $\mathbf{A}$, and curl your fingers towards the direction of $\mathbf{B}$, your thumb points in the direction of $\mathbf{n}$.

The cross product is anti-commutative, which means that

$$(\mathbf{A} \times \mathbf{B}) = -(\mathbf{B} \times \mathbf{A}) \tag{D.1.5}$$

The cross product can be computed from the determinant of

$$\begin{aligned} \mathbf{A} \times \mathbf{B} &= \begin{vmatrix} \mathbf{i} & \mathbf{j} & \mathbf{k} \\ a_1 & a_2 & a_3 \\ b_1 & b_2 & b_3 \end{vmatrix} \\ &= (a_2 b_3 - a_3 b_2)\mathbf{i} + (a_3 b_1 - a_1 b_3)\mathbf{j} + (a_1 b_2 - a_2 b_1)\mathbf{k} \end{aligned} \tag{D.1.6}$$

D.2 Linear Algebra

Matrix Multiplication

Matrix multiplication can be performed using Algorithm D.2.12.

Algorithm D.2.12 Matrix Multiplication

```
function matrix_multiplication
    Input
        A                                              Matrix of size m × n
        B                                              Matrix of size n × p
    if columns(A) ≠ rows(B) then
        throw invalid_argument
    end if
    Initialize matrix C of size m × p with zeros
    for i ← 0 to m − 1 do
        for j ← 0 to p − 1 do
            for k ← 0 to n − 1 do
                C[i][j] ← C[i][j] + A[i][k] · B[k][j]
            end for
        end for
    end for
    return C
end function
```

Inverse of a 2 × 2 Matrix

Let $[\mathbf{A}]$ be a 2×2 matrix of real numbers given by

$$[\mathbf{A}] = \begin{bmatrix} a & b \\ c & d \end{bmatrix} \tag{D.2.1}$$

The inverse of [**A**] is

$$[\mathbf{A}]^{-1} = \begin{bmatrix} a & b \\ c & d \end{bmatrix}^{-1} = \frac{1}{ad - bc}\begin{bmatrix} d & -b \\ -c & a \end{bmatrix} \tag{D.2.2}$$

where det [**A**] $\neq 0$.

Inverse of a 3 × 3 Matrix

Let [**A**] be a 3×3 matrix of real numbers given by

$$[\mathbf{A}] = \begin{bmatrix} a_{11} & a_{12} & a_{13} \\ a_{21} & a_{22} & a_{23} \\ a_{31} & a_{32} & a_{33} \end{bmatrix} \tag{D.2.3}$$

The inverse of a 3×3 matrix is

$$[\mathbf{A}]^{-1} = \frac{1}{\det[\mathbf{A}]}[\tilde{\mathbf{A}}] \tag{D.2.4}$$

where det [**A**] $\neq 0$ and is given by

$$\det[\mathbf{A}] = a_{11}a_{22}a_{33} + a_{12}a_{23}a_{31} + a_{13}a_{21}a_{32} - a_{13}a_{22}a_{31} - a_{12}a_{21}a_{33} - a_{11}a_{23}a_{32} \tag{D.2.5}$$

and $[\tilde{\mathbf{A}}]$ is the adjugate matrix of [**A**] given by

$$[\tilde{\mathbf{A}}] = \begin{bmatrix} \begin{vmatrix} a_{22} & a_{23} \\ a_{32} & a_{33} \end{vmatrix} & \begin{vmatrix} a_{13} & a_{12} \\ a_{33} & a_{32} \end{vmatrix} & \begin{vmatrix} a_{12} & a_{13} \\ a_{22} & a_{23} \end{vmatrix} \\ \begin{vmatrix} a_{23} & a_{21} \\ a_{33} & a_{31} \end{vmatrix} & \begin{vmatrix} a_{11} & a_{13} \\ a_{31} & a_{33} \end{vmatrix} & \begin{vmatrix} a_{13} & a_{11} \\ a_{23} & a_{21} \end{vmatrix} \\ \begin{vmatrix} a_{21} & a_{22} \\ a_{31} & a_{32} \end{vmatrix} & \begin{vmatrix} a_{12} & a_{11} \\ a_{32} & a_{31} \end{vmatrix} & \begin{vmatrix} a_{11} & a_{12} \\ a_{21} & a_{22} \end{vmatrix} \end{bmatrix}$$

$$= \begin{bmatrix} a_{22}a_{33} - a_{23}a_{32} & a_{13}a_{32} - a_{12}a_{33} & a_{12}a_{23} - a_{13}a_{22} \\ a_{23}a_{31} - a_{21}a_{33} & a_{11}a_{33} - a_{13}a_{31} & a_{13}a_{21} - a_{11}a_{23} \\ a_{21}a_{32} - a_{22}a_{31} & a_{12}a_{31} - a_{11}a_{32} & a_{11}a_{22} - a_{12}a_{21} \end{bmatrix} \tag{D.2.6}$$

D.3 Taylor-Series Expansion

The Taylor series expansion of a function $\phi(x)$ about the point x and evaluated at $x + \Delta x$ is given by

$$\phi(x + \Delta x) = \phi(x) + \left.\frac{\partial \phi}{\partial x}\right|_x \frac{\Delta x}{1!} + \left.\frac{\partial^2 \phi}{\partial x^2}\right|_x \frac{\Delta x^2}{2!} + \left.\frac{\partial^3 \phi}{\partial x^3}\right|_x \frac{\Delta x^3}{3!} + O(\Delta x^4) + \cdots \tag{D.3.1}$$

This equation can be written in several discrete forms in order to yield useful approximations for values and derivatives near a point. The order of the method depends on the accuracy needed. Second-order approximations are typically sufficient for modeling most physical systems. However, in some cases, such as the case of modeling turbulence properties near a smooth wall, higher-order approximations are needed [86].

D.4 Differentiation Using Finite-Differences

Various methods can be used to evaluate the derivative of a function at a point. The methods presented in this section are based on the Taylor-series expansion of the function. Verification of the order of accuracy of the approximations developed in this section has been performed as published by Hunsaker [86].

First Derivative Approximations

Second-Order Solutions

The second-order finite-difference approximation for the first derivative at a point can be developed by considering the Taylor-series expansion of the function about the point x_i. Given the location of two discrete points at distances Δx_1, and Δx_2, along with the values of the function at those points, ϕ_1 and ϕ_2, the system of Taylor-series expansions can be written

$$\phi_1 = \phi_i + \left.\frac{\partial \phi}{\partial x}\right|_i \Delta x_1 + \left.\frac{\partial^2 \phi}{\partial x^2}\right|_i \frac{\Delta x_1^2}{2} + O(\Delta x_1^3) + \cdots \tag{D.4.1}$$

$$\phi_2 = \phi_i + \left.\frac{\partial \phi}{\partial x}\right|_i \Delta x_2 + \left.\frac{\partial^2 \phi}{\partial x^2}\right|_i \frac{\Delta x_2^2}{2} + O(\Delta x_2^3) + \cdots \tag{D.4.2}$$

Retaining only the first three terms in the Taylor series and rearranging the system of equations gives

$$\begin{bmatrix} \left.\frac{\partial \phi}{\partial x}\right|_i \\ \left.\frac{\partial^2 \phi}{\partial x^2}\right|_i \end{bmatrix} = \begin{bmatrix} \Delta x_1 & \frac{\Delta x_1^2}{2} \\ \Delta x_2 & \frac{\Delta x_2^2}{2} \end{bmatrix}^{-1} \begin{bmatrix} \phi_1 - \phi_i \\ \phi_2 - \phi_i \end{bmatrix} \tag{D.4.3}$$

The analytic inverse of a 2×2 matrix can be used to yield

$$\begin{bmatrix} \left.\frac{\partial \phi}{\partial x}\right|_i \\ \left.\frac{\partial^2 \phi}{\partial x^2}\right|_i \end{bmatrix} = \frac{1}{\Delta x_1 \Delta x_2^2 - \Delta x_1^2 \Delta x_2} \begin{bmatrix} \Delta x_2^2 & -\Delta x_1^2 \\ -2\Delta x_2 & 2\Delta x_1 \end{bmatrix} \begin{bmatrix} \phi_1 - \phi_i \\ \phi_2 - \phi_i \end{bmatrix} \tag{D.4.4}$$

Retaining only the first solution gives the **second-order finite-difference approximation for the first derivative of a function at the location x_i**

$$\left.\frac{\partial \phi}{\partial x}\right|_i = \frac{\Delta x_2^2 \phi_1 + (\Delta x_1^2 - \Delta x_2^2)\phi_i - \Delta x_1^2 \phi_2}{\Delta x_1 \Delta x_2^2 - \Delta x_1^2 \Delta x_2} \tag{D.4.5}$$

Equation (D.4.5) is a general solution and is not constrained to uniform spacing. It can be used to obtain the forward-difference, backward-difference, or central-difference formulas. For example, the forward-difference formula can be obtained by using $\Delta x_1 = x_{i+1} - x_i$, $\Delta x_2 = x_{i+2} - x_i$, $\phi_1 = \phi_{i+1}$, and $\phi_2 = \phi_{i+2}$. The backward-difference formula can be obtained by using $\Delta x_1 = x_{i-1} - x_i$, $\Delta x_2 = x_{i-2} - x_i$, $\phi_1 = \phi_{i-1}$, and $\phi_2 = \phi_{i-2}$. The central-difference formula can be obtained by using $\Delta x_1 = x_{i-1} - x_i$, $\Delta x_2 = x_{i+1} - x_i$, $\phi_1 = \phi_{i-1}$, and $\phi_2 = \phi_{i+1}$. These solutions are general and do not require uniform spacing.

For the special case of uniform spacing in x, all the Δx values in Eq. (D.4.5) are related to a common Δx spacing. The forward-difference formula for this case can be obtained by applying $\Delta x_1 = \Delta x$, $\Delta x_2 = 2\Delta x$, $\phi_1 = \phi_{i+1}$ and $\phi_2 = \phi_{i+2}$. Using this in Eq. (D.4.5) gives the **second-order forward-difference approximation for the first derivative of a function at the location x_i with uniform spacing**

$$\left.\frac{\partial \phi}{\partial x}\right|_i = \frac{-3\phi_i + 4\phi_{i+1} - \phi_{i+2}}{2\Delta x} \tag{D.4.6}$$

The backward-difference formula for uniform spacing can be obtained by applying $\Delta x_1 = -\Delta x$, $\Delta x_2 = -2\Delta x$, $\phi_1 = \phi_{i-1}$ and $\phi_2 = \phi_{i-2}$. Using this in Eq. (D.4.5) gives the **second-order backward-difference approximation for the first derivative of a function at the location x_i with uniform spacing**

$$\left.\frac{\partial \phi}{\partial x}\right|_i = \frac{3\phi_i - 4\phi_{i-1} + \phi_{i-2}}{2\Delta x} \tag{D.4.7}$$

The central-difference formula for uniform spacing can be obtained by applying $\Delta x_1 = -\Delta x$, $\Delta x_2 = \Delta x$, $\phi_1 = \phi_{i-1}$ and $\phi_2 = \phi_{i+1}$. Using this in Eq. (D.4.5) gives the **second-order central-difference approximation for the first derivative of a function at the location x_i with uniform spacing**

$$\left.\frac{\partial \phi}{\partial x}\right|_i = \frac{\phi_{i+1} - \phi_{i-1}}{2\Delta x} \tag{D.4.8}$$

Higher-Order Solutions

In general, the nth-order approximation for the first derivative can be developed by retaining $n+1$ terms in the Taylor series at n points about the point of interest. These can be written as

$$\begin{aligned}
\phi_1 &= \phi_i + \left.\frac{\partial \phi}{\partial x}\right|_i \Delta x_1 + \cdots + \left.\frac{\partial^n \phi}{\partial x^n}\right|_i \frac{\Delta x_1^n}{n!} + O(\Delta x_1^{n+1}) + \cdots \\
\phi_2 &= \phi_i + \left.\frac{\partial \phi}{\partial x}\right|_i \Delta x_2 + \cdots + \left.\frac{\partial^n \phi}{\partial x^n}\right|_i \frac{\Delta x_2^n}{n!} + O(\Delta x_2^{n+1}) + \cdots \\
&\vdots \\
\phi_{n-1} &= \phi_i + \left.\frac{\partial \phi}{\partial x}\right|_i \Delta x_{n-1} + \cdots + \left.\frac{\partial^n \phi}{\partial x^n}\right|_i \frac{\Delta x_{n-1}^n}{n!} + O(\Delta x_{n-1}^{n+1}) + \cdots \\
\phi_n &= \phi_i + \left.\frac{\partial \phi}{\partial x}\right|_i \Delta x_n + \cdots + \left.\frac{\partial^n \phi}{\partial x^n}\right|_i \frac{\Delta x_n^n}{n!} + O(\Delta x_n^{n+1}) + \cdots
\end{aligned} \tag{D.4.9}$$

Retaining $n+1$ terms in each Taylor series of this system of equations and rearranging yields

$$\begin{bmatrix} \left.\frac{\partial \phi}{\partial x}\right|_i \\ \left.\frac{\partial^2 \phi}{\partial x^2}\right|_i \\ \vdots \\ \left.\frac{\partial^{n-1} \phi}{\partial x^{n-1}}\right|_i \\ \left.\frac{\partial^n \phi}{\partial x^n}\right|_i \end{bmatrix} = \begin{bmatrix} \Delta x_1 & \frac{\Delta x_1^2}{2!} & \cdots & \frac{\Delta x_1^{n-1}}{(n-1)!} & \frac{\Delta x_1^n}{n!} \\ \Delta x_2 & \frac{\Delta x_2^2}{2!} & \cdots & \frac{\Delta x_2^{n-1}}{(n-1)!} & \frac{\Delta x_2^n}{n!} \\ \vdots & \vdots & \ddots & \vdots & \vdots \\ \Delta x_{n-1} & \frac{\Delta x_{n-1}^2}{2!} & \cdots & \frac{\Delta x_{n-1}^{n-1}}{(n-1)!} & \frac{\Delta x_{n-1}^n}{n!} \\ \Delta x_n & \frac{\Delta x_n^2}{2!} & \cdots & \frac{\Delta x_n^{n-1}}{(n-1)!} & \frac{\Delta x_n^n}{n!} \end{bmatrix}^{-1} \begin{bmatrix} \phi_1 - \phi_i \\ \phi_2 - \phi_i \\ \vdots \\ \phi_{n-1} - \phi_i \\ \phi_n - \phi_i \end{bmatrix} \tag{D.4.10}$$

The solution to the first equation in this series gives the nth-order approximation for the first derivative at a point. This solution can be used for forward-differencing, backward-differencing, or central-differencing, and is not constrained to uniform spacing.

Tabulated Solutions for Uniform Spacing

For the special case of uniform spacing in x, the solution for the first derivative from Eq. (D.4.10) can be written as the summation of influences from neighboring points as shown in Eqs. (D.4.6)–(D.4.8). These solutions can be expressed in a tabulated form denoting the coefficients to be multiplied by each term. For example, Eqs. (D.4.6)–(D.4.8) can be expressed in tabulated form as shown in Table D.4.1. Fourth-order approximations for the first derivative can be obtained from Table D.4.2. Sixth-order approximations for the first derivative can be obtained from Table D.4.3.

Table D.4.1. Second-order approximations for the first derivative. Coefficients are to be multiplied by the corresponding function values and divided by the specified denominator.

	Forward Difference	Central Difference	Backward Difference
ϕ_{i-2}			1
ϕ_{i-1}		-1	-4
ϕ_i	-3		3
ϕ_{i+1}	4	1	
ϕ_{i+2}	-1		
Denominator		$2\,\Delta x$	

Table D.4.2. Fourth-order approximations for the first derivative. Coefficients are to be multiplied by the corresponding function values and divided by the specified denominator.

	Forward Difference		Central Difference	Backward Difference	
ϕ_{i-4}					3
ϕ_{i-3}				-1	-16
ϕ_{i-2}			1	6	36
ϕ_{i-1}		-3	-8	-18	-48
ϕ_i	-25	-10	0	10	25
ϕ_{i+1}	48	18	8	3	
ϕ_{i+2}	-36	-6	-1		
ϕ_{i+3}	16	1			
ϕ_{i+4}	-3				
Denominator			$12\Delta x$		

Table D.4.3. Sixth-order approximations for the first derivative. Coefficients are to be multiplied by the corresponding function values and divided by the specified denominator.

	Forward Difference			Central Difference	Backward Difference		
ϕ_{i-6}							10
ϕ_{i-5}						-2	-72
ϕ_{i-4}					1	15	225
ϕ_{i-3}				-1	-8	-50	-400
ϕ_{i-2}			2	9	30	100	450
ϕ_{i-1}		-10	-24	-45	-80	-150	-360
ϕ_i	-147	-77	-35	0	35	77	147
ϕ_{i+1}	360	150	80	45	24	10	
ϕ_{i+2}	-450	-100	-30	-9	-2		
ϕ_{i+3}	400	50	8	1			
ϕ_{i+4}	-225	-15	-1				
ϕ_{i+5}	72	2					
ϕ_{i+6}	-10						
Denominator	$60\Delta x$						

Second Derivative Approximations

Second-Order Solutions

The second-order finite-difference approximation for the second derivative at a point can be developed by considering the Taylor-series expansion of the function about the point x_i. Given the location of three discrete points at distances Δx_1, Δx_2, and Δx_3, along with the values of the function at those points, ϕ_1, ϕ_2, and ϕ_3, the system of Taylor-series expansions can be written

$$\phi_1 = \phi_i + \left.\frac{\partial\phi}{\partial x}\right|_i \Delta x_1 + \left.\frac{\partial^2\phi}{\partial x^2}\right|_i \frac{\Delta x_1^2}{2} + \left.\frac{\partial^3\phi}{\partial x^3}\right|_i \frac{\Delta x_1^3}{6} + O(\Delta x_1^4) + \cdots \tag{D.4.11}$$

$$\phi_2 = \phi_i + \left.\frac{\partial\phi}{\partial x}\right|_i \Delta x_2 + \left.\frac{\partial^2\phi}{\partial x^2}\right|_i \frac{\Delta x_2^2}{2} + \left.\frac{\partial^3\phi}{\partial x^3}\right|_i \frac{\Delta x_2^3}{6} + O(\Delta x_2^4) + \cdots \tag{D.4.12}$$

$$\phi_3 = \phi_i + \left.\frac{\partial\phi}{\partial x}\right|_i \Delta x_3 + \left.\frac{\partial^2\phi}{\partial x^2}\right|_i \frac{\Delta x_3^2}{2} + \left.\frac{\partial^3\phi}{\partial x^3}\right|_i \frac{\Delta x_3^3}{6} + O(\Delta x_3^4) + \cdots \tag{D.4.13}$$

Retaining the first four terms in the series and rearranging yields

$$\begin{bmatrix} \left.\frac{\partial\phi}{\partial x}\right|_i \\ \left.\frac{\partial^2\phi}{\partial x^2}\right|_i \\ \left.\frac{\partial^3\phi}{\partial x^3}\right|_i \end{bmatrix} = \begin{bmatrix} \Delta x_1 & \frac{\Delta x_1^2}{2} & \frac{\Delta x_1^3}{6} \\ \Delta x_2 & \frac{\Delta x_2^2}{2} & \frac{\Delta x_2^3}{6} \\ \Delta x_3 & \frac{\Delta x_3^2}{2} & \frac{\Delta x_3^3}{6} \end{bmatrix}^{-1} \begin{bmatrix} \phi_1 - \phi_i \\ \phi_2 - \phi_i \\ \phi_3 - \phi_i \end{bmatrix} \tag{D.4.14}$$

Solving this system of equations and retaining only the second solution yields the **second-order finite-difference approximation for the second derivative of a function at the location x_i**

$$\left.\frac{\partial^2\phi}{\partial x^2}\right|_i = 2\frac{\Delta\phi_1(\Delta x_2^2 - \Delta x_3^2)/\Delta x_1 + \Delta\phi_2(\Delta x_3^2 - \Delta x_1^2)/\Delta x_2 + \Delta\phi_3(\Delta x_1^2 - \Delta x_2^2)/\Delta x_3}{\Delta x_1\Delta x_2\Delta x_3[(\Delta x_3 - \Delta x_2)/\Delta x_1 - (\Delta x_3 - \Delta x_1)/\Delta x_2 + (\Delta x_2 - \Delta x_1)/\Delta x_3]} \tag{D.4.15}$$

where $\Delta\phi_1 = \phi_1 - \phi_i$, $\Delta\phi_2 = \phi_2 - \phi_i$, and $\Delta\phi_3 = \phi_3 - \phi_i$. Equation (D.4.15) is a general solution for the second-order approximation for a second-derivative at a point and is not constrained to uniform spacing. It can be used to develop forward-difference, central-difference, and backward-difference formulas. For example, a forward-difference formula can be obtained by using $\Delta x_1 = x_{i+1} - x_i$, $\Delta x_2 = x_{i+2} - x_i$, and $\Delta x_3 = x_{i+3} - x_i$, with function values $\phi_1 = \phi_{i+1}$, $\phi_2 = \phi_{i+2}$, and $\phi_3 = \phi_{i+3}$ in Eq. (D.4.15). A backward-difference formula can be obtained by using $\Delta x_1 = x_{i-1} - x_i$, $\Delta x_2 = x_{i-2} - x_i$, and $\Delta x_3 = x_{i-3} - x_i$, with function values $\phi_1 = \phi_{i-1}$, $\phi_2 = \phi_{i-2}$, and $\phi_3 = \phi_{i-3}$ in Eq. (D.4.15). A central-difference formula can be obtained by using $\Delta x_1 = x_{i-1} - x_i$, $\Delta x_2 = x_{i+1} - x_i$, and $\Delta x_3 = x_{i+2} - x_i$, with function values $\phi_1 = \phi_{i-1}$, $\phi_2 = \phi_{i+1}$, and $\phi_3 = \phi_{i+2}$ in Eq. (D.4.15). These solutions are general and do not require uniform spacing.

For the special case of uniform spacing in x, all the Δx values in Eq. (D.4.15) are related to a common Δx spacing. For example, a forward-difference formula for this case can be obtained by applying $\Delta x_1 = \Delta x$, $\Delta x_2 = 2\Delta x$, and $\Delta x_3 = 3\Delta x$, with function values $\phi_1 = \phi_{i+1}$, $\phi_2 = \phi_{i+2}$, and $\phi_3 = \phi_{i+3}$. Using this in Eq. (D.4.15) gives a **second-order forward-difference approximation for the second derivative of a function at the location x_i with uniform spacing**

$$\left.\frac{\partial^2\phi}{\partial x^2}\right|_i = \frac{2\phi_i - 5\phi_{i+1} + 4\phi_{i+2} - \phi_{i+3}}{\Delta x^2} \tag{D.4.16}$$

A backward-difference formula for uniform spacing can be obtained by applying $\Delta x_1 = -\Delta x$, $\Delta x_2 = -2\Delta x$, and $\Delta x_3 = -3\Delta x$, with function values $\phi_1 = \phi_{i-1}$, $\phi_2 = \phi_{i-2}$, and $\phi_3 = \phi_{i-3}$. Using this in Eq. (D.4.15) gives a **second-order backward-difference approximation for the second derivative of a function at the location x_i with uniform spacing**

$$\left.\frac{\partial^2\phi}{\partial x^2}\right|_i = \frac{2\phi_i - 5\phi_{i-1} + 4\phi_{i-2} - \phi_{i-3}}{\Delta x^2} \tag{D.4.17}$$

The central-difference formula for uniform spacing can be obtained by applying $\Delta x_1 = -\Delta x$, $\Delta x_2 = \Delta x$, and $\Delta x_3 = 2\Delta x$, with function values $\phi_1 = \phi_{i-1}$, $\phi_2 = \phi_{i+1}$, and $\phi_3 = \phi_{i+2}$. Using this in Eq. (D.4.15) gives the **second-order central-difference approximation for the second derivative of a function at the location x_i with uniform spacing**

$$\left.\frac{\partial^2\phi}{\partial x^2}\right|_i = \frac{\phi_{i-1} - 2\phi_i + \phi_{i+1}}{\Delta x^2} \tag{D.4.18}$$

Higher-Order Approximations

In general, the nth-order approximation for the second derivative can be developed by retaining $n+2$ terms in the Taylor series at $n+1$ points. These can be written as

$$\begin{aligned}
\phi_1 &= \phi_i + \left.\frac{\partial \phi}{\partial x}\right|_i \Delta x_1 + \cdots + \left.\frac{\partial^{n+1}\phi}{\partial x^{n+1}}\right|_i \frac{\Delta x_1^{n+1}}{(n+1)!} + O(\Delta x_1^{n+2}) + \cdots \\
\phi_2 &= \phi_i + \left.\frac{\partial \phi}{\partial x}\right|_i \Delta x_2 + \cdots + \left.\frac{\partial^{n+1}\phi}{\partial x^{n+1}}\right|_i \frac{\Delta x_2^{n+1}}{(n+1)!} + O(\Delta x_2^{n+2}) + \cdots \\
&\vdots \\
\phi_n &= \phi_i + \left.\frac{\partial \phi}{\partial x}\right|_i \Delta x_n + \cdots + \left.\frac{\partial^{n+1}\phi}{\partial x^{n+1}}\right|_i \frac{\Delta x_n^{n+1}}{(n+1)!} + O(\Delta x_n^{n+2}) + \cdots \\
\phi_{n+1} &= \phi_i + \left.\frac{\partial \phi}{\partial x}\right|_i \Delta x_{n+1} + \cdots + \left.\frac{\partial^{n+1}\phi}{\partial x^{n+1}}\right|_i \frac{\Delta x_{n+1}^{n+1}}{(n+1)!} + O(\Delta x_{n+1}^{n+2}) + \cdots
\end{aligned} \tag{D.4.19}$$

Retaining $n+2$ terms in each Taylor series of this system of equations and rearranging yields

$$\begin{bmatrix} \left.\frac{\partial \phi}{\partial x}\right|_i \\ \left.\frac{\partial^2 \phi}{\partial x^2}\right|_i \\ \vdots \\ \left.\frac{\partial^n \phi}{\partial x^n}\right|_i \\ \left.\frac{\partial^{n+1} \phi}{\partial x^{n+1}}\right|_i \end{bmatrix} = \begin{bmatrix} \Delta x_1 & \frac{\Delta x_1^2}{2!} & \cdots & \frac{\Delta x_1^n}{n!} & \frac{\Delta x_1^{n+1}}{(n+1)!} \\ \Delta x_2 & \frac{\Delta x_2^2}{2!} & \cdots & \frac{\Delta x_2^n}{n!} & \frac{\Delta x_2^{n+1}}{(n+1)!} \\ \vdots & \vdots & \ddots & \vdots & \vdots \\ \Delta x_n & \frac{\Delta x_n^2}{2!} & \cdots & \frac{\Delta x_n^n}{n!} & \frac{\Delta x_n^{n+1}}{(n+1)!} \\ \Delta x_{n+1} & \frac{\Delta x_{n+1}^2}{2!} & \cdots & \frac{\Delta x_{n+1}^n}{n!} & \frac{\Delta x_{n+1}^{n+1}}{(n+1)!} \end{bmatrix}^{-1} \begin{bmatrix} \phi_1 - \phi_i \\ \phi_2 - \phi_i \\ \vdots \\ \phi_n - \phi_i \\ \phi_{n+1} - \phi_i \end{bmatrix} \tag{D.4.20}$$

The solution to the second equation in this series gives the nth-order approximation for the second derivative at a point. This solution can be used for forward-differencing, backward-differencing, or central-differencing, and is not constrained to uniform spacing.

Tabulated Solutions for Uniform Spacing

For the special case of uniform spacing in x, the solution for the second derivative from Eq. (D.4.20) can be written as the summation of influences from neighboring points as shown in Eqs. (D.4.16)–(D.4.18). These solutions can be easily obtained through a tabulated form. For example, Eqs. (D.4.16)–(D.4.18) can be expressed in tabulated form as shown in Table D.4.4. Fourth-order approximations for the second derivative can be obtained from Table D.4.5. Sixth-order approximations for the second derivative can be obtained from Table D.4.6.

Table D.4.4. Second-order approximations for the second derivative. Coefficients are to be multiplied by the corresponding function values and divided by the specified denominator.

	Forward Difference	Central Difference	Backward Difference
ϕ_{i-3}			-1
ϕ_{i-2}			4
ϕ_{i-1}		1	-5
ϕ_i	2	-2	2
ϕ_{i+1}	-5	1	
ϕ_{i+2}	4		
ϕ_{i+3}	-1		
Denominator		Δx^2	

Table D.4.5. Fourth-order approximations for the second derivative. Coefficients are to be multiplied by the corresponding function values and divided by the specified denominator.

	Forward Difference		Central Difference	Backward Difference	
ϕ_{i-5}					-10
ϕ_{i-4}				1	61
ϕ_{i-3}				-6	-156
ϕ_{i-2}			-1	14	214
ϕ_{i-1}		10	16	-4	-154
ϕ_i	45	-15	-30	-15	45
ϕ_{i+1}	-154	-4	-16	10	
ϕ_{i+2}	214	14	-1		
ϕ_{i+3}	-156	-6			
ϕ_{i+4}	61	1			
ϕ_{i+5}	-10				
Denominator			$12\Delta x^2$		

Table D.4.6. Sixth-order approximations for the second derivative. Coefficients are to be multiplied by the corresponding function values and divided by the specified denominator.

	Forward Difference			Central Difference	Backward Difference		
ϕ_{i-7}							-126
ϕ_{i-6}						11	1019
ϕ_{i-5}					-2	-90	-3618
ϕ_{i-4}					16	324	7380
ϕ_{i-3}				2	-54	-670	-9490
ϕ_{i-2}			-11	-27	85	855	7911
ϕ_{i-1}		126	214	270	130	-486	-4014
ϕ_{i}	938	-70	-378	-490	-378	-70	938
ϕ_{i+1}	-4014	-486	130	270	214	126	
ϕ_{i+2}	7911	855	85	-27	-11		
ϕ_{i+3}	-9490	-670	-54	2			
ϕ_{i+4}	7380	324	16				
ϕ_{i+5}	-3618	-90	-2				
ϕ_{i+6}	1019	11					
ϕ_{i+7}	-126						
Denominator				$180\Delta x^2$			

D.5 Integration

The integral of a Taylor series expansion for a function, $\phi(x)$ about a point i from the point x_a to x_b can be written

$$\int_{x_a}^{x_b} \phi(x)dx = \int_{x_i+\Delta x_a}^{x_i+\Delta x_b} \left[\phi_i + \left.\frac{\partial \phi}{\partial x}\right|_i (x - x_i) + \left.\frac{\partial^2 \phi}{\partial x^2}\right|_i \frac{(x - x_i)^2}{2} + O(\Delta x^3) + \cdots \right] dx \tag{D.5.1}$$

Performing the integration gives

$$\int_{x_a}^{x_b} \phi(x)dx = \phi_i(\Delta x_b - \Delta x_a) + \left.\frac{\partial \phi}{\partial x}\right|_i \frac{(\Delta x_b^2 - \Delta x_a^2)}{2} + \left.\frac{\partial^2 \phi}{\partial x^2}\right|_i \frac{(\Delta x_b^3 - \Delta x_a^3)}{6} + O(\Delta x^4) + \cdots \tag{D.5.2}$$

Global error for integration is one order less than local truncation error. Therefore, for uniform spacing, only the first term on the right-hand side of the Eq. (D.5.2) must be retained for second-order accuracy. For an integration step size of Δx with the value specified at the midpoint i, the integral over the step can be evaluated by applying the definitions $\Delta x_b = \Delta x/2$, and $\Delta x_a = -\Delta x/2$. Using these definitions in Eq. (D.5.2) gives the second-order midpoint rule

$$\int_{-\Delta x/2}^{\Delta x/2} \phi(x)dx \approx \phi_i \Delta x \tag{D.5.3}$$

Fourth-Order Runge-Kutta

The fourth-order Runge-Kutta integration method is outlined in Algorithm D.5.13.

Algorithm D.5.13 Fourth-Order Runge-Kutta Method (RK4)

Input
- $f(t, y)$ — The derivative function
- t_0 — Initial time
- y_0 — Initial value of function
- Δt — Time step

$k_1 \leftarrow f(t_0, y_0)$
$k_2 \leftarrow f\left(t_0 + \frac{\Delta t}{2},\ y_0 + \frac{\Delta t}{2} \cdot k_1\right)$
$k_3 \leftarrow f\left(t_0 + \frac{\Delta t}{2},\ y_0 + \frac{\Delta t}{2} \cdot k_2\right)$
$k_4 \leftarrow f(t_0 + \Delta t,\ y_0 + \Delta t \cdot k_3)$
$y_1 \leftarrow y_0 + \frac{\Delta t}{6}(k_1 + 2k_2 + 2k_3 + k_4)$
return y_1

D.6 Linear Interpolation

Interpolation is widely used to estimate the value of a function when only neighboring information is known. The Taylor series expansion at points a and b about point i where $\Delta x_a = x_a - x$ and $\Delta x_b = x_b - x$ gives

$$\begin{aligned}\phi_a &= \phi(x_i + \Delta x_a) \\ &= \phi_i + \left.\frac{\partial \phi}{\partial x}\right|_i \Delta x_a + \left.\frac{\partial^2 \phi}{\partial x^2}\right|_i \frac{\Delta x_a^2}{2} + \left.\frac{\partial^3 \phi}{\partial x^3}\right|_i \frac{\Delta x_a^3}{6} + O(\Delta x_a^4) + \cdots\end{aligned} \tag{D.6.1}$$

$$\begin{aligned}\phi_b &= \phi(x_i + \Delta x_b) \\ &= \phi_i + \left.\frac{\partial \phi}{\partial x}\right|_i \Delta x_b + \left.\frac{\partial^2 \phi}{\partial x^2}\right|_i \frac{\Delta x_b^2}{2} + \left.\frac{\partial^3 \phi}{\partial x^3}\right|_i \frac{\Delta x_b^3}{6} + O(\Delta x_b^4) + \cdots\end{aligned} \tag{D.6.2}$$

Subtracting the product of Eq. (D.6.2) and Δx_a from the product of Eq. (D.6.1) and Δx_b gives

$$\phi_a \Delta x_b - \phi_b \Delta x_a = \phi_i(\Delta x_b - \Delta x_a) + \left.\frac{\partial^2 \phi}{\partial x^2}\right|_i \frac{\Delta x_b \Delta x_a^2 - \Delta x_a \Delta x_b^2}{2} + O(\Delta x^4) + \cdots \tag{D.6.3}$$

Solving for ϕ_i gives

$$\phi_i = \frac{\phi_a \Delta x_b - \phi_b \Delta x_a}{\Delta x_b - \Delta x_a} - \left.\frac{\partial^2 \phi}{\partial x^2}\right|_i \left(\frac{\Delta x_b \Delta x_a^2 - \Delta x_a \Delta x_b^2}{2(\Delta x_b - \Delta x_a)} \right) + \cdots \tag{D.6.4}$$

Retaining only the first term on the right-hand side yields the second-order approximation for the value at point i

$$\phi_i = \frac{\phi_a \Delta x_b - \phi_b \Delta x_a}{\Delta x_b - \Delta x_a} \tag{D.6.5}$$

This relationship is called *linear interpolation* if the point i lies between points a and b. Otherwise it is called *linear extrapolation*.

Bibliography

[1] des Poids et Mesures, B. I., "General Conference on Weights and Measures," *Proceedings of the 26th Meeting of the CGPM*, Bureau International des Poids et Mesures, 2018, p. 472.

[2] Phillips, W. F., Hailey, C. E., and Gebert, G. A., "Review of Attitude Representations Used for Aircraft Kinematics," *Journal of Aircraft*, Vol. 38, 2001, pp. 718–737. https://doi.org/10.2514/2.2824.

[3] Phillips, W. F., "Aircraft Flight Simulation," *Mechanics of Flight*, John Wiley & Sons, Inc., Hoboken, NJ, 2010, Chap. 11, 2nd ed., pp. 989–1068.

[4] Phillips, W. F., "Position and Orientation: The Euler Angle Formulation," *Mechanics of Flight*, John Wiley & Sons, Inc., Hoboken, NJ, 2010, Chap. 7, 2nd ed., pp. 735–752.

[5] Euler, L., "Formulae Generales pro Translatione Quacunque Corporum Rigidorum," *Novi Commentari Academiae Scientiarum Imperialis Petropolitanae*, Vol. 20, 1775.

[6] Hamilton, W. R., "On Quaternions: Or a New System of Imaginaries in Algebra," *Philosophical Magazine*, Vol. 25, No. 3, 1844. https://doi.org/10.1080/14786444408644923.

[7] Hamilton, W. R., *Lectures on Quaternions*, Hodges and Smith, Dublin, Ireland, 1853.

[8] Hamilton, W. R., *Elements of Quaternions*, Longmans-Green, London, 1866. https://doi.org/10.1017/CBO9780511707162.

[9] Lanham, C., "Inertia Calculation Procedure for Preliminary Design," Tech. Rep. ASD-TR-79-5004, Aeronautical Systems Division Wright-Patterson AFB OH, 1979. https://doi.org/10.21236/ADA072679.

[10] Moulton, B. C., and Hunsaker, D. F., "Simplified Mass and Inertial Estimates for Aircraft with Components of Constant Density," *AIAA SciTech 2023 Forum*, 2023, pp. AIAA–2023–2432. https://doi.org/10.2514/6.2023-2432.

[11] Hunsaker, D. F., Reid, J. T., and Joo, J. J., "Geometric Definition and Ideal Aerodynamic Performance of Parabolic Trailing-Edge Flaps," *International Journal of Astronautics and Aeronautical Engineering*, Vol. 4, 2019, pp. 1–17. https://doi.org/10.35840/2631-5009/7526.

[12] Selig, M., "UIUC Airfoil Coordinates Database," https://m-selig.ae.illinois.edu/ads/coord_database.html, 2023.

[13] Moulton, B. C., and Hunsaker, D. F., "Analytic Solutions for Volume, Mass, Center of Gravity, and Inertia of Wing Segments and Rotors of Constant Density," *Aerospace*, Vol. 11, No. 6, 2024, p. 492. https://doi.org/10.3390/aerospace11060492.

[14] Phillips, W. F., "Aircraft Equations of Motion: Inertial and Gyroscopic Coupling," *Mechanics of Flight*, John Wiley & Sons, Inc., Hoboken, NJ, 2010, Chap. 7, 2nd ed., pp. 805–806.

[15] National Aeronautics and Space Administration, "U.S. Standard Atmosphere, 1976," Tech. Rep. NASA-TM-X-74335, National Aeronautics and Space Administration, 1976.

[16] Phillips, W. F., "Limits of Continuum Aerodynamics," *Journal of Aircraft*, Vol. 47, No. 1, 2010, pp. 225–228. https://doi.org/10.2514/1.44755.

[17] Phillips, W. F., "Aircraft Equations of Motion: Introduction," *Mechanics of Flight*, John Wiley & Sons, Inc., Hoboken, NJ, 2010, Chap. 7, 2nd ed., pp. 715–725.

[18] Phillips, W. F., "Aircraft Controls and Maneuverability: Lateral Control and Maneuverability," *Mechanics of Flight*, John Wiley & Sons, Inc., Hoboken, NJ, 2010, Chap. 6, 2nd ed., pp. 666–679.

[19] Phillips, W. F., Alley, N. R., and Niewoehner, R. J., "Effects of Nonlinearities on Subsonic Aerodynamic Center," *Journal of Aircraft*, Vol. 45, 2008, pp. 1244–1255. https://doi.org/10.2514/1.34241.

[20] Phillips, W. F., and Snyder, D. O., "Modern Adaptation of Prandtl's Classic Lifting-Line Theory," *Journal of Aircraft*, Vol. 37, No. 4, 2000, pp. 662–670. https://doi.org/10.2514/2.2649.

[21] Munk, M. M., "General Theory of Thin Wing Sections," Tech. Rep. NACA-TR-142, National Advisory Committee for Aeronautics, 1922.

[22] Glauert, H., *The Elements of Aerofoil and Airscrew Theory*, Cambridge University Press, London, 1926.

[23] Hunsaker, D. F., Pope, O. D., Taylor, J. D., and Hodson, J. J., "Aerodynamic Center of Arbitrary Airfoils Below Stall," *Journal of Aircraft*, Vol. 56, 2019, pp. 2158–2171. https://doi.org/10.2514/1.C035579.

[24] Karamcheti, K., "Problem of the Airfoil: Moment on the Airfoil," *Principles of Ideal-Fluid Aerodynamics*, John Wiley & Sons, Inc., New York, NY, 1966, Chap. 16, pp. 471–474.

[25] Hoerner, S. F., *Fluid-Dynamic Drag*, Hoerner, Midland Park, New Jersey, 1958.

[26] Hoerner, S. F., and Borst, H. V., *Fluid-Dynamic Lift*, Hoerner, Midland Park, New Jersey, 1985.

[27] Finck, R. D., "USAF (United States Air Force) Stability and Control DATCOM (Data Compendium)," Tech. Rep. AFWAL-TR-83-3048, United States Air Force, 1978.

[28] White, F. M., *Viscous Fluid Flow*, McGraw Hill, 2006.

[29] Phillips, W. F., "Overview of Aerodynamics: Incompressible Flow over Finite Wings," *Mechanics of Flight*, John Wiley & Sons, Inc., Hoboken, NJ, 2010, Chap. 1, 2nd ed., pp. 46–94.

[30] Phillips, W. F., Hunsaker, D. F., and Niewoehner, R. J., "Estimating the Subsonic Aerodynamic Center and Moment Components for Swept Wings," *Journal of Aircraft*, Vol. 45, No. 3, 2008, pp. 1033–1043. https://doi.org/10.2514/1.33445.

[31] Phillips, W. F., "Overview of Aerodynamics: Wing Aerodynamic Center and Pitching Moment," *Mechanics of Flight*, John Wiley & Sons, Inc., Hoboken, NJ, 2010, Chap. 1, 2nd ed., pp. 120–130.

[32] Polhamus, E. C., "A Simple Method of Estimating the Subsonic Lift and Damping in Roll of Sweptback Wings," Tech. Rep. NACA-TN-1862, National Advisory Committee for Aeronautics, 1949.

[33] Phillips, W. F., "Overview of Aerodynamics: Trailing-Edge Flaps and Section Flap Effectiveness," *Mechanics of Flight*, John Wiley & Sons, Inc., Hoboken, NJ, 2010, Chap. 1, 2nd ed., pp. 39–46.

[34] Pope, O. D., "Aerodynamic Centers of Arbitrary Airfoils," Master's thesis, Utah State University, 2017. https://doi.org/10.26076/1a04-e36c.

[35] Bowman, W. J., and Snyder, D. O., "A Minimalistic Approach to Teaching Aircraft Design," *43rd AIAA Aerospace Sciences Meeting and Exhibit, Reno, Nevada*, 2005, pp. AIAA–2005–0282. https://doi.org/10.2514/6.2005-282.

[36] Bowman, W. J., and Snyder, D. O., "A Less Minimalistic Approach to Teaching Aircraft Design," *44th AIAA Aerospace Sciences Meeting and Exhibit, Reno, Nevada*, 2006, pp. AIAA–2006–94. https://doi.org/10.2514/6.2006-94.

[37] Beard, R. W., and McLain, T. W., *Small Unmanned Aircraft: Theory and Practice*, Princeton University Press, 2012.

[38] Phillips, W. F., "Aircraft Equations of Motion: Nondimensional Linearized Equations of Motion," *Mechanics of Flight*, John Wiley & Sons, Inc., Hoboken, NJ, 2010, Chap. 7, 2nd ed., pp. 788–797.

[39] Nguyen, L. T., Ogburn, M. E., Gilbert, W. P., Kibler, K. S., Brown, P. W., and Deal, P. L., "Simulator Study of Stall/Post-Stall Characteristics of a Fighter Airplane With Relaxed Static Stability," Tech. Rep. NASA-TP-1538, NASA Langley Research Center, Hampton, 1979.

[40] Phillips, W. F., *Mechanics of Flight*, 2nd ed., John Wiley & Sons, Inc., Hoboken, NJ, 2010.

[41] Phillips, W. F., "Longitudinal Static Stability and Trim," *Mechanics of Flight*, John Wiley & Sons, Inc., Hoboken, NJ, 2010, Chap. 4, 2nd ed., pp. 377–496.

[42] Phillips, W. F., "Aircraft Controls and Maneuverability: Longitudinal Control and Maneuverability," *Mechanics of Flight*, John Wiley & Sons, Inc., Hoboken, NJ, 2010, Chap. 6, 2nd ed., pp. 605–623.

[43] Phillips, W. F., "Longitudinal Static Stability and Trim: Effects of the Fuselage, Nacelles, and External Stores," *Mechanics of Flight*, John Wiley & Sons, Inc., Hoboken, NJ, 2010, Chap. 4, 2nd ed., pp. 472–475.

[44] Phillips, W. F., "Lateral Static Stability and Trim," *Mechanics of Flight*, John Wiley & Sons, Inc., Hoboken, NJ, 2010, Chap. 5, 2nd ed., pp. 497–604.

[45] Phillips, W. F., "Lateral Static Stability and Trim: Estimating the Sidewash Gradient on a Vertical Tail," *Mechanics of Flight*, John Wiley & Sons, Inc., Hoboken, NJ, 2010, Chap. 5, 2nd ed., pp. 518–525.

[46] Phillips, W. F., "Lateral Static Stability and Trim: Roll Control and Trim Requirements," *Mechanics of Flight*, John Wiley & Sons, Inc., Hoboken, NJ, 2010, Chap. 5, 2nd ed., pp. 567–574.

[47] Phillips, W. F., "Aircraft Controls and Maneuverability: Stall Recovery," *Mechanics of Flight*, John Wiley & Sons, Inc., Hoboken, NJ, 2010, Chap. 6, 2nd ed., pp. 661–666.

[48] Margolis, K., "Supersonic Wave Drag of Sweptback Tapered Wings at Zero Lift," Tech. Rep. NACA-TN-1448, National Advisory Committee for Aeronautics, 1947.

[49] Fisher, L. R., "Approximate Corrections for the Effects of Compressibility on the Subsonic Stability Derivatives of Swept Wings," Tech. Rep. NACA-TN-1854, National Advisory Committee for Aeronautics, 1949.

[50] Loftin, L. K., Jr, *Quest for Performance: The Evolution of Modern Aircraft*, NASA-SP-468, Scientific and Technical Information Branch, National Aeronautics and Space Administration, 1985.

[51] Hoerner, S., and Borst, H., "Lift of Airplane Configurations," *Fluid Dynamic Lift*, Hoerner Fluid Dynamics, Bricktown, NJ, 1975, Chap. 2, pp. 1–22.

[52] Raymer, D. P., "Ground Effect," *Aircraft Design: A Conceptual Approach*, AIAA, 2012, 5th ed., p. 463. https://doi.org/10.2514/4.869112.

[53] McCormick, B. W., "Ground Roll," *Aerodynamics, Aeronautics, and Flight Mechanics*, John Wiley & Sons, Inc., 1979, 1st ed., pp. 419–424.

[54] Anderson, J. D., "Calculation of Ground Roll," *Aircraft Performance & Design*, McGraw-Hill, 1999, pp. 355–363.

[55] Anderson, J. D., "Takeoff Performance," *Introduction to Flight*, McGraw-Hill, 2012, 7th ed., pp. 520–526.

[56] Phillips, W. F., "Aircraft Performance: Takeoff and Landing Performance," *Mechanics of Flight*, John Wiley & Sons, Inc., Hoboken, NJ, 2010, Chap. 3, 2nd ed., pp. 337–353.

[57] Torenbeek, E., "Ground Effects," *Synthesis of Subsonic Airplane Design*, Delft University Press, 1982, pp. 551–554.

[58] Phillips, W. F., and Hunsaker, D. F., "Lifting-Line Predictions for Induced Drag and Lift in Ground Effect," *Journal of Aircraft*, Vol. 50, No. 4, 2013, pp. 1226–1233. https://doi.org/10.2514/1.C032152.

[59] Delaunay, B., "Sur la sphère vide. A la mémoire de Georges Voronoï," *Izv. Math.*, 1934.

[60] Adams, S. J., Hunsaker, D. F., Fagley, C., Lewis, J. B., and Graves, R. E., "Comparison and Combination of Aerodynamic Data Sources for a Modern Fighter Aircraft," *AIAA SCITECH 2025 Forum*, 2025, pp. AIAA–2025–0366. https://doi.org/10.2514/6.2025-0366.

[61] Bolander, C. R., Kohler, A. J., Hunsaker, D. F., Myszka, D., and Joo, J. J., "Static Trim of a Bio-Inspired Rotating Empennage for a Fighter Aircraft," *AIAA SciTech 2023 Forum*, 2023, pp. AIAA–2023–0624. https://doi.org/10.2514/6.2023-0624.

[62] Bolander, C. R., "Aerodynamic Implications of a Bio-Inspired Rotating Empennage Design for Control of a Fighter Aircraft," Ph.D. thesis, Utah State University, 2023. https://doi.org/10.26076/90ec-d95e.

[63] Phillips, W. F., "Overview of Propulsion," *Mechanics of Flight*, John Wiley & Sons, Inc., Hoboken, NJ, 2010, Chap. 2, 2nd ed., pp. 153–258.

[64] Mattingly, J. D., *Elements of Propulsion: Gas Turbines and Rockets*, American Institute of Aeronautics and Astronautics, Inc, 2006.

[65] McCormick, B. W., *Aerodynamics, Aeronautics, and Flight Mechanics*, 2nd ed., John Wiley & Sons, Inc., 1995.

[66] Raffaelli, I., "Effect of Altitude on Power of Aviation Engines," Tech. Rep. NACA-TM-287, National Advisory Committee for Aeronautics, 1924.

[67] Phillips, W. F., "Overview of Propulsion: Propeller Blade Theory," *Mechanics of Flight*, John Wiley & Sons, Inc., Hoboken, NJ, 2010, Chap. 2, 2nd ed., pp. 173–202.

[68] Maoquan, Z., and Haixin, C., "Modeling and validation of propulsion system for small battery powered electric aircraft," *7th European Conference for Aeronautics and Space Sciences (EUCASS)*, 2017. https://doi.org/10.13009/EUCASS2017-701.

[69] Corbett, J. P., and Wright, F. B., "Stabilization of computer circuits," Tech. Rep. WADC-TR-57-425, Wright Air Development Center, 1957.

[70] Robinson, A. C., "On the use of quaternions in simulation of rigid-body motion," Tech. Rep. WADC-TR-58-17, Wright Air Development Center, 1958.

[71] Mitchell, E. E. L., and Rogers, A. E., "Quaternion Parameters in the Simulation of a Spinning Rigid Body," *Simulation*, Vol. 4, No. 6, 1965, pp. 390–396. https://doi.org/10.1177/003754976500400610.

[72] Phillips, W. F., "Aircraft Flight Simulation: Numerical Integration of the Quaternion Formulation," *Mechanics of Flight*, John Wiley & Sons, Inc., Hoboken, NJ, 2010, Chap. 11, 2nd ed., pp. 1021–1036.

[73] Pope, D. A., "An Exponential Method of Numerical Integration of Ordinary Differential Equations," *Communications of the ACM*, Vol. 6, No. 8, 1963, pp. 491–493.

[74] Hoffman, J. D., "Ordinary Differential Equations: Finite Difference Equations," *Numerical Methods for Engineers and Scientists*, McGraw-Hill, Inc., New York, NY, 1992, Chap. 7, pp. 239–266.

[75] Abraham, T. A., "Linearized Rigid-Body Dynamics of Asymmetric Aircraft," Ph.D. thesis, Utah State University, 2025. https://doi.org/10.26076/42c2-2f26.

[76] Phillips, W. F., "Aircraft Equations of Motion: Linearized Equations of Motion," *Mechanics of Flight*, John Wiley & Sons, Inc., Hoboken, NJ, 2010, Chap. 7, 2nd ed., pp. 754–767.

[77] Press, W. H., Flannery, B. P., Teukolsky, S. A., and Vetterling, W. T., *Numerical Recipes - The Art of Scientific Computing*, Cambridge University Press, 1986.

[78] Phillips, W. F., "Aircraft Flight Simulation: Aircraft Position in Geographic Coordinates," *Mechanics of Flight*, John Wiley & Sons, Inc., Hoboken, NJ, 2010, Chap. 11, 2nd ed., pp. 1044–1063.

[79] U.S. Department of Defense, "Flying Qualities of Piloted Airplanes," Military Specification MIL-F-8785C, U.S. Department of Defense, 1980.

[80] U.S. Department of Defense, "Flying Qualities of Piloted Aircraft," Department of Defense Interface Standard MIL-STD-1797A, U.S. Department of Defense, 1995.

[81] Beal, T. R., "Digital Simulation of Atmospheric Turbulence for Dryden and von Karman Models," *Journal of Guidance, Control, and Dynamics*, Vol. 16, 1993, pp. 132–138. https://doi.org/10.2514/3.11437.

[82] Schmidt, D. K., "Models of Atmospheric Turbulence," *Modern Flight Dynamics*, McGraw-Hill, 2012, 1st ed., pp. 839–851.

[83] Mechtley, E. A., "Physical Constants and Conversion Factors," Tech. Rep. NASA-SP-7012, National Aeronautics and Space Administration, 1973.

[84] Fox, M. C., and Forrest, D. K., "Supersonic Aerodynamic Characteristics of an Advanced F-16 Derivative Aircraft Configuration," Tech. Rep. NASA-TP-3355, National Aeronautics and Space Administration, 1993.

[85] Butcher, D., "Non-honeycomb F-16 horizontal stabilizer, structural design," *ICAS Proc. 1982, AIAA Aircraft Syst. And Technol. Conf. Seattle*, 1982, pp. 586–592.

[86] Hunsaker, D. F., "Evaluation of an Incompressible Energy-Vorticity Turbulence Model for Fully Rough Pipe Flow," Ph.D. thesis, Utah State University, 2011. https://doi.org/10.26076/0620-8e52.

Nomenclature

Acronyms

ABM Adams-Bashforth-Moulton

ADS air data system

ASCII American Standard Code for Information Interchange

CAD computer-aided design

CAS calibrated airspeed

CFD computational fluid dynamics

CG center of gravity

CSV comma-separated values

DC direct current

EAS equivalent airspeed

EPP expanded polypropylene

ESC electronic speed controller

FAR Federal Aviation Regulations

FIFO first-in, first-out

GPS global positioning system

HTTP hypertext transfer protocol

HUD heads-up display

IAS indicated airspeed

IGES Initial Graphics Exchange Specification

IMU inertial measurement unit

IP Internet Protocol

JSON JavaScript Object Notation

LCD liquid-crystal display

LU lower-upper decomposition

MSL mean sea level

NACA National Advisory Committee for Aeronautics

NASA National Aeronautics and Space Administration

PSD power spectral density

PWM pulse-width modulation

RADAR radio detection and ranging

RAM random access memory

RH right-hand

LH left-hand

RK4 fourth-order Runge-Kutta

RMS root mean square

RPM revolutions per minute

SI International System of Units

STEP Standard for the Exchange of Product Data

STL stereolithography

TAS true airspeed

TCP transmission control protocol

UDP user datagram protocol

UIUC University of Illinois Urbana-Champaign

USAF United States Air Force

VTK Visualization Toolkit

WMM World Magnetic Model

Mathematical Symbols

$\dot{}$ a dot over a symbol denotes the time rate of change

, a comma between subscripts denotes a derivative with respect to the following symbol. For example, a first derivative of C_N with respect to α is expressed as $C_{N,\alpha} = dC_N/d\alpha$, and the second derivative of C_N with respect to α is expressed as $C_{N,\alpha,\alpha} = (d^2C_N)/(d\alpha^2)$

$\otimes$ quaternion product

* quaternion conjugate

C_{angle} cosine of angle

S_{angle} sine of angle

Latin Symbols

A aerodynamic axial force in the $-x_b$ direction

A_c area of a thin cylinder

A_i amplitude of the ith sinusoidal component

a speed of sound

a_c centrifugal acceleration used for gravity relief

a_x x-component of acceleration measured by accelerometer

a_y y-component of acceleration measured by accelerometer

a_z z-component of acceleration measured by accelerometer

b wing span

b_h horizontal tail span

b_w main wing span

C_A dimensionless axial force coefficient, Eq. (3.4.26)

C_b battery C-rating

C_D dimensionless drag force coefficient, Eq. (3.4.26)

$C_{D,L}$ constant for linear term in drag polar

$C_{D,L\overline{q}}$ change in C_D with $C_{L_1}\overline{q}$

$C_{D,L\delta_e}$ change in C_D with $C_{L_1}\delta_e$

C_{D,L^2} constant for quadratic term in drag polar

$C_{D,\overline{p}}$ change in C_D with $\overline{p}$

$C_{D,\overline{q}}$ change in C_D with $\overline{q}$

$C_{D,\overline{r}}$ change in C_D with $\overline{r}$

C_{D,S^2} change in C_D with C_S^2

$C_{D,\alpha}$ change in C_D with α

$C_{D,\hat{\alpha}}$ change in C_D with $\hat{\alpha}$

$C_{D,\alpha\overline{q}}$ change in C_D with $\alpha\overline{q}$

$C_{D,\alpha\delta_e}$ change in C_D with $\alpha\delta_e$

$C_{D,\beta}$ change in C_D with β

$C_{D,\hat{\beta}}$ change in C_D with $\hat{\beta}$

C_{D,δ_a} change in C_D with δ_a

C_{D,δ_e} change in C_D with δ_e

C_{D,δ_e^2} change in C_D with δ_e^2

C_{D,δ_r} change in C_D with δ_r

$C_{D_{L0}}$ C_D at $C_L = C_S = 0$

C_{D_N} C_D from Newtonian theory

$C_{D_{ss}}$ sub-stall C_D

C_{D_0} C_D at $\alpha = \beta = \overline{p} = \overline{q} = \overline{r} = \hat{\alpha} = \hat{\beta} = \mathrm{U} = 0$

C_{D0} cylinder drag coefficient constant

$C_{D\mathrm{M}0}$ C_D at $\mathrm{M} = 0$

$C_{\ell\mathrm{M}0}$ C_ℓ at $\mathrm{M} = 0$

C_I motor torque conversion factor

C_L dimensionless lift force coefficient, Eq. (3.4.26)

$C_{L,\overline{p}}$ change in C_L with $\overline{p}$

$C_{L,\overline{q}}$ change in C_L with $\overline{q}$

$C_{L,\overline{r}}$ change in C_L with $\overline{r}$

$C_{L,\alpha}$ change in C_L with α

$C_{L,\hat{\alpha}}$ change in C_L with $\hat{\alpha}$

$C_{L,\beta}$ change in C_L with β

$C_{L,\hat{\beta}}$ change in C_L with $\hat{\beta}$

C_{L,δ_a} change in C_L with δ_a

C_{L,δ_e} change in C_L with δ_e

C_{L,δ_r} change in C_L with δ_r

C_{L_g} lift coefficient in ground effect

C_{L_h} lift coefficient of horizontal tail

$C_{L_h,\alpha}$ lift slope of horizontal tail

C_{L_N} C_L from Newtonian theory

$C_{L_{ss}}$ sub-stall C_L

$C_{L_v,\alpha}$ lift slope of vertical tail

C_{L_w} lift coefficient of main wing

$C_{L_w,\alpha}$ lift slope of main wing

C_{L_0} C_L at $\alpha = \beta = \overline{p} = \overline{q} = \overline{r} = \hat{\alpha} = \hat{\beta} = \mathrm{U} = 0$

C_{L_1} pseudo lift coefficient defined in Eq. (3.7.20)

$C_{L0,\alpha}$ lift slope at zero angle of attack

C_{LM0} C_L at M = 0

C_ℓ dimensionless rolling-moment coefficient, Eq. (3.4.33)

$C_{\ell,\overline{p}}$ change in C_ℓ with $\overline{p}$

$C_{\ell,\overline{q}}$ change in C_ℓ with $\overline{q}$

$C_{\ell,\overline{r}}$ change in C_ℓ with $\overline{r}$

$C_{\ell,\alpha}$ change in C_ℓ with α

$C_{\ell,\hat{\alpha}}$ change in C_ℓ with $\hat{\alpha}$

$C_{\ell,\alpha\overline{r}}$ change in C_ℓ with $\alpha\overline{r}$

$C_{\ell,\beta}$ change in C_ℓ with β

$C_{\ell,\hat{\beta}}$ change in C_ℓ with $\hat{\beta}$

C_{ℓ,δ_a} change in C_ℓ with δ_a

C_{ℓ,δ_e} change in C_ℓ with δ_e

C_{ℓ,δ_r} change in C_ℓ with δ_r

C_{ℓ_0} C_ℓ at $\alpha = \beta = \overline{p} = \overline{q} = \overline{r} = \hat{\alpha} = \hat{\beta} = \mathbf{U} = 0$

C_m dimensionless pitching-moment coefficient, Eq. (3.4.33)

$C_{m,A}$ constant used in general inviscid airfoil theory, Eq. (3.5.19)

$C_{m,N}$ constant used in general inviscid airfoil theory, Eq. (3.5.19)

$C_{m,\overline{p}}$ change in C_m with $\overline{p}$

$C_{m,\overline{q}}$ change in C_m with $\overline{q}$

$C_{m,\overline{r}}$ change in C_m with $\overline{r}$

$C_{m,\alpha}$ change in C_m with α

$C_{m,\hat{\alpha}}$ change in C_m with $\hat{\alpha}$

$C_{m,\beta}$ change in C_m with β

$C_{m,\hat{\beta}}$ change in C_m with $\hat{\beta}$

C_{m,δ_a} change in C_m with δ_a

C_{m,δ_c} change in C_m with control-surface deflection

C_{m,δ_e} change in C_m with δ_e

C_{m,δ_r} change in C_m with δ_r

$C_{m_{ac}}$ C_m at the aerodynamic center

$C_{m_{c/4}}$ C_m at the airfoil quarter-chord

C_{m_h,δ_e} change in horizontal tail pitching moment with δ_e

C_{m_O} dimensionless C_m at the origin

C_{m_N} C_m from Newtonian theory

$C_{m_{ss}}$ sub-stall C_m

C_{m_v,δ_r} change in vertical tail pitching moment with δ_r

C_{m_w} main wing pitching moment

C_{m_0} C_m at $\alpha = \beta = \overline{p} = \overline{q} = \overline{r} = \hat{\alpha} = \hat{\beta} = \mathbf{U} = 0$

$C_{m_{0h}}$ horizontal tail pitching moment at zero angle of attack and elevator deflection

C_{M_1} coefficient used for compressibility corrections

C_{M_2} coefficient used for compressibility corrections

$C_{m0,\alpha}$ constant used in general inviscid airfoil theory, Eq. (3.5.19)

$C_{\mathcal{P}M0}$ pressure coefficient at zero Mach number

C_{mM0} C_m at M = 0

C_N dimensionless normal force coefficient, Eq. (3.4.26)

C_n dimensionless yawing-moment coefficient, Eq. (3.4.33)

$C_{n,\overline{p}}$ change in C_n with $\overline{p}$

$C_{n,\overline{q}}$ change in C_n with $\overline{q}$

$C_{n,\overline{r}}$ change in C_n with $\overline{r}$

$C_{N,\alpha}$ change in C_N with α

$C_{n,\alpha}$ change in C_n with α

$C_{n,\hat{\alpha}}$ change in C_n with $\hat{\alpha}$

$C_{n,\alpha\overline{p}}$ change in C_n with $\alpha\overline{p}$

$C_{n,\alpha\delta_a}$ change in C_n with $\alpha\delta_a$

$C_{n,\beta}$ change in C_n with β

$C_{n,\hat{\beta}}$ change in C_n with $\hat{\beta}$

C_{n,δ_a} change in C_n with δ_a

C_{n,δ_e} change in C_n with δ_e

C_{n,δ_r} change in C_n with δ_r

C_{n_0} C_n at $\alpha = \beta = \overline{p} = \overline{q} = \overline{r} = \hat{\alpha} = \hat{\beta} = \mathbf{U} = 0$

C_{N1} linear coefficient in normal-force-coefficient relation given in Eq. (4.5.11)

C_{n1} linear coefficient in yawing-moment-coefficient relation given in Eq. (4.5.12)

C_{N2} quadratic coefficient in normal-force-coefficient relation given in Eq. (4.5.11)

C_{n2} quadratic coefficient in yawing-moment-coefficient relation given in Eq. (4.5.12)

C_{N3} cubic coefficient in normal-force-coefficient relation given in Eq. (4.5.11)

C_{n3} cubic coefficient in yawing-moment-coefficient relation given in Eq. (4.5.12)

$C_{n\mathrm{M}0}$ C_n at M = 0

$C_{\mathcal{P}}$ pressure coefficient, Eq. (3.3.9

C_{P_b} rotor brake-power coefficient

C_{P0} constant coefficient in power-coefficient relation given in Eq. (4.5.10)

C_{P1} linear coefficient in power-coefficient relation given in Eq. (4.5.10)

C_{P2} quadratic coefficient in power-coefficient relation given in Eq. (4.5.10)

C_S dimensionless side-force coefficient, Eq. (3.4.26)

$C_{S,\overline{p}}$ change in C_S with $\overline{p}$

$C_{S,\overline{q}}$ change in C_S with $\overline{q}$

$C_{S,\overline{r}}$ change in C_S with $\overline{r}$

$C_{S,\alpha}$ change in C_S with α

$C_{S,\hat{\alpha}}$ change in C_S with $\hat{\alpha}$

$C_{S,\alpha\overline{p}}$ change in C_S with $\alpha\overline{p}$

$C_{S,\beta}$ change in C_S with β

$C_{S,\hat{\beta}}$ change in C_S with $\hat{\beta}$

C_{S,δ_a} change in C_S with δ_a

C_{S,δ_e} change in C_S with δ_e

C_{S,δ_r} change in C_S with δ_r

C_{S_0} C_S at $\alpha = \beta = \overline{p} = \overline{q} = \overline{r} = \hat{\alpha} = \hat{\beta} = \mathbf{U} = 0$

$C_{S\mathrm{M}0}$ C_S at M = 0

C_T thrust coefficient

C_{T0} constant coefficient in thrust-coefficient relation given in Eq. (4.5.9)

C_{T1} linear coefficient in thrust-coefficient relation given in Eq. (4.5.9)

C_{T2} quadratic coefficient in thrust-coefficient relation given in Eq. (4.5.9)

C_x dimensionless force coefficient in the x direction, Eq. (3.4.26)

C_Y dimensionless side-force coefficient in body-fixed or stability coordinates

C_y dimensionless force coefficient in the y direction, Eq. (3.4.26)

C_z dimensionless force coefficient in the z direction, Eq. (3.4.26)

c airfoil section chord

c_f flap chord

c_f chord length of fuselage

c_r wing root chord

c_t wing tip chord

$\overline{c}$ wing mean chord length

$\overline{c}_h$ horizontal tail mean chord length

$\overline{c}_v$ vertical tail mean chord length

$\overline{c}_w$ main wing mean chord length

D aerodynamic drag force in the $-x_w$ direction

D_s aerodynamic drag force in the $-x_s$ direction

d horizontal step distance or navigation distance

d_f fuselage equivalent diameter

d_r rotor diameter

d_{vp} distance from the center of projection to the view plane

$[\mathbf{E}]$ identity matrix

$\mathbf{E}$ Euler axis vector

E_A battery voltage model constant

E_B battery voltage model constant

E_b battery terminal voltage

E_{b_0} no-load battery voltage

$E_{b_{0F}}$ no-load battery voltage at full capacity

E_C battery voltage model constant

E_m	voltage applied to motor
E_x	x-component of the Euler axis vector
E_{xx}	$E_x E_x (1 - C_\Theta)$
E'_{xx}	$-E_x E_x C_{\Theta/2}/S_{\Theta/2}$
E_{xy}	$E_x E_y (1 - C_\Theta)$
E'_{xy}	$-E_x E_y C_{\Theta/2}/S_{\Theta/2}$
E_{xz}	$E_x E_z (1 - C_\Theta)$
E'_{xz}	$-E_x E_z C_{\Theta/2}/S_{\Theta/2}$
E_y	y-component of the Euler axis vector
E_{yy}	$E_y E_y (1 - C_\Theta)$
E'_{yy}	$-E_y E_y C_{\Theta/2}/S_{\Theta/2}$
E_{yz}	$E_y E_z (1 - C_\Theta)$
E'_{yz}	$-E_y E_z C_{\Theta/2}/S_{\Theta/2}$
E_z	z-component of the Euler axis vector
E_{zz}	$E_z E_z (1 - C_\Theta)$
E'_{zz}	$-E_z E_z C_{\Theta/2}/S_{\Theta/2}$
$\{\mathbf{e}\}$	Euler-Rodrigues quaternion
$\{\mathbf{e}\}_r$	normalized Euler-Rodrigues quaternion
e	specific energy
e_O	Oswald Efficiency
e_x	x-component of the Euler-Rodrigues quaternion
e_y	y-component of the Euler-Rodrigues quaternion
e_z	z-component of the Euler-Rodrigues quaternion
e_0	scalar component of the Euler-Rodrigues quaternion
$\overline{\mathbf{F}}$	pseudo aerodynamic force vector including aerodynamic and thrust effects
$\mathbf{F}$	aerodynamic force vector
$\mathbf{F}_b$	aerodynamic force vector in body-fixed coordinate system
$\mathbf{F}_c$	component aerodynamic force vector in component coordinate system
$\mathbf{F}_{\text{ref}}$	aerodynamic force vector at the reference location
$\mathbf{F}_s$	net surface force vector on the vehicle
$\mathbf{F}_T$	thrust component of the pseudo aerodynamic force vector
F_N	aerodynamic force normal to surface
F_{x_b}	x_b-component of aerodynamic force vector
F_{x_s}	x_s-component of aerodynamic force vector
F_{x_w}	x_w-component of aerodynamic force vector
F_{y_b}	y_b-component of aerodynamic force vector
F_{y_s}	y_s-component of aerodynamic force vector
F_{y_w}	y_w-component of aerodynamic force vector
F_{z_b}	z_b-component of aerodynamic force vector
F_{z_s}	z_s-component of aerodynamic force vector
F_{z_w}	z_w-component of aerodynamic force vector
$\mathbf{f}_b$	fluid element body forces
$\mathbf{f}_s$	fluid element surface forces
f	frequency
f_i	frequency of the ith sinusoidal component
f_k	the kth frequency bin
f_s	sampling frequency
$f'(x)$	derivative of scalar function $f(x)$
$f(x)$	scalar function used in Newton's method
G_m	motor gear ratio
g	acceleration of gravity
g_{ssl}	standard acceleration of gravity at sea level
H	geometric altitude above sea level
$[\mathbf{h}]$	gyroscopic tensor
$\mathbf{h}$	angular momentum vector
$\mathbf{h}_b$	angular momentum vector in the body-fixed coordinate system
$\mathbf{h}_c$	angular momentum vector of a component in the component coordinate system
$\mathbf{h}_v$	angular momentum vector in the vehicle coordinate system
h	height above ground
h_c	cylinder height
h_h	rotor hub height
h_v	distance above CG of vertical tail AC
h_{vp}	view-plane height
$[\mathbf{I}]$	mass moment of inertia tensor

$[\mathbf{I}]_b$ inertia tensor of rotor blades

$[\mathbf{I}]_c$ component inertia tensor about vehicle center of gravity

$[\mathbf{I}]_{cc}$ component inertia tensor about component center of gravity

$[\mathbf{I}]_{CG}$ inertia tensor about the center of gravity

$[\mathbf{I}]_h$ inertia tensor of rotor hub

$[\mathbf{I}]_v$ vehicle inertia tensor about vehicle origin

I_b battery current

$I_{b\max}$ maximum current at which a battery can be safely discharged

$I_{c\max}$ maximum current that can be passed through the ESC

I_m motor current

I_{m0} no-load motor current

I_{xx} moment of inertia, Eq. (2.2.9)

I_{xy} product of inertia, Eq. (2.2.12)

I_{xz} product of inertia, Eq. (2.2.13)

I_{yx} product of inertia, Eq. (2.2.12)

I_{yy} moment of inertia, Eq. (2.2.10)

I_{yz} product of inertia, Eq. (2.2.14)

I_{zx} product of inertia, Eq. (2.2.13)

I_{zy} product of inertia, Eq. (2.2.14)

I_{zz} moment of inertia, Eq. (2.2.11)

$\mathbf{i}_x$ unit vector in the x-direction

$\mathbf{i}_y$ unit vector in the y-direction

$\mathbf{i}_z$ unit vector in the z-direction

$[\mathbf{J}]$ Jacobian matrix

J rotor advance ratio

J_{ij} Jacobian entry $\partial R_i/\partial x_j$

j imaginary unit, $j \equiv \sqrt{-1}$

K Knudsen number, Eq. (3.3.1)

K_c rotor pitch-to-diameter ratio

K_S constant in Sutherland's law

K_v motor speed constant

k frequency-bin index

L aerodynamic lift force in the $-z_s = -z_w$ direction

L_b roll-disturbance length scale based on wingspan

L_u x-component turbulence length scale

L_v y-component turbulence length scale

L_w z-component turbulence length scale

ℓ aerodynamic rolling moment about the x axis

ℓ_b aerodynamic rolling moment about the x_b axis

ℓ_m motor torque

ℓ_r rotor torque

ℓ_s motor shaft torque

ℓ_s aerodynamic rolling moment about the x_s axis

ℓ_w aerodynamic rolling moment about the x_w axis

$\mathbf{l}_{ca}$ line-of-sight vector from $\mathbf{P}_c$ to $\mathbf{P}_a$

l_f distance aft of CG to center of pressure of fuselage

l_h distance aft of CG to AC of horizontal tail

l_{ref} characteristic length of the geometry of interest

l_v distance aft of CG to AC of vertical tail

l_w distance aft of CG to AC of main wing

l_{wt} distance defined in Eq. (3.7.36)

l_x length in the x direction

l_y length in the y direction

l_z length in the z direction

$\overline{\mathbf{M}}$ pseudo aerodynamic moment vector including aerodynamic and thrust effects

$\mathbf{M}$ aerodynamic moment vector

$\mathbf{M}_b$ aerodynamic moment vector in body-fixed coordinate system

$\mathbf{M}_c$ component aerodynamic moment vector in component coordinate system

$\mathbf{M}_O$ moment vector about the origin

$\mathbf{M}_{\text{ref}}$ aerodynamic moment vector about the reference location

$\mathbf{M}_s$ net surface moment vector on the vehicle about the vehicle center of gravity

$\mathbf{M}_T$ thrust component of the pseudo aerodynamic moment vector

$\mathbf{M}_1$ moment vector about point 1

$\mathcal{M}$ mass of an object or the complete vehicle

$\mathcal{M}_b$ mass of the rotor blades

$\mathcal{M}_c$ mass of component

$\mathcal{M}_h$ mass of the rotor hub

M Mach number, Eq. (3.3.4)

M_∞ freestream Mach number

M_{x_b} x_b-component of aerodynamic moment vector

M_{x_s} x_s-component of aerodynamic moment vector

M_{x_w} x_w-component of aerodynamic moment vector

M_{y_b} y_b-component of aerodynamic moment vector

M_{y_s} y_s-component of aerodynamic moment vector

M_{y_w} y_w-component of aerodynamic moment vector

M_{z_b} z_b-component of aerodynamic moment vector

M_{z_s} z_s-component of aerodynamic moment vector

M_{z_w} z_w-component of aerodynamic moment vector

$\dot{\mathcal{M}}_e$ mass flow rate entering or exiting the vehicle, positive for mass exiting

m aerodynamic pitching moment about the y axis

m_b aerodynamic pitching moment about the y_b axis

m_s aerodynamic pitching moment about the y_s axis

m_w aerodynamic pitching moment about the y_w axis

N aerodynamic normal force in the $-x_b$ direction

N number of equations in a nonlinear system

N_b number of blades on a rotor

N_b bit count of a binary signal

N_g number of polygons in a VTK geometry definition

N_i number of integers in a VTK polygon or line definition

N_l number of lines in a VTK line definition

N_m motor rotations per minute

N_p number of vertices or points in a polygon or polyline

N_r rotor rotations per minute

N_s motor shaft rotations per minute

N_s number of discrete signal values

$\mathbf{n}$ normal vector to a surface

$\mathbf{n}_{vp}$ view-plane normal vector

n aerodynamic yawing moment about the z axis

n_b aerodynamic yawing moment about the z_b axis

n_s aerodynamic yawing moment about the z_s axis

n_w aerodynamic yawing moment about the z_w axis

$\bar{\mathbf{n}}$ load factor vector

$\bar{n}_L$ load factor in the direction of lift

$\bar{n}_N$ normal load factor

$\mathbf{P}$ component location relative to vehicle CG

$\mathbf{P}_a$ point of interest to be projected

$\mathbf{P}_b$ point along the line of sight, including the projected point on the view plane

$\mathbf{P}_{b_c}$ camera or center-of-projection location in body-fixed coordinates

$\mathbf{P}_c$ camera or center-of-projection point

$\mathbf{P}_{CG}$ location of the center of gravity

$\mathbf{P}_{f_c}$ camera or center-of-projection location in Earth-fixed coordinates

$\mathbf{P}_{\mathrm{IMU}}$ IMU position vector

$\mathbf{P}_{\text{ref}}$ aerodynamic reference location relative to the vehicle center of gravity

$\mathbf{P}_0$ center point of the view plane

$\mathbf{P}_1$ location of point 1 relative to an origin

$\mathbf{P}_1, \mathbf{P}_2$ neighboring view-plane corner points used to define the view-plane normal

$\mathcal{P}$ pressure

$\mathcal{P}_i$ pressure at the base of atmosphere range i

$\mathcal{P}_\infty$ freestream fluid pressure

$\mathcal{P}_{\text{ssl}}$ standard sea-level pressure

$\mathcal{P}_0$ stagnation pressure

P_a propulsive power or power available

P_b rotor brake power

P_E probability of exceedance

$P_{\max}$ maximum power at sea level

P_{xx} periodogram estimate of the power spectral density

p angular rate about the x axis

$\overline{p}$ traditional dimensionless angular rate about the x axis, Eq. (3.4.23)

$\tilde{p}$ turbulent angular-rate fluctuation about x axis

$\breve{p}$ dynamic dimensionless angular rate about the x axis, Eq. (3.4.24)

$\{\mathbf{Q}\}$ arbitrary quaternion

Q dynamic pressure, Eq. (3.3.7)

Q_b battery capacity remaining

Q_{b_0} total battery capacity

Q_x x-component of an arbitrary quaternion

Q_y y-component of an arbitrary quaternion

Q_z z-component of an arbitrary quaternion

Q_0 scalar component of an arbitrary quaternion

$\mathbf{q}$ heat flux vector

q angular rate about the y axis

$\overline{q}$ traditional dimensionless angular rate about the y axis, Eq. (3.4.23)

$\tilde{q}$ turbulent angular-rate fluctuation about y axis

$\breve{q}$ dynamic dimensionless angular rate about the y axis, Eq. (3.4.24)

$[\mathbf{R}]$ rotation matrix

$\mathbf{R}$ residual vector

$\mathcal{R}$ gas constant

R_A wing aspect ratio

R_{A_h} horizontal tail aspect ratio

$R_{A_{vp}}$ view-plane aspect ratio

R_{A_w} main wing aspect ratio

R_b battery internal resistance

R_c ESC resistance

R_D induced-drag influence ratio

R_E mean radius of the Earth at sea level

R_e equatorial radius of the Earth

R_e Reynolds number, Eq. (3.3.3)

R_{E_Z} radius of the Earth used to compute geopotential altitude

R_i the ith component of the residual vector

R_L lift influence ratio

R_m motor resistance

R_p polar radius of the Earth

R_x Earth ellipsoid primary radius of curvature in the north-south direction

R_y Earth ellipsoid primary radius of curvature in the east-west direction

$\mathbf{r}_{ac}$ location of wing aerodynamic center relative to vehicle origin

$\mathbf{r}_b$ location of vehicle CG relative to vehicle origin

$\mathbf{r}_c$ location of component relative to vehicle origin

$\mathbf{r}_{\text{IMU}}$ IMU location relative to the vehicle center of gravity

$\mathbf{r}_o$ component origin relative to vehicle origin

r radial coordinate in cylindrical and spherical coordinate systems

r angular rate about the z axis

r_h rotor hub radius

r_r rotor radius

r_1 inner radius

r_2 outer radius

$\overline{r}$ dimensionless angular rate about the z axis, Eq. (3.4.23)

$\tilde{r}$ turbulent angular-rate fluctuation about r axis

$\check{r}$ dynamic dimensionless angular rate about the z axis, Eq. (3.4.24)

S aerodynamic side force in the y_w direction

S_f maximum cross-sectional area of fuselage

S_h horizontal tail planform area

$S_{i,j,k}$ wing segment volume integral define in Eq. (2.4.20)

S_{ref} reference area

S_v vertical tail planform area

S_w wing planform area

$\mathbf{s}$ vector between two points

s signal value

$s_{\max}$ maximum signal value

$s_{\min}$ minimum signal value

s_q quantized signal value

s_x x component of $\mathbf{s}$

s_y y component of $\mathbf{s}$

s_z z component of $\mathbf{s}$

$\{\mathbf{T}\}$ temporary quaternion

$\mathcal{T}$ temperature

$\mathcal{T}_i$ temperature at the base of atmosphere range i

$\mathcal{T}_i'$ temperature gradient within atmosphere range i

$\mathcal{T}_\infty$ freestream temperature

$\mathcal{T}_S$ temperature constant in Sutherland's law

$\mathcal{T}_{\text{ssl}}$ standard sea-level temperature

$\mathcal{T}_0$ stagnation temperature

T thrust

T_a exponential coefficient in thrust equation

T_x x-component of a temporary quaternion

T_y y-component of a temporary quaternion

T_z z-component of a temporary quaternion

T_0 scalar component of a temporary quaternion

T_0 constant coefficient in thrust equation

T_1 linear coefficient in thrust equation

T_2 quadratic coefficient in thrust equation

t time

$t_{\max}$ airfoil section maximum thickness

$\mathbf{U}$ control vector

U_e''' volumetric heating

$\mathbf{u}_c$ component velocity unit vector

$\mathbf{u}_D$ unit vector in the direction of drag

$\mathbf{u}_L$ unit vector in the direction of lift

$\mathbf{u}_m$ unit vector about axis of pitching moment

$\mathbf{u}_N$ unit vector in the direction of normal force

$\mathbf{u}_w$ wing velocity unit vector

u x_b vehicle velocity component relative to the surrounding air

u_{c_x} x component of $\mathbf{u}_c$

u_{c_y} y component of $\mathbf{u}_c$

u_{c_z} z component of $\mathbf{u}_c$

u_{w_x} x component of $\mathbf{u}_w$

u_{w_z} z component of $\mathbf{u}_w$

$\tilde{u}$ x-component of turbulent velocity fluctuation

$\mathbf{V}$ velocity vector

$\mathbf{V}_c$ component velocity vector

$\mathbf{V}_e$ velocity of mass entering or exiting the vehicle relative to the inertial coordinate system

$\mathbf{V}_g$ ground velocity vector

$\mathbf{V}_h$ horizontal vehicle velocity vector relative to the Earth-fixed coordinate system

$\mathbf{V}_w$ wing velocity vector

$\mathbf{V}_{w_f}$ wind velocity vector relative to the Earth-fixed coordinate system

$\mathbf{V}_{wh}$ horizontal wind velocity vector

$\forall$ volume of an object

V velocity magnitude

$\forall_b$ volume of rotor blades

V_c component velocity magnitude

V_{CAS} calibrated airspeed

V_{climb} climb rate, positive upward

V_{EAS} equivalent airspeed

V_f magnitude of the vehicle velocity relative to the Earth

V_g ground speed

$\forall_h$ volume of rotor hub

V_h horizontal vehicle speed

V_{IAS} indicated airspeed

V_∞ freestream velocity magnitude

V_N velocity magnitude normal to component

V_t horizontal or tangential velocity about Earth's center

V_{TAS} true airspeed

V_w wing velocity magnitude

V_{wh} horizontal wind speed

V_{20} mean wind speed at 20 ft above ground

$\mathbf{v}_L$ vector in the direction of lift

$\mathbf{v}_N$ vector in the direction of normal force

v y_b vehicle velocity component relative to the surrounding air

$\tilde{v}$ y-component of turbulent velocity fluctuation

$\tilde{v}_{CG}$ lateral turbulent velocity fluctuation at the vehicle center of gravity

$\tilde{v}_v$ lateral turbulent velocity fluctuation at the vertical tail

$\mathbf{W}$ weight vector

W weight of an object or the complete vehicle

W_{x_b} x_b-component of weight vector

W_{x_f} x_f-component of weight vector

W_{x_v} x_v-component of weight vector

W_{y_b} y_b-component of weight vector

W_{y_f} y_f-component of weight vector

W_{y_v} y_v-component of weight vector

W_{z_b} z_b-component of weight vector

W_{z_f} z_f-component of weight vector

W_{z_v} z_v-component of weight vector

w z_b vehicle velocity component relative to the surrounding air

w_{vp} view-plane width

$\tilde{w}$ z-component of turbulent velocity fluctuation

$\tilde{w}_{CG}$ vertical turbulent velocity fluctuation at the vehicle center of gravity

$\tilde{w}_h$ vertical turbulent velocity fluctuation at the horizontal tail

$\mathbf{x}$ vector of independent variables in Newton's method

$\mathbf{x}_{c_{vp}}$ array of view-plane corner-point x-coordinates in the camera coordinate system

$\mathbf{x}_{f_{vp}}$ array of view-plane corner-point x-coordinates in the Earth-fixed coordinate system

x Cartesian coordinate

x_a atmosphere-fixed coordinate aligned with x_f, Fig. 1.1.1

x_{ac} aerodynamic center x coordinate

x_b vehicle body-fixed coordinate pointing forward from the CG, Figs. 1.1.1 and 3.4.1

x_{b_h} axial location of horizontal tail AC in body-fixed coordinate system

$x_{b_{\mathrm{wingtip}}}$ axial location of main wing tip in body-fixed coordinate system

x_{b_v} axial location of vertical tail AC in body-fixed coordinate system

x_c axial component-fixed coordinate

x_f Earth-fixed coordinate pointing northward, Fig. 1.1.1

x_{ff} one-dimensional location in the frozen turbulence field

x_i independent-variable value at Newton iteration i

x_i the ith sampled value of a signal

x_{i+1} updated independent-variable value in Newton's method

x_{root} root of a scalar equation

x_s vehicle stability x coordinate pointing forward, Fig. 3.4.1

x_v vehicle-fixed coordinate pointing forward, Fig. 1.1.1

x_w vehicle wind x coordinate pointing into freestream, Fig. 3.4.1

$\bar{x}$ x coordinate of the center of gravity

$\bar{x}$ sample mean of a signal

$\hat{x}$ dimensionless airfoil chordwise coordinate, $\hat{x} \equiv x/c$

$\hat{x}_{\text{max}}$ dimensionless chordwise location of maximum thickness

$\tilde{x}_i$ deviation of the ith sample from the mean

$x(t)$ continuous signal as a function of time

Y aerodynamic side force in the $y_b = y_s$ direction

$\mathbf{y}$ array of vehicle states

$\mathbf{y}_{c_{vp}}$ array of view-plane corner-point y-coordinates in the camera coordinate system

$\mathbf{y}_{f_{vp}}$ array of view-plane corner-point y-coordinates in the Earth-fixed coordinate system

y Cartesian coordinate

y_a atmosphere-fixed coordinate aligned with y_f, Fig. 1.1.1

y_{ac} aerodynamic center y coordinate

y_b vehicle body-fixed coordinate pointing rightward from the CG, Figs. 1.1.1 and 3.4.1

y_c component-fixed coordinate

y_f Earth-fixed coordinate pointing eastward, Fig. 1.1.1

y_s vehicle stability y coordinate, Fig. 3.4.1

y_v vehicle-fixed coordinate pointing rightward, Fig. 1.1.1

y_w vehicle wind y coordinate, Fig. 3.4.1

$\bar{y}$ y coordinate of the center of gravity

$\hat{y}$ dimensionless airfoil normal coordinate, $\hat{y} \equiv y/c$

Z geopotential altitude

Z_i geopotential altitude at the base of atmosphere range i

$\mathbf{z}_{c_{vp}}$ array of view-plane corner-point z-coordinates in the camera coordinate system

$\mathbf{z}_{f_{vp}}$ array of view-plane corner-point z-coordinates in the Earth-fixed coordinate system

z Cartesian coordinate

z_a atmosphere-fixed coordinate aligned with z_f, Fig. 1.1.1

z_{ac} aerodynamic center z coordinate

z_b vehicle body-fixed coordinate pointing downward from the CG, Figs. 1.1.1 and 3.4.1

z_{b_v} z_b location of vertical tail AC in body-fixed coordinate system

z_c component-fixed coordinate

z_f Earth-fixed coordinate pointing downward, Fig. 1.1.1

z_s vehicle stability z coordinate, Fig. 3.4.1

z_v vehicle-fixed coordinate pointing downward, Fig. 1.1.1

z_w vehicle wind z coordinate, Fig. 3.4.1

$\bar{z}$ z coordinate of the center of gravity

Greek Symbols

α angle of attack

α_b blending angle

α_{b0} minimum-value blending angle

α_{L0} zero-lift angle of attack

α_{L0_h} horizontal tail zero-lift angle of attack

α_{L0_w} main wing zero-lift angle of attack

α_0 mounting angle of wing

α_{0_f} mounting angle of fuselage

α_{0_h} horizontal tail mounting angle

α_{0_w} main wing mounting angle

$\hat{\alpha}$ dimensionless form of $\dot{\alpha}$, Eq. (3.4.20)

β traditional sideslip angle

β_D ground effect drag factor

β_f flank angle

β_L ground effect lift factor

$\hat{\beta}$ dimensionless form of $\dot{\beta}$, Eq. (3.4.21)

Γ dihedral angle

Γ relaxation factor

γ heat capacity ratio

γ_a rotor blade inertia parameter, Eq. (2.5.22)

γ_b rotor blade inertia parameter, Eq. (2.5.23)

γ_c rotor blade inertia parameter, Eq. (2.5.24)

γ_d rotor blade inertia parameter, Eq. (2.5.25)

γ_e rotor blade inertia parameter, Eq. (2.5.26)

γ_f rotor blade inertia parameter, Eq. (2.5.27)

γ_f Earth-fixed climb angle

γ_g rotor blade inertia parameter, Eq. (2.5.28)

γ_h rotor blade inertia parameter, Eq. (2.5.29)

γ_i rotor blade inertia parameter, Eq. (2.5.30)

γ_j rotor blade inertia parameter, Eq. (2.5.31)

γ_k rotor blade inertia parameter, Eq. (2.5.32)

γ_l rotor blade inertia parameter, Eq. (2.5.33)

γ_m rotor blade inertia parameter, Eq. (2.5.34)

γ_n rotor blade inertia parameter, Eq. (2.5.35)

γ_o rotor blade inertia parameter, Eq. (2.5.36)

γ_r relative climb angle measured with respect to the surrounding air mass

Δ change or increment in a quantity

δ integer indicating right (1) or left (−1) wing

δ integer indicating right-handed (1) or left-handed (−1) rotor

δ finite-difference step size

δ actuator state

δ_a aileron deflection

δ_{a_c} commanded aileron deflection

δ_c commanded actuator state

δ_c control-surface deflection

δ_D ground effect drag factor

δ_e elevator deflection

δ_{e_c} commanded elevator deflection

δ_L ground effect lift factor

δ_{lef} leading-edge flap deflection

$\delta_{\max}$ maximum actuator-state limits

$\delta_{\min}$ minimum actuator-state limits

δ_r rudder deflection

δ_{r_c} commanded rudder deflection

δ_{sb} speed-brake deflection

δ_0 initial actuator state

$\dot{\delta}$ actuator rate

$\dot{\delta}_{\max}$ maximum actuator-rate limits

$\dot{\delta}_{\min}$ minimum actuator-rate limits

$\dot{\delta}_0$ initial actuator rate

$\ddot{\delta}$ actuator acceleration

$\ddot{\delta}_{\max}$ maximum actuator-acceleration limits

$\ddot{\delta}_{\min}$ minimum actuator-acceleration limits

$\Delta\mathcal{P}_C$ calibrated Pitot-static pressure difference

$\Delta\mathcal{P}_I$ indicated Pitot-static pressure difference

$\Delta\psi_g$ geodesic change in bearing

Δs signal resolution or quantization step size

Δt integration time-step size

$\Delta\mathbf{x}$ Newton step vector

Δx Newton step for a scalar independent variable

Δx_{ff} step size in the frozen turbulence field

ε error

ε_c control-surface effectiveness

ε_d downwash at horizontal tail

$\varepsilon_{d,\alpha}$ downwash gradient

ε_{d0} downwash angle at zero angle of attack

ε_E Earth-ellipsoid eccentricity

ε_e elevator effectiveness

ε_r rudder effectiveness

$\varepsilon_{s,\beta}$ sidewash gradient

ε_t	tolerance used for real-number comparisons
$\tilde{\varepsilon}_a$	aileron section effectiveness
ζ	damping ratio
η_c	ESC efficiency
η_f	flap efficiency
η_g	motor gearbox efficiency
η_h	horizontal tail efficiency
η_P	propulsive efficiency
η_p	white-noise sample for $\tilde{p}$ turbulence filter
η_u	white-noise sample for $\tilde{u}$ turbulence filter
η_v	vertical tail efficiency
η_v	white-noise sample for $\tilde{v}$ turbulence filter
η_w	white-noise sample for $\tilde{w}$ turbulence filter
Θ	angle associated with rotation about the Euler axis
θ	Euler elevation angle
θ_f	flap-chord fraction change of variables
θ_{vp}	view-plane angle
κ_a	wing volume integral parameter, Eq. (2.4.22)
κ_b	wing volume integral parameter, Eq. (2.4.27)
κ_c	wing volume integral parameter, Eq. (2.4.28)
κ_d	wing volume integral parameter, Eq. (2.4.41)
κ_e	wing volume integral parameter, Eq. (2.4.42)
κ_f	wing volume integral parameter, Eq. (2.4.43)
κ_g	wing volume integral parameter, Eq. (2.4.44)
κ_Γ	dihedral factor
κ_ℓ	rolling-moment factor
$\kappa_{\ell\overline{p}}$	rolling-moment factor
κ_1	constant defined in Eq. (3.5.31)
κ_2	constant defined in Eq. (3.5.31)
$\Lambda_{c/4}$	quarter-chord sweep angle
Λ_{le}	leading-edge sweep angle
Λ_{te}	trailing-edge sweep angle
λ	molecular mean-free path
λ	eigenvalues of the second-order actuator model
λ	line parameter used to locate the intersection of a line with the view plane
λ_b	blending-rate parameter
μ	dynamic viscosity
μ_a	airfoil section dimensionless thickness distribution
μ_S	constant in Sutherland's law
π	ratio of the circumference to diameter of a circle, $2\cos^{-1}(0)$
ρ	density
ρ_∞	freestream flow density
ρ_{ssl}	sea-level density
ρ_0	stagnation density
$[\sigma]$	fluid stress tensor
σ	blending parameter
σ	actuator damping rate
σ	standard deviation or RMS turbulence amplitude
σ_{n_p}	white-noise standard deviation for $\tilde{p}$ turbulence filter
σ_{n_u}	white-noise standard deviation for $\tilde{u}$ turbulence filter
σ_{n_v}	white-noise standard deviation for $\tilde{v}$ turbulence filter
σ_{n_w}	white-noise standard deviation for $\tilde{w}$ turbulence filter
σ_p	standard deviation of $\tilde{p}$ fluctuations
σ_u	standard deviation of $\tilde{u}$ fluctuations
σ_v	standard deviation of $\tilde{v}$ fluctuations
σ_w	standard deviation of $\tilde{w}$ fluctuations
τ	throttle setting, expressed as a fraction of maximum thrust or power
τ_c	commanded throttle setting
$\tau_{\max}$	airfoil section percent maximum thickness

τ_r wing root percent maximum thickness

τ_t wing tip percent maximum thickness

$\tau(\hat{x})$ airfoil thickness distribution as a function of dimensionless chordwise coordinate

υ_0 airfoil section area integral defined in Eq. (2.4.8)

υ_1 airfoil section area integral defined in Eq. (2.4.9)

υ_2 airfoil section area integral defined in Eq. (2.4.10)

υ_3 airfoil section area integral defined in Eq. (2.4.11)

Φ power spectral density

Φ_n white-noise power spectral density

Φ_p power spectral density of $\tilde{p}$ fluctuations

Φ_q power spectral density of $\tilde{q}$ fluctuations

Φ_r power spectral density of $\tilde{r}$ fluctuations

Φ_u power spectral density of $\tilde{u}$ fluctuations

Φ_v power spectral density of $\tilde{v}$ fluctuations

Φ_w power spectral density of $\tilde{w}$ fluctuations

$\mathit{\Phi}$ geographic latitude

$\mathit{\Phi}_d$ destination latitude

ϕ Euler bank angle

ϕ_i phase of the ith sinusoidal component

φ azimuth angle in cylindrical coordinates

φ rotation angle between two coordinate systems

$\mathit{\Psi}$ geographic longitude

$\mathit{\Psi}_d$ destination longitude

ψ Euler azimuth angle

ψ_g ground-track angle or bearing

ψ_h heading angle

ψ_w wind direction angle

$\mathbf{\Omega}$ vorticity vector

ω angular rotation rate vector

ω spatial frequency or wave number

ω_m motor angular rate

ω_n undamped natural frequency

ω_r rotor angular rate

ω_s motor shaft angular rate

ω_x angular rate about the x axis

ω_y angular rate about the y axis

ω_z angular rate about the z axis

$\dot{\omega}$ angular acceleration vector

Index

F

G

H

I

K

L

S

T

U

V

W

www.ingramcontent.com/pod-product-compliance
Ingram Content Group UK Ltd.
Pitfield, Milton Keynes, MK11 3LW, UK
UKHW020222290926
15470UKWH00001B/337